Chemical Engineering Fluid Mechanics

THIRD EDITION

Chemical Engineering Fluid Mechanics

THIRD EDITION

Ron Darby
Texas A&M University, College Station, Texas

Raj P. Chhabra
Indian Institute of Technology, Kanpur, India

CRC Press
Taylor & Francis Group
Boca Raton London New York

CRC Press is an imprint of the
Taylor & Francis Group, an **informa** business

CRC Press
Taylor & Francis Group
6000 Broken Sound Parkway NW, Suite 300
Boca Raton, FL 33487-2742

ISBN 13: 978-1-4987-2442-5 (hbk)

Library of Congress Cataloging-in-Publication Data

Names: Darby, Ron, 1932- | Chhabra, R. P.
Title: Chemical engineering fluid mechanics.
Description: Third edition / Ron Darby and Raj P. Chhabra. | Boca Raton :
CRC Press, 2017. | Includes bibliographical references and index.
Identifiers: LCCN 2016025950| ISBN 9781498724425 (hardback : alk. paper) |
ISBN a9781498724449 (e-book) | ISBN 9781498724456 (e-book)
Subjects: LCSH: Chemical processes. | Fluid mechanics.
Classification: LCC TP155.7 .D28 2017 | DDC 660/.284--dc23
LC record available at https://lccn.loc.gov/2016025950

This book is dedicated to the memory and honor of Isaac Newton, for without whom our vast scientific and engineering achievements may never have happened.

Isaac Newton (1643–1727)

And to Dr. Theodor Seuss Geisel (1904–1991), for his humor, insight, and inspiration.

"You have brains in your head. You have feet in your shoes. You can steer yourself in any direction you choose. You're on your own, and you know what you know. And you are the guy who'll decide where to go."

—*Dr. Seuss*[*]

---

[*] http://www.brainyquote.com/quotes/quotes/d/drseuss414097.html.

Contents

Preface

The objective of this edition remains the same as that of the first two editions, namely, to present the fundamental (macroscopic) conservation equations that govern the behavior of fluids and to apply these to a wide variety of practical situations of interest to the chemical engineer. Where purely theoretical expressions are not completely adequate to describe the actual fluid system of interest, empirical information is utilized and generalized by principles of similitude where possible.

The general approach to problem solving espoused here is represented by the "baby versus the bathwater" philosophy illustrated in the book. That is, it is believed that the most useful approach to engineering problem solving is to concentrate on those factors that have a major (or significant) influence on the desired solution (i.e., the "baby") and neglect those that have a negligible or insignificant influence on the answer (i.e., the "bathwater"). The best engineer is the one who is able to recognize the difference between these effects and make the appropriate assumptions in order to achieve the simplest answer that is the most realistic.

Just as in the first two editions, the scope of this edition includes both Newtonian and non-Newtonian incompressible fluids and compressible fluids in subsonic through choked flow. Both internal flows (in pipes, conduits, fittings, etc.) and external flows (fluid drag on particles, sedimentation, fluidization, cyclone separation, etc.) are included.

The chapter on flow measurement has been expanded, the information on drag on nonspherical particles has been expanded, the material on control valves has been completely revised and updated, and a new chapter has been added that includes the sizing of safety relief valves for both single- and two-phase fluids.

A number of examples have been added throughout the book to illustrate the application of the principles. However, it is the authors' experience that too many students tend to rely on finding a worked problem that represents the problem they are trying to solve and just changing the numbers, rather than starting with the fundamental principles and deriving the solution, which is the desired approach. Thus, a few examples may be good, but too many may actually be detrimental.

This book has found extensive use as both a textbook for students in applied fluid mechanics courses and a useful reference for practicing engineers. It is hoped that this revised edition will also find similar applications for a greater number of readers.

"Engineering is all about tradeoffs."

"You cannot teach a man anything; you can only help him discover it in himself."

—Galileo

"In theory there is no difference between theory and practice. In practice there is."

—Yogi Berra

About Professor Darby

Professor Ronald Darby was born in La Veta, Colorado on September 12, 1932. He earned his BS (1955) and PhD (1963) in chemical engineering from Rice University, Houston, Texas. After working briefly for Ling-Temco-Vought (LTV) in Dallas, Texas, he returned to academia and joined the Department of Chemical Engineering at Texas A&M University (TAMU) in 1965 as an assistant professor. He remained a faculty member at TAMU until his formal retirement in 2001. Thereafter, he was Professor Emeritus until his death on July 14, 2017.

During his long career at Texas A&M, Ron taught a variety of courses in chemical engineering. His research interests were in the areas of non-Newtonian and viscoelastic fluid mechanics and two phase flows in the context of safety relief valves, and he published extensively on these topics. In addition to a large number of research publications in archival journals, Ron authored two books, *Viscoelastic Fluids: An Introduction to their Properties and Behavior* (Dekker, 1976) and *Chemical Engineering Fluid Mechanics* (3rd ed., CRC Press, 2017).

He was elected fellow of the AIChE in 1994, and he was honored with numerous awards for his research contributions. These included the Faculty Distinguished Achievement Award (1971) from his department and the Publication Award from AIChE in 1984. He was also the recipient of the Distinguished Service Award (2003) of the Texas AIChE and the Former Students Award from Texas A&M for excellence in teaching. In 1981, he was designated "Minnie Stevens Piper Professor of Chemical Engineering."

Ron was, in addition, a captain in the U.S. navy. In 1975, he earned "First commanding officer," USNR's outstanding leadership award.

Acknowledgments

The authors would like to recognize, and give special thanks, to the following people for their contributions to this edition of the book:

Ms. Swati Patel and Anoop Gupta, graduate students, Chemical Engineering Department, IIT Kanpur, India, for assistance in preparing the illustrations in the book and typing of the solutions manual.

Dr. Leonid B. Korelstein, Cofounder and VP on Science of Piping Systems Research & Engineering Co. (NTP Truboprovod) and Associate Professor in D. Mendeleev University of Chemical Technology of Russia, Moscow, Russia, for special contributions to Chapters 9 and 16 and review of other sections of this edition.

Mr. Freeman Self, Bechtel Distinguished Engineer, Bechtel Oil Gas & Chemicals, Houston, TX, for extensive review and contribution to Chapter 11 of this edition of the book and for the cover design.

Unit Conversion Factors

Dimension	Equivalent Units
Mass	$1 \text{ kg} = 1000 \text{ g} = 0.001$ metric tons (tonne) $= 2.20462 \text{ lb}_m = 35.274 \text{ oz}$
	$1 \text{ lb}_m = 453.59237 \text{ g} = 0.45359237 \text{ kg} = 5 \times 10^{-4} \text{ ton} = 16 \text{ oz}$
Force	$1 \text{ N} = 1 \text{ kg m/s}^2 = 10^5 \text{ dynes} = 10^5 \text{ g cm/s}^2 = 0.22418 \text{ lb}_f$
	$1 \text{ lb}_f = 32.174 \text{ lb}_m \text{ ft/s}^2 = 4.4482 \text{ N} = 4.4482 \times 10^5 \text{ dynes}$
Length	$1 \text{ m} = 100 \text{ cm} = 10^6 \text{ microns } (\mu\text{m}) = 10^{10} \text{ Å} = 39.37 \text{ in} = 3.2808 \text{ ft} = 1.0936 \text{ yards} = 0.0006214 \text{ miles}$
	$1 \text{ ft} = 12 \text{ in} = 1/3 \text{ yard} = 0.3048 \text{ m} = 30.48 \text{ cm}$
Volume	$1 \text{ m}^3 = 1000 \text{ L} = 10^6 \text{ cm}^3 = 35.3145 \text{ ft}^3 = 264.17 \text{ gal}$
	$1 \text{ ft}^3 = 1728 \text{ in.}^3 = 7.4805 \text{ gal} = 0.028317 \text{ m}^3 = 28.317 \text{ L} = 28{,}317 \text{ cm}^3$
Pressure	$1 \text{ atm} = 1.01325 \times 10^5 \text{ N/m}^2 \text{ (Pa)} = 1.01325 \text{ bar} = 1.01325 \times 10^6 \text{ dyne/cm}^2 = 760 \text{ mm Hg @ } 0°\text{C}$ (torr) $= 10.333 \text{ m H}_2\text{O @ } 4°\text{C} = 33.9 \text{ ft H}_2\text{O @ } 4°\text{C} = 29.921 \text{ in Hg @ } 0°\text{C} = 14.696 \text{ lb}_f/\text{in}^2 \text{ (psi)}$
Energy	$1 \text{ J} = 1 \text{ N m} = 10^7 \text{ ergs} = 10^7 \text{ dyne cm} = 2.667 \times 10^{-7} \text{ kW h} = 0.23901 \text{ cal} = 0.7376 \text{ ft lb}_f = 9.486 \times 10^{-4} \text{ btu } (550 \text{ ft lb}_f/\text{hp s})$
Power	$1 \text{ W} = 1 \text{ J/s} = 0.23901 \text{ cal/s} = 0.7376 \text{ ft lb}_f/\text{s} = 9.486 \times 10^{-4} \text{ btu/s} = 1 \times 10^{-3} \text{ kW} = 1.341 \times 10^{-3} \text{ hp}$
Flow rate	$1 \text{ m}^3/\text{s} = 35.3145 \text{ ft}^3/\text{s} = 264.17 \text{ gal/s} = 1.585 \times 10^4 \text{ gal/min (GPM)} = 10^6 \text{ cm}^3/\text{s}$
	$1 \text{ GPM} = 6.309 \times 10^{-5} \text{ m}^3/\text{s} = 2.228 \times 10^{-3} \text{ ft}^3/\text{s} = 63.09 \text{ cm}^3/\text{s}$
Viscosity	$1 \text{ Poise} = 1 \text{ dyn s/cm}^2 = 100 \text{ cP} = 0.1 \text{ N s/m}^2 = 0.1 \text{ Pa s} = 1 \text{ gm/(cm s)} = 0.0672 \text{ lb}_m/(\text{ft s}) = 2.09 \times 10^{-3} \text{ lb}_f \text{ s/ft}^2$

Example: The factor to convert Pa to psi is $14.696 \text{ psi}/(1.01325 \times 10^5 \text{ Pa})$.

Some values of
the gas constant

$$R = 8.314 \times 10^3 \text{ kg m}^2/\text{s}^2 \text{ kgmol K}$$
$$= 8.314 \times 10^7 \text{ g cm}^2/\text{s}^2 \text{ gmol K}$$
$$= 82.05 \text{ cm}^3 \text{ atm/gmol K}$$
$$= 1.987 \text{ cal/gmol K or btu/lbmol R}$$
$$= 1545 \text{ ft lb}_f/\text{lbmol R}$$
$$= 10.73 \text{ ft}^3 \text{ psi/lbmol R}$$
$$= 0.730 \text{ ft}^3 \text{ atm/lbmol R}$$

1 Basic Concepts

"Engineering is learning how to make a decision with insufficient information."

—**Anonymous**

I. FUNDAMENTALS

A. INTRODUCTION AND SCOPE

The understanding of, and ability to predict, the flow behavior of fluids is fundamental to all aspects of the chemical engineering profession. Such behavior includes the relationship between the (driving) forces and the flow rates of various classes of fluids, and the characteristics of the equipment used to contain/handle/process these fluids. A wide variety of fluids with different properties may be encountered by the chemical engineer in the industrial applications that may be of concern in various chemical or petroleum process industries, biological or food and pharmaceutical industries, polymer and materials processing industries, etc. These properties range from common simple incompressible (Newtonian) liquids, that is, fluids such as water, oils, various petroleum fractions, etc., to complex nonlinear (non-Newtonian) fluids such as pastes, emulsions, foams, suspensions, high-molecular-weight polymeric fluids, biological fluids (e.g., blood), etc. as well as compressible (gaseous) fluids such as air, N_2, CO_2, etc. This book is concerned with the fundamental principles that govern the flow behavior of each of these types of fluids, as well as the resulting basic relationships needed to predict their behavior (relations between flow rates, pressures, forces on solid boundaries, etc.) and the corresponding analysis of a wide variety of equipment and practical situations commonly found in various industries. These principles are also applicable to such situations as the flow of blood in vessels, transport of sludge and silt in rivers and streams, pneumatic conveying of particles, the trajectory of a pitched baseball, etc.

B. BASIC LAWS

The fundamental principles that apply to the analysis of fluid flows are the three "conservation laws":

1. Conservation of mass
2. Conservation of energy (the first law of thermodynamics)
3. Conservation of momentum (Newton's second law of motion)

To which may be added

4. The second law of thermodynamics (i.e., will it work or not?)
5. Conservation of dimensions (e.g., the "fruit salad" law*)
6. Conservation of dollars (economics)

Although the second law of thermodynamics (#4) is not strictly a "conservation law," it provides a practical limitation on what processes are possible or are most likely. It states that a process can occur spontaneously only if it goes from a state of higher energy to the one of lower energy (e.g., water will flow downhill by itself, but it must be "pushed" by expending energy in order to make it flow uphill).

---

* The "fruit salad law" states that you "can't add apples and oranges unless you want fruit salad."

These basic *conservation laws* provide relations between various fluid properties and operating conditions at different points within a *system* (see Section IV). In addition, appropriate *rate* or *transport* models are required that describe the *rate* at which these conserved quantities are transported from one part of the system to another. For example, if the *mass* of a given fluid element is m (e.g., kg or lb_m), the rate at which that mass is transported is the *mass flow rate*, $\dot{m}$ (e.g., kg/s or lb_m/s). These *conservation* and *rate* laws are the starting point for the solution to every engineering problem.

The conservation of energy, for example, is often expressed in terms of the thermodynamic (equilibrium) properties of the system. This implies a system that is in a state of *static equilibrium*. However, most systems of interest to us are not at equilibrium but are *dynamic* (i.e., in motion) and it is the *rate* of transport of energy, mass, or momentum which is of interest. In order to transport energy at a finite *rate*, this equilibrium must be disturbed and additional "nonequilibrium" energy is required that depends upon the *rate* of transport. This "extra" energy is expended (e.g., dissipated or "lost") by transformation from useful mechanical energy to low-grade thermal energy in order to transport mass, energy, or momentum at a *finite rate*. The farther from equilibrium the system is (i.e., the faster it goes), the greater is the resistance to motion ("friction") and the greater is the energy that is "lost" or dissipated to a less useful low-grade thermal energy in order to overcome this resistance. In more mundane terms, this tells us that useful energy must be expended (or dissipated) in order to "push" water through a pipe (e.g., by a pump) at a rate that increases with the water flow rate. Furthermore, the faster the water is "pushed," the further from "equilibrium" it is and the greater the amount of energy that must be supplied. This "extra energy" is typically described as the "flow resistance" or "friction loss" (although the energy is not really "lost"—it is transformed from higher-order mechanical energy to lower-order thermal energy).

C. Experience

Engineering is much more than just "applied science and math." Although science and math are important "tools of the trade," it is the engineer's ability to use these tools (and others), along with considerable *judgment* and *experience* to "make things work"—that is, to make it possible to get reasonable answers to real problems with (sometimes) limited or incomplete information. A key aspect of "judgment and experience" is the ability to organize and utilize information obtained from one "system" and apply it to analyze or design similar systems on a different scale or in a different setting. For example, the "conservation of dimensions" (or *fruit salad law*) allows us to design experiments and to acquire and organize data (e.g., "experience") obtained in a lab test or model system in the most general and efficient form so that it can be applied to problems in similar systems that may involve different properties and/or a different scale.

Another aspect of this is the "baby and the bathwater" principle. This relates to the ability to recognize and account for the most important or controlling factors in a given problem (the "baby") while being able to ignore the factors that have only minor influence and can be neglected (the "bathwater"). Because the vast majority of engineering problems in fluid mechanics cannot be solved without resort to experience (e.g., data or empirical knowledge), this very important principle (i.e., dimensionless variables) will be used extensively (see Chapter 2).

"Simple laws can very well describe complex structures. The miracle is not the complexity of our world, but the simplicity of the equations describing that complexity."

—Sander Bais, B. 1945, *Theoretical Physicist*

II. OBJECTIVE

It is the intent of this book to show how basic laws and concepts, along with pertinent knowledge of the relevant fluid properties, operating conditions, engineering data, and suitable assumptions (e.g., judgment), can be applied to the analysis and design of a wide variety of practical engineering systems involving the flow of fluids. It is the authors' belief that engineers are much more effective if they approach the problem-solving process systematically from a basic perspective, starting with first principles to develop a solution, rather than looking for a "similar" solved problem (that may or may not be applicable) as an example to follow. It is this philosophy, along with the objective of arriving at reasonable workable solutions to practical problems, upon which this work is based.

A. A NOTE ON PROBLEM SOLVING

Because engineers are primarily "problem solvers" (a major objective of this book!), it is appropriate to offer some guidelines for the problem-solving process. Of course, the primary objective of the process is to arrive at the correct solution to the problem, but equally important is the ability to arrive at the solution through an organized and logical process that is clear and can easily be verified by others. A useful mnemonic that represents some guidelines that are helpful in achieving a good problem solution is the "4 C's":

- *Clear*—All symbols and reference points should be obvious or clearly defined, and the procedure should be organized so that all assumptions, sources of data, basic principles and equations, method of solution, etc. are clearly evident.
- *Complete*—The "knowns" and "unknowns" should be identified, and all pertinent steps connecting the basic principles to the problem solution should be included. All calculations should be shown, with numerical values accompanied by their respective units, and conversion factors included as needed (see Sections II and III of Chapter 2).
- *Correct*—Of course, the answer should be correct, with numerical answers expressed in terms of the appropriate number of significant digits (see Section VII of Chapter 2). This also means that the answer should make sense physically.
- *Concise*—All equations and calculations needed to clearly follow the solution procedure should be included, but extraneous side calculations or details not directly needed to follow the procedure should be omitted.

Such a systematic approach also enables one to revisit each step and/or question the validity of assumptions made in order to improve the solution in a systematic manner. Many of these guidelines can be satisfied by initially drawing a diagram or sketch that represents the physical layout of the problem, labeling the diagram with all the variable symbols and reference points, and identifying all known values and unknown quantities. These guidelines are offered as an aid in effectively addressing a problem and organizing the solution process but are not "cast in stone." They are in no way intended as a substitution for the *thinking* and *analysis* required for effective problem solving.

Example 1.1 Problem Solving Illustration

To illustrate the problem-solving process, we apply the guidelines to determine the pressure at a depth of 3000 ft below the surface of the ocean.
Draw a sketch:

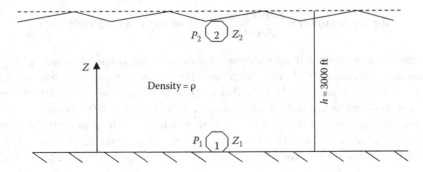

Note that the diagram has been labeled with the pertinent reference points (1, 2), pressures (P) and elevations (Z) at these points, the depth (h), and the water density (ρ).

Assumptions: We will make the following assumptions:

1. The sea water is an incompressible (constant density) liquid with $\rho = 64$ lb_m/ft^3.
2. The surface of the ocean and the sea bottom are flat.
3. The barometric pressure is exactly 1 atm.

It should be noted that none of these assumptions is exact, but they should be adequate to obtain a reasonable answer.

Theory: The theoretical expression that relates the pressure in an incompressible fluid to the elevation in the fluid is derived in Chapter 4 and is expressed as

$$\Phi = \text{constant}, \quad \text{where } \Phi = P + \rho gZ \text{ by definition.}$$

Note that once the correct theoretical expression is identified, pertinent variables and parameters can be identified from the terms in the equation and used to complete the labeling of the diagram.

Apply theory: We must tailor the general theoretical expression to the specific problem at hand:

$$\Phi_2 = \Phi_1 \quad \text{or} \quad P_1 + \rho gZ_1 = P_2 + \rho gZ_2$$

Complete the solution: Now solve the equation for the desired quantity, insert values of known quantities with conversion factors as required (noting that $h = Z_2 - Z_1$), and calculate the answer:

$$P_1 = P_2 + \rho g(Z_2 - Z_1) = 1\,\text{atm} + \frac{(64\,lb_m/ft^3)(32.17\,ft/s^2)(3000\,ft)}{(32.17\,lb_m\,ft/lb_f\,s^2)(14.7\,lb_f/in.^2\,atm)(144\,in.^2/ft^2)}$$

$$= 91.7\,\text{atm}(14.7\,\text{psi/atm}) = \boxed{1350\,\text{psi}}$$

Note that the answer has been rounded to three digits or significant figures. This should be consistent with the probable precision of the data (e.g., ρ and h) used to calculate the answers, in light of the assumptions made. Conversion factors and significant digits are discussed in detail in Chapter 2.

III. PHENOMENOLOGICAL RATE OR TRANSPORT LAWS

In addition to the laws for the conservation of mass, energy, momentum, etc. there are additional physical laws that govern the *rate* at which these conserved quantities are transported from one region to another in a continuous medium. These are called *phenomenological* laws because they are based on observable phenomena and logical deduction but they cannot be derived from first principles. These "rate" or "transport" models can be written for any conserved quantity (mass, energy, momentum, electric charge, etc.) and can be expressed in a general form as

$$Rate\ of\ transport = \frac{Driving\ force}{Resistance} = (Conductance) \times (Driving\ force) \qquad (1.1)$$

This expression can be applied to the transport of any conserved (extensive) quantity "Q," for example, mass, energy, momentum, electric charge, etc. The rate of transport of Q per unit area normal to the direction of transport is called the *flux* of Q, with dimensions of "Q"/(time × area). This transport equation can be applied on a microscopic scale in a stationary medium or to a fluid in motion. The moving fluid can be in *laminar flow* in which the mechanism for the transport of Q is the intermolecular forces of attraction between molecules or groups of molecules, or in *turbulent flow* in which the transport mechanism is the result of the motion of turbulent eddies which move in three dimensions and carry Q along with them. The *resistance* or *conductance* term in Equation 1.1 is also called the *transport coefficient*. For laminar or stationary media, the *transport coefficient* is a fluid or material property, but for turbulent flows it also depends upon the degree of turbulence in the medium.

On the microscopic or molecular level (e.g., stationary media or laminar flow), the *driving force* for the transport of Q in any direction (e.g., the y direction) is the negative of the *concentration gradient of Q* in that (y) direction. That is, Q flows "downhill," from a region of high concentration to a region of low concentration, and the rate of transport of Q is proportional to the change in the concentration of Q divided by the distance over which it changes. At any specific point in the system, this can be expressed as

$$\text{Flux of } Q \text{ in the } y \text{ direction} = q_y = K_t \left[-\frac{dC_Q}{dy} \right] \tag{1.2}$$

where K_t is the *transport coefficient* for the quantity Q and C_Q is the concentration of Q. For microscopic (molecular) transport, K_t is a property only of the medium, which is assumed to be a continuum, that is, all relevant physical properties can be defined at each and every point within the medium. This means that the smallest region of practical interest must be very large relative to the size of the individual molecules (or the distance between them) or any substructure of the medium such as particles, drops, bubbles, etc. It is further assumed that these properties are homogeneous and isotropic (i.e., they are the same at all points within the medium and are independent of direction). For macroscopic systems involving turbulent convective transport, the *driving force* is a representative difference in the concentration of Q, and the transport coefficient includes the effective distance (δ) over which this concentration changes, for example,

$$\text{Turbulent flux of } Q \text{ in the } y \text{ direction} = q_y = \frac{K_t}{\delta} \Delta C_Q \tag{1.3}$$

where the *transport coefficient* is now (K_t/δ) and is dependent upon the flow conditions as well as properties of the medium. Note the absence of the minus sign on the right-hand side (*RHS*) of Equation 1.3. This is because the concentration change (ΔC_Q) is generally interpreted as the *magnitude* of the change in concentration along the path of increasing y.

Example 1.2

What are the dimensions of the transport coefficient K_t?

Analysis: "Dimensions" represent the "physical character" of a given quantity (e.g., mass [M], length [L], time [t], etc.) (a complete discussion of units and dimensions is given in detail in Chapter 2). The "conservation of dimensions" or *fruit salad law* requires that all terms in any valid equation must have the same net dimensions. By applying this "law" to Equation 1.2 that defines the coefficient K_t, we can deduce the dimensions of K_t from the dimensions of the other terms in the equation.

Solution:

Denoting dimensions by square brackets (i.e., "the dimensions of x = [x]"), the dimensions of each term in Equation 1.2 are

$$[\text{Flux of } Q] = [q_y] = [Q]/(L^2\, t), \quad [C_Q] = [Q]/L^3, \quad [y] = L$$

Thus, the "dimensional" form of Equation 1.2 is

$$[q_y] = \frac{[Q]}{L^2 t} = [K_t] \frac{[Q]}{L^4}$$

Since the dimensions of Q cancel out, the dimensions of K_t are L^2/t, or

$$[K_t] = \frac{L^2}{t}$$

By similar reasoning, the dimensions of the coefficient (K_t/δ) are seen to be L/t.

A.   FOURIER'S LAW OF HEAT CONDUCTION

Figure 1.1 illustrates two horizontal parallel plates with a uniform "medium" (either solid or fluid) in between. If the top plate is kept at a temperature T_1 that is higher than the temperature T_0 of the bottom plate, there will be transport of thermal energy (heat) from the upper plate to the lower plate through the medium, in the $-y$ direction. If the flux of heat in the y direction is denoted by q_y, then the corresponding transport law can be written as

$$q_y = -\alpha_T \frac{d(\rho c_v T)}{dy} \tag{1.4}$$

where
α_T is the *thermal diffusion coefficient*, with dimensions of $L^2\, t^{-1}$
($\rho c_v T$) is the "concentration of thermal energy (heat)"

Because the density (ρ) and the heat capacity (c_v) are assumed to be independent of position, this equation can be rewritten in the following form:

$$q_y = -k \frac{dT}{dy} \doteq -k \frac{\Delta T}{\Delta y} = -k \frac{(T_1 - T_0)}{(y_1 - y_0)} \tag{1.5}$$

where $k = \alpha_T \rho c_v$ is the *thermal conductivity* of the medium. The two forms of the equation on the right follow from the fact that the gradient (derivative) term is constant in the geometry of Figure 1.1. This is not true in general but is valid for this geometry and for any continuum in which the reference locations are sufficiently close together. Note that q_y is negative because heat is being transported in the "$-y$" direction, from high temperature to low temperature. This law was formulated by Fourier in 1822 and is known as *Fourier's law* of heat conduction. This law applies to stationary solids or fluids and to fluids moving in the x direction with straight stream-lines (e.g., laminar flow).

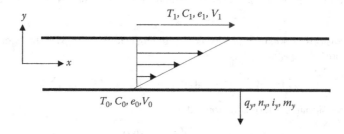

FIGURE 1.1   Transport of energy, mass, charge, and momentum from the upper surface to the lower surface.

B.   Fick's Law of Diffusion

An analogous situation can be envisioned if the medium is stationary (or is a fluid in laminar flow in the x direction) and the temperature difference $(T_1 - T_o)$ is replaced by the concentration difference $(C_1 - C_o)$ of a substance that is soluble in the fluid (e.g., a top plate made of pure salt in contact with water). If the soluble species (e.g., the salt) is denoted by A, it will diffuse through the medium (B) from the high concentration C_1 at the top plate to the low concentration C_o at the bottom plate. If the flux of species A in the y direction is denoted by n_{Ay}, then the transport law is given by

$$n_{Ay} = -D_{AB}\frac{dC_A}{dy} \doteq -D_{AB}\frac{\Delta C_A}{\Delta y} = -D_{AB}\frac{(C_{A1} - C_{A0})}{(y_1 - y_0)} \tag{1.6}$$

where D_{AB} is the molecular diffusivity of species A in the medium B. Here, n_{Ay} is negative because species A is diffusing in the $-y$ direction. Equation 1.6 is known as *Fick's law of diffusion* and was formulated in 1855 (note that it is exactly the same as Fourier's law, with just the symbols changed). In fact, both the molecular diffusivity D_{AB} and the thermal diffusivity α_T have identical dimensions of $L^2\,t^{-1}$.

C.   Ohm's Law of Electrical Conductivity

The same transport law can be written for electric charge (which is another conserved quantity). In this case, the top plate in Figure 1.1 is assumed to be at an electric potential of e_1 which is higher than the potential e_0 at the bottom plate (note: "electric potential" is equivalent to "concentration of electric charge" or voltage). The corresponding "charge flux," i.e., "current density," in the y direction is i_y, which is negative because the charge transport is in the $-y$ direction. The corresponding expression for the charge flux is known as *Ohm's law* (1827) and is written as

$$i_y = -k_e\frac{de}{dy} \doteq -k_e\frac{\Delta e}{\Delta y} = -k_e\frac{(e_1 - e_0)}{(y_1 - y_0)} \tag{1.7}$$

where k_e is the electrical conductivity of the medium between the plates, with dimensions of $L^2\,t^{-1}$.

D.   Newton's Law of Viscosity

Momentum is also a conserved quantity, and an equivalent transport law can be written for the transport (flux) of momentum. We must be careful here, however, because velocity and momentum are vector quantities as opposed to mass, energy, and charge, which are all scalars. Even though we may draw some analogies between these scalar quantities and the one-dimensional transport of vector components, these analogies do not hold in general for multidimensional systems or for complex geometries. However, the conditions depicted in Figure 1.1 are for simple one-dimensional transport, in which case we can include the transport of the x-component of linear momentum by analogy with the transport of scalar quantities.

In this case, we consider the top plate to be subject to a force in the x direction, causing it to move with a constant velocity V_1 in the x direction, while the bottom plate is stationary $(V_0 = 0)$. There is x momentum associated with any given element of fluid of mass m moving in the x direction with velocity v_x. This x momentum is mv_x and the corresponding concentration of x momentum (i.e., momentum per unit volume) must be ρv_x. If we denote the flux of this x momentum in the y direction by $(\tau_{yx})_{mf}$, the transport expression is

$$(\tau_{yx})_{mf} = -v\frac{d(\rho v_x)}{dy} \tag{1.8}$$

where ν is called the *kinematic* viscosity, with dimensions of $L^2\,t^{-1}$. It should be evident that $(\tau_{yx})_{mf}$ is negative because the faster moving fluid at the top drags the slower moving fluid along with it so that x momentum is transported in the $-y$ direction by virtue of this fluid "drag." Because the fluid is assumed to be homogeneous, the density is independent of position and the flux of x momentum in the $-y$ direction can therefore be written as

$$(\tau_{yx})_{mf} = -\mu\frac{dV_x}{dy} \doteq -\mu\frac{\Delta V_x}{\Delta y} = -\mu\frac{(V_1 - V_0)}{(y_1 - y_0)} \tag{1.9}$$

where $\mu = \rho\nu$ is the *viscosity* (or sometimes the *dynamic viscosity*). Equation 1.9 is *Newton's law of viscosity* and was formulated in 1687! It applies to so-called Newtonian fluids in laminar flow (non-Newtonian fluids and turbulent flows will be considered in later chapters).

1. Momentum Flux and Shear Stress

Both Newton's law of viscosity and the conservation of momentum are related to Newton's second law of motion, which is commonly written as $F_x = ma_x = d(mV_x)/dt$, that is, force is equivalent to a rate of change of momentum. For a fluid in steady flow (constant velocity), this is equivalent to $F_x = \dot{m}v_x$ where $\dot{m}$ is the fluid mass flow rate. F_x is the force acting in the x direction on the top plate in Figure 1.1, which makes it move, and it is also the *driving force* for the rate of transport of x momentum $(\dot{m}V_x)$ from the faster to the slower moving fluid, in the $-y$ direction. The *flux* of x momentum in the y direction $(\tau_{yx})_{mf}$ is therefore equivalent to the force F_x acting on unit area A_y. Note that "area" has both magnitude and direction due to the orientation of the surface and is thus a vector quantity with three components. The direction of the area vector is defined by the unit vector that is the outward normal to the surface that bounds the fluid volume of interest (i.e., the "*system*"). Because a surface has two sides and hence has both a "positive" and "negative" direction, the convention is that the surface bounding the volume element whose *outward normal vector* points in the positive direction is *positive* and the surface whose *outward normal vector* points in the *negative* direction is *negative*. In Figure 1.1, $+A_y$ is the area of the surface bounding the fluid element in contact with the top plate (the "system"), which has an outward normal vector in the $+y$ direction. The quantity F_x/A_y is also known as the *shear* stress, τ_{yx}, that acts on the fluid (i.e., it is the $+F_x$ force, acting in the $+x$ direction, on the area A_y of $+y$ surface). It should be evident that, with reference to Figure 1.1, the momentum flux $(\tau_{yx})_{mf}$ is negative ($+x$ momentum being transported in the $-y$ direction), whereas the shear stress (τ_{yx}) is positive ($+F_x$ force acting on $+A_y$ surface), thus

$$\tau_{yx} = -\left(\tau_{yx}\right)_{mf} \tag{1.10}$$

That is, the momentum flux is always the negative of the corresponding shear stress. In the field of mechanics, the "shear stress" sign convention is customary, as opposed to the "momentum flux" sign convention, and that is the one we will follow in this book. It is important to recognize the difference between these two sign conventions, because references that use the momentum flux convention define the viscosity of a Newtonian fluid by Equation 1.9, whereas if the shear stress convention is used, viscosity is defined by the equation

$$\tau_{yx} = \mu\frac{dV_x}{dy} \cong \mu\frac{\Delta V_x}{\Delta y} = \mu\frac{\left(V_1 - V_0\right)}{\left(y_1 - y_0\right)} \tag{1.11}$$

2. Vectors and Dyads

The equations representing each of the preceding one-dimensional transport laws are all identical, except for the notation (e.g., "the game is the same—only the color of the jerseys is different"), and all of the transport coefficients (D_{AB}, α_T, k_e, and ν have the same dimensions, i.e., L^2/t).

However, there are some features of Newton's law of viscosity that distinguish it from the other laws, which are important when applying this law. First, as previously mentioned, momentum is fundamentally different from the other conserved quantities in that it is a *vector* (with directional properties), whereas the others (mass, energy, electric charge) are *scalars*, independent of direction. Now, the gradient operator or "directional derivative" ($d()/dy$ in one dimension or $\nabla()$ in more general 3D notation) is also a vector. It thus follows that the gradient of a scalar is a vector (i.e., the flux of heat, mass, or charge), and the gradient of a vector (e.g., velocity or momentum), and hence the flux of momentum (or stress), is a *dyad* or *second-order tensor*. Dyads are associated with two directions: the direction of the vector quantity and the direction in which it varies. Thus, $(\tau_{yx})_{mf}$ represents the flux of x momentum in the y direction. Likewise, the shear stress components (e.g., τ_{yx}), which represent a force (F_x) acting on an area component (A_y) oriented normal to the y direction, are *tensor components* or *dyads*, with nine possible components. Again, recall that area is a vector because of its orientation, with a positive area being that which bounds the fluid element of interest with its *outward normal vector* pointing in the positive coordinate direction. In general, a 3D (x, y, z) space vector can be represented by three components, whereas a *dyad* or *second-order tensor* has nine possible components. This difference between momentum flux and the flux of a scalar quantity is very significant when applying the equations to 3D space or to path lines, which are not straight. In such cases, the analogy between Newton's law of viscosity and the other transport laws is lost.

3. Newtonian and Non-Newtonian Fluids

Fluids that obey Newton's law of viscosity are characterized by the fluid property "viscosity" (μ), which is assumed to be constant (i.e., independent of position, time, or flow conditions). This is true of the other transport properties for heat, mass, and electric charge as well, all of which are assumed to be dependent upon the thermodynamic state of the material (e.g., temperature, pressure, density) but independent of the "dynamic" state, that is, the state of stress or deformation. Many common fluids are Newtonian, although there are many fluids for which this assumption is not true with respect to viscosity, and for these the viscosity does depend upon the state of stress or deformation and hence is a *function* of the dynamic state of the fluid. Gases, low molecular weight simple inorganic and organic liquids and solutions of relatively low-molecular-weight compounds, molten metals, and salts all are examples of Newtonian fluid behavior. "Non-Newtonian" fluids are typically complex materials with significant "substructure," such as melts or solutions of high-molecular-weight polymers, suspensions of solids in liquids, emulsions of liquids in liquids, foam suspensions of gas bubbles in liquids, surfactant solutions, etc. Typical examples are paint, ink, pastes, liquid soaps, blood, mud, ketchup, mayonnaise, mustard, milk shakes, creams, lotions, etc.

The characteristics of linear, nonlinear viscous, and viscoelastic fluids are discussed in Chapter 3, and an analysis of the flow characteristics of a variety of "purely viscous" non-Newtonian fluids is included in later chapters. However, a completely general treatment of the flow and deformation behavior of non-Newtonian (including viscoelastic) fluids is beyond the scope of this book, and for this the reader is referred to the more advanced literature (e.g., Darby, 1971; Carreau et al., 1997; Chhabra and Richardson, 2008). Nevertheless, quite a bit can be learned and many practical problems solved by considering relatively simple models for the non-Newtonian properties of some complex fluids, provided the complexities of viscoelastic behavior can be avoided. These properties can be measured in the laboratory, with proper attention to data interpretation, and can be represented by any of several relatively simple mathematical expressions such as those illustrated in Chapter 3 and the books cited above. We do not attempt to delve in detail into the molecular or structural properties of these complex fluids but will make use of information that can be readily obtained through routine measurements. In this context, the flow behavior of non-Newtonian fluids will be considered in parallel with Newtonian fluids wherever possible and within the framework of the continuum hypothesis.

Example 1.3

With respect to Figure 1.1, consider the two parallel flat surfaces spaced 1 mm apart, with a Newtonian grease between them that has a viscosity of 200 cP. If the area of the plates in contact with the grease is 1 m², determine the force (in lb$_f$) required to make the top plate move at a velocity of 1 ft/s when the bottom plate is stationary.

Analysis: Equation 1.11 applies here

$$\tau_{yx} = \mu \frac{dV_x}{dy} \cong \mu \frac{\Delta V_x}{\Delta y} = \mu \frac{(V_1 - V_0)}{(y_1 - y_0)}$$

We use the form on the far RHS because the gradient (dV_x/dy) is constant, that is, the velocity varies linearly in the gap. Setting V_0 and y_0 equal to zero and $\tau_{yx} = F_x/A_y$, the equation can be rearranged to solve for F_x:

$$F_x = \tau_{yx} A_y = \frac{A_1 \mu V_1}{y_1}$$

Inserting the given quantities with units and conversion factors, and noting that 200 cP = 2 P = 2 (g/cm s) = 2 (dyn s/cm²), gives

$$F_x = (1\,m^2)(100\,cm/m)^2(2\,dyn\,s/cm^2)(1\,ft/s)(30.48\,cm/ft)\left(\frac{1}{1\,mm}\right)(10\,mm/cm)$$

$$= 6.096 \times 10^6 \, dyn$$

$$= 60.96\,N\,(0.2242\,lb_f/N) = \boxed{13.7\,lb_f}$$

Note that the actual number of significant digits is indeterminate because all given values are whole numbers, which implies either "infinitely exact" values or unspecified precision. When in doubt about the number of significant digits to report in the answer, the rule of thumb is *a maximum of three digits, unless more can be justified*. Note that three significant digits implies a precision somewhere between 0.05% and 0.5%, which is better than that of most engineering data.

"I was told to measure twice with a micrometer, mark it with chalk, and then cut it with an axe."

—Anonymous

IV. THE "SYSTEM"

The basic conservation laws, as well as the transport models, are generally applied to a specified region or volume of fluid designated as "the system" (sometimes called a "control volume"). The *system* is not actually the volume of the fluid container (e.g., a tank, pipe, pump, etc.) but is the *fluid within a defined region*. For flow problems, there may be one or more streams entering and/or leaving the system, each of which can carry or transport a conserved quantity (e.g., Q) into and/or out of the system at a defined rate (Figure 1.2). Q may also be transported into or out of the system through the system boundaries by other means in addition to being carried by the in/out streams. The general conservation law for a flow problem with a defined system with respect to any conserved quantity Q can be written as

$$\left\{ \begin{matrix} \text{Rate of } Q \\ \text{into the system} \end{matrix} \right\} - \left\{ \begin{matrix} \text{Rate of } Q \\ \text{out of the system} \end{matrix} \right\} = \left\{ \begin{matrix} \text{Rate of accumulation of } Q \\ \text{within the system} \end{matrix} \right\} \quad (1.12)$$

If Q can be produced or consumed within the system (e.g., through a chemical or nuclear reaction, speeds approaching the speed of light, etc.), then a "rate of generation" term can be included on the

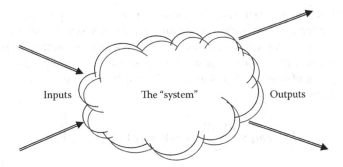

FIGURE 1.2 The "system."

left of Equation 1.12. However, these effects will not be significant in the systems with which we are concerned. The system in Figure 1.2, for example, would be the material (fluid) between the parallel plates. There are no "in" or "out" streams in this example, but the conserved quantity is transported into the system by microscopic (molecular) interactions through the upper boundary of the system. These and related concepts will be expanded upon in Chapter 5 and succeeding chapters.

V. TURBULENT MACROSCOPIC (CONVECTIVE) TRANSPORT MODELS

The transport expressions illustrated in Equations 1.5 through 1.7 and 1.9 apply to the transport of heat, mass, charge, or momentum from one region of the continuum to another by virtue of molecular interactions only. That is, there is no actual bulk motion of the material in the transport direction (y), which means that the medium must be stationary or moving only in the direction normal to the direction of the transport (x). Consequently, the flow must be *laminar*, that is, all elements move in straight parallel streamlines in the x direction. This occurs only if the velocity is sufficiently low so that it is dominated by stabilizing viscous forces. However, as the velocity increases, destabilizing inertial forces eventually overcome the viscous forces and the flow becomes *turbulent*. Turbulence produces a 3D fluctuating flow field dominated by "eddies" that produce a high degree of mixing or "convection" due to the bulk motion of these eddies. As a result, the flow is highly mixed, except for a region in the immediate vicinity of the solid boundary that is called the *boundary layer* (thickness δ). As the fluid velocity must be zero at a stationary boundary, there must be a region (boundary layer) adjacent to the wall which is laminar. Consequently, the major resistance to transport in turbulent (convective) flows is within this boundary layer, the size of which depends upon the dynamic state of the flow field, as well as the fluid and solid properties. As the distance from the wall increases, this laminar boundary layer will transition to a turbulent boundary layer with a thickness that depends on the intensity of the turbulence in the main flow field. This turbulent boundary layer thickness is typically small compared to the total dimensions of the flow field (see Chapter 6 for more detail on this subject).

The general turbulent transport models for the transport of heat and mass are

$$\text{Heat flux:} \quad q_y = k_e \frac{\Delta T}{\delta} = h \Delta T \tag{1.13}$$

$$\text{Mass flux:} \quad n_{Ay} = D_e \frac{\Delta C_A}{\delta} = K_m \Delta C_A \tag{1.14}$$

where
  k_e is a turbulent or "eddy thermal conductivity"
  h is the *heat transfer coefficient*
  D_e is the "eddy diffusivity"
  K_m is the *mass transfer coefficient*

The situation with regard to convective (turbulent) momentum transport is somewhat more complex because of the tensorial character of momentum flux (or stress). As previously explained, Newton's second law provides a correspondence between a force F_x (acting in the x direction) and the rate of transport of x momentum. For continuous steady flow in the x direction at a constant velocity V_x in a conduit of cross-sectional area A_x, there is transport of x momentum in the x direction, which is given by

$$F_x = ma_x = \frac{d(mV_x)}{dt} = \dot{m}V_x = (\rho V_x A_x)V_x = \rho V_x^2 A_x \tag{1.15}$$

The corresponding flux of x momentum in the x direction is $F_x/A_x = \rho V_x^2$. This is also the driving force for the transport of x momentum in the $-y$ direction (toward the wall), that is, $\tau_{yx} = F_x/A_y$. Therefore, the convective flux of x momentum from the fluid to the wall (or the equivalent stress exerted by the fluid on the wall) can be expressed as

$$\text{Momentum flux from fluid to wall:} \left(\tau_{yx}\right)_{mf \to wall} = \tau_w = \frac{f}{2}\rho V_x^2 \tag{1.16}$$

where f is the *Fanning friction factor*. The factor of 2 is arbitrary, and various other definitions of the friction factor vary from this by factors of 2 or 4 (see Chapter 6). Although Equation 1.16 is the counterpart of Equations 1.13 and 1.14 for heat and mass, it appears somewhat different as a result of the correspondence between force and the rate of momentum, and the multidimensional (tensor) nature of stress and momentum flux. It is seen from Equation 1.11 that laminar flows are dominated by the fluid viscosity (which is stabilizing), and from Equation 1.16 that turbulent flows are dominated by the fluid density (i.e., inertial forces), which is destabilizing. A fundamental definition of f and its dependence on flow conditions and fluid properties, which is consistent for either laminar or turbulent flow, are discussed in Chapters 5 and 6. Because of the highly complex unsteady fluctuations in turbulent flows, the corresponding turbulent transport coefficients (i.e., h, K_m, and f) cannot be predicted from fundamental principles and must be determined by experimental investigation and correlation.

Example 1.4

It can be argued that all physical quantities have the mathematical properties of a "tensor," when the latter is defined in terms of the number of directions associated with the quantity. For example, a scalar can be described as a "zero-order tensor," as it has magnitude only, but zero directions associated with it; a vector is a "first-order tensor" as it has a magnitude and one direction; a dyad is a "second-order tensor," with a magnitude and two directions, etc.

(a) With regard to the fluid contained between two parallel plates illustrated in Figure 1.1, identify the "tensorial order" and directional character of each of the transported quantities associated with the fluid.
(b) Considering an extension of the "vector" and "tensor (dyad)" character of these quantities, can you think of a physical quantity that can be described as a third-order tensor (e.g., a triad)?

Solution:

(a) The transported quantities are heat, mass, electric charge, and momentum. Each of these quantities is measured by an intensive variable that indicates the "concentration" of that quantity, per unit mass of the medium (e.g., temperature, concentration, voltage or potential, and velocity). All of these quantities, and their intensive variables, are scalars ("zero-order tensors") except momentum (velocity), which is a "first-order tensor" (vector) with magnitude and one direction.

(b) The rate of transport of these quantities is proportional to the rate at which the concentration of that quantity (as represented by its intensive property) changes with direction, that is, it is "concentration gradient." It can be observed that the "gradient" of a quantity adds an additional direction to that quantity, as the gradient represents the magnitude and direction of the change of that quantity and hence is a "gradient vector." Hence, the gradient of each of the scalars (heat/temperature, mass/concentration, electric charge/potential) is a vector, and the gradient of a vector (e.g., momentum/velocity) is a second-order tensor (dyad). By extension, the gradient of a dyad (momentum flux or stress) must be a "triad" or "third-order tensor," that is, the rate at which "stress" (components) varies with position, or the "stress gradient" must be a triad. In principle, this concept can be extended indefinitely.

SUMMARY

The key points covered in this chapter include:

- A basic understanding of quantities that are "conserved" in engineering systems (e.g., mass, energy, momentum, electric charge, "dimensions," and "dollars");
- The equations defining fundamental laws that govern the *rate of transport* of these conserved quantities in a dynamic system by both microscopic (molecular) and macroscopic (convective) mechanisms (e.g., Fourier's law, Fick's law, Ohm's law, Newton's law of viscosity, and the turbulent convective transport coefficients for mass, heat, and momentum);
- The concepts of *force* and *momentum* and their relationship to *stress* and *momentum flux*;
- The concept of a *system* as the reference or control volume to which the conservation equations are applied;
- The significance of *tensors* versus *vectors* and *scalars* with respect to conserved and transported quantities;
- The "4 C's" guidelines for effective problem solving and the importance of "experience" and "judgment" in the practice of engineering.

PROBLEMS

1. Write the general 3D equations for the definitions of each of the following laws: Fick's, Fourier's, Ohm's, and Newton's laws. Identify the conserved quantity in each of these laws. Can you represent all of these laws by the same generalized expression? If so, does that mean that all of these processes are always analogous? If not, why not?

2. The general expression for the conservation of any (conserved) quantity Q can be written in the form of Equation 1.12. We have said that "dollars" are also a conserved quantity, so Equation 1.12 should apply. Substituting "$" for Q in this equation and assuming that the "system" is your bank account:
    (a) Identify specific items that correspond to "rate in" and "rate out" and "rate of accumulation" with respect to your "system" (i.e., each term should correspond to the rate at which dollars move into or out of your account).
    (b) Identify one or more "driving force" effects that govern each of these rate terms, that is, the things that influence how fast the dollars go in or out. Use this to define respective "transport coefficients" relating the "driving force" and the "rate" for each term in the equation.

3. A dimensionless grouping of variables and parameters that are important in pipe flow is called the Reynolds number:

$$N_{Re} = \frac{DV\rho}{\mu} = \frac{\rho V^2}{\mu V/D}$$

where
  V is the average velocity over the cross section in a pipe of diameter D
  ρ is the fluid density
  μ is the fluid viscosity

Comparing the numerator and denominator of the RHS of the equation with Equations 1.16 and 1.11 shows that this group is a ratio of the convective (turbulent) momentum flux to the laminar (viscous) momentum flux, or the ratio of (destabilizing) inertia forces to (stabilizing) viscous forces. Thus, at low values of the Reynolds number, viscous forces dominate and the flow is smooth and stable (laminar), and at high Reynolds numbers, inertial forces dominate and the flow is unstable (turbulent). It has been found that laminar flow occurs in a pipe only when the Reynolds number is less than about 2000. Calculate the maximum velocity and the corresponding volumetric flow rate (in cm^3/s) at which laminar flow of air and water is possible in pipes with the following diameters:

$$D = 0.25, 0.5, 1.0, 2.0, 4.0, 6.0, \text{ and } 10.0 \text{ in.}$$

4. A layer of water is flowing down a flat plate that is inclined at an angle of 20° to the vertical. If the depth of the layer is ¼ in., what is the shear stress exerted by the plate on the water? (*Remember*: stress is a *dyad*, with two directions associated with each component.)

5. The relationship between the forces and fluid velocity can easily be illustrated by considering the simple one-dimensional steady flow between two parallel plates, Figure 1.1. The area of each plate $A_y = 1 \text{ m}^2$ and the gap between the two plates is 0.1 mm (very thin). Fluids of different viscosities, as listed below, are contained in between the two plates. A tangential force of 1 N is applied to the top plate moving in the x direction at a fixed velocity V_0 while the bottom plate is held stationary. Calculate the value of V_0 for each of the following fluids:

Substance	Viscosity (cP)
Olive oil	100
Castor oil	600
Pure glycerin (293 K)	1500
Honey	10^4
Corn syrup	10^5
Bitumen	10^{11}
Molten glass	10^{15}

In view of these results, what is the meaning of the phrase from the Bible: "*the mountains flowed before the Lord*" and the significance of the phrase: "*Everything flows*"!

6. A slider bearing consists of a sleeve surrounding a cylindrical shaft that is free to move axially within the sleeve. A lubricant (e.g., grease) is in the gap between the shaft and the sleeve to isolate the metal surfaces and support the stress resulting from the motion of the shaft. The diameter of the shaft is 1 in. and the inside diameter of the sleeve is 1.02 in., and its length is 2 in.
   (a) If you want to limit the total force on the sleeve to less than 0.5 lb_f when the shaft is moving with a velocity of 20 ft/s, what should the viscosity of the grease be, in centipoise?
   (b) If the grease has a viscosity of 400 cP (centipoise), what is the force exerted on the sleeve when the shaft is moving at a velocity of 20 ft/s?

(c) The sleeve is cooled to a temperature of 150°F, and it is desired to keep the shaft tempera-
     ture below 200°F. What is the cooling rate (i.e., the rate at which heat must be removed
     through the sleeve by the coolant, in Btu/h), to achieve this temperature? The properties
     of the grease are as follows: specific heat = 0.5 Btu/lb$_m$ °F, specific gravity (SG) = 0.85,
     thermal conductivity = 0.06 Btu/h ft °F.

(d) If the grease becomes contaminated, it could be corrosive to the shaft metal. Assume that
     this occurs and the surface of the shaft starts to corrode at a rate of 0.1 μm/year. If this cor-
     rosion rate is constant, determine the maximum concentration of metal ions in the grease
     when the ions from the shaft just reach the sleeve. The properties of the shaft metal are
     as follows: molecular weight (MW) = 65, SG = 8.5, diffusivity of metal ions in grease =
     8.5×10^{-5} cm^2/s.

7. By making use of the analogies between the various transport models for molecular transport
    of conserved quantities, describe how you would set up an experiment to solve each of the fol-
    lowing problems by making electrical measurements of voltage and current (e.g., describe the
    design of the experiment, how and where you would measure current and/or voltage, and how
    the measurements would be related to the desired quantity).

(a) Determine the rate of heat transfer from a long cylinder to a fluid flowing normal to the
     cylinder axis, if the surface temperature of the cylinder is T_0 and the fluid temperature far
     away from the cylinder is T_1. Determine also the temperature distribution within the fluid
     and the cylinder.

(b) Determine the rate at which a (spherical) mothball evaporates when it is immersed in stag-
     nant air, and also the concentration distribution of the evaporating compound in the air.

(c) Determine the local stress as a function of position on the surface of a wedge-shaped body
     immersed in a fluid that is flowing slowly parallel to the surface. Determine also the local
     velocity distribution in the fluid as a function of position.

8. What is the "system" in Example 1.3?

9. Explain the "fruit salad law," and how does it apply to the formulation of equations describing
    engineering systems?

NOTATION*

A	Area, [L^2]
A_y	Area component whose normal vector is in the y direction, [L^2]
C	Concentration of soluble species, [M/L^3]
C_A	Concentration of species A, [M/L^3]
c_v	Heat capacity at constant volume, [H/MT]
D	Pipe diameter, [L]
D_{AB}	Diffusivity of species A in medium B, [L^2/t]
D_e	Eddy diffusivity, [L^2/t]
e	Electric potential (concentration of charge), [C/L^2]
f	Pipe Fanning friction factor, [—]
F_x	Component of force in x direction, [ML/t^2]
g	Acceleration due to gravity, [L/t^2]
h	Depth, [L]
h	Heat transfer coefficient, [H/L^{2}tT]
k	Thermal conductivity, [H/LtT]

---

* Dimensions given in brackets are as follows: L = length, M = mass, t = time, T = temperature, C = charge, H = "heat" =
thermal energy = ML^2/T^2 (see Chapter 2 for discussion of units and dimensions).

k_e	Electrical conductivity, [L^2/t]
K_m	Mass transfer coefficient, [L/t]
K_T	General transport coefficient, [L^2/t]
i_y	Flux of charge (current density), [C/L^{2}t]
k	Thermal conductivity, [H/LTt]
L	Length, [L]
m	Mass, [M]
$\dot{m}$	Mass flow rate, [M/t]
n_{Ay}	Flux of species A in the y direction, [M/L^{2}t]
N_{Re}	Reynolds number, [—]
P	Pressure, [M/Lt^2]
Q	Generic notation for any conserved (or transported) quantity, [varies]
q_y	Flux of heat in the y direction, per unit area normal to the y direction, [H/L^{2}t]
t	Time, [t]
T	Temperature, [K or °R]
v_x	Local or point velocity in the x direction, [L/t]
V	Spatial average velocity or velocity of a boundary, [L/t]
X	Coordinate direction, [L]
x, y, z	Coordinate directions, [L]
z	Elevation [L]

GREEK

α_T	Thermal diffusivity (= $\rho c_v T$), [L^2/t]
δ	Boundary layer thickness, [L]
ρ	Density, [M/L^3]
μ	Viscosity, [M/Lt] = [Ft/L^2]
$(\tau_{yx})_{mf}$	Flux of x momentum in the y direction (= $-\tau_{yx}$), [M/Lt^2] = [F/L^2]
τ_{yx}	Shear stress component = (force in x direction)/(area of y surface) = $-(\tau_{yx})_{mf}$, [M/Lt^2] = [F/L^2]
ν	Kinematic viscosity (=μ/ρ), [L^2/t]
Φ	Potential, [F/L^2]

SUBSCRIPTS

mf	Momentum flux, [M/Lt^2] = [F/L^2]
0,1	locations 0,1

REFERENCES

Carreau, P.J., D. Dekee, and R.P. Chhabra, *Rheology of Polymeric Systems: Principles and Applications*, Hanser, New York, 1997.

Chhabra, R.P. and J.F. Richardson, *Non-Newtonian Flow and Applied Rheology*, 2nd edn., Butterworth-Heinemenn, Oxford, U.K., 2008.

Darby, R., *Viscoelastic Fluids*, Marcel Dekker, New York, 1976.

2 Dimensional Analysis and Scale-Up

> *"Today's scientists have substituted mathematics for experiments, and they wander off through equation after equation, and eventually build a structure which has no relation to reality."*
>
> **—Nikola Tesla, 1856–1943, Inventor**

I. INTRODUCTION

In this chapter, we consider various aspects of *dimensions* and *units* and the different systems in use for describing these quantities. In particular, the distinction between *scientific* and *engineering* systems of dimensions is explained, and the various *metric* and *English** units used in each system are discussed. It is important that the engineer be familiar with both English and metric systems, as they are all in common use in various fields of engineering and will continue to be for the indefinite future. It is common to encounter a variety of units in different systems during the analysis of a given problem, and the engineer must be adept at dealing with all of them.

The concept of *conservation of dimensions* will be applied to the dimensional analysis and scale-up of engineering systems. It will be shown how these principles may be used in the design and interpretation of laboratory experiments on "model" or "prototype" systems to predict the behavior of large-scale ("field" or "real") systems (this process is also known as *similitude*). These concepts are presented early on, because we shall make frequent use of them in describing the results of both theoretical and experimental analyses of engineering systems in a form that is the most concise, general, and useful. These methods can also provide guidance in choosing the best approach to follow in the solution of many complex problems.

II. UNITS AND DIMENSIONS

A. DIMENSIONS

The *dimensions* of a quantity identify the *physical character* of that quantity, for example, force (F), mass (M), length (L), time (t), temperature (T), electric charge (e), etc. On the other hand, *units* identify the reference scale by which the magnitude of the respective physical quantity is measured. Many different reference scales (units) can be defined for a given dimension; for example, the dimension of length can be measured in units of miles, centimeters, inches, meters, yards, angstroms, furlongs, light years, kilometers, etc. Dimensions can be classified as either *fundamental* or *derived*. *Fundamental* dimensions are those that cannot be expressed in terms of any other dimensions and include length (L), time (t), temperature (T), mass (M), and/or force (F) (depending upon the *system* of dimensions used). *Derived* dimensions can be expressed in terms of fundamental dimensions. Using the notation that "$[x]$ = the dimensions of x," we see, for example, that area ($[A] = L^2$), volume ($[V] = L^3$), energy ($[E] = FL = ML^2/t^2$), power ($[HP] = FL/t = ML^2/t^3$), viscosity

---

* Although this system of units originated in England, the English have since adopted the SI (metric) system. However, this system is still in use by most companies and the public in the United States and hence is sometimes called the American system.

($[\mu] = Ft/L^2 = M/Lt$), thermal conductivity ($[k] = ML/t^3T$), etc. are all "derived" dimensions that can be expressed in terms of fundamental dimensions.

There are two systems of fundamental dimensions in use (with their associated systems of units). These can be referred to as *scientific* and *engineering* systems. These systems differ basically in the manner in which the dimension of force is defined. In both systems, mass, length, and time are fundamental dimensions. Furthermore, Newton's second law provides a relation between the dimensions of force, mass, length, and time:

$$\text{Force} = \text{mass} \times \text{acceleration}$$

that is,

$$F = ma \tag{2.1}$$

or

$$[F] = [ma] = ML/t^2$$

In *scientific systems*, this is accepted as the definition of force; that is, force is a derived dimension, being identical to ML/t^2.

In *engineering systems*, however, force is considered in a more practical or "pragmatic" context. This is because the mass of a body is not usually measured directly but is instead determined by its "weight" (W), that is, the gravitational force resulting from the mutual attraction between two bodies of mass m_1 and m_2:

$$W = G\frac{m_1 m_2}{r^2} \tag{2.2}$$

where
G is a constant having a value of 6.67×10^{-11} N m^2/kg^2
r is the distance between the centers of masses m_1 and m_2

If m_2 is the mass of the earth, for example, and r is its radius at a certain location on earth, then W is the *weight* of mass m_1 at that location:

$$W = m_1 g \tag{2.3}$$

The quantity g is called the *acceleration due to gravity* and is equal to Gm_2/r^2 where G is a constant. On the earth at sea level and 45° latitude (the condition for *standard gravity*), the value of g is 32.174 ft/s^2 or 9.806 m/s^2. The value of g is obviously different on the moon (different r and m_2), and also varies slightly over the surface of the earth as well (since the radius of the earth varies with both elevation and latitude).

Example 2.1

The average radius of the earth is about 3963 miles. Determine the mass of the earth in kg and lb$_m$.

Analysis:

The acceleration due to gravity, g, is defined by the relation between the earth's mass and radius and the weight of an object. These parameters are embodied in the constant G in Equation 2.2. Combining Equations 2.2 and 2.3, we get

$$g = \frac{Gm_2}{r^2},$$

where m_2 is the mass of the earth and r is its radius. Solving for m_2, and noting from Equation 2.1 that $N = kg\ m/s^2$, we get

$$m_2 = \frac{gr^2}{G} = (9.806\ \text{m/s}^2)(3963\ \text{miles})^2 \left(5280\ \text{ft/miles}\ \frac{1}{3.281}\ \text{m/ft}\right)^2 \left(\frac{1}{6.67 \times 10^{-11}}\ \text{kg}^2/\text{Nm}^2\right)$$

$$= 6.07 \times 10^{24}\ \text{kg} \times 2.205\ \text{lb}_\text{m}/\text{kg},$$

$$\boxed{m_2 = 1.34 \times 10^{25}\ \text{lb}_\text{m}.}$$

Since the mass of a body is determined indirectly by measuring its weight (i.e., the gravitational force acting on the mass) under specified gravitational conditions, engineers decided that it would be more practical and convenient if a system of dimensions were defined in which "what you see is what you get," that is, the numerical magnitudes of mass and weight are equal under standard conditions. This must not violate Newton's law, however, so that both Equations 2.1 and 2.3 are valid. Since the value of g is not unity when expressed in common units of length and time, the only way to have the numerical values of weight and mass be the same under any conditions is to introduce a "conversion factor" that forces this equivalence. This factor is designated g_c and is incorporated into Newton's second law for engineering systems (sometimes referred to as "gravitational systems") as follows:

$$F = \frac{ma}{g_c}, \quad W = \frac{mg}{g_c} \tag{2.4}$$

This additional definition of force is equivalent to treating F as a *fundamental dimension*, in addition to mass (m), the redundancy being accounted for by the conversion factor g_c. Thus, if a unit for the *weight* of mass m is defined so that the numerical values of F and m are identical under standard gravity conditions (i.e., $a = g_{std} = 32.174\ \text{ft/s}^2$), it follows that the numerical magnitude of g_c must be identical to that of g_{std} so that their *magnitudes* cancel out in Equation 2.4. However, it is important to distinguish between g (ft/s^2) and g_c (ft lb$_\text{m}$/lb$_\text{f}$ s^2), because they are fundamentally different quantities. As explained earlier, g is not a constant—it is a *variable* that depends on both m_2 and r (Equation 2.2). However, g_c is a *constant*, because it is merely a conversion factor that is defined by the value of standard gravity. Note that these two quantities are also physically different, because they have different dimensions as well as different units:

$$[g] = \frac{L}{t^2}, \quad [g_c] = \frac{ML}{Ft^2} \tag{2.5}$$

The factor g_c is the conversion factor that relates equivalent force and mass units in engineering systems. In these systems, both *force* and *mass* can be considered *fundamental* dimensions, because they are related by two separate (but compatible) definitions: Newton's second law and the definition of weight. The conversion factor g_c thus accounts for the redundancy in these two definitions.

B. Units

Several different sets of units are in use in both scientific and engineering systems of dimensions. These can be classified as either metric (SI and cgs) or English (fps). Although the internationally accepted standard is the SI scientific system, English engineering* units are still very common in the United States and will probably remain so for the foreseeable future. Therefore, the reader

---

* English engineering units are often referred to as American engineering units.

TABLE 2.1

Systems of Dimensions/Units

Dimension	Scientific				Engineering			
	L	M	F	g_c	L	M	F	g_c
English	ft	lb_m	poundal	1	ft	lb_m	lb_f	32.174
	ft	slug	lb_f	1				
Metric								
(SI)	m	kg	N	1	m	kg_m	kg_f	9.806
(cgs)	cm	g	dyn	1	cm	g_m	g_f	980.6

Conversion factors: g_c [ML/Ft^2], $F = ma/g_c$

$g_c = 32.174$ (lb_m ft/lb_f s^2) $= 9.806$ (kg_m m/kg_f s^2) $= 980.6$ (g_m cm/g_f s^2),

$= 1$ (kg m/N s^2) $= 1$ (g cm/dyn s^2) $= 1$ (slug ft/lb_f s^2) $= 1$ (lb_m ft/poundal s^2),

$= 12$ (in./ft) $= 60$ (s/min) $= 30.48$ (cm/ft) $= 778$ (ft lb_f/Btu) $=$ etc. $= 1\left[0\right]$.

should master at least these two systems and become adept at converting between them. These systems are illustrated in Table 2.1.

Note that there are two different English scientific systems, one in which M, L, and t are fundamental and F is derived, and another in which F, L, and t are fundamental and M is derived. In the system in which mass is fundamental, the unit for mass is the *slug*, and for the system in which force is fundamental, the force unit is the *poundal*. However, these systems are archaic and rarely used in practice. Also, the metric engineering systems with units of kg_f and g_f have generally been replaced by the SI system, although they are still in use in some places. The most common systems in general use are the scientific metric (e.g., SI) and the English engineering (or American) systems.

Since Newton's second law is satisfied identically in scientific units with no conversion factor (i.e., $g_c = 1$), the following identities hold:

$$g_c = 1 \text{ kg m/N s}^2 = 1 \text{ g cm/dyn s}^2 = 1 \text{ slug ft/lb}_f \text{ s}^2 = 1 \text{ lb}_m \text{ ft/poundal s}^2,$$

$$g_c = 1 \text{ kg m/N s}^2 = 1.$$

In summary, in engineering systems both F and M are considered fundamental dimensions because of the engineering definition of weight, in addition to Newton's second law. However, this apparent redundancy is rectified by the conversion factor g_c. The value of this conversion factor in the various engineering units provides the following identities:

$$g_c = 9.806 \text{ kg}_m \text{ m/kg}_f \text{ s}^2 = 980.6 \text{ g}_m \text{ cm/g}_f \text{ s}^2 = 32.174 \text{ lb}_m \text{ ft/lb}_f \text{ s}^2.$$

C.   CONVERSION FACTORS

Conversion factors relate the magnitudes of different units with common dimensions and are actually identities; that is, 1 ft is identical to 12 in., 1 Btu is identical to 778 ft lb_f, etc. Because any identity can be expressed as a ratio with a magnitude and no net dimensions, the same holds for any conversion factor, that is,

$$12 \text{ in./ft} = 778 \text{ ft lb}_f \text{/Btu} = 30.48 \text{ cm/ft} = 14.7 \text{ psi/atm} = 10^3 \text{ dyn/N} = 1\left[0\right].$$

A table of commonly encountered conversion factors is included in the "Unit Conversion Factors" front matter of this book. The value of any quantity expressed in a given set of units can be converted to any other equivalent set of units by multiplying or dividing by the appropriate conversion factor in the table to cancel the unwanted units.

Example 2.2

To convert a quantity X measured in feet to the equivalent in miles:

$$X \text{ ft} = (X \text{ ft})/(5280 \text{ ft/miles}) = (X/5280) \text{ miles}.$$

Note that the conversion factor relating mass units in scientific systems to those in engineering systems can be obtained by equating the appropriate values of g_c from the two systems, for example,

$$g_c = 1 \text{ slug ft/lb}_f \text{ s}^2 = 32.174 \text{ lb}_m \text{ ft/lb}_f \text{ s}^2.$$

Thus, after canceling common units, the conversion factor relating *slugs* to lb_m is $\boxed{32.174 \ lb_m/slug.}$

III.   CONSERVATION OF DIMENSIONS

Physical laws, theories, empirical relations, etc., are expressed by equations that relate the various variables and parameters. These equations usually contain a number of terms—for example, when a projectile is fired from a gun, the relation between the vertical elevation (z) of the projectile and its horizontal distance (x) from the gun at any time can be expressed in the form

$$z = ax + bx^2 \tag{2.6}$$

This equation can be derived from the laws of physics, and in the process the parameters a and b are related to such factors as the muzzle velocity, projectile mass, angle of inclination of the gun, wind resistance, etc. The equation may also be empirical if measured values of z versus x can be fit by an equation of this form, with no reference to the laws of physics. For any equation to be valid, every term in the equation must have the same physical character, that is, the same net dimensions, and consequently the same units in any consistent system of units. This is known as the *law of conservation of dimensions* (otherwise known as the "fruit salad law"—"you can't add apples and oranges, unless you are making fruit salad"). Let us look further at Equation 2.6. Since both z and x have dimensions of length, for example, $[x] = L$, $[z] = L$, it follows from the fruit salad law that the dimensions of a and b must be

$$[a] = 0, \quad [b] = 1/L$$

that is, a has no dimensions (it is dimensionless) and the dimensions of b are 1/L, or L^{-1}. For the sake of argument, let us assume that x and z are measured in feet and that the values of a and b in the equation are 5 and 10 ft^{-1}, respectively. Thus, if $x = 1$ ft,

$$z = (5)(1 \text{ ft}) + (10 \text{ ft}^{-1})(1 \text{ ft})^2 = 15 \text{ ft} \tag{2.7}$$

On the other hand, if we choose to measure x and z in inches, the value of z for $x = 1$ in. is

$$z = (5)(1 \text{ in.}) + (10 \text{ ft}^{-1})\left(\frac{1}{12} \text{ ft/in.}\right)(1 \text{ in.})^2 = 5.83 \text{ in.} \tag{2.8}$$

This is still in the form of Equation 2.6, that is,

$$z = ax + bx^2,$$

but now $a = 5$ and $b = 10/12 = 0.833$ in.$^{-1}$. Thus, the magnitude of a has not changed, but the magnitude of b *has* changed. This simple example illustrates two important principles:

1. *Conservation of dimensions* ("*fruit salad*" *law*): All terms in a given equation must have the same net dimensions and consistent units for the equation to be valid.
2. *Scaling*: The fact that the value of the dimensionless parameter a is the same regardless of the units (e.g., scale) used in the problem illustrates the universal nature of dimensionless quantities. That is, the magnitude of any dimensionless quantity will always be independent of the scale of the problem or the units used. This is the basis for the application of *dimensional analysis*, which permits information and relationships determined in a small-scale system (e.g., a prototype or "model") to be applied directly to a *similar system* of a different size or scale if the system variables are expressed in dimensionless form. This process is known as "scale-up."

The universality of certain dimensionless quantities is often taken for granted. For example, the exponent 2 in the last term of Equation 2.6 has no dimensions and hence has the same magnitude regardless of the scale or units used for measurement. Likewise, the kinetic energy per unit mass of a body moving with a velocity V is given by

$$ke = \frac{V^2}{2} \tag{2.9}$$

Both of the numerical quantities in this equation, 1/2 and 2, are dimensionless so they always have the same magnitude regardless of the units used to measure V.

A.   NUMERICAL VALUES

Ordinarily, numerical quantities that appear in equations that have a theoretical basis (such as the ke equation above) are dimensionless and hence "universal." However, many valuable engineering relations have an empirical rather than a theoretical basis, in which case this conclusion does not always hold. For example, a very useful expression for the (dimensionless) friction loss coefficient (K_f) for valves and fittings that we will use later is

$$K_f = \frac{K_1}{N_{Re}} + K_i \left(1 + \frac{K_d}{D_{n(in.)}^{0.3}} \right) \tag{2.10}$$

Here, N_{Re} is the Reynolds number, which is dimensionless, as are K_f and the constants K_1 and K_i. However, the term $D_{n(in.)}$ is the nominal diameter of the fitting in inches, which has dimensions of length. According to the "fruit salad" law, the constant K_d in the term $K_d/D_{n(in.)}^{0.3}$ must therefore have dimensions of $L^{0.3}$ and so is not independent of scale, that is, its magnitude is defined only in specific units. In fact, as the magnitude of $D_{n(in.)}$ is defined with units of *in.*, K_d must also be expressed in units of *in.*, for example, *in.*$^{0.3}$. If $D_{n(in.)}$ were to be measured in centimeters, for example, the value of K_d would be 1.32 times as large, because $[(1 \text{ in.})(2.54 \text{ cm/in.})]^{0.3} = (2.54 \text{ cm})^{0.3} = 1.32 \text{ cm}^{0.3}$. Therefore, caution must be exercised when using such empirical relations that are not in a dimensionless form or are not dimensionally consistent.

B. Consistent Units

The conclusion that dimensionless numerical values are universal is valid only if a consistent system of units is used for all quantities in a given expression. That is, units for all quantities with the same dimensions should always be the same and not "mixed," that is, always using *ft* for the unit of length and not mixing *ft* and *m* in the same expression. If such is not the case, then the numerical quantities may include conversion factors that relate the different units. For example, the velocity (V) of a fluid flowing in a pipe is related to the volumetric flow rate (Q) and the internal pipe diameter (D) by the following "universal" equation with consistent units:

$$V = \frac{4}{\pi} \frac{Q}{D^2} \tag{2.11}$$

The dimensions must be the same on both sides of the equation, that is, $[V] = L/t$, $[Q/D^2] = (L^3/t)$ $(1/L^2) = L/t$, and if the units are consistent (e.g., all length dimensions in the same units), then the dimensionless numerical values (4 and π) are "universal," with the same values regardless of the system of dimensions or units chosen. However, the following equations for V can also be found in various engineering reference books:

$$V = 183.3 \frac{Q}{D^2} \tag{2.12}$$

$$V = 0.408 \frac{Q}{D^2} \tag{2.13}$$

$$V = 0.286 \frac{Q}{D^2} \tag{2.14}$$

Although the dimensions of V and Q/D^2 are the same in each of these equations, it is evident that the numerical coefficient is not universal despite the fact that it must be dimensionless. This is because a consistent system of units is *not* used in these three equations. In each equation, the unit of V is *ft/s* and D is in *in*. However, in Equation 2.12, Q is in *ft³/s*, whereas in Equation 2.13, Q is in *gallons per min* (*gpm*), and in Equation 2.14, it is in barrels per hour (*bbl/h*). Thus, although the dimensions are consistent, the units are not, and therefore the numerical coefficients must include unit conversion factors. Only in Equation 2.11 are all the units assumed to be from the same consistent set (e.g., Q in *ft³/s* and D in *ft*), so that the factor $4/\pi$ is both dimensionless and unitless and is thus *universal*. It is always advisable to write equations in a universally valid form, where possible, to avoid confusion; that is, all quantities should be expressed in consistent units, and numerical values should be both dimensionless and unitless.

Let us consider another example to illustrate this point. The well-known Hazen–Williams formula for the flow of water in smooth pipes relates the volumetric flow rate (Q, in gpm) to the applied pressure gradient ($\Delta P/L$, in psi/ft) and the pipe diameter (D, in in.) is

$$Q = 61.9 D^{2.63} \left(\frac{\Delta P}{L} \right)^{0.54} \tag{2.15}$$

When Equation 2.15 is rewritten in terms of the dimensions of each quantity, we get

$$\left[L^3 t^{-1} \right] = \left[L \right]^{2.63} \left[ML^{-2} t^{-2} \right]^{0.54}$$

or

$$\left[L^3 t^{-1} \right] = \left[M \right]^{0.54} \left[L \right]^{1.55} \left[t \right]^{-1.08} \tag{2.16}$$

The equation is clearly not dimensionally homogeneous and, therefore, the constant factor of 61.9 is not dimensionless. Thus, its value will change from one system of units to another, and this clearly does not make good sense.

IV. DIMENSIONAL ANALYSIS

The law of conservation of dimensions can be applied to arrange all of the variables and parameters that are involved in a given problem into a set of *dimensionless groups*. The original set of (dimensional) variables can then be replaced by the resulting set of dimensionless groups as a new set of variables, and these can then be used in the equations or correlations that completely define the system behavior. That is, any valid relationship (theoretical or empirical) between the original variables can be expressed in terms of these dimensionless groups. This has two important advantages:

1. Dimensionless quantities are universal (as we have seen), so any relationship involving dimensionless variables is independent of the size or scale of the system (including the fluid properties, as long as the fluid has the same rheological behavior, as discussed later, i.e., Newtonian or non-Newtonian [and described by the same rheological model], compressible or incompressible, etc.). Consequently, information obtained from a model (small-scale) system that can be represented in *dimensionless* form can be applied directly to geometrically and dynamically similar systems of any size or scale. This allows us to translate information directly from laboratory models to large-scale equipment or plant operations (scale-up). Geometrical similarity requires that the two systems have the same shape (geometry), and dynamic similarity requires them to be operating in the same dynamic regime (i.e., both must be in either laminar or turbulent flow). This will be expanded upon later.
2. The number of dimensionless groups is always less than the number of original variables involved in the problem. Thus, the relations that define the behavior of a given system are much simpler when expressed in terms of the dimensionless groups as the variables, because fewer variables are required. In other words, the amount of effort required to represent a relationship between the dimensionless groups is much less than that required to relate each of the original variables independently, and so the resulting relation will be simpler in form. For example, a relation between two variables (*x* vs. *y*) requires two dimensions, whereas a relation between three variables (*x* vs. *y* vs. *z*) requires three dimensions, or a family of two-dimensional curves (e.g., a set of *x* vs. *y* curves, with each curve for a different *z*). This is equivalent to the difference between one page and a book of many pages. Relating four variables would obviously require many books or volumes and relating five variables would necessitate the entire library. Thus, reducing the number of variables from, say, four to two would dramatically simplify any problem involving these variables.

It is important to realize that the process of dimensional analysis only replaces the set of original (dimensional) variables with an equivalent (smaller) set of dimensionless variables (i.e., the dimensionless groups). It does not tell how these variables are related; the relationship must be determined either theoretically, by application of basic scientific principles, or empirically by measurements and data analysis. However, dimensional analysis is a very powerful tool in that it can provide a direct guide for experimental design and scale-up and for expressing operating relationships in the most general, simplified, and useful form in almost all branches of engineering, including living systems.

There are a number of different approaches to dimensional analysis. A classical method is the "Buckingham Π theorem," so called because Buckingham used the symbol Π to represent the dimensionless groups.* Another classical approach, which involves a more direct application of the law of conservation of dimensions, is attributed to Lord Rayleigh. Numerous variations of these methods have also been presented in the literature. The one thing all of these methods have in common is that they require knowledge of the variables and parameters that are important in the problem as a starting point. This can be determined through common sense, logic, intuition, experience, physical reasoning, or by asking someone who is more experienced or knowledgeable. They can also be determined from knowledge of the physical principles that govern the system (e.g., the equations for the conservation of mass, energy, momentum, etc.), as written for the specific system to be analyzed. These equations may be macroscopic or microscopic (e.g., coupled sets of partial differential equations, along with their boundary conditions). The variables and parameters that appear in these governing equations, as well as in any relevant boundary conditions, constitute the set of relevant variables. However, this knowledge often requires as much (or more) insight, intuition, and/or experience than is required to compose the list of variables from logical deduction or intuition.

The analysis of any engineering problem requires key assumptions to distinguish those factors that are important in the problem from those that are insignificant. This can be referred to as the "bathwater" rule—it is necessary to separate the "baby" from the "bathwater" in any problem, that is, to retain the most significant elements (the "baby") and discard the insignificant ones (the "bathwater"), and not vice versa! The talent required to do this depends much more upon sound understanding of fundamentals and the exercise of good judgment and intuition than upon mathematical facility, and the best engineer is most often the one who is able to make the most appropriate assumptions to simplify a problem (i.e., to discard the "bathwater" and retain the "baby"). Many problem statements, as well as solutions, involve assumptions that are implied but not stated. One should always be on the lookout for such implicit assumptions and try to identify them wherever possible since they may set corresponding limits on the applicability of the results.

The method we will use to illustrate the dimensional analysis process is one that involves a minimum of manipulations. It does require, as do all methods, an initial knowledge of the variables (and parameters) that are important in the system and the dimensions of these variables. At this stage, the omission of an important variable from the list of variables or inclusion of an irrelevant variable will obviously lead to an erroneous solution. Thus, compiling the list of relevant variables is germane to the success of this method. This step thus necessitates a good physical understanding of the problem at hand. The objective of the process is to determine an appropriate limited set of dimensionless groups of these variables that can then be used in place of the larger number of original individual variables for the purpose of simplifying the description of the system behavior. The process will be explained by means of an example, and the results will be used to illustrate the application of dimensional analysis to experimental design and scale-up.

A. PIPELINE ANALYSIS

The procedure for performing a dimensional analysis will be illustrated by means of an example concerning the flow of a liquid through a circular pipe. In this example, we will determine an appropriate set of dimensionless groups that can be used to represent the relationship between the flow rate of an incompressible fluid in a pipeline, the properties of the fluid, the dimensions of the pipeline, and the driving force (pressure drop) for moving the fluid, as illustrated in Figure 2.1.

---

* We use the recommended notation of the AIChE for dimensionless groups that are named after their originator, that is, a capital N with a subscript identifying the person the group is named for, such as N_{Re} for Reynolds number, N_{Ar} for Archimedes number, etc. However, there are a variety of other dimensionless quantities that are identified by other symbols.

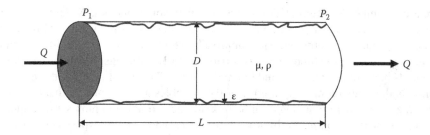

FIGURE 2.1 Flow in a pipeline.

The stepwise procedure is as follows:

Step 1: Identify the important variables. Most of the important *fundamental* variables in this
system should be obvious and are labeled in Figure 2.1. By *fundamental* we mean that none
of the variables should be related to others by definition. For example, the flow rate can be
represented by either the total volumetric flow rate (Q) or the average velocity in the pipeline (V). However, these are related by the definition $Q = \pi D^2 V/4$, so if D is chosen as an
important variable, then either V or Q can be chosen to represent the flow rate, but not both.
We shall choose V. The driving force can be represented by ΔP, the difference between the
pressure at the upstream end of the pipe (P_1) and that at the downstream end (P_2) ($\Delta P =
P_1 - P_2$), assuming the pipe to be horizontal. The pipe dimensions are the diameter (D)
and length (L), and the fluid properties are the density (ρ) and viscosity (μ), assuming the
fluid to be Newtonian. It is also possible that the "texture" of the pipe wall (i.e., the surface
roughness ε) is important. This identification of the pertinent variables is the most important step in the process and can be done by using experience, judgment, brainstorming,
and intuition or by examining the basic equations that describe the fundamental physical
principles governing the system along with appropriate boundary conditions. For example,
one might ask if surface tension and gravity are relevant here. It is also important to include
only "fundamental" variables, that is, those that are not derivable from others through basic
definitions. For example, as pointed out earlier, the fluid velocity (V), the pipe diameter (D),
and the volumetric flow rate (Q) are related by the definition $Q = \pi D^2 V/4$. Thus, these three
variables are not independent, since any one of them can be derived from the other two by
this definition. It would therefore be necessary to include only two of the three.

Step 2: List all the problem variables and parameters, along with their dimensions. The
procedure is simplest if the most fundamental dimensions in a scientific system (i.e., M, L, t)
are used (e.g., force dimensions can be eliminated by utilizing a scientific system, energy
can be converted to FL = ML^2/t^2, etc.):

Variable	Dimensions
V	L/t
$\Delta P = (P_1 - P_2)$	$F/L^2 = M/(Lt^2)$
D	L
L	L
ε	L
ρ	M/L^3
μ	M/(Lt)
Seven variables	Three dimensions

The number of dimensionless groups that will result from the analysis is equal to the number of variables less the minimum number of fundamental dimensions involved in these

variables. Thus, there will be $7 - 3 = 4$ groups in this problem. This is important, but this method does not tell us anything about what these groups are.

Step 3: Choose a set of reference variables. The choice of variables is arbitrary, except that the following criteria must be satisfied:

1. The number of reference variables must be equal to the minimum number of fundamental dimensions in the problem (in this case, three).
2. No two reference variables should have exactly the same dimensions.
3. All the dimensions that appear in the problem variables must also be represented among the dimensions of the reference variables.

In general, the procedure is easiest if the reference variables chosen have the simplest combination of dimensions, consistent with the preceding criteria. In this problem, we have three dimensions (M, L, t), so we need three reference variables. The variables D, ε, and L all have the dimension of length, so we can choose only one of these. We will choose D (arbitrarily) as one reference variable:

$$[D] = \text{L}.$$

The dimension t appears in V, ΔP, and μ, but V has the simplest combination of dimensions, so we choose it as our second reference variable:

$$[V] = \text{L}/\text{t}.$$

We also need a reference variable containing the dimension M, which could be either ρ or μ. Since ρ has the simplest dimensions, we choose it for the third reference variable:

$$[\rho] = \text{M}/\text{L}^3.$$

Our three reference variables are therefore D, V, and ρ, and they contain all three dimensions: M, L, and t.

Step 4: Solve the foregoing "dimensional equations" for the dimensions (L, t, M) in terms of the reference variables (D, V, ρ), that is,

$$\text{L} = [D], \quad \text{t} = [D/V], \quad \text{M} = [\rho D^3].$$

Step 5: Write the dimensional equations for each of the remaining variables. Substitute the results of step 4 for the dimensions in terms of the reference variables:

$$[\varepsilon] = \text{L} = [D],$$

$$[L] = \text{L} = [D],$$

$$[\mu] = \frac{\text{M}}{\text{Lt}} = \left[\frac{\rho D^3}{D(D/V)}\right] = [\rho V D],$$

$$[\Delta P] = \frac{\text{M}}{\text{Lt}^2} = \left[\frac{\rho D^3}{D(D/V)^2}\right] = [\rho V^2].$$

Step 6: Each of these equations is a dimensional identity, so dividing one side by the other results in one dimensionless group from each equation:

$$N_1 = \frac{\varepsilon}{D} \quad \text{or} \quad \frac{D}{\varepsilon},$$

$$N_2 = \frac{L}{D} \quad \text{or} \quad \frac{D}{L},$$

$$N_3 = \frac{\mu}{DV\rho} \quad \text{or} \quad \frac{DV\rho}{\mu},$$

$$N_4 = \frac{\Delta P}{\rho V^2} \quad \text{or} \quad \frac{\rho V^2}{\Delta P}.$$

These four dimensionless groups can now be used as the primary variables to define the behavior of the system in place of the original seven variables. This can be done by grouping the various variables and parameters from a theoretical solution of the problem, or by the grouping of measured quantities from experimental testing.

B. Uniqueness

The results of the foregoing procedure are not unique, because the reciprocal (or any power) of any group is just as valid as the initial group. In fact, any combination of these groups will be dimensionless and will be just as valid as any other combination, as long as all of the original variables are represented among the final set of groups. For example, these four groups can be replaced by any other four groups formed by combining these groups, and, indeed, a different set of (related) groups would have resulted if we had used a different set of reference variables. Also, any set of groups derived by combination of any other set would be valid just as long as all of the initial variables appear somewhere in the set. As we shall see, the set of groups that is most appropriate will depend on the particular problem to be solved, that is, which of the variables are known (independent) and which are unknown (dependent). Specifically, it is most convenient to arrange the groups so that the unknown (dependent) variables each appear in only one group, if possible. It should be noted that the variables that were *not* chosen as a reference variables each appear in only one group.

C. Dimensionless Variables

The original seven variables in this problem may now be replaced by an equivalent set of four dimensionless groups of variables. For example, if it is desired to determine the driving force required to transport a given fluid at a given rate through a given pipe, the relation could be represented as

$$\Delta P = fn(V, D, L, \varepsilon, \rho, \mu)$$

or, in terms of the equivalent dimensionless variables (groups),

$$N_4 = fn(N_1, N_2, N_3).$$

Note that the number of variables has been reduced from the original seven to four (groups). At this point, we can conjecture whether additional variables, such as gravity (g) or surface tension (σ), should be included. This would obviously result in two additional groups, but would they be relevant? As this is a closed system with no fluid–fluid interface, it is expected that surface tension would not be significant (surface tension between the fluid and the pipe wall is also negligible compared with the other forces at the wall, unless the pipe diameter is very small). Also, if one end of the pipe is at a different elevation than the other, gravity will contribute to the driving force, but should be a constant for all systems on the earth's surface. As we shall see later, this gravitational driving force can be combined with the ΔP driving force and hence no additional variables would be introduced. Thus, selection of the significant variables is extremely important. Furthermore, the relationship between these dimensionless variables or groups is independent of scale (i.e., the "size" of L, D, or ε) and the fluid properties (assumed to be Newtonian and incompressible, thus characterized by constant values of ρ and μ). The functional relationship between these groups is always the same for *similar* systems, regardless of size or scale. Thus, any two *similar* systems will be exactly equivalent if the values of all dimensionless variables or groups are the same in each. By *similar* we mean that both systems must have the same geometry or shape (which is actually another dimensionless variable), and both must be operating under comparable dynamic conditions (e.g., either laminar or turbulent flow, as will be expanded on later). Also, the fluids must be rheologically similar (e.g., Newtonian, or the same class of non-Newtonian. The difference between Newtonian and non-Newtonian fluids will be discussed in detail in Chapter 3.). For the present, a Newtonian fluid is one that requires only one rheological property, the viscosity (μ), to determine its flow behavior, whereas a non-Newtonian fluid requires a rheological *function* that contains two or more parameters. Each of these parameters is a rheological property, so in place of the viscosity for a Newtonian fluid, the non-Newtonian fluid would require two or more *rheological properties* or *parameters*, depending upon the specific model that describes the fluid viscosity function, and a corresponding increase in the number of dimensionless groups.

D.   PROBLEM SOLUTION

It should be emphasized that the specific relationship between the variables or groups that is implied in the foregoing discussion is not determined by dimensional analysis. This must be determined from theoretical or experimental analysis. Dimensional analysis gives only an appropriate set of independent dimensionless groups that can be used as generalized variables in these relationships, and which minimizes the number of variables needed to characterize a system. However, because of the universal generality of the dimensionless groups, any functional relationship between them that is valid in any system must also be valid in any other *similar* (geometrically and dynamically) system.

E.   ALTERNATE GROUPS

The preceding set of dimensionless groups is convenient for representing the behavior of a pipeline if it is desired to determine the driving force (ΔP) required to move a given fluid at a given rate through a given pipeline, because the unknown quantity (ΔP) appears in only one group (N_4), which can be considered the *dependent variable*. However, the same variables apply if the driving force (e.g., ΔP) is known and it is desired to determine the flow rate (Q or V) that would result for a given fluid through a given pipe. In this case, V is the *dependent (unknown) variable*, but it appears in more than one group (N_3 and N_4). Therefore, there is no single dependent group containing the unknown variable. However, this set of groups is not unique, so we can rearrange the groups into another equivalent set in which the unknown velocity appears in only one group. This can easily be done, for example, by combining groups N_3 and N_4 to form a new group (N_5) that does not contain V:

$$N_5 = N_3^2 N_4 = \left(\frac{DV\rho}{\mu}\right)^2 \left(\frac{\Delta P}{\rho V^2}\right) = \frac{\Delta P D^2 \rho}{\mu^2} \tag{2.17}$$

This new group can then be used in place of either N_3 or N_4, along with N_1 and N_2 to complete the required set of four groups in which the unknown V appears in only one group. If we replace N_4 by N_5, the implied relation can be expressed as

$$N_3 = fn\left(N_1, N_2, N_5\right) \quad \text{or} \quad \frac{DV\rho}{\mu} = \left(\frac{\varepsilon}{D}, \frac{L}{D}, \frac{\Delta P D^2 \rho}{\mu^2}\right) \tag{2.18}$$

in which the unknown (V) appears only in the group on the left.

Let us reexamine our original problem for a moment. If the pipeline is relatively long and is operating at steady state and the fluid is incompressible, then the conditions over any given length of the pipe will be the same as any other segment of the same length, except for the regions very near the entrance and exit of the pipe. If these regions are short relative to the rest of the pipe (e.g., $L \gg D$), their effect is negligible and therefore the pressure drop *per unit length of pipe* should be the same over all segments of the pipe. Thus, the only significance of the pipe length is to spread the total pressure drop over the entire length, so that the two variables ΔP and L are not independent because $\Delta P \propto L$. They can therefore be combined into one variable—the *pressure gradient*, $\Delta P/L$. This reduces the total number of variables from seven to six and the number of groups from four to three. These three groups can be derived by following the original procedure. However, because ΔP and L each appear in only one of the original groups (N_2 and N_4 respectively), dividing one of these by the other will automatically produce a new group (N_6) with the desired variable (the pressure gradient) in the resulting group, which will then replace N_2 and N_4:

$$N_6 = \frac{N_4}{N_2} = \frac{D\Delta P/L}{\rho V^2} \tag{2.19}$$

The three groups are now N_1, N_3, and N_6:

$$N_1 = \frac{\varepsilon}{D}, \quad N_3 = \frac{DV\rho}{\mu}, \quad N_6 = \frac{D\Delta P/L}{\rho V^2} \tag{2.20}$$

Group N_6 (or some multiple thereof) is also known as a *friction factor* (f), because the driving force (ΔP) is required to overcome "friction" (i.e., the energy dissipated) in the pipeline (assuming it to be horizontal), and N_3 is known as the *Reynolds number* (N_{Re}). There are various definitions of the pipe friction factor, each of which is some multiple of N_6; for example, the *Fanning friction factor* is $N_6/2$ and the *Darcy friction factor* is $2N_6$. These are discussed in further detail in Chapter 5. The group N_4 is also known as the *Euler number*.

V. SCALE-UP

We have stated that dimensional analysis results in an appropriate set of groups that can be used to describe the behavior of a system independent of its scale, but it does not tell how these groups are related. In fact, dimensional analysis does not result in any *numbers* related to the groups (except for exponents on the variables). The relationship between the groups that represents the system behavior must be determined by either theoretical analysis or experimentation. Even when theoretical results are possible, however, it is often necessary to obtain data to evaluate or confirm the adequacy of the theory. Because relationships between dimensionless variables are independent of scale, the groups also provide a guide for the proper design of a small-scale experiment that is intended to simulate a larger-scale similar system and for scaling up the results of model measurements to the full-scale system. For example, the operation of our pipeline can now be described by a functional relationship of the form

$$N_6 = fn(N_1, N_3)$$

or

$$\frac{D\Delta P/L}{\rho V^2} = fn\left(\frac{\varepsilon}{D}, \frac{DV\rho}{\mu}\right) \tag{2.21}$$

This is valid for any Newtonian fluid in any (circular) pipe of any size (scale) under given dynamic conditions (e.g., laminar or turbulent). Conversely, for given values of $N_1 = \varepsilon/D$ and $N_3 = DV\rho/\mu$, there will always be a unique value of N_6 regardless of the size (L, D) of the pipe and the fluid properties (ρ, μ). Thus, if the values of N_3 (the Reynolds number, N_{Re}) and N_1 (the relative roughness, ε/D) for an experimental model (prototype) are identical to the values for a full-scale system, it follows that the value of N_6 (the friction factor) must also be the same in the two systems under the same dynamic conditions. In such a case, the *model* is said to be dynamically similar to the full-scale (*field*) system, and measurements of the variables in N_6 can be translated (scaled) directly from the *model* to the *field* system. In other words, the equality between the groups $N_3(N_{Re})$ and $N_1(\varepsilon/D)$ in the model and in the field is a necessary condition for the dynamic similarity of the two systems.

Example 2.3 Laminar Flow of a Newtonian Fluid in a Pipe

It turns out (for reasons that will be explained later) that if the Reynolds number in pipe flow has a value less than about 2000, the fluid elements follow a smooth, straight pattern called *laminar flow*. In this case, the "friction loss" (i.e., the pressure drop) does not depend upon the pipe wall roughness (ε) or the density (ρ) (the reason for this will also become clear when we examine the mechanism of pipe flow in Chapter 6). With two fewer variables we would have two fewer groups, so that for a "long" pipe ($L \gg D$) the system can be described completely by only one group (that does not contain either ε or ρ). The form of this group can be determined by repeating the dimensional analysis procedure or simply by eliminating these two variables from the original three groups. This is easily done by multiplying the friction factor (f) by the Reynolds number (N_{Re}) to get the required group that is independent of the fluid density (ρ), that is,

$$N_7 = fN_{Re} = \left(\frac{DV\rho}{\mu}\right)\left(\frac{D\Delta P/L}{\rho V^2}\right) = \frac{\Delta P D^2}{LV\mu} = \text{Constant} \tag{2.22}$$

Because this is the only "variable" that is needed to describe this system, it follows that the value of this group must be the same for *all laminar flows* of any Newtonian fluid at any flow rate in any pipe. This is in contrast to *turbulent pipe flow* (which occurs for $N_{Re} > 4000$) in long pipes, for which three groups (e.g., f, N_{Re}, and ε/D) are required to describe the flow completely. That is, turbulent flow in two different pipes must satisfy the same functional relationship between these three groups even though the actual values of the individual groups may be quite different in the two systems. However, for laminar pipe flow, since only one group (fN_{Re}) is required, the value of that group must be the same in *all* Newtonian laminar pipe flows, regardless of the values of the individual variables. The numerical value of this group will be derived theoretically in Chapter 6 (where it is shown that the value is 16 when f is the *Fanning friction factor*).

As an example of the application of dimensional analysis to experimental design and scale-up, consider the following problem.

Example 2.4 Scale-Up of Pipe Flow

We would like to know the total pressure driving force (ΔP) required to pump an oil ($\mu = 30$ cP, $\rho = 0.85$ g/cm^3) through a horizontal pipeline with a diameter (D) of 48 in. and a length (L) of 700 miles, at a flow rate (Q) of 1 million barrels/day. The pipe is to be commercial steel, which has an equivalent roughness (ε) of 0.0018 in. To get this information, we want to design a laboratory experiment in which the laboratory model (m) and the full-scale field pipeline (f) are operating under dynamically similar conditions so that measurements of ΔP in the model can be scaled up directly to find ΔP in the field. The necessary conditions for dynamical similarity for this system are

$$\left(N_3\right)_m = \left(N_3\right)_f \quad \text{or} \quad \left(\frac{DV\rho}{\mu}\right)_m = \left(\frac{DV\rho}{\mu}\right)_f$$

and

$$\left(N_1\right)_m = \left(N_1\right)_f \quad \text{or} \quad \left(\frac{\varepsilon}{D}\right)_m = \left(\frac{\varepsilon}{D}\right)_f,$$

from which it follows that

$$\left(N_6\right)_m = \left(N_6\right)_f \quad \text{or} \quad \left(\frac{\Delta PD}{L\rho V^2}\right)_m = \left(\frac{\Delta PD}{L\rho V^2}\right)_f.$$

Since the volumetric flow rate (Q) is specified instead of the velocity (V), we can make the substitution $V = 4Q/\pi D^2$ to get the following equivalent groups:

$$\left(\frac{\varepsilon}{D}\right)_m = \left(\frac{\varepsilon}{D}\right)_f \tag{2.23}$$

$$\left(\frac{4Q\rho}{\pi D\mu}\right)_m = \left(\frac{4Q\rho}{\pi D\mu}\right)_f \tag{2.24}$$

$$\left(\frac{\pi^2 \Delta PD^5}{16L\rho Q^2}\right)_m = \left(\frac{\pi^2 \Delta PD^5}{16L\rho Q^2}\right)_f \tag{2.25}$$

Note that all the numerical coefficients cancel out. By substituting the known values for the pipeline variables into Equation 2.24, we find that the value of the Reynolds number for this flow is 54,500, as follows:

$$N_{Re} = \frac{4Q\rho}{\pi D\mu} = \frac{4(10^6 \text{ bbl/day})(0.85 \text{ g/cm}^3)}{\pi(48 \text{ in.})(30 \text{ cP})}$$

$$\times \frac{(42 \text{ gal/bbl})(28{,}317 \text{ cm}^3/\text{ft}^3)(100 \text{ cP/P})}{(7.48 \text{ gal/ft}^3)(86{,}400 \text{ s/day})(2.54 \text{ cm/in.})(\text{g/(cm s P)})}$$

$$= 54{,}500.$$

This indicates turbulent flow (see Chapter 6), so we will assume that all three of these groups are important.

We now identify the knowns and unknowns in the problem. The knowns obviously include all of the field variables except $(\Delta P)_f$. Because we will measure the pressure drop in the lab model, $(\Delta P)_m$, after specifying the model test conditions that simulate the field conditions, this will also be known. This value of $(\Delta P)_m$ will then be scaled up to find the unknown pressure drop in the field $(\Delta P)_f$. Thus,

$$\text{Knowns (7):} \quad (D, L, \varepsilon, Q, \mu, \rho)_f, \quad (\Delta P)_m$$

$$\text{Unknowns (7):} \quad (D, L, \varepsilon, Q, \mu, \rho)_m, \quad (\Delta P)_f.$$

There are seven unknowns but only three equations that relate these quantities. Therefore, four of the unknowns can be chosen "arbitrarily." This process is not really arbitrary, however, because we are constrained by certain practical considerations such as a lab model that must be smaller than the field pipeline, and test materials that are convenient, inexpensive, and readily available. For example, the diameter of the pipe to be used in the model could, in principle, be chosen arbitrarily. However, it is related to the field pipe diameter by Equation 2.23:

$$D_m = D_f \left(\frac{\varepsilon_m}{\varepsilon_f} \right).$$

Thus, if we were to use the same pipe material (e.g., commercial steel) for the model as in the field, the roughness values would have to be the same so we would also have to use the same diameter (48 in.) in our lab tests. This is obviously not practical, but a smaller diameter for the model would obviously require a much smoother material in the lab (because $D_m \ll D_f$ requires $\varepsilon_m \ll \varepsilon_f$) to satisfy Equation 2.23. The smoothest material we can find would be glass or plastic or smooth drawn tubing such as copper or stainless steel, all of which have equivalent roughness values of the order of $\varepsilon = 0.00006$ in. (see Table 6.1). If we choose one of these (e.g., plastic), then the required lab diameter is set by Equation 2.23:

$$D_m = D_f \left(\frac{\varepsilon_m}{\varepsilon_f} \right) = (48 \text{ in.}) \left(\frac{0.00006}{0.0018} \right) = 1.6 \text{ in.}$$

Since the roughness values are only approximate, so is this value of D_m. Thus, we could choose a convenient size pipe for the model with a diameter as close as possible to 1.6 in. (e.g., from Appendix F, we see that a *Schedule 40, 1½ in.* pipe has an internal diameter of 1.61 in., which is fortuitous).

We now have five remaining unknowns—$(Q, \rho, \mu, L)_m$ and $(\Delta P)_f$—and only two remaining equations, so we still have three "arbitrary" choices. Of course, we will choose a pipe length for the model that is much less than the 700 miles in the field, but it only has to be much longer than its diameter to avoid end effects. This is inherent in the assumption that the variable $\Delta P/L$ is uniform over the entire pipe. Thus, we can choose any convenient length that will fit into the lab (say, $L = 50$ ft). Actually, as we have seen, ΔP and L are not independent but constitute only one variable, $\Delta P/L$, so that the only restriction on this variable in the lab is that the pipe be "long," that is, $L_m \gg D_m$. This still leaves two "arbitrary" unknowns to specify. Since there are two fluid properties to specify (μ and ρ), this means that we can choose (arbitrarily) any (Newtonian) fluid for the lab test. Water is the most convenient, readily available, and inexpensive fluid, so if we use it $\mu = 1$ cP, $\rho = 1.0$ g/cm³ we will have used all our "arbitrary" choices. The remaining two unknowns, Q_m and $(\Delta P)_f$, are determined by the two remaining equations, Equations 2.24 and 2.25:

$$Q_m = Q_f \left(\frac{\rho_f}{\rho_m} \right) \left(\frac{D_m}{D_f} \right) \left(\frac{\mu_m}{\mu_f} \right) = \left(10^6 \text{ bbl/day} \right) \left(\frac{0.85}{1.0} \right) \left(\frac{1.6}{48} \right) \left(\frac{1.0}{30} \right) = 944 \text{ bbl/day}$$

or

$$Q_m = (944 \text{ bbl/day})(42 \text{ gal/bbl}) \left(\frac{1}{1440} \text{ day/min} \right) = 27.5 \text{ gpm.}$$

Note that if the same units are used for the variables in both the model and the field, no conversion factors are needed because only ratios are involved so that the units cancel out.

Now our experiment has been designed: We will use 1½ in. Sch. 40 plastic pipe (with a wall roughness $\varepsilon = 0.00006$ in.) and an inside diameter of 1.6 in. and a length of 50 ft, and pump water through it at a rate of 27.5 gpm. Then we measure the pressure drop through this pipe in the lab and use our final equation to scale up this value to find the field pressure drop. If the measured pressure drop with this system in the lab is, say, $\Delta P_m = 1.2\,\text{psi}$, then the pressure drop in the field pipeline, from Equation 2.25, would be

$$\Delta P_f = \Delta P_m \left(\frac{D_m}{D_f}\right)^5 \left(\frac{L_f}{L_m}\right)\left(\frac{\rho_f}{\rho_m}\right)\left(\frac{Q_f}{Q_m}\right)^2,$$

$$= (1.2\,\text{psi})\left(\frac{1.6}{48}\right)^5 \left(\frac{700\,\text{miles}}{50\,\text{ft}}\,5280\,\text{ft/miles}\right)\left(\frac{0.85}{1.0}\right)\left(\frac{10^6}{944}\right)^2 = 3480\,\text{psi}.$$

This total pressure driving force would probably not be produced by a single pump but would be apportioned among several pumps spaced along the pipeline.

This example illustrates the power of dimensional analysis as an aid in experimental design and the scale-up of lab measurements to field conditions. We have actually determined the pumping requirements for a large pipeline by applying the results of dimensional analysis to select laboratory conditions and design a laboratory test model that simulates the field pipeline, making one measurement in the lab and scaling up this value to determine the field performance. We have not used any scientific principles or engineering correlations other than the principle of conservation of dimensions and the exercise of logic and judgment. However, we shall see later that information is available to us, based upon similar experiments that have been conducted by others, with the results presented in dimensionless form so that we can use them to solve this and similar problems without conducting any additional experiments.

VI. DIMENSIONLESS GROUPS IN FLUID MECHANICS

Table 2.2 lists some dimensionless groups that are commonly encountered in fluid mechanics problems. The name of the group and its symbol, definition, significance, and most common area of application are given in the table. Wherever feasible, it is desirable to express basic relations (either theoretical or empirical) in dimensionless form, with the variables being dimensionless groups, because this represents the most general way of presenting results and is independent of scale or specific system properties. We shall follow this guideline insofar as is practical in this book.

VII. ACCURACY AND PRECISION

First of all, we should make a clear distinction between *accuracy* and *precision*. *Accuracy* is a measure of how close a given value is to the "true" value, whereas *precision* is a measure of the uncertainty in the value or how "reproducible" that value is. For example, if we were to measure the width of a standard piece of paper using a ruler, we might find that it is 21.5 cm, give or take 0.1 cm. The "give or take" (i.e., the uncertainty) value of 0.1 cm is the precision of the measurement, which is determined by how close we are able to reproduce the measurement with the ruler. However, it is possible that when the ruler is compared with a "standard" unit of measure, it is found to be in error by, say, 0.2 cm. Thus, the "accuracy" of the ruler is limited, which contributes to the uncertainty of the measurement. However, we may not know what this limitation is unless we can compare our "instrument" to one that we know to be true. The accuracy of a given value may be difficult to determine, but the precision of a measurement can be determined by evaluating the *reproducibility,* if multiple repetitions of the measurement are made. Unfortunately, when using values or data provided by others from handbooks, textbooks, journals, etc., we do not usually have access to either the "true" value or information on the reproducibility of the measured values. However, we can make use of both common sense (i.e., reasonable judgment) and convention to

TABLE 2.2
Dimensionless Groups in Fluid Mechanics

Name	Symbol	Formula	Notation	Significance	Application
Archimedes number	N_{Ar}	$N_{Ar} = \dfrac{\rho_f g \Delta\rho d^3}{\mu^2}$	ρ_f = Fluid density $\Delta\rho$ = Solid density − fluid density	(Buoyant × inertial)/ (viscous forces)	Settling particles, fluidization
Bingham number	N_{Bi}	$N_{Bi} = \dfrac{\tau_o D}{\mu_\infty V}$	τ_o = Yield stress μ_∞ = Limiting viscosity	(Yield)/(viscous stresses)	Flow of Bingham plastics
Bond number	N_{Bo}	$N_{Bo} = \dfrac{\Delta\rho d^2 g}{\sigma}$	σ = Surface tension	(Gravity)/(surface tension forces)	Rise or fall of drops or bubbles
Cauchy number	N_c	$N_c = \dfrac{\rho V^2}{K}$	K = Bulk modulus	(Inertial)/ (compressible forces)	Compressible flow
Euler number	N_{Eu}	$N_{Eu} = \dfrac{\Delta P}{\rho V^2}$	ΔP = Pressure drop in pipe	Pressure energy/ kinetic energy	Flow in closed conduits
Drag coefficient	C_D	$C_D = \dfrac{F_D}{\rho V^2 A/2}$	F_D = Drag force A = Cross-sectional area normal to flow	(Drag stress)/(1/2 momentum flux)	External flows
Fanning (Darcy) friction factor	f (f or f_D)	$f = \dfrac{e_f D}{2V^2 L}$ $f_D = 4f$ $f = \dfrac{\tau_w}{\rho V^2/2}$	e_f = Friction loss (energy/mass) τ_w = Wall stress	(Energy dissipated)/ (K.E. of flow × $4L/D$) or (wall stress)/ (momentum flux)	Flow in pipes, channels, fittings, etc.
Froude number	N_{Fr}	$N_{Fr} = \dfrac{V^2}{gL}$	L = Characteristic length	(Inertial)/(gravity forces)	Free surface flows
Hedstrom number	N_{He}	$N_{He} = \dfrac{\tau_o D^2 \rho}{\mu_\infty^2}$	τ_o = Yield stress μ_∞ = Limiting visc.	(Yield × inertia)/ (viscous) stresses	Flow of Bingham plastics
Reynolds number	N_{Re}	$N_{Re} = \dfrac{DV\rho}{\mu} = \dfrac{\rho V^2}{\mu V/D}$ $= \dfrac{4Q\rho}{\pi D\mu} = \dfrac{\rho V^2}{\tau_w/8}$	τ_w = wall stress	Inertial momentum flux/viscous momentum flux	Pipe/internal flows (equivalent forms for external flows)
Mach number	N_{Ma}	$N_{Ma} = \dfrac{V}{c}$	c = Speed of sound	Gas velocity/speed of sound	High speed compressible flow

estimate the implied precision of a given value. If the value is represented in scientific notation, the number of digits to the right of the decimal indicates the implied precision (the uncertainty being one-half of the digit to the far right of the decimal point). For example, if the distance from Dallas to Houston is stated as being 250 miles and we drive at 60 miles/h, should we say that it would take us 4.166667 (= 250/60) h for the trip? This number implies that we can determine the answer to a precision of 0.0000005 h, which is one part in 10^7, or less than 2 ms! This is obviously ludicrous, because the mileage value is nowhere near that precise (is it ±1 mile, ±5 miles?—*exactly* where did we start and end?), nor can we expect to drive at a speed having this degree of precision (e.g., 60 ± 0.000005 mph, or about ±20 μm/s!). It is conventional to assume that the precision of a given number is comparable to the magnitude of one-half of the last digit in that number. That is, we assume that the value of 250 miles implies 250 ± 0.5 miles (or perhaps ± 1 mile). However, unless the numbers are given in scientific notation, so that the least significant digit can be associated with the last digit to the right of the decimal place, there will be some uncertainty, in which case common sense (judgment) should prevail.

For example, if the diameter of a tank is specified to be 10.32 ft, we could assume that this value has a precision (or uncertainty) of about 0.005 ft (or 0.06 in., or 1.5 mm). However, if the diameter is said to be 10 ft, the number of digits cannot provide an accurate guide to the precision of the number. It is unlikely that a tank of that size would be constructed to the precision of 1.5 mm, so we would probably assume (optimistically!) that the uncertainty is about 0.5 in., or that the measurement is "roughly 10.0 ft" (i.e., assume three significant digits). However, if I say that I have five fingers on my hand this means *exactly* five, no more, no less (i.e., an "infinite" number of "significant digits").

In general, the number of decimal digits that are included in reported data, or the precision to which values can be read from graphs or plots, should be consistent with the precision of the data. Therefore, answers calculated from data with limited precision will likewise be limited in precision (computer people have an acronym for this, i.e., "GIGO," which stands for "garbage in, garbage out"). When the actual precision of data or other information is uncertain, a general "rule of thumb" is to report numbers to no more than three "significant digits." This corresponds to an uncertainty of somewhere between 0.05% and 0.5% (which is actually greater precision than can be justified by most engineering data). Inclusion of more than three digits in your answer implies a greater precision than this and should be justified. Those who report values with a large number of digits that cannot be justified are usually making the implied statement "I just wrote down the numbers—I really didn't think about it." This is most unfortunate, because if these people don't think about the numbers they write down, how can we be sure that they are thinking about other critical aspects of the problem?

Example 2.5

Our vacation time accrues by the hour, a certain number of hours of vacation time being credited per month worked. When we request leave or vacation, we are likewise expected to report it in increments of 1 h. We received a statement from the accountants that we have accrued "128.00 h of vacation time." What is the precision of this number?

The precision is implied by half of the digit furthest to the right of the decimal point, that is, 0.005 h, or 18 s. Does this imply that we must report leave taken to the closest 18 s? (We think not. It takes at least a minute to fill out the leave request form—would this time be charged against our accrued leave? The accountant just "wasn't thinking" when these numbers were reported. Actually, the "two decimal points" were automatically set in her computer.)

When combining values, each of which has a finite precision or uncertainty, it is important to be able to estimate the corresponding uncertainty of the result. Although there are various "rigorous" ways of doing this, a very simple method that gives good results as long as the relative uncertainty is a small fraction of the value is to use the approximation (which is really just the first term of a Taylor series expansion):

$$A(1 \pm a)^x \cong A(1 \pm xa + \cdots),$$

which is valid for any value of x if $a < 0.1$ (about). This assumes that the relative uncertainty of each quantity is expressed as a fraction of the given value, for example, the fractional uncertainty in the value A is a or, equivalently, the percentage error in A is $100a$.

Example 2.6

Suppose we wish to calculate the shear stress on the bob surface in a cup and bob viscometer from a measured value of the torque or moment on the bob (see Chapter 3 for the details). The equation for this is

$$\tau_{r\theta} = \frac{T}{2\pi R_i^2 L}.$$

If the torque (T) can be measured to $\pm5\%$, the bob radius (R_i) is known to $\pm1\%$, and the length (L) is known to $\pm3\%$, the corresponding uncertainty in the shear stress can be determined as follows:

$$\tau_{r\theta} = \frac{T(1\pm0.05)}{2\pi R_i^2(1\pm0.01)^2 L(1\pm0.03)},$$

$$= \frac{T}{2\pi R_i^2 L}[1\pm0.05\pm2(0.02\pm0.03)]$$

$$\boxed{\tau_{r\theta} = \frac{T}{2\pi R_i^2 L}(1\pm0.12).}$$

That is, there would be a 12% error, or uncertainty, in the answer (which we would probably round off to 10%). Note that even though terms in the denominator have a negative exponent, the maximum error due to these terms is still cumulative, because a given error may be either positive or negative, that is, errors may either accumulate (giving rise to the maximum possible error) or all cancel out (we should be so lucky!).

SUMMARY

The key points covered in this chapter include the following:

- The concept of *dimensions* as related to the physical character of a quantity and *units* as a measure of the *magnitude* of the dimensions.
- The distinction between *scientific* and *engineering* systems of dimensions (with associated *metric* and *English* units in both systems) and the definition of *weight* as a gravitational force, being integral to the definition of *engineering* systems of dimensions.
- A clear understanding of the distinction between mass and force.
- The concept that all conversion factors are ratios with a net magnitude of unity and zero dimensions.
- The "fruit salad law" that says that all terms in a given equation *must* have the *same net dimensions (and the same units)* if the equation is valid.
- Numerical values that arise in theoretical expressions are normally dimensionless, but often numbers or parameters in empirical expressions are not dimensionless, or may include conversion factors that are applicable only to specific units.
- The value of any quantity that is dimensionless is the same regardless of the scale or units used in any similar system.
- The *universality* of dimensionless quantities that permits them to be used to *scale-up* results from a small-scale system to apply to a larger similar system.
- Replacing the collection of variables (and parameters) that govern a particular system with the appropriate *dimensionless groups* of these variables results in a fewer number of variables necessary to describe the system, and renders relationships between these variables independent of scale or size of the system.

- Understanding that dimensional analysis merely replaces the collection of individual variables in a given problem with fewer corresponding dimensionless variables (groups), but does not provide a relation between these variables.
- Understanding the difference between *accuracy* and *precision* and methods for determining each.

PROBLEMS

UNITS AND DIMENSIONS

1. Determine the weight of 1 g mass at sea level in units of:
   (a) dynes
   (b) lb_f
   (c) g_f
   (d) poundals

2. One cubic foot of water weighs 62.4 lb_f under conditions of standard gravity.
   (a) What is its weight in dynes, poundals, and g_f?
   (b) What is its density in lb_m/ft^3 and slugs/ft³?
   (c) What is its weight on the moon ($g = 5.4$ ft/s²) in lb_f?
   (d) What is its density on the moon?

3. The acceleration due to gravity on the moon is about 5.4 ft/s². If your weight is 150 lb_f on the earth:
   (a) What is your mass on the moon, in slugs?
   (b) What is your weight on the moon, in SI units?
   (c) What is your weight on earth, in poundals?

4. You weigh a body with a mass m on an electronic scale, which is calibrated with a known mass.
   (a) What does the scale actually measure, and what are its dimensions?
   (b) If the scale is calibrated in the appropriate system of units, what would the scale reading be if the mass of m is (1) 1 slug; (2) 1 lb_m (in scientific units); (3) 1 lb_m (in engineering units); (3) 1 g_m (in scientific units); (4) 1 g_m (in engineering units).

5. Explain why the gravitational "constant" (g) is different at Reykjavik, Iceland, than it is at Quito, Ecuador. At which location is it greatest, and why? If you could measure the value of g at these two locations, what would this tell you about the earth?

6. You have purchased a 5 oz bar of gold (100% pure), at a cost of $400/oz. Because the bar was weighed in air, you conclude that you got a bargain, because its true mass is greater than 5 oz due to the buoyancy of air. If the true density of the gold is 1.9000 g/cm³, what is the actual value of the bar based upon its true mass?

7. You purchased 5 oz of gold in Quito, Ecuador ($g = 977.110$ cm/s²), for $400/oz. You then took the gold and the same spring scale on which you weighed it in Quito to Reykjavik, Iceland ($g = 983.06$ cm/s²), where you weighed it again and sold it for $400/oz. How much money did you make or lose, or did you break even?

8. Calculate the pressure at a depth of 2 miles below the surface of the ocean. Explain and justify any assumptions you make. The physical principle that applies to this problem can be described by the equation

$$\Phi = \text{constant},$$

   where

   $\Phi = P + \rho gz$

   z is the vertical distance measured upward from any horizontal reference plane

   Express your answer in units of (a) atm, (b) psi, (c) Pa, (d) poundal/ft², (e) dyn/cm², (f) kg_f/m^2, (g) N/m².

9. (a) Use the principle in Problem 8 to calculate the pressure at a depth of 1000 ft below the surface of the ocean (in psi, Pa, and atm). Assume that the ocean water density is 64 lb_m/ft^3.

  (b) If this ocean were on the moon, what would be the answer to (a)? Use the following information to solve this problem: The diameter of the moon is 2160 miles, the diameter of the earth is 8000 miles, and the density of the earth is 1.6 times that of the moon.

10. The following formula for the pressure drop through a valve was found in a design manual:

$$h_L = \frac{522\,K q^2}{d^4},$$

where
 h_L is the "head loss" in feet of fluid flowing through the valve
 K is the dimensionless resistance coefficient for the valve
 q is the flow rate through the valve, in ft^3/s
 d is the diameter of the valve, in in.

  (a) Can this equation be used without changing anything if SI units are used for the variables? Explain.
  (b) What are the dimensions of "522" in this equation? What are its units?
  (c) Determine the pressure drop through a 2 in. valve with a K of 4 for water at 20°C flowing at a rate of 50 gpm, in units of: (1) feet of water, (2) psi, (3) atm, (4) Pa, (5) dyn/cm^2; and (6) inches of mercury.

11. When the energy balance on the fluid in a stream tube is written in the following form, it is known as the Bernoulli equation:

$$\frac{P_2 - P_1}{\rho} + g(z_2 - z_1) + \frac{\alpha}{2}\left(V_2^2 - V_1^2\right) + e_f + w = 0,$$

where
 w is the work done on a unit mass of fluid
 e_f is the energy per unit mass dissipated by friction in the fluid, including all thermal energy effects due to heat transfer or internal generation
 α is equal to either 1 or 2 for turbulent or laminar flow, respectively

If $P_1 = 25$ psig, $P_2 = 10$ psig, $z_1 = 5$ m, $z_2 = 8$ m, $V_1 = 20$ ft/s, $V_2 = 5$ ft/s, $\rho = 62.4\ lb_m/ft^3$, $\alpha = 1$, and $w = 0$, calculate the value of e_f in each of the following systems of units:
  (a) SI
  (b) mks engineering (e.g., metric engineering)
  (c) English engineering
  (d) English scientific (with M as a fundamental dimension)
  (e) English thermal units (e.g., Btu)
  (f) Metric thermal units (e.g., calories)

CONVERSION FACTORS, PRECISION

12. Determine the value of the gas constant, R, in units of ft^3 atm/lb mol °R), starting with the value of the standard molar volume of a perfect gas (i.e., 22.4 m^3/kg mol).

13. Calculate the value of the Reynolds number for sodium flowing at a rate of 50 gpm through a tube with a 1/2 in. ID, at 400°F.

14. The conditions at two different positions along a pipeline (at points 1 and 2) are related by the Bernoulli equation (see Problem 11). For flow in a pipe,

$$e_f = \left(\frac{4fL}{D}\right)\left(\frac{V^2}{2}\right),$$

where
   D is the pipe diameter
   L is the pipe length between points 1 and 2

If the flow is laminar ($N_{Re} < 2000$), the value of $\alpha = 2$ and $f = 16/N_{Re}$, but for turbulent flow in a smooth pipe, $\alpha = 1$ and $f = 0.0791/N_{Re}^{1/4}$. The work done by a pump on the fluid ($-w$) is related to the power delivered to the fluid (HP) and the mass flow rate of the fluid ($\dot{m}$) by $HP = -w\dot{m}$. Consider water ($\rho = 1$ g/cc, $\mu = 1$ cP) being pumped at a rate of 150 gpm through a 2000 ft long 3 in. diameter pipe. The water is transported from a reservoir ($z = 0$) at atmospheric pressure to a condenser at the top of a column that is at an elevation of 30 ft and a pressure of 5 psig. Determine

(a) The value of the Reynolds number in the pipe
(b) The value of the friction factor in the pipe (assuming that it is smooth)
(c) The power that the pump must deliver to the water, in horsepower (hp)

15. The Peclet number (N_{Pe}) is defined as

$$N_{Pe} = N_{Re}N_{Pr} = \left(\frac{DV\rho}{\mu}\right)\left(\frac{c_p\mu}{k}\right) = \frac{DGc_p}{\mu},$$

where
   D is the pipe diameter
   G is the mass flux $= \rho V$
   c_p is the specific heat
   k is the thermal conductivity
   μ is the viscosity

Calculate the value of N_{Pe} for water at 60°F flowing through a 1 cm. diameter tube at a rate of 100 lb$_m$/h. (Use the most accurate data you can find, and state your answer in the appropriate number of digits consistent with the data you use.)

16. The heat transfer coefficient (h) for a vapor bubble rising through a boiling liquid is given by

$$h = A\left(\frac{kV\rho c_p}{d}\right)^{1/2} \quad \text{where} \quad V = \left(\frac{\Delta\rho g\sigma}{\rho_v^2}\right),$$

where
   h is the heat transfer coefficient [e.g., Btu/(h °F ft^2)]
   c_p is the liquid heat capacity [e.g., cal/(g °C)]
   k is the liquid thermal conductivity [e.g., J/(s K m)]
   σ is the liquid/vapor surface tension [e.g., dyn/cm]
   $\Delta\rho = \rho_{liquid} - \rho_{vapor} = \rho_l - \rho_v,$
   d is the bubble diameter
   g is the acceleration due to gravity

(a) What are the fundamental dimensions of V and A?

(b) If the value of h is 1000 Btu/(h ft^2 °F) for a 5 mm diameter steam bubble rising in boiling water at atmospheric pressure, determine the corresponding values of V and A in SI units. You must look up values for the other quantities you need, and be sure to cite the sources you use for these data.

17. Determine the value of the Reynolds number for SAE 10 lube oil at 100°F flowing at a rate of 2000 gpm through a 10 in. Schedule 40 pipe. The oil SG is 0.92, and its viscosity can be found in Appendix A. If the pipe is made of commercial steel ($\varepsilon = 0.0018$ in.), use the Moody diagram (see Figure 6.4) to determine the friction factor f for this system. Estimate the precision of your answer, based upon the information and procedure you use to determine it (i.e., tell what the reasonable upper and lower bounds, or the corresponding percentage variation, should be for the value of f based on the information you used).

18. Determine the value of the Reynolds number for water flowing at a rate of 0.5 gpm through a 1 in. ID pipe. If the diameter of the pipe is doubled at the same flow rate, how much will each of the following change:
(a) The Reynolds number
(b) The pressure drop
(c) The friction factor

19. The pressure drop for a fluid with a viscosity of 5 cP and a density of 0.8 g/cm^3 flowing at a rate of 30 g/s in a 50 ft long 1/4 in. diameter pipe is 10 psi. Use this information to determine the pressure drop for water at 60°F flowing at 0.5 gpm in a 2 in. diameter pipe. What is the value of the Reynolds number for each of these cases?

DIMENSIONAL ANALYSIS AND SCALE-UP

20. In the steady flow of a Newtonian fluid through a long uniform circular tube, if $N_{Re} < 2000$, the flow is laminar and the fluid elements move in smooth straight parallel lines. Under these conditions, it is known that the relationship between the flow rate and the pressure drop in the pipe does not depend upon the fluid density or the pipe wall material.
(a) Perform a dimensional analysis of this system to determine the dimensionless groups that apply. Express your result in a form in which the Reynolds number can be identified.
(b) If water is flowing at a rate of 0.5 gpm through a pipe with an ID of 1 in., what is the value of the Reynolds number? If the diameter is doubled at the same flow rate, what will be the effect on the Reynolds number and on the pressure drop?

21. Perform a dimensional analysis to determine the groups relating the variables that are important in determining the settling rate of very small solid particles falling in a liquid. Note that the driving force for moving the particles is gravity and the corresponding *net* weight of the particle. At very slow settling velocities, it is known that the velocity is independent of the fluid density. Show that this also requires that the velocity be inversely proportional to the fluid viscosity.

22. A simple pendulum consists of a small, heavy ball of mass m at the end of a long string of length L. The period of the pendulum should depend on these factors, as well as gravity, which is the driving force for making it move. What information can you get about the relationship between these variables from a consideration of their dimensions? Suppose you measured the period, T_1, of a pendulum with mass m_1 and length l_1. How could you use this to determine the period of a different pendulum with a different mass and length? What would be the ratio of the pendulum period on the moon to that on the earth? How could you use the pendulum to determine the variation of g on the earth's surface?

23. An ethylene storage tank in your plant explodes. The distance that the blast wave travels from the blast site (R) depends upon the energy released in the blast (E), the density of the air (ρ), and time (t). Use dimensional analysis to determine:
    (a) The dimensionless group(s) that can be used to describe the relationship between the variables in the problem;
    (b) The ratio of the velocity of the blast wave at a distance of 2000 ft from the blast site to the velocity at a distance of 500 ft from the sites;
         The pressure difference across the blast wave (ΔP) also depends upon the blast energy (E), the air density (ρ), and time (t). Use this information to determine:
    (c) The ratio of the blast pressure at a distance of 500 ft from the blast site to that at a distance of 2000 ft from the site.

24. It is known that the power required to drive a fan depends upon the impeller diameter (D), the impeller rotational speed (ω), the fluid density (ρ), and the volume flow rate (Q). (Note that the fluid viscosity is not important for gases under normal conditions.)
    (a) What is the minimum number of fundamental dimensions required to define all of these variables?
    (b) How many dimensionless groups are required to determine the relationship between the power and all the other variables? Find these groups by dimensional analysis, and arrange the results so that the power and the flow rate each appear in only one group.

25. A centrifugal pump with an 8 in. diameter impeller operating at a rotational speed of 1150 rpm requires 1.5 hp to deliver water at a rate of 100 gpm and a pressure of 15 psi. Another pump for water, which is geometrically similar but has an impeller diameter of 13 in., operates at a speed of 1750 rpm. Estimate the pump pressure, flow capacity, and power requirements of this second pump. (Under these conditions, the performance of both pumps is independent of the fluid viscosity.)

26. A gas bubble of diameter D rises with a velocity V in a liquid of density ρ and viscosity μ.
    (a) Determine the dimensionless groups that include the effects of all of the significant variables, in such a form that the liquid viscosity appears in only one group. Note that the driving force for the bubble motion is buoyancy, which is equal to the weight of the displaced fluid.
    (b) You want to know how fast a 5 mm diameter air bubble will rise in a liquid with a viscosity of 20 cP and a density of 0.85 g/cm³. You want to simulate this system in the laboratory using water ($\mu = 1$ cP, $\rho = 1$ g/cm³) and air bubbles. What size (diameter) air bubble should you use?
    (c) You perform the experiment, and measure the velocity of the air bubble in water (V_m). What is the ratio of the velocity of the 5 mm diameter bubble in the field liquid (V_f) to that in the lab (V_m)?

27. You must predict the performance of a large industrial mixer under various operating conditions. To obtain the necessary data, you decide to run a laboratory test on a small-scale model of the unit. You have deduced that the power (P) required to operate the mixer depends upon the following variables:

Tank diameter, D
Impeller diameter, d
Impeller rotational speed, N
Fluid density, ρ
Fluid viscosity, μ

    (a) Determine the minimum number of fundamental dimensions involved in these variables and the number of dimensionless groups that can be defined by them.
    (b) Find an appropriate set of dimensionless groups such that D and N each appear in only one group. If possible, identify one or more of the groups with groups commonly encountered in other systems.

(c) You want to know how much power would be required to run a mixer in a large tank 6 ft in diameter, using an impeller with a diameter of 3 ft operating at a speed of 10 rpm, when the tank contains a fluid with a viscosity of 25 cP and a specific gravity of 0.85. To do this, you run a lab test on a model of the system, using a scale model of the impeller that is 10 in. in diameter. The only appropriate fluid you have in the lab has a viscosity of 15 cp and a specific gravity of 0.75. Can this fluid be used for the test? Explain.

(d) If the preceding lab fluid is used, what size of the tank should be used in the lab, and how fast should the lab impeller be rotated?

(e) With the lab test properly designed and the proper operating conditions chosen, you run the test and find that it takes 150 W to operate the lab test model. How much power would be required to operate the larger field mixer under the plant operating conditions?

28. When an open tank with a free surface is stirred with an impeller, a vortex will form around the shaft. It is important to prevent this vortex from reaching the impeller, because entrainment of air in the liquid tends to cause foaming. The shape of the free surface depends upon (among other things) the fluid properties, the speed and size of the impeller, the size of the tank, and the depth of the impeller below the free surface.

(a) Perform a dimensional analysis of this system to determine an appropriate set of dimensionless groups that can be used to describe the system performance. Arrange the groups so that the impeller speed appears in only one group.

(b) In your plant you have a 10 ft diameter tank containing a liquid that is 8 ft deep. The tank is stirred by an impeller that is 6 ft in diameter and is located 1 ft from the tank bottom. The liquid has a viscosity of 100 cP and a specific gravity of 1.5. You need to know the maximum speed at which the impeller can be rotated without entraining the vortex in the impeller blades. To find this out, you design a laboratory test using a scale model of the impeller that is 8 in. in diameter. What, if any, limitations are there on your freedom to select a fluid for use in the lab test?

(c) Select an appropriate fluid for the lab test and determine how large the tank used in the lab should be and where in the tank the impeller should be located.

(d) The lab impeller is run at such a speed that the vortex just reaches the impeller. What is the relation between this speed and that at which entrainment would occur in the tank in the plant?

29. The variables involved in the performance of a centrifugal pump include the fluid properties (μ and ρ), the impeller diameter (d), the casing diameter (D), the impeller rotational speed (N), the volumetric flow rate of the fluid (Q), the head (H) developed by the pump ($\Delta P = \rho g H$), and the power required to drive the pump (HP).

(a) Perform a dimensional analysis of this system to determine an appropriate set of dimensionless groups that would be appropriate to characterize the pump. Arrange the groups so that the fluid viscosity and the pump power each appear in only one group.

(b) You want to know what pressure a pump will develop with a liquid that has a specific gravity of 1.4 and a viscosity of 10 cP, at a flow rate of 300 gpm. The pump has an impeller with a diameter of 12 in., which is driven by a motor running at 1100 rpm. (It is known that the pump performance is independent of fluid viscosity unless the viscosity is greater than about 50 cP.) You want to run a lab test that simulates the operation of the larger field pump using a similar (scaled) pump with an impeller that has a diameter of 6 in. and a 3600 rpm motor. Should you use the same liquid in the lab as in the field, or can you use a different liquid? Why?

(c) If you use the same liquid, what flow rate should be used in the lab to simulate the operating conditions of the field pump?

(d) If the lab pump develops a pressure of 150 psi at the proper flow rate, what pressure will the field pump develop with the field fluid?

(e) What pressure would the field pump develop with water at a flow rate of 300 gpm?

30. The purpose of a centrifugal pump is to increase the pressure of a liquid in order to move it through a piping system. The pump is driven by a motor, which must provide sufficient power to operate the pump at the desired conditions. You wish to find the pressure developed by a pump operating at a flow rate of 300 gpm with an oil having a specific gravity (SG) of 0.8 and a viscosity of 20 cP, and the required horsepower for the motor to drive the pump. The pump has an impeller diameter of 10 in., and the motor runs at 1200 rpm.
   (a) Determine the dimensionless groups that would be needed to completely describe the performance of the pump.
   (b) You want to determine the pump pressure and motor horsepower by measuring these quantities in the lab on a smaller-scale model of the pump that has a 3 in. diameter impeller and a 1800 rpm motor, using water as the test fluid. Under the operating conditions for both the lab model and the field pump, the value of the Reynolds number is very high, and it is known that the pump performance is independent of the fluid viscosity under these conditions. Determine the proper flow rate at which the lab pump should be tested and the ratio of the pressure developed by the field pump to that of the lab pump operating at this flow rate as well as the ratio of the required motor power in the field to that in the lab.
   (c) The pump efficiency (η_e) is the ratio of the power delivered by the pump to the fluid (as determined by the pump pressure and flow rate) to the power delivered to the pump by the motor. Because this is a dimensionless number, it should also have the same value for both the lab and field pumps when they are operating under equivalent conditions. Is this condition satisfied?

31. When a ship moves through the water, it causes waves. The energy and momentum in these waves must come from the ship, which is manifested as a "wave drag" force on the ship. It is known that this drag force (F) depends upon the ship speed (V), the fluid properties (ρ, μ), the length of the waterline (L), and the beam width (W) as well as the shape of the hull, among other things. (There is at least one important "other thing" that relates to the "wave drag," i.e., the energy required to create and sustain the waves from the bow and the wake. What is this additional variable?) Note that "shape" is a dimensionless parameter, which is implied by the requirement of geometrical similarity. If two geometries have the same shape, the ratio of each corresponding dimension of the two will also be the same.
   (a) Perform a dimensional analysis of this system to determine a suitable set of dimensionless groups that could be used to describe the relationship between all of the variables. Arrange the groups such that viscous and gravitational parameters each appear in separate groups.
   (b) It is assumed that "wave drag" is independent of viscosity and that "hull drag" is independent of gravity. You wish to determine the drag on a ship having a 500 ft long waterline moving at 30 mph through seawater (SG = 1.1). You can make measurements on a scale model of the ship, 3 ft long, in a towing tank containing fresh water. What speed should be used for the model to simulate the wave drag and the hull drag?

32. You want to find the force exerted on an undersea pipeline by a 10 mph current flowing normal to the axis of the pipe. The pipe is 30 in. in diameter, and the density of seawater is 64 lb_m/ft^3 and its viscosity is 1.5 cP. To determine this, you test a 1½ in. diameter model of the pipe in a wind tunnel at 60°F. What velocity should you use in the wind tunnel in order to scale the measured force to the conditions in the sea? What is the ratio of the force on the pipeline in the sea to that on the model measured in the wind tunnel?

33. You want to determine the thickness of the film when a Newtonian fluid flows uniformly down an inclined plane at an angle θ with the horizontal at a specified flow rate. To do this, you design a laboratory experiment from which you can scale up measured values to any other Newtonian fluid under corresponding conditions.

(a) List all the independent variables that are important in this problem, with their dimensions. If there are any variables that are not independent but act only in conjunction with one another, list only the net combination that is important.

(b) Determine an appropriate set of dimensionless groups for this system, in such a way that the fluid viscosity and the plate inclination each appear in only one group.

(c) Decide what variables you would choose for convenience, what variables would be specified by the analysis, and what you would measure in the lab.

34. You would like to know the thickness of a syrup film as it drains at a rate of 1 gpm down a flat surface that is 6 in. wide and is inclined at an angle of 30° from the vertical. The syrup has a viscosity of 100 cP and a SG of 0.9. In the laboratory, you have a fluid with a viscosity of 70 cP and a SG of 1.0 and a 1 ft wide plane inclined at an angle of 45° from the vertical.

(a) At what flow rate, in gpm, would the laboratory conditions simulate the specified conditions?

(b) If the thickness of the film in the laboratory is 3 mm at the proper flow rate, what would the thickness of the film be for the 100 cP fluid at the specified conditions?

35. The size of liquid droplets produced by a spray nozzle depends upon the nozzle diameter, the fluid velocity, and the fluid properties (which may, under some circumstances, include surface tension).

(a) Determine an appropriate set of dimensionless groups for this system.

(b) You want to know what size droplets will be generated by a fuel oil nozzle with a diameter of 0.5 mm at an oil velocity of 10 m/s. The oil has a viscosity of 10 cP, a SG of 0.82, and a surface tension of 35 dyn/cm. You have a nozzle in the lab with a nozzle diameter of 0.2 mm that you want to use in a lab experiment to find the answer. Can you use the same fuel oil in the lab test as in the field? If not, why not?

(c) If the only fluid you have is water, tell how you would design the lab experiment. Note: water has a viscosity of 1 cP and a SG of 1, but its surface tension can be varied by adding small amounts of surfactant that does not affect the viscosity or density.

(d) Determine what conditions you would use in the lab, what you would measure, and the relationship between the measured and unknown droplet diameters.

36. Small solid particles of diameter d and density ρ_s are carried horizontally by an air stream moving at velocity V. The particles are initially at a distance h above the ground, and you want to know how far they will be carried horizontally before they settle to the ground. To find this out, you decide to conduct a lab experiment using water as the test fluid.

(a) Determine what variables you must set in the lab and how the value of each of these variables is related to the corresponding variable in the air system. You should note that several forces act on the particle: the drag force due to the moving fluid, which depends on the fluid and solid properties, the size of the particle and the relative velocity, and the gravitational force, which is directly related to the densities of both the solid and the fluid in a known manner.

(b) Is there any reason why this experiment might not be feasible in practice?

37. You want to find the wind drag on a new automobile design at various speeds. To do this, you test a 1/30 scale model of the car in the lab. You must design an experiment whereby the drag force measured in the lab can be scaled up directly to find the force on the full-scale car at a given speed.

(a) What is the minimum number of (dimensionless) variables required to completely define the relationship between all the important variables in the problem? Determine the appropriate variables (e.g., the dimensionless groups).

(b) The only fluids you have available in the lab are air and water. Could you use either one of these, if you wanted to? Why (or why not)?

(c) Tell which of these fluids you would use in the lab, and then determine what the velocity of this fluid past the model car would have to be so that the experiment would simulate the drag on the full-scale car at 40 mph. If you decide that it is possible to use either one of the two fluids, determine the answer for each of them.

(d) What is the relationship between the measured drag force on the model and the drag force on the full-scale car? If possible, determine this relationship for the other fluid, as well. Repeat this for a speed of 70 mph.

(e) It turns out that for very high values of the Reynolds number, the drag force is independent of the fluid viscosity. Under these conditions, if the speed of the car doubles, by what factor does the power required to overcome wind drag change?

38. The power required to drive a centrifugal pump and the pressure that the pump will develop depends upon the size (diameter) and speed (angular velocity) of the impeller, the volumetric flow rate through the pump, and the fluid properties. However, if the fluid is not too viscous (e.g., less than about 100 cP), the pump performance is essentially independent of the fluid viscosity. Under these conditions:

(a) Perform a dimensional analysis to determine the dimensionless groups that would be required to define the pump performance. Arrange the groups so that the power and pump pressure each appear in only one group.

   You have a pump with an 8 in. diameter impeller that develops a pressure of 15 psi and requires 1.5 hp to operate when running at 1150 rpm with water at a flow rate of 100 gpm. You also have a similar pump with a 13 in. diameter impeller, driven by a 1750 rpm motor, and you would like to know what pressure this pump would develop with water and what power would be required to drive it.

(b) If the second pump is to be operated under equivalent (similar) conditions to the first one, what should the flow rate be?

(c) If this pump is operated at the proper flow rate, what pressure will it develop, and what power will be required to drive it when pumping water?

39. In a distillation column, vapor is bubbled through the liquid to provide good contact between the two phases. The bubbles are formed when the vapor passes upward through a hole (orifice) in a plate (tray) that is in contact with the liquid. The size of the bubbles depends upon the diameter of the orifice, the velocity of the vapor through the orifice, the viscosity and density of the liquid, and the surface tension between the vapor and the liquid.

(a) Determine the dimensionless groups required to completely describe this system, in such a manner that the bubble diameter and the surface tension do not appear in the same group.

(b) You want to find out what size bubbles would be formed by a hydrocarbon vapor passing through a 1/4 in. orifice at a velocity of 2 ft/s, in contact with a liquid having a viscosity of 4 cP and a density of 0.95 g/cm^3 (the surface tension is 30 dyn/cm). To do this, you run a lab experiment using air and water (with a surface tension of 60 dyn/cm).

   (i) What size orifice should you use, and what should the air velocity through the orifice be?

   (ii) You design and run this experiment and find that the air bubbles are 0.1 in. in diameter. What size would the vapor bubbles be in the organic fluid above the 1/4 in. orifice?

40. A flag will flutter in the wind at a frequency that depends upon the wind speed, the air density, the size of the flag (length and width), gravity, and the "area density" of the cloth (i.e., the mass per unit area). You have a very large flag (40 ft long and 30 ft wide) that weighs 240 lb, and you want to find the frequency at which it will flutter in a wind of 20 mph.

(a) Perform a dimensional analysis to determine an appropriate set of dimensionless groups that could be used to describe this problem.

(b) To find the flutter frequency you run a test in a wind tunnel (at normal atmospheric temperature and pressure) using a flag made from a cloth that weighs 0.05 lb/ft². Determine (1) the size of the flag and the wind speed that you should use in the wind tunnel and (2) the ratio of the flutter frequency of the big flag to that which you observe for the model flag in the wind tunnel.

41. If the viscosity of the liquid is not too high (e.g., less than about 100 cP), the performance of many centrifugal pumps is not very sensitive to the fluid viscosity. You have a pump with an 8 in. diameter impeller that develops a pressure of 15 psi and consumes 1.5 hp when running at 1150 rpm pumping water at a rate of 100 gpm. You also have a similar pump with a 13 in. diameter impeller, driven by a 1750 rpm motor, and you would like to know what pressure that pump would develop with water, and how much power it would take to drive it.
    (a) If the second pump is to be operated under conditions similar to that of the first, what should the flow rate be?
    (b) When operated at this flow rate with water, what pressure should it develop and what power would be required to drive it?

42. The pressure developed by a centrifugal pump depends on the fluid density, the diameter of the pump impeller, the rotational speed of the impeller, and the volumetric flow rate through the pump (centrifugal pumps are not recommended for highly viscous fluids, so viscosity is not commonly an important variable). Furthermore, the pressure developed by the pump is commonly expressed as the "pump head," which is the height of a column of the fluid in the pump that exerts the same pressure as the pump pressure.
    (a) Perform a dimensional analysis to determine the minimum number of variables required to represent the pump performance characteristic in the most general (dimensionless) form.
    (b) The power delivered to the fluid by the pump is also important. Should this be included in the list of important variables, or can it be determined from the original set of variables? Explain.
         You have a pump in the field that has a 1.5 ft diameter impeller that is driven by a motor operating at 750 rpm. You want to determine what head the pump will develop when pumping a liquid with density of 50 lb$_m$/ft³ at a rate of 1000 gpm. You do this by running a test in the lab on a scale model of the pump that has a 0.5 ft diameter impeller using water (at 70°F) and a motor that runs at 1200 rpm.
    (c) At what flow rate of water should the lab pump be operated (in gpm)?
    (d) If the lab pump develops a head of 85 ft at this flow rate, what head would the pump in the field develop with the operating fluid at the specified flow rate?
    (e) How much power (in horsepower) is transferred to the fluid in both the lab and the field cases?
    (f) The pump efficiency is defined as the ratio of the power delivered to the fluid to the power of the motor that drives the pump. If the lab pump is driven by a 2 hp motor, what is the efficiency of the lab pump? If the efficiency of the field pump is the same as that of the lab pump, what power motor (horsepower) would be required to drive it?

NOTATION

D Inside pipe diameter, [L]
f Friction factor, [—]
F Dimension of force
G Gravitational constant = 6.67×10^{-11} N m²/kg², Equation 2.2, [FL²/M²] = [L³/Mt²]
g Acceleration due to gravity, [L/t²]

g_c Conversion factor, $[ML/(Ft^2)]$

ID   Inside diameter of pipe, [L]

K_1 Loss parameter (see Chapter 7), [—]

K_d Loss parameter (see Chapter 7), [—]

K_f Loss coefficient (see Chapter 7), [—]

K_i Loss parameter (see Chapter 7), [—]

ke Kinetic energy per unit mass, $[FL/M = L^2/t^2]$

L   Dimension of length

L Length, [L]

M   Dimension of mass

m Mass, [M]

N_{Re} Reynolds number, [—]

N_x Dimensionless group x [—]

P Pressure, $[F/L^2 = M/(Lt^2)]$

Q Volumetric flow rate, $[L^3/t]$

R Radius, [L]

T Torque, $[FL = ML^2/t^2]$

t   Dimension of time

t Time [t]

V Spatial average velocity, [L/t]

v   Local velocity, [L/t]

W Weight, $[F = ML/t^2]$

x Coordinate direction, [L]

z Coordinate direction (measured upward), [L]

GREEK

ε   Roughness, [L]

μ   Viscosity, [M/Lt]

ρ   Density, $[M/L^3]$

τ_{yx} Shear stress, force in x direction on y surface, $[F/L^2 = ML/t^2]$

SUBSCRIPTS

1   Upstream reference point 1

2   Downstream reference point 2

m "Model"

f "Field"

x, y, r, θ Coordinate directions

3 Fluid Properties in Perspective

I. CLASSIFICATION OF MATERIALS AND FLUID PROPERTIES

What is a fluid? It isn't a solid, but what is a solid? Perhaps it is easier to define these materials in terms of how they respond (i.e., deform or flow) when subjected to an applied force in a specific situation, such as *simple shear* illustrated in Figure 3.1 (which is virtually identical to Figure 1.1). We envision the material contained between two large parallel plates of area A_y, the bottom one being fixed and the upper one subject to a tangential force F_x in the x direction. This force causes the upper plate to be displaced in the x direction by an amount U_x and/or to move with a constant velocity V_x after reaching a steady state.

The material is assumed to adhere to the plates (known as the "no-slip" condition), and its deformation properties can be classified by the way the top plate responds when the force is applied. The mechanical behavior of a material, and its corresponding mechanical or rheological* properties, can be defined in terms of how the *shear stress* (τ_{yx}) (force per unit area) and the *shear strain* (γ_{yx}) (which is a relative displacement and a measure of deformation) are related. These are defined, respectively, in terms of the total force (F_x) acting on area A_y of the plate and the relative displacement (U_x/h_y) of the plates, that is,

$$\tau_{yx} = \frac{F_x}{A_y} \tag{3.1}$$

and

$$\gamma_{yx} = \frac{U_x}{h_y} = \frac{du_x}{dy} \tag{3.2}$$

The manner in which the shear strain, or displacement, responds to an applied shear stress, or force, (or vice versa) in this situation defines the mechanical or rheological characteristics of the material. The parameters in any quantitative functional relation between the stress and strain are the *rheological properties* of the material. It is noted that the shear stress has dimensions of force per unit area (with units of, e.g., Pa, dyn/cm², or lb$_f$/ft²) and that shear strain is dimensionless (it has no units).

For example, if the material between the plates is a perfectly *rigid solid* (e.g., a brick), it will not move at all, no matter how much force is applied (unless it breaks). Thus, the quantitative relation that defines the behavior of this material is

$$\gamma_{yx} = 0 \tag{3.3}$$

---

* "Rheology" is the study of the deformation and flow behavior of materials, both fluids and solids. See for example: Darby (1976), Barnes et al. (1989), Boger and Walters (1992), etc.

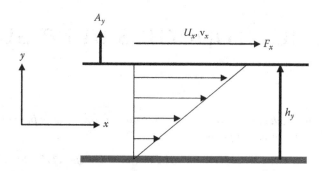

FIGURE 3.1 Simple shear.

However, if the top plate moves a distance that is in proportion to the applied force and then stops, the material is called a *linear elastic (Hookean) solid* (e.g., rubber). The quantitative relation that defines such a material is

$$\tau_{yx} = G\gamma_{yx} \tag{3.4}$$

where G is a constant called the *shear modulus*. Note also that if the force (stress) is removed, the displacement (strain) also goes to zero, that is, the material reverts to its original undeformed state. Such an ideal elastic material is thus said to have a "perfect memory."

On the other hand, if the top plate moves but its displacement is not directly proportional to the applied force (it may be either more or less than proportional to the force), the material is said to be a *nonlinear* (i.e., *non-Hookean*) *elastic solid*. It can be represented by an equation of the form

$$G = \frac{\tau_{yx}}{\gamma_{yx}} = fn\left(\tau \text{ or } \gamma\right) \tag{3.5}$$

Here, G is still called the shear modulus but it is no longer a constant. It is, instead, a *function* of either how far the plate moves (γ_{yx}) or the magnitude of the applied force (τ_{yx}), that is, $G(\gamma)$ or $G(\tau)$. The particular form of the function will depend upon the specific nature of the material. Note, however, that such a material still exhibits a "perfect memory" because it returns to its undeformed state when the force (stress) is removed.

At the other extreme, if the molecules of the material are so far apart that they exert negligible attraction on each other (e.g., a gas under very low pressure), the plate can be moved by the application of a negligible force. The equation that describes this material is

$$\tau_{yx} = 0 \tag{3.6}$$

Such an ideal material is called an *inviscid (Pascalian) fluid*. However, if the molecules do exhibit a significant mutual attraction such that the force (e.g., the shear stress) is proportional to the *relative rate* of movement (i.e., the velocity gradient), the material is known as a *Newtonian fluid*. The equation that describes this behavior is

$$\tau_{yx} = \mu\dot{\gamma}_{yx} \tag{3.7}$$

where $\dot{\gamma}_{yx}$ is the *rate of shear strain*, or *shear rate* for short:

$$\dot{\gamma}_{yx} = \frac{d\gamma_{yx}}{dt} = \frac{dv_x}{dy} = \frac{V_x}{h_y} \tag{3.8}$$

and μ is the fluid *viscosity*. Note that Equation 3.7 defines the viscosity, that is, $\mu \equiv \tau_{yx}/\dot{\gamma}_{yx}$, which has dimensions of Ft/L^2 (with units of $(Pa\ s)$, $(dyn\ s/cm^2 = gm/cm\ s = Poise)$, $(lb_f\ s/ft^2)$, etc.).* Note that when the stress is removed from this fluid, the strain *rate* goes to zero, that is, the motion stops, but there is no "memory" or tendency to return to any past state.

If the properties of the fluid are such that the shear stress and shear rate are not directly proportional but are instead related by some more complex function, the fluid is said to be *non-Newtonian* (similar to a *non-Hookean* solid). For such fluids, the viscosity is still defined as $\tau_{yx}/\dot{\gamma}_{yx}$, but it is no longer a constant, being instead a function of either the shear rate or shear stress. It is called the *apparent viscosity* (*function*) and is designated by η:

$$\eta = \frac{\tau_{yx}}{\dot{\gamma}_{yx}} = fn(\tau \text{ or } \dot{\gamma}) \tag{3.9}$$

The actual mathematical form of this function will depend upon the nature (i.e., the "constitution") of the particular material. The most common fluids of simple structure (water, air, glycerin, oils, etc.) are Newtonian. However, fluids with complex structure (high-molecular-weight polymer melts or solutions, suspensions of fine particles, emulsions, foams, soap and surfactant solutions, etc.) are generally non-Newtonian. Some very common examples of non-Newtonian fluids are mud, paint, ink, mayonnaise, shaving cream, dough, mustard, yoghurt, catsup, toothpaste, blood, synovial fluid, sludge, etc.

Actually, some substances exhibit both elastic (solid) properties and viscous (fluid) properties under appropriate conditions. For instance, a ball made from silly putty bounces off the floor like a rubber ball (i.e., elastic), whereas when left alone it relaxes into a "puddle" on its own over a long time scale (i.e., viscous). These materials are said to be *viscoelastic* and are most notably materials composed of high-molecular-weight polymers and/or complex molecular structure. The complete description of the rheological properties of these materials may involve a function relating the stress and strain as well as derivatives or integrals of these with respect to time. Because the elastic properties of these materials (both fluids and solids) impart "memory" to the material (as described previously), which results in a tendency to recover to a preferred state upon the removal of the force (stress), they are often termed "memory materials" and exhibit time-dependent properties.

This classification of material behavior is summarized in Table 3.1 (in which the subscripts (x, y) have been omitted for simplicity). Since we are concerned with fluids, we will concentrate primarily on the flow behavior of Newtonian and non-Newtonian fluids. However, we will later illustrate some of the unique characteristics of viscoelastic fluids, such as the ability of solutions of certain high polymers to flow through pipes in turbulent flow with much less energy expenditure than the solvent alone (see Section VIII of Chapter 6).

II. DETERMINATION OF FLUID VISCOUS (RHEOLOGICAL) PROPERTIES

As previously discussed, the flow behavior of fluids is determined by their rheological properties, which govern the relationships between shear stress and shear rate. In principle, these properties could be determined by measurements in a "simple shear" test as illustrated in Figure 3.1. One would put the "unknown fluid" in the gap between the plates, subject the upper plate to a specified velocity (V), and measure the required force (F) (or vice versa). The shear stress (τ) would be determined by F/A, the shear rate $(\dot{\gamma})$ is given by V/h, and the viscosity (η) by $\tau/\dot{\gamma}$.

The experiment is repeated for different combinations of V and F to determine the viscosity at various shear rates (or shear stresses). However, this geometry is not convenient to work with,

---

* Conversion Factors for viscosity include: 1 poise = 1 dyn s/cm² = 100 cP = 0.1 N s/m² = 0.1 Pa s = 1 gm/(cm s) = 0.0672 lb$_m$/(ft s) = 2.09 × 10^{-3} lb$_f$ s/ft².

TABLE 3.1

Classification of Materials

Rigid Solid (Euclidean)	Linear Elastic Solid (Hookean)	Nonlinear Elastic Solid (Non-Hookean)	Viscoelastic Fluids and Solids (Nonlinear)	Nonlinear Viscous Fluid (Non-Newtonian)	Linear Viscous Fluid (Newtonian)	Inviscid Fluid (Pascalian)
$\gamma = 0$	$\tau = G\gamma$	$\tau = fn(\gamma)$		$\tau = fn(\dot\gamma)$	$\tau = \mu\,\dot\gamma$	$\tau = 0$
	or	or		or	or	or
	$G = \tau/\gamma$	$G = \tau/\gamma$		$\eta = \tau/\dot\gamma$	$\mu = \tau/\dot\gamma$	$\mu = 0$
	Shear Modulus— (Constant)	Modulus— *Function* of γ or τ		Viscosity— *Function* of $\dot\gamma$ or τ	Viscosity— (Constant)	

$\longleftarrow$Purely elastic solids $\longrightarrow$ $\tau = fn(\gamma, \dot\gamma, ...)$ $\longleftarrow$Purely viscous fluids $\longrightarrow$

Elastic deformations—store energy   Viscous deformations—dissipate energy

Source: Darby, R., *Viscoelastic Fluids*, Marcel Dekker, New York, 1976.

because it is hard to keep the fluid in the gap with no confining walls, and correction for the effect of the walls is not simple. However, there are more convenient geometries for measuring viscous properties, as described in the following sections. The working equations used to obtain viscosity from measured quantities will be given here, although the development of these equations will be delayed until after the appropriate fundamental principles have been discussed.

A. Cup and Bob (Couette) Viscometer

As the name implies, the cup and bob (Couette) viscometer consists of two concentric cylinders, the outer "cup" and the inner "bob," with the test fluid in the annular gap (see Figure 3.2).

One cylinder (preferably the cup) is rotated at a fixed angular velocity (Ω). The force is transmitted to the sample causing it to deform and is then transferred by the fluid to the other cylinder (i.e., the bob). This force results in a torque (T) that can be measured by a torsion spring, for example. Thus, the known quantities are the radius of the inner bob (R_i) and the outer cup (R_o), the length

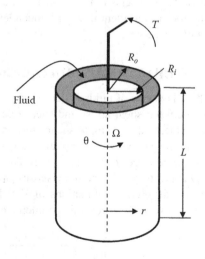

FIGURE 3.2   Cup and bob (Couette) viscometer.

of bob surface in contact with the fluid sample (L), and the measured angular velocity (Ω) and torque (T). From these quantities, we must determine the corresponding shear stress and shear rate to find the fluid viscosity. The shear stress is determined by a balance of moments on a cylindrical surface within the sample (at a distance r from the center), and the torsion spring

$$T = \text{Force} \times \text{Lever arm} = \text{Shear stress} \times \text{Surface area} \times \text{Radius}$$

or

$$T = \tau_{r\theta}(2\pi rL)(r) \tag{3.10}$$

where the subscripts on the shear stress (r, θ) represent the force in the θ direction acting on the r surface (the cylindrical surface perpendicular to r). Solving for the shear stress:

$$\tau_{r\theta} = \frac{T}{2\pi r^2 L} \tag{3.11}$$

Setting $r = R_i$ gives the stress on the bob surface (τ_i), and $r = R_o$ gives the stress on the cup (τ_o). If the gap is small [i.e., $(R_o - R_i)/R_i \leq 0.02$], the curvature can be neglected and the flow in the gap is equivalent to the flow between flat parallel plates, similar to that shown in Figure 3.1. In this case, an average shear stress should be used [i.e., $(\tau_i + \tau_o)/2$], and the average shear rate is given by

$$\dot{\gamma}_{r\theta} = \frac{dv_\theta}{dr} \cong \frac{\Delta V}{\Delta r} = \frac{R_o\Omega - 0}{R_o - R_i} = \frac{\Omega}{1 - R_i/R_o}$$

or

$$\dot{\gamma} = \frac{\Omega}{1-\beta} \tag{3.12}$$

where $\beta = R_i/R_o$. However, if the gap is not small, the shear rate must be corrected for the curvature in the velocity profile. This can be done by applying the following approximate expression for the shear rate at the bob (which is accurate to $\pm 1\%$ for most conditions and is better than $\pm 5\%$ for the worst case; see, e.g., Darby, 1985):

$$\dot{\gamma}_i = \frac{2\Omega}{n'(1-\beta^{2/n'})} \tag{3.13}$$

where

$$n' = \frac{d(\log T)}{d(\log \Omega)} \tag{3.14}$$

is the point slope of the plot of $\log T$ versus $\log \Omega$, at the value of Ω (or T) in Equation 3.12. Thus, a series of data points of T versus Ω must be obtained in order to determine the value of the slope (n') at each point, which is needed to determine the corresponding values of the shear rate. If $n' = 1$ (i.e., $T \propto \Omega$), the fluid is Newtonian. The viscosity at each shear rate (or shear stress) is then determined by dividing the shear stress at the bob (Equation 3.11 with $r = R_i$) by the shear rate at the bob (Equation 3.13), for each data point.

Example 3.1

The following data were taken in a cup and bob viscometer, with a bob radius of 2 cm, a cup radius of 2.05 cm, and a bob length of 15 cm. Determine the viscosity of the sample, and the equation for the model that best represents this viscosity.

Torque, T (dyn cm)	Speed, Ω (rpm)
2,000	2
3,500	4
7,200	10
12,500	20
20,000	40

The viscosity is the shear stress at the bob, as given by Equation 3.11, divided by the shear rate at the bob, as given by Equation 3.13.

Sample Calculation:

For the first data point, the shear stress at the bob is given by

$$\tau_{r\theta} = \frac{T}{2\pi r_i^2 L} = \frac{2000 \text{ dyn cm}}{2\pi (2 \text{ cm})^2 15 \text{ cm}} = \boxed{5.31 \text{ dyn/cm}^2}$$

and the shear rate is given by Equation 3.13. The value of β for this system is $\beta = 2.0/2.05 = 0.976$. Although this is a small gap, we will assume it is not small enough to use Equation 3.12 for the shear rate and will illustrate the use of Equation 3.13 instead. The value of n' in Equation 3.13 is determined from the *point slope* of the (log T) versus (log *rpm*) plot at each data point, as illustrated by the plot shown in Figure E3.1. The line through the data is the best fit of all data points by linear least squares (this is easily found by using a spreadsheet) and is found to have a slope of 0.77 (with $r^2 = 0.999$). In general, if the data do not fall on a straight line as in this example, the point slope (tangent) must be determined at each data point, resulting in a different value of n' for each data point. Using 0.77 for n' in Equation 3.13 gives

$$\dot{\gamma}_i = \frac{2\Omega}{n'\left(1-\beta^{2/n'}\right)} = \frac{2(2 \text{ rev/min})(2\pi \text{ rad/rev})}{0.77\left(1-0.976^{2/0.77}\right)(60 \text{ s/min})} = \boxed{8.90 \text{ s}^{-1}}$$

$$\eta = \frac{\tau_{r\theta}}{\dot{\gamma}_{r\theta}} = \frac{5.31 \text{ dyn/cm}^2}{8.90 \text{ s}^{-1}} = 0.61 \text{ dyn s/cm}^2 = 0.61 \text{ poise}$$

Shear Stress at Bob (dyn/cm²)	Shear Rate at Bob (s⁻¹)	Viscosity (Poise)
5.31	8.90	0.61
9.28	17.8	0.53
19.1	44.5	0.43
33.2	89.0	0.37
53.1	178	0.31

The plot of viscosity versus shear rate for this material is shown in Figure E3.1b.

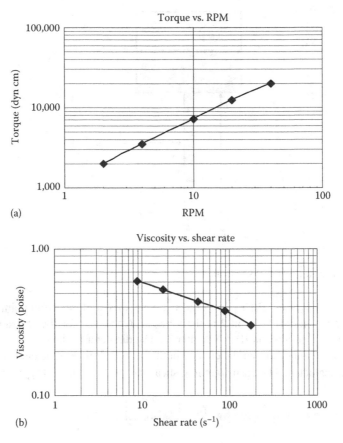

FIGURE E3.1 Examples of (a) torque versus speed and (b) the corresponding calculated viscosity versus shear rate for data in Example 3.1.

It should be noted that in practice some corrections arising from the edge effects (or end effects), secondary flows (Taylor vortices) at high rotational speeds, etc. may be required depending upon the specific designs of cup and bob viscometers and the properties of the fluid. These aspects are described in detail in Darby (1976) and Chhabra and Richardson (2008) and generally limit the maximum reliable shear rate attainable in cup and bob viscometers.

B. TUBE FLOW (POISEUILLE) VISCOMETER

Another common method of determining viscosity is by measuring the total driving force or "pressure drop" ($\Delta\Phi = \Delta P + \rho g \Delta z$) and flow rate ($Q$) in steady laminar flow through a straight, uniform circular tube of length L and diameter D (this is called *Poiseuille flow*). This can be done by using pressure taps through the tube wall to measure the pressure difference directly or by measuring the total pressure difference from a reservoir to the end of the tube, as illustrated in Figure 3.3. The latter is more common because tubes of very small diameter are usually used, but this arrangement requires that correction factors be applied for the static head of the fluid in the reservoir and the entrance pressure loss from the reservoir to the tube (detailed descriptions of the relevant corrections can be found in Darby (1976) and Chhabra and Richardson (2008)). As will be shown later, a momentum (force) balance on the fluid in the tube provides a relation between the shear stress at the tube wall (τ_w) and the measured pressure drop:

$$\tau_w = \frac{-\Delta\Phi}{4L/D} \tag{3.15}$$

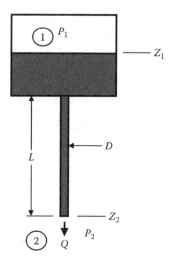

FIGURE 3.3 Tube flow (Poiseuille) viscometer.

where $\Delta\Phi = \Phi_2 - \Phi_1 = (P_2 - P_1) + \rho g(z_2 - z_1)$ is the net driving force from point 1 in the reservoir (at the surface of the liquid) to point 2 at the tube exit (corrected for the "entrance loss" from the reservoir to the tube (see Section III of Chapter 7). The diameter of the reservoir is normally much larger than that of the tube, so that the change in the head and the kinetic energy of the fluid in the reservoir can be neglected. The corresponding shear rate at the tube wall ($\dot\gamma_w$) is given by

$$\dot\gamma_w = \Gamma\left(\frac{3n'+1}{4n'}\right) \tag{3.16}$$

where

$$\Gamma = \frac{32Q}{\pi D^3} = \frac{8V}{D} \tag{3.17}$$

is the wall shear rate for a Newtonian ($n' = 1$) fluid, and

$$n' = \frac{d\log\tau_w}{d\log\Gamma} = \frac{d\log(-\Delta\Phi)}{d\log Q} \tag{3.18}$$

is the point slope of the log–log plot of $\Delta\Phi$ versus Q, evaluated at each data point. This n' is the same as that determined from the cup and bob viscometer, for a given fluid. As before, if $n' = 1$ (i.e., $\Delta\Phi \propto Q$), the fluid is Newtonian. The apparent viscosity is given by $\eta = \tau_w/\dot\gamma_w$.

Example 3.2

The following data were obtained for a 5% aqueous polymer solution in a horizontal pipe of 42 mm internal diameter. The pressure difference was measured by two pressure transducers mounted 63 cm apart in the middle section of the pipe so that end effects are negligible.

V (m/s)	3.28	2.96	2.56	2.01	1.8	1.43	1.15	0.98	0.77	0.60	0.35
$-\Delta P$ (kPa)	6.28	6.00	5.44	4.80	4.32	3.74	3.25	2.96	2.53	2.16	1.53

Obtain the true shear stress–shear rate data for this polymer solution.

Sample Calculation:

The values of Γ are calculated from Equation 3.17. For the first data point, the calculation is

$$\Gamma = \frac{32Q}{\pi D^3} = \frac{8V}{D} = \frac{8(3.28\ \text{m/s})(1000\ \text{mm/m})}{42\ \text{mm}} \boxed{= 625\ \text{s}^{-1}}$$

$\Gamma = \dfrac{8V}{D}\ (\text{s}^{-1})$	$\tau_w = \dfrac{D(-\Delta P)}{4L}\ (\text{Pa})$	$\dot{\gamma}_w = \Gamma\left(\dfrac{3n'+1}{4n'}\right)(\text{s}^{-1})$	$\eta = \dfrac{\tau_w}{\dot{\gamma}_w}\ (\text{Poise})$
625	104	714	1.47
564	100	643	1.56
488	90.7	557	1.63
383	80.0	437	1.83
343	72.0	391	1.84
272	62.3	310	2.00
219	54.2	250	2.17
187	49.3	213	2.31
147	42.2	167	2.52
114	36	130	2.77
66.7	25.5	76	2.36

As the tube in this case is horizontal, the driving force is simply the pressure difference, $-\Delta\Phi = -\Delta P$, so the wall shear stress becomes

$$\tau_w = \frac{D(-\Delta P)}{4L} = \frac{(42\ \text{mm})(6.28\ \text{kPa})(1000\ \text{Pa/kPa})}{4(63\ \text{cm})(10\ \text{mm/cm})} = \boxed{104\ \text{Pa}}$$

In order to evaluate the point slope, $n' = d\log\tau_w/d\log(8V/D)$, these data are plotted in Figure E3.2 on log–log coordinates. The value of the slope of the plot, n', is seen to be constant at $n' = 0.64$. Therefore, the true shear rate at the wall $\dot{\gamma}_w$ can be calculated using Equation 3.16:

$$\dot{\gamma}_w = \Gamma\left(\frac{3n'+1}{4n'}\right) = (625\ \text{s}^{-1})\left(\frac{3(0.64)+1}{4(0.64)}\right) = 714\ \text{s}^{-1}$$

and the apparent viscosity at this shear rate is

$$\eta = \frac{\tau_w}{\dot{\gamma}_w} = \frac{104\ \text{Pa}\left[10\ \text{dyn/)(cm}^2/\text{Pa})\right]\left[1\ \text{poise/(dyn s/cm}^2)\right]}{714\ \text{s}^{-1}} = 1.47\ \text{poise}$$

These values are also summarized in the preceding table. The viscosity versus shear rate curve is also shown in Figure E3.2, where this polymer solution is seen to exhibit shear thinning behavior.

III.   TYPES OF NON-NEWTONIAN FLUID BEHAVIOR

When the measured values of shear stress or viscosity are plotted versus shear rate, various types of behavior may be observed depending upon the fluid properties, as shown in Figures 3.4 and 3.5. It should be noted that the shear stress and shear rate can both be either positive or negative, depending upon the direction of motion or the applied force, the reference frame, etc. (however, by our

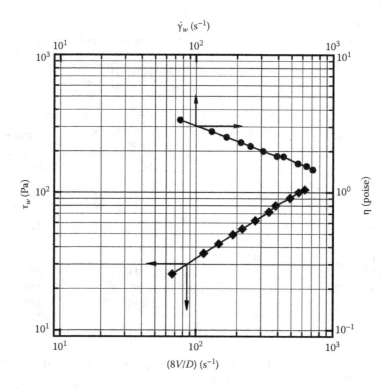

FIGURE E3.2 Plot of τ_w versus $\Gamma(8V/D)$ to evaluate n' and viscosity versus shear rate.

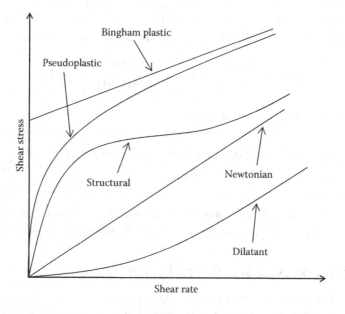

FIGURE 3.4 Shear stress versus shear rate for various types of fluid behavior.

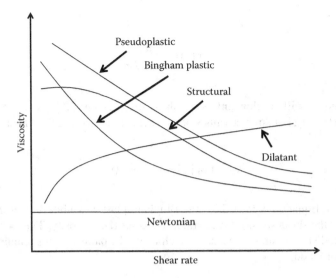

FIGURE 3.5 Viscosity versus shear rate for fluids in Figure 3.4.

convention, they are always of the same sign). Because the viscosity must always be positive, the shear rate (or shear stress) argument in the viscosity function for a non-Newtonian fluid should be the absolute magnitude regardless of the actual sign of the shear rate and shear stress.

A. NEWTONIAN FLUID

If the shear stress versus shear rate plot is a straight line through the origin (or a straight line with a slope of unity on a log–log plot), the fluid is Newtonian:

$$\text{\textit{Newtonian}:} \quad \tau = \mu \dot{\gamma} \tag{3.19}$$

where μ is the viscosity.

B. BINGHAM PLASTIC MODEL

If the data appear to be linear but do not extrapolate through the origin, intersecting the shear stress (τ) axis instead at a shear stress value of τ_o, the material is called a Bingham plastic:

$$\text{\textit{Bingham plastic}:} \quad \text{\textit{For}} \ |\tau| > \tau_o: \tau = \pm \, \tau_o + \mu_\infty \dot{\gamma} \tag{3.20}$$

The *yield stress*, τ_o, and the *high shear limiting* (or plastic) *viscosity*, μ_∞, are the two rheological properties required to determine the flow behavior of a Bingham plastic. The positive sign is used when τ and $\dot{\gamma}$ are positive, and the negative sign when they are negative. The viscosity function for the Bingham plastic is

$$\eta(\dot{\gamma}) = \frac{\tau_o}{|\dot{\gamma}|} + \mu_\infty \tag{3.21}$$

or

$$\eta(\tau) = \frac{\mu_\infty}{1 - \tau_o / |\tau|} \tag{3.22}$$

Because this material will not flow unless the shear stress exceeds the yield stress, these equations apply only when $|\tau| > \tau_o$. For smaller values of the shear stress, the material behaves as a rigid solid, that is,

$$\text{For } |\tau| < \tau_o: \quad \dot{\gamma} = 0 \tag{3.23}$$

As is evident from Equation 3.21 or 3.22, the Bingham plastic exhibits a *shear thinning* viscosity, that is, the larger the shear stress or shear rate, the lower the viscosity. This behavior is typical of many concentrated slurries and suspensions such as muds, paints, foams, emulsions (e.g., mayonnaise), ketchup, or blood.

C. POWER LAW MODEL

If the data (either shear stress or viscosity) exhibit a straight line on a log–log plot, the fluid is said to follow the *power law* model, which can be represented as

$$\textit{Power Law}: \tau = m |\dot{\gamma}|^{(n-1)} \dot{\gamma}$$

or

$$\tau = m\dot{\gamma}^n \quad \textit{if both } \tau \textit{ and } \dot{\gamma} \textit{ are } (+) \tag{3.24}$$

$$\tau = -m(-\dot{\gamma})^n \quad \textit{if both } \tau \textit{ and } \dot{\gamma} \textit{ are } (-)$$

The two viscous rheological properties are m, the *consistency coefficient*, and n, the *flow index*. The apparent viscosity function for the power law model in terms of shear rate is

$$\eta(\dot{\gamma}) = m |\dot{\gamma}|^{n-1} \tag{3.25}$$

or in terms of shear stress

$$\eta(\tau) = m^{1/n} |\tau|^{(n-1)/n} \tag{3.26}$$

Note that n is dimensionless but m has dimensions of Ft^n/L^2. However, m is also equal to the apparent viscosity of the fluid at a shear rate of 1 s^{-1}, so it is a "viscosity" parameter with equivalent units. It is evident that if $n = 1$ the power law model reduces to a Newtonian fluid with $m = \mu$. If $n < 1$, the fluid is shear thinning (or *pseudoplastic*), and if $n > 1$ the model represents shear thickening (or *dilatant*) behavior, as illustrated in Figures 3.5 and 3.6. Most non-Newtonian fluids are shear thinning, whereas shear thickening behavior is relatively rare, being observed primarily for some concentrated suspensions of very small particles (e.g., starch suspensions) and some unusual polymeric fluids. The power law model

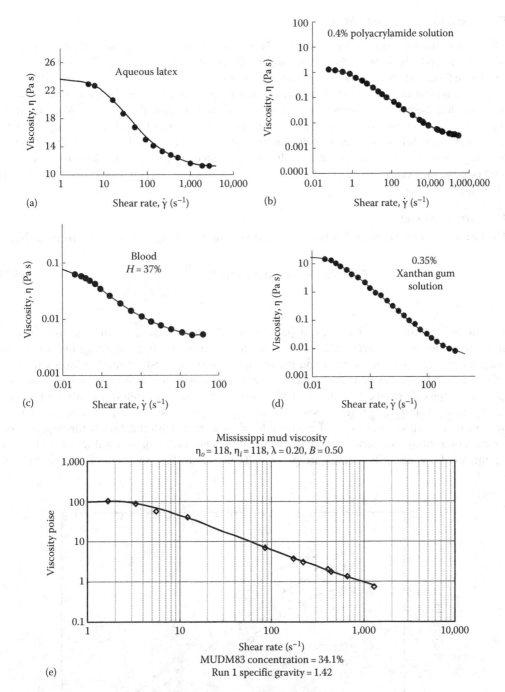

FIGURE 3.6 (a–e) Some examples of structural viscosity behavior for a range of fluids.

is very popular for curve fitting viscosity data for many fluids over one to three decades of shear rate. However, it is dangerous to extrapolate beyond the range of measurements using this model, because for $n < 1$ it predicts a viscosity that increases without bound as the shear rate decreases and a viscosity that decreases without bound as the shear rate increases, both of which are physically unrealistic.

D. STRUCTURAL VISCOSITY MODELS

The typical viscous behavior for many non-Newtonian fluids (e.g., polymeric fluids, flocculated suspensions, colloids, foams, gels) is illustrated by the curves labeled "structural" in Figures 3.5 and 3.6. These fluids exhibit Newtonian behavior at very low and very high shear rates, with shear thinning or pseudoplastic behavior at intermediate shear rates. In some materials, this can be attributed to a reversible "structure" or network that forms in the "rest" or equilibrium state. When the material is sheared, the structure breaks down, resulting in a shear-dependent (shear thinning) behavior. Some real examples of this type of behavior are shown in Figure 3.6. These show that structural viscosity behavior is exhibited by fluids as diverse as polymer solutions, blood, latex emulsions, and mud (sediment). Equations (i.e., models) that represent this type of behavior are described in the following sections.

1. Carreau Model

The *Carreau model* (Carreau, 1972) is very useful for describing the viscosity of structural fluids:

$$\textit{Carreau model:} \quad \eta(\dot{\gamma}) = \eta_\infty + \frac{\eta_o - \eta_\infty}{\left[1 + \left(\lambda^2 \dot{\gamma}^2\right)\right]^p} \tag{3.27}$$

This model contains four rheological parameters: the *low shear limiting viscosity* (η_o), the *high shear limiting viscosity* (η_∞), a *time constant* (λ), and the *shear thinning index* (p). This is a very general viscosity model, and it can represent the viscosity function for a wide variety of materials. However, it may require data over a range of six to eight *decades* of shear rate to completely define the shape of the curve (and hence to determine all four parameters). As an example, Figure 3.7 shows viscosity data for several polyacrylamide solutions over a range of about six orders of magnitude of shear rate, with the curves through the data points representing the Carreau model fit to the data. The corresponding values of the Carreau parameters for each of the curves are given in Table 3.2. In fact, over certain ranges of shear

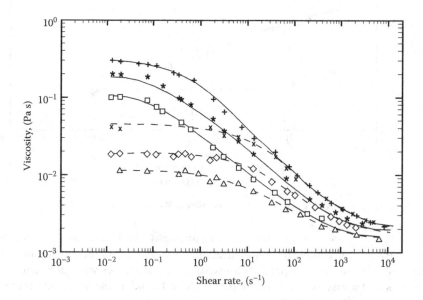

FIGURE 3.7 Viscosity data and Carreau model fit of polyacrylamide solutions. (From Darby, R. and Pvisa-Art, S., *Can. J. Chem. Eng.*, 69, 1395, 1991.)

TABLE 3.2

Values of Carreau Parameters for Model Fit in Figure 3.7

Solution	η_o [(Pa s) × 10]	λ (s)	p	η_∞ [(Pa s) × 10⁴]
100 (mg/kg) (fresh)	1.113	11.89	0.266	1.30
250 (mg/kg) (fresh)	1.714	6.67	0.270	1.40
500 (mg/kg) (fresh)	3.017	3.53	0.300	1.70
100 (mg/kg) (sheared)	0.098	0.258	0.251	1.30
250 (mg/kg) (sheared)	0.169	0.167	0.270	1.40
500 (mg/kg) (sheared)	0.397	0.125	0.295	1.70

Source: Darby, R. and Pvisa-Art, S., *Can. J. Chem. Eng.*, 69, 1395, 1991.

rate, the Carreau model reduces to various other popular models for special cases (including the Bingham plastic and power law models), as shown below.

a. *Low to Intermediate Shear Rate Range*

If $\eta_\infty \ll (\eta, \eta_o)$, the Carreau model reduces to a three-parameter model (η_o, λ, and p) that is equivalent to a power law model with a low shear limiting viscosity of η_o, also known as the *Ellis model*, which can be written as follows:

$$\text{Ellis model:} \quad \eta(\dot{\gamma}) = \frac{\eta_o}{\left[1 + (\lambda\dot{\gamma})^2\right]^p} \tag{3.28}$$

b. *Intermediate to High Shear Rate Range*

If $\eta_o \gg (\eta, \eta_\infty)$ and $(\lambda\dot{\gamma})^2 \gg 1$, the Carreau model reduces to the equivalent of a power law model with a high shear limiting viscosity, called the *Sisko model*:

$$\text{Sisko model:} \quad \eta(\dot{\gamma}) = \eta_\infty + \frac{\eta_o}{\left(\lambda^2\dot{\gamma}^2\right)^p} \tag{3.29}$$

Although this appears to have four parameters, it is really a three-parameter model because the combination η_o/λ^{2p} is a single parameter, along with the two parameters p and η_∞.

c. *Intermediate Shear Rate Behavior*

For $\eta_\infty \ll \eta \ll \eta_o$ and $(\lambda\dot{\gamma})^2 \gg 1$, the Carreau model reduces to the *power law model*:

$$\text{Power law model:} \quad \eta(\dot{\gamma}) = \frac{\eta_o}{\left(\lambda^2\dot{\gamma}^2\right)^p} \tag{3.30}$$

In this case, the power law parameters m and n are equivalent to the following combinations of Carreau parameters:

$$m = \frac{\eta_o}{\lambda^{2p}}, \quad n = 1 - 2p \tag{3.31}$$

d. Bingham Plastic Behavior

If the value of p is set equal to 1/2 in the Sisko model, the result is equivalent to the *Bingham plastic* model:

$$\text{Bingham plastic model:} \quad \eta(\dot{\gamma}) = \eta_\infty + \frac{\eta_o}{\lambda|\dot{\gamma}|} \tag{3.32}$$

where

the yield stress τ_o is equivalent to η_o/λ

η_∞ is the limiting (high shear) viscosity

2. Other Models

A variety of more complex models have been proposed to fit a wider range and variety of viscosity data. Three of these are presented here.

a. Meter Model

A stress-dependent viscosity model, which has the same general characteristics as the *Carreau* model except that it uses shear stress as the independent variable, is the *Meter model* (Meter, 1964):

$$\text{Meter model:} \quad \eta(\tau) = \eta_\infty + \frac{\eta_o - \eta_\infty}{1 + (\tau^2/\sigma^2)^a} \tag{3.33}$$

where

σ is a characteristic stress parameter

a is the shear thinning index

This form is particularly suited for conditions when the shear stress is known.

b. Yasuda Model

The *Yasuda model* (Yasuda et al., 1981) is a modification of the Carreau model with one additional parameter a (a total of five parameters):

$$\text{Yasuda model:} \quad \eta = \eta_\infty + \frac{\eta_o - \eta_\infty}{\left[1 + \left(\lambda^2\dot{\gamma}^2\right)^a\right]^{p/a}} \tag{3.34}$$

which reduces to the Carreau model for $a = 1$ (this is also sometimes called the *Carreau–Yasuda* model). This model is particularly useful for representing melt viscosity data for polymers with a broad molecular weight distribution, for which the zero-shear viscosity is approached very gradually. Both of these models reduce to Newtonian behavior at very low and very high shear rates and to power law behavior at intermediate shear rates.

IV. TEMPERATURE DEPENDENCE OF VISCOSITY

All fluid properties are dependent upon temperature. For most fluids, the viscosity is the property that is most sensitive to temperature changes. The following relations apply to Newtonian fluids, although for non-Newtonian fluids represented by any of the above models, any model parameter having dimensions of viscosity will generally follow similar temperature dependence.

A. LIQUIDS

For liquids, as the temperature increases the degree of molecular motion increases, reducing the short-range attractive forces between molecules and thereby lowering the viscosity of the fluid. The viscosity of various liquids is shown as a function of temperature in Appendix A. For many liquids, this temperature dependence can be represented reasonably well by the Arrhenius equation:

$$\mu = A \exp\left(\frac{B}{T}\right) \tag{3.35}$$

where T is the absolute temperature. If the viscosity of a liquid is known at two different temperatures, this information can be used to evaluate the parameters A and B, which then permits the calculation of the viscosity at any other temperature. If the viscosity is known at only one temperature, this value can be used as a reference point to establish the temperature scale for Figure A.2, which can then be used to estimate the viscosity at any other temperature. Viscosity data for a large number of Newtonian liquids have been fitted by Yaws et al. (1994) by the equation

$$\log_{10}\mu = A + \frac{B}{T} + CT + DT^2 \tag{3.36}$$

where
   T is in Kelvin
   the viscosity μ is in centipoise

The values of the correlation parameters A, B, C, and D are tabulated by Yaws et al. (1994).

For non-Newtonian fluids, any model parameter with the dimensions or physical significance of viscosity (e.g., the power law consistency, m, or the Carreau parameters η_∞ and η_o) will depend on temperature in a manner similar to the viscosity of a Newtonian fluid (e.g., Equation 3.35 or 3.36).

B. GASES

In contrast to the behavior of liquids, the viscosity of a gas increases with increasing temperature. This is because gas molecules are much farther apart, so the short-range attractive forces are very small. However, as the temperature is increased, the molecular kinetic energy increases, resulting in a greater exchange of momentum between the molecules and consequently resulting in a higher viscosity. The viscosity of gases is not as sensitive to temperature as that of liquids, however, and can often be represented by the equation

$$\mu = aT^b \tag{3.37}$$

The parameters a and b can be evaluated from a knowledge of the viscosity at two different temperatures, and the equation can then be used to calculate the viscosity at any other temperature. The value of the parameter b is often close to 1.5. In fact, if the viscosity (μ_1) of a gas is known at only one temperature (T_1), the following equation can be used to estimate the viscosity at any other temperature:

$$\mu = \mu_1 \left(\frac{T}{T_1}\right)^{3/2} \left(\frac{T_1 + 1.47T_B}{T + 1.47T_B}\right) \tag{3.38}$$

where
   the temperatures are in Rankine
   T_B is the boiling point of the gas

V. DENSITY

In contrast to viscosity, the density of both liquids and gases decreases with increasing temperature, and the density of gases is much more sensitive to temperature than that of liquids. If the density of a liquid (ρ) and its vapor (ρ_v) are known at 60°F, the density at any other temperature can be estimated by the equation

$$\frac{(\rho-\rho_v)_T}{(\rho-\rho_v)_{60°F}} = \left(\frac{T_c-T}{T_c-519.67}\right)^{1/N} \tag{3.39}$$

where
  the temperatures are in degrees Rankine
  T_c is the critical temperature of the substance

The value of N is given in Table 3.3 for various liquids. This method yields the value of $(\rho-\rho_v)_T$ at the temperature of interest (T) and one must still estimate the value of ρ_v to evaluate the density of the liquid, ρ.

The specific gravity of hydrocarbon liquids at 60°F is also often represented by the API gravity, or °API:

$$SG_{60°F} = \frac{141.5}{131.5+°API} \tag{3.40}$$

For gases, if the temperature is well above the critical temperature and the pressure below the critical pressure, the ideal gas law usually applies:

$$\rho = \frac{PM}{RT} = \frac{M}{\tilde{V}_m}\left(\frac{T_{ref}}{T}\right)\left(\frac{P}{P_{ref}}\right) \tag{3.41}$$

where
  M is the gas molecular weight
  Temperatures and pressures are absolute
  $\tilde{V}_m$ is the *standard molar volume (22.4 m³/(kg mol) at 273 K and 1 atm, 359 ft³/(lb mol) at 492°R and 1 atm, or 379.4 ft³/lb mol at 520°R (60°F) and 1 atm)*

The notation *SCF* (which stands for "standard cubic feet") is often used for hydrocarbon gases to represent the volume in ft³ that would be occupied by the gas at *60°F* and *1 atm* pressure and is thus actually a measure of the mass of the gas.

For other methods of predicting fluid properties and their temperature dependence, the reader is referred to the book by Reid et al. (1977).

TABLE 3.3
Parameter N in Equation 3.39

Liquid	N
Water and alcohols	4
Hydrocarbons and ethers	3.45
Organics	3.23
Inorganics	3.03

VI. SURFACE TENSION

Surface tension is a physical property of an interface between two phases such as a liquid–liquid, liquid–gas, or liquid–solid interface. The unique orientation of the fluid molecules at such an interface gives rise to unique forces and energy at the interface. For instance, it is the surface tension of water that tends to keep the small air bubbles spherical! Similarly, it is the surface tension that makes the flow of oil in narrow pores and passages of the rocks rather difficult. Surface tension arises from the cohesion between molecules at the fluid interface. A molecule surrounded by identical molecules experiences a zero net force, as it is pulled equally in all directions. In contrast, a molecule close to a wall, or an interface with another fluid, experiences a net force due to this orientation. From a fluid mechanics standpoint, surface tension can complicate the analysis of problems involving the formation and/or motion of bubbles and drops in another liquid such as in ink-jet printers, or when the flow passage is of the order of micro- and nanometers such as in microfluidic devices. In contrast to the forces such as pressure or shear stresses that act on areas and the body forces acting on the volume of a fluid, surface tension is a force per unit length of the interface with the units of dyn/cm or N/m. Note that surface tension also represents a concentration of surface energy, with dimensions of energy/area (e.g., N m/m^2). Thus, in microfluidic and nanodevices, the characteristic dimension (l) is of the order of micrometers or nanometers and the corresponding surface areas and volumes are O(l^2) and O(l^3), respectively. Thus, in such situations, the surface tension force can be a significant factor. Finally, there are situations when the driving force for the flow arises from the surface tension gradient itself, such as the flow in capillaries. While none of these situations are addressed in depth in this text, in view of the growing importance of nano- and microscale engineering, the readers must be aware of these issues and one must look up specialized books on this subject (e.g., Probstein, 2003).

SUMMARY

The key points covered in this chapter include:

- The understanding and description of linear and nonlinear deformation and flow of solids and fluids.
- The measurement of fluid viscous properties by Couette and Poiseuille viscometers.
- The types of fluid materials that exhibit nonlinear viscous (non-Newtonian) behavior.
- The various mathematical models that are capable of describing different classes of non-Newtonian behavior.
- Temperature dependency of the viscosity of liquids and gases.
- Nature of the temperature dependency of density for liquids and gases.
- Awareness of the importance of surface tension in systems of small dimensions.

PROBLEMS

RHEOLOGICAL PROPERTIES

1. (a) Using tabulated data for the viscosity of water and SAE 10 lube oil as a function of temperature, plot the data in a form that is consistent with each of the following equations:
    - (i) $\mu = A \exp(B/T)$
    - (ii) $\mu = aT^b$
    (b) Arrange the equations in (a) in such a form that you can use linear regression analysis to determine the values of A, B and a, b that give the best fit to the data for each fluid (a spreadsheet is useful for this). (Note that T is the absolute temperature here.)

2. The viscosity of a fluid sample is measured in a cup and bob viscometer. The bob is 15 cm long with a diameter of 9.8 cm, and the cup has a diameter of 10 cm. The cup rotates, and the torque is measured on the bob. The following data were obtained:

Ω (rpm)	2	4	10	20	40
T (dyn cm)	3.6×10^5	3.8×10^5	4.4×10^5	5.4×10^5	7.4×10^5

(a) Determine the viscosity of the sample.

(b) What viscosity model equation would be the most appropriate for describing the viscosity of this sample? Convert the data to corresponding values of viscosity versus shear rate, and plot them on appropriate axes consistent with the data and your equation. Use linear regression analysis in a form that is consistent with the plot to determine the values of each of the parameters in your equation.

(c) What is the viscosity of this sample at a cup speed of 100 rpm in the viscometer?

3. A fluid sample is contained between two parallel plates separated by a distance of 2 ± 0.1 mm. The area of the plates is 100 ± 0.01 cm^2. The bottom plate is stationary, and the top plate moves with a velocity of 1 cm/s when a force of 315 ± 25 dyn is applied to it, and at 5 cm/s when the force is 1650 ± 25 dyn.

(a) Is the fluid Newtonian?

(b) What is its viscosity?

(c) What is the range of uncertainty to your answer to (b)?

4. The following materials exhibit flow properties that can be described by models that include a yield stress (e.g., Bingham plastic): (a) catsup, (b) toothpaste, (c) paint, (d) coal slurries, and (e) printing ink. In terms of typical applications of these materials, describe how the yield stress is beneficial to their behavior, in contrast to how they would behave if they were Newtonian.

5. Consider each of the fluids for which the viscosity is shown in Figure 3.7, all of which exhibit a "structural viscosity" characteristic. Explain how the "structure" of each of these fluids influences the nature of the viscosity curve for that fluid.

6. Starting with the equations for $\tau = fn(\dot{\gamma})$ that define the power law and Bingham plastic fluids, derive the equations for the viscosity functions for these models as a function of shear stress, that is, $\eta = fn(\tau)$.

7. A paint sample is tested in a Couette (cup and bob) viscometer that has an outer radius of 5 cm, an inner radius of 4.9 cm, and a bob length of 10 cm. When the outer cylinder is rotated at a speed of 4 rpm, the torque on the bob is 0.0151 N m, and at a speed of 20 rpm, the torque is 0.0226 N m.

(a) What are the corresponding values of shear stress and shear rate for these two data points (in cgs units)?

(b) What can you conclude about the viscous properties of the paint sample?

(c) Which of the following models could be used to describe the paint:
   (i)   Newtonian
   (ii) Bingham plastic
   (iii) Power law?
   Explain why.

(d) Determine the values of the fluid properties (i.e., parameters) of the models in (c) that could be used.

(e) What would the viscosity of the paint be at a shear rate of 500 s^{-1} (in poise)?

8. The quantities that are measured in a cup and bob viscometer are the rotation rate of the cup (rpm) and the corresponding torque transmitted to the bob. These quantities are then converted to corresponding values of shear rate ($\dot{\gamma}$) and shear stress (τ), which in turn can be converted to corresponding values of viscosity (η).

(a) Show what a log–log plot of τ versus $\dot{\gamma}$ and η versus $\dot{\gamma}$ would look like for materials that follow the following models: (i) Newtonian, (ii) power law (shear thinning), (iii) power law (shear thickening), (iv) Bingham plastic, and (v) structural.

(b) Show how the values of the parameters for each of the models listed in (a) can be evaluated from the respective plot of η versus $\dot{\gamma}$. That is, relate each model parameter to some characteristic or combination of characteristics of the plot such as the slope, specific values read from the plot, intersection of tangent lines, etc.

9. What is the difference between shear stress and momentum flux? How are they related? Illustrate each one in terms of the angular flow in the gap in a cup and bob viscometer, in which the outer cylinder (cup) is rotated and the torque is measured at the stationary inner cylinder (bob).

10. A fluid is contained in the annulus in a cup and bob viscometer. The bob has a radius of 50 mm and a length of 10 cm and is made to rotate inside the cup by application of a torque on a shaft attached to the bob. If the cup inside radius is 52 mm and the applied torque is 0.03 ft lb$_f$, what is the shear stress in the fluid at the bob surface and at the cup surface? If the fluid is Newtonian with a viscosity of 50 cP, how fast will the bob rotate (in rpm) with this applied torque?

11. You measure the viscosity of a fluid sample in a cup and bob viscometer. The radius of the cup is 2 in. and that of the bob is 1.75 in., and the length of the bob is 3 in. At a speed of 10 rpm, the measured torque is 500 dyn cm, and at 50 rpm, it is 1200 dyn cm. What is the viscosity of the fluid? What can you deduce about the properties of the fluid?

12. A sample of coal slurry is tested in a Couette (cup and bob) viscometer. The bob has a diameter of 10.0 cm and a length of 8.0 cm, and the cup has a diameter of 10.2 cm. When the cup is rotated at a rate of 2 rpm, the torque on the bob is 6.75×10^4 dyn cm, and at a rate of 50 rpm, it is 2.44×10^6 dyn cm. If the slurry follows the power law model, what are the values of the flow index and consistency coefficient? If the slurry follows the Bingham plastic model, what are the values of the yield stress and the limiting viscosity? What would the viscosity of this slurry be at a shear rate of 500 s^{-1} as predicted by each of these models? Which number would you be more likely to believe, and why?

13. You must analyze the viscous properties of blood. Its measured viscosity is 6.49 cP at a shear rate of 10 s^{-1} and 4.66 cP at a shear rate of 80 s^{-1}.
    (a) How would you describe these viscous properties?
    (b) If the blood is subjected to a shear stress of 50 dyn/cm^2, what would its viscosity be if it is described by (i) the power law model and (ii) the Bingham plastic model? Which answer do you think would be better, and why?

14. The following data were measured for the viscosity of a 500 ppm polyacrylamide solution in distilled water:

Shear Rate (s^{-1})	Viscosity (cP)	Shear Rate (s^{-1})	Viscosity (cP)
0.015	300	15	30
0.02	290	40	22
0.05	280	80	15
0.08	270	120	11
0.12	260	200	8
0.3	200	350	6
0.4	190	700	5
0.8	180	2,000	3.3
2	100	4,500	2.2
3.5	80	7,000	2.1
8	50	20,000	2

Find the model that best represents these data, and determine the values of the model parameters by fitting the model to the data. (This can be done most easily by trial and error, using a spreadsheet.)

15. What viscosity model best represents the following data? Determine the values of the parameters in the model. Show a plot of the data together with the line that represents the model, to show how well the model works. (*Hint*: The easiest way to do this is by trial and error, fitting the model equation to the data using a spreadsheet.)

Shear Rate (s^{-1})	Viscosity (Poise)
0.007	7745
0.01	7690
0.02	7399
0.05	6187
0.07	5488
0.1	4705
0.2	3329
0.5	2033
0.7	1692
1	1392
2	952
5	576
7	479
10	394
20	270
50	164
100	113
200	77.9
500	48.1
700	40.4
1,000	33.6
2,000	23.8
5,000	15.3
7,000	13.2
10,000	11.3
20,000	8.5
50,000	6.1
100,000	5.5

16. You would like to determine the pressure drop in a slurry pipeline. To do this, you need to know the rheological properties of the slurry. To evaluate these properties, you test the slurry by pumping it through a 1/8 in. ID tube that is 10 ft long. You find that it takes a 5 psi pressure drop to produce a flow rate of 100 cm³/s in the tube and that a pressure drop of 10 psi results in a flow rate of 300 cm³/s. What can you deduce about the rheological characteristics of the slurry from these data? If it is assumed that the slurry can be adequately described by the power law model, what would be the values of the appropriate fluid properties (i.e., the flow index and consistency parameter) for the slurry?

17. A film of paint, 3 mm thick, is applied to a flat surface that is inclined to the horizontal by an angle θ. If the paint is a Bingham plastic, with a yield stress of 150 dyn/cm², a limiting viscosity of 65 cP, and a SG of 1.3, how large would the angle θ have to be before the paint would start to run? At this angle, what would the shear rate be if the paint follows

the power law model instead, with a flow index of 0.6 and a consistency coefficient of 215 (in cgs units)?

18. A thick suspension is tested in a Couette (cup and bob) viscometer that has a cup radius of 2.05 cm, a bob radius of 2.00 cm, and a bob length of 15 cm. The following data are obtained:

Cup Speed (rpm)	Torque on Bob (dyn cm)
2	2,000
4	6,000
10	19,000
20	50,000
50	150,000

What can you deduce about (a) the viscous properties of this material, and (b) the best model to use to represent these data?

19. You have obtained data for a viscous fluid in a cup and bob viscometer that has the following dimensions: cup radius = 2 cm, bob radius = 1.5 cm, bob length = 3 cm. The data are tabulated in the following, where n' is the point slope of the log T versus log N curve:

N (rpm)	T (dyn cm)	n'	N (rpm)	T (dyn cm)	n'
1	1.13×10^4	0.01	100	1.25×10^4	0.6
2	1.13×10^4	0.02	200	1.42×10^4	0.7
5	1.13×10^4	0.05	500	1.93×10^4	0.8
10	1.13×10^4	0.1	1000	2.73×10^4	0.9
20	1.14×10^4	0.2	2000	4.31×10^4	1.0
50	1.16×10^4	0.5			

(a) Determine the viscosity of the fluid. How would you describe its viscosity?
(b) What kind of viscous model (equation) would be appropriate to describe this fluid?
(c) Use the data to determine the values of the fluid properties that are defined by the model.

20. A sample of a viscous fluid is tested in a cup and bob viscometer that has a cup radius of 2.1 cm, a bob radius of 2.0 cm, and a bob length of 5 cm. When the cup is rotated at 10 rpm, the torque measured at the bob is 6,000 dyn cm, and at 100 rpm, the torque is 15,000 dyn cm.
(a) What is the viscosity of this sample?
(b) What can you conclude about the viscous properties of the sample?
(c) If the cup is rotated at 500 rpm, what will be the torque on the bob and the fluid viscosity? Clearly explain any assumptions you make to answer this question, and tell how you might check the validity of these assumptions.

21. You have a sample of a sediment that is non-Newtonian. You measure its viscosity in a cup and bob viscometer having a cup radius of 3.0 cm, a bob radius of 2.5 cm, and a length of 5 cm. At a rotational speed of 10 rpm, the torque transmitted to the bob is 700 dyn cm, and at 100 rpm, the torque is 2500 dyn cm.
(a) What is the viscosity of the sample?
(b) Determine the values of the model parameters that represent the sediment viscous properties if it is represented by (i) the power law model or (ii) the Bingham plastic model.
(c) What would the flow rate of the sediment be (in cm³/s) in a 2 cm diameter tube, 50 m long, that is subjected to a differential pressure driving force of 25 psi assuming that (i) the power law model and (ii) the Bingham plastic model apply?
Which of these two answers do you think is best, and why?

22. Acrylic latex paint can be described by the Bingham plastic model with a yield stress of 112 dyn/cm², a limiting viscosity of 80 cP, and a density of 0.95 g/cm³. What is the maximum thickness of this paint that can be applied to a vertical wall without it running?

23. Santa Claus and his loaded sleigh are sitting on your roof, which is covered with snow. The sled's two runners each have a length L and width W, and the roof is inclined at an angle θ to the horizontal. The thickness of the snow between the runners and the roof is H. If the snow has properties of a Bingham plastic, derive an expression for the total mass (m) of the loaded sleigh at which it will just start to slide on the roof if it is pointed straight downhill. If the actual mass is twice this minimum mass, determine an expression for the speed at which the sled will slide. (*Note*: Snow does not actually behave as a Bingham plastic!)

24. You must design a piping system to handle a sludge waste product. However, you don't know the properties of the sludge, so you test it in a cup and bob viscometer with a cup diameter of 10 cm, a bob diameter of 9.8 cm, and a bob length of 8 cm. When the cup is rotated at 2 rpm, the torque on the bob is 2.4×10^4 dyn cm, and at 20 rpm, it is 6.5×10^4 dyn cm.
    (a) If you use the power law model to describe the sludge, what are the values of the flow index and consistency?
    (b) If you use the Bingham plastic model instead, what are the values of the yield stress and limiting viscosity?

25. A fluid sample is tested in a cup and bob viscometer that has a cup diameter of 2.25 in., a bob diameter of 2 in., and length of 3 in. The following data are obtained:

Rotation Rate (rpm)	Torque (dyn cm)
20	2,500
50	5,000
100	8,000
200	10,000

    (a) Determine the viscosity of this sample.
    (b) What model would provide the best representation of this viscosity function, and why?

26. You test a sample in a cup and bob viscometer to determine the viscosity. The diameter of the cup is 55 mm, that of the bob is 50 mm, and the length is 65 mm. The cup is rotated and the torque on the bob is measured, giving the following data:

Cup Speed (rpm)	Torque on Bob (dyn cm)
2	3,000
10	6,000
20	11,800
30	14,500
40	17,800

    (a) Determine the viscosity of this sample.
    (b) How would you describe the viscosity of this material?
    (c) What model would be the most appropriate to represent this viscosity?
    (d) Determine the values of the parameters in the model that fit the model to the data.

27. Consider each of the fluids for which the viscosity is shown in Figure 3.7, all of which exhibit a typical "structural viscosity" characteristic. Explain why this is a logical consequence of the composition or "structural makeup" for each of these fluids.

28. You are asked to measure the viscosity of an emulsion, so you use a tube flow viscometer similar to that illustrated in Figure 3.4, with the container open to the atmosphere. The length of the tube is 10 cm, its diameter is 2 mm, and the diameter of the container is 3 in. When the level of the sample is 10 cm above the bottom of the container, the emulsion drains through the tube at a rate of 12 cm³/min, and when the level is 20 cm, the flow rate is 30 cm³/min. The emulsion density is 1.3 g/cm³.
    (a) What can you tell from the data about the viscous properties of the emulsion?
    (b) Determine the viscosity of the emulsion.
    (c) What would the sample viscosity be at a shear rate of 500 s⁻¹?

29. You must determine the horsepower required to pump a coal slurry through an 18 in. diameter pipeline, 300 miles long, at a rate of 5 million tons/year. The slurry can be described by the Bingham plastic model, with a yield stress of 75 dyn/cm², a limiting viscosity of 40 cP, and a density of 1.4 g/cm³. For non-Newtonian fluids, the flow is not sensitive to the wall roughness.
    (a) Determine the dimensionless groups that characterize this system. Use these to design a lab experiment, from which you can scale up measurements to find the desired horsepower.
    (b) Can you use the same slurry in the lab as in the actual pipeline?
    (c) If you use a slurry in the lab that has a yield stress of 150 dyn/cm², a limiting viscosity of 20 cP, and a density of 1.5 g/cm³, what size pipe and what flow rate (in gpm) should you use in the lab?
    (d) If you run the lab system as designed and measure a pressure drop ΔP (psi) over a 100 ft length of pipe, show how you would use this information to determine the required horsepower for the actual pipeline.

30. You want to determine how fast a rock will settle in mud, which behaves like a Bingham plastic. The first step is to perform a dimensional analysis of the system.
    (a) List the important variables that have an influence on this problem, with their dimensions (give careful attention to the factors that cause the rock to fall when listing these variables), and determine the appropriate dimensionless groups.
    (b) Design an experiment in which you measure the velocity of a solid sphere falling in a Bingham plastic in the lab, and use the dimensionless variables to scale the answer to find the velocity of a 2 in. diameter rock, with a density of 3.5 g/cm³, falling in a mud with a yield stress of 300 dyn/cm², a limiting viscosity of 80 cP, and a density of 1.6 g/cm³. Should you use this same mud in the lab, or can you use a different material that is also a Bingham plastic but with a different yield stress and limiting viscosity?
    (c) If you use a suspension in the lab with a yield stress of 150 dyn/cm², a limiting viscosity of 30 cP, and a density of 1.3 g/cm³ and a solid sphere, how big should the sphere be and how much should it weigh?
    (d) If the sphere in the lab falls at a rate of 4 cm/s, how fast will the 2 in. diameter rock fall in the other mud?

31. A pipeline has been proposed to transport a coal slurry 1200 miles from Wyoming to Texas, at a rate of 50 million tons/year, through a 36 in. diameter pipeline. The coal slurry has the properties of a Bingham plastic, with a yield stress of 150 dyn/cm², a limiting viscosity of 40 cP, and a SG of 1.5. You must conduct a lab experiment in which the measured pressure gradient can be used to determine the total pressure drop in the pipeline.

(a) Perform a dimensional analysis of the system to determine an appropriate set of dimensionless groups to use (you may neglect the effect of wall roughness for this fluid).

(b) For the lab test fluid, you have available a sample of the above coal slurry and three different muds with the following properties:

	Yield Stress (dyn/cm²)	Limiting Viscosity (cP)	Density (g/cm³)
Mud 1	50	80	1.8
Mud 2	100	20	1.2
Mud 3	250	10	1.4

Which of these would be the best to use in the lab, and why?

(c) What size pipe and what flow rate (in lb_m/min) should you use in the lab?

(d) If the measured pressure gradient in the lab is 0.016 psi/ft, what is the total pressure drop in the pipeline?

32. A fluid sample is subjected to a "sliding plate" (simple shear) test. The area of the plates is 100 ± 0.01 cm² and the spacing between them is 2 ± 0.1 mm. When the moving plate travels at a speed of 0.5 cm/s, the force required to move it is measured to be 150 dyn, and at a speed of 3 cm/s, the force is 1100 dyn. The force transducer has a sensitivity of 50 dyn. What can you deduce about the viscous properties of the sample?

33. You want to predict how fast a glacier that is 200 ft thick will flow down a slope inclined 25° to the horizontal. Assume that the glacier ice can be described by the Bingham plastic model with a yield stress of 50 psi, a limiting viscosity of 840 poise, and a SG of 0.98. The following materials are available to you in the lab, which also may be described by the Bingham plastic model:

	Yield Stress (dyn/cm²)	Limiting Viscosity (cP)	SG
Mayonnaise	300	130	0.91
Shaving cream	175	15	0.32
Catsup	130	150	1.2
Paint	87	95	1.35

You want to set up a lab experiment to measure the velocity at which the model fluid flows down an inclined plane and scale this value to find the velocity of the glacier.

(a) Determine the appropriate set of dimensionless groups.

(b) Which of the above materials would be the best to use in the lab? Why?

(c) What is the film thickness that you should use in the lab, and at what angle should the plane be inclined?

(d) What would be the scale factor between the measured velocity in the lab and the glacier velocity?

(e) What problems might you encounter when conducting this experiment?

34. Your boss gives you a sample of "gunk" and asks you to measure its viscosity. You do this in a cup and bob viscometer that has an outer (cup) diameter of 2 in., an inner (bob) diameter of 1.75 in., and a bob length of 4 in. You run the viscometer at three speeds and record the following data:

Rotational Velocity, Ω (rpm)	Torque on Bob, T (dyn cm)
1	10,500
10	50,000
100	240,000

    (a) How would you classify the viscous properties of this material?

    (b) Calculate the viscosity of the sample in cP.

    (c) What viscosity model best represents these data, and what are the values of the viscous properties (i.e., the model parameters) for the model?

35. The dimensions and measured quantities in the viscometer in Problem 34 are known to the following precision:

T:	$\pm 1\%$ of full scale (full scale = 500,000 dyn cm)
Ω:	$\pm 1\%$ of reading
D_o, D_i, and L:	± 0.002 in.

Estimate the maximum percentage uncertainty in the measured viscosity of the sample for each of the three data points.

36. A concentrated slurry is prepared in an open 8 ft diameter mixing tank, using an impeller with a diameter of 6 ft located 3 ft below the free surface. The slurry is non-Newtonian and can be described as a Bingham plastic with a yield stress of 50 dyn/cm², a limiting viscosity of 20 cP, and a density of 1.5 g/cm³. A vortex is formed above the impeller, and if the speed is too high the vortex can reach the blades of the impeller entraining air and causing problems. Since this condition is to be avoided, you need to know how fast the impeller can be rotated without entraining the vortex. To do this, you conduct a lab experiment using a scale model of the impeller that is 1 ft in diameter. You must design the experiment so that the critical impeller speed can be measured in the lab and scaled up to determine the critical speed of the larger mixer.

    (a) List all the variables that are important in this system, and determine an appropriate set of dimensionless groups.

    (b) Determine the diameter of the tank that should be used in the lab, and the depth below the surface at which the impeller should be located.

    (c) Should you use the same slurry in the lab model as in the field? If not, what properties should the lab slurry have?

    (d) If the critical speed of the impeller in the lab system is ω (rpm), what is the critical speed of the impeller in the large tank?

37. You would like to know the thickness of a paint film as it drains at a rate of 1 gpm down a flat surface that is 6 in. wide and is inclined at an angle of 30° to the vertical. The paint is non-Newtonian and can be described as a Bingham plastic with a limiting viscosity of 100 cP, a yield stress of 60 dyn/cm², and a density of 0.9 g/cm³. You have data from the laboratory for the film thickness of a Bingham plastic that has a limiting viscosity of 70 cP, a yield stress of 40 dyn/cm², and a density of 1 g/cm³ flowing down a plane 1 ft wide inclined at an angle of 45° to the vertical, at various flow rates.

    (a) At what flow rate (in gpm) will the laboratory system correspond to the conditions of the other system?

    (b) If the film thickness of the laboratory fluid is 3 mm at these conditions, what would the film thickness be for the other system?

38. The following data were obtained for a proprietary salad dressing tested at 22°C in a cup and bob viscometer (cup diameter = 4.2 cm, bob diameter = 4.01 cm, length of 6 cm):

rpm	2	4	8	32	64	256
Torque (N m) × 10³	3.98	4.84	6.06	8.48	10.6	19.7

Calculate the shear stress, shear rate, and viscosity for this material, and fit a suitable viscosity model to this data set.

39. A kaolin-in-water suspension was tested in a 13 mm internal diameter horizontal tube, and the following data were reported:

Γ (s^{-1})	660	803	978	1208	1518	1790	2081	2300	2629	2988
τ_w (Pa)	37.7	38.4	40.1	41.7	43.3	44.8	46.7	48.2	62.7	73.6

Obtain the true shear stress, shear rate, and viscosity for this suspension. Does this suspension exhibit a yield stress? Fit a suitable viscosity model to approximate the rheology of this suspension.

40. The same suspension as that in Problem 39 above was subsequently tested in a 28 mm internal diameter pipe, and the following data reported:

Γ (s^{-1})	127	200	289	406	557	744	951	1079	1409	1610
τ_w (Pa)	31	32.8	33.6	34.2	36.4	37.8	40.0	43.4	66.0	79.25

Obtain the true shear stress, shear rate, and viscosity for this material using this information. Is this consistent with the results obtained in Problem 39? If not, give possible reasons.

NOTATION

A_y — Area whose outward normal vector is in the y direction, [L^2]
F_x — Force component in the x direction, [$F = M\,L/t^2$]
$fn()$ — A function of whatever is in the ()
G — Shear modulus, [$F/L^2 = M/Lt^2$]
g — Acceleration due to gravity, [L/t^2]
h_y — Distance between plates in the y direction, [L]
L — Length, [L]
m — Power law consistency parameter, [$M/(Lt^{2-n})$]
n — Power law flow index, [—]
n' — Variable defined by Equation 3.14 or 3.18, [—]
P — Pressure, [$F/L^2 = M/Lt^2$]
p — Parameter in Carreau model, [—]
Q — Volumetric flow rate, [L^3/t]
R — Radius, [L]
r — Radial direction or coordinate, [L]
SG — Specific gravity, [—]
T — Temperature (absolute), [T]
T — Torque or moment, [$FL = M\,L^2/t^2$]
U_x — Displacement of boundary in the x direction, [L]
u_x — Local displacement in the x direction, [L]
v_x — Local velocity component in x direction, [L/t]
V — Bulk or average velocity, [L/t]
z — Vertical direction measured upward, [L]

GREEK

β — (R_i/R_o) ratio of inner to outer radius in Couette viscometer
Γ — Shear rate at tube wall for Newtonian fluid, Equation 3.16, [1/t]
γ_{yx} — Gradient of x displacement in y direction (shear strain, or γ), [—]
$\dot{\gamma}_{yx}$ — Gradient of x velocity in y direction (shear rate, or $\dot{\gamma}$), [1/t]
$\Delta()$ — Value of $()_2 - ()_1$

λ Fluid time constant parameter, [t]

μ Viscosity (Newtonian), [M/Lt]

μ_∞ Bingham limiting viscosity, [M/Lt]

η Viscosity function (non-Newtonian), [M/Lt]

ρ Density, [M/L^3]

θ Angular displacement, [—]

τ_o Yield stress, [F/L^2 = M/Lt^2]

τ_{yx} Force in x direction on y surface (shear stress, or τ), [F/L^2 = M/Lt^2]

$\tau_{r\theta}$ Force in θ direction on r surface (shear stress), [F/L^2 = M/Lt^2]

Φ Static potential = $P + \rho gz$, [F/L^2 = M/L t^2]

Ω Angular velocity of boundary, [1/t]

SUBSCRIPTS

1	Reference point 1
2	Reference point 2
i	Inner
o	Outer
o	Zero shear rate parameter
w	Value at wall
x, y, r, θ	Coordinate directions
∞	High shear limiting parameter

REFERENCES

Barnes, H.A., J.F. Hutton, and K. Walters, *An Introduction to Rheology*, Elsevier, New York, 1989.

Boger, D.V. and K. Walters, *Rheological Phenomena in Focus*, Elsevier, Amsterdam, the Netherlands, 1992.

Carreau, P.J., Rheological equations from molecular network theories, *Trans. Soc. Rheol.*, 16, 99–127, 1972.

Chhabra, R.P. and J.F. Richardson, *Non-Newtonian Flow and Applied Rheology*, 2nd edn., Butterworth-Heinemann, Oxford, U.K., 2008.

Darby, R., *Viscoelastic Fluids*, Marcel Dekker, New York, 1976.

Darby, R., Couette viscometer data reduction for materials with a yield stress, *J. Rheology*, 29, 369–378, 1985.

Darby, R. and S. Pivsa-Art, An improved correlation for turbulent drag reduction in dilute polymer solutions, *Can. J. Chem. Eng.*, 69, 1395–1400, 1991.

Meter, D.M., Tube flow of non-Newtonian polymer solutions: Part I, Laminar flow and rheological models, *AIChE J.*, 10, 878–881, 1964.

Probstein, R.F., *Physico-Chemical Hydrodynamics: An Introduction*, Wiley, New York, 2003.

Reid, R.C., J.M. Prausnitz, and T.K. Sherwood, *The Properties of Gases and Liquids*, 3rd edn., McGraw-Hill, New York, 1977.

Yasuda, K., R.C. Armstrong, and R.E. Cohen, Shear flow properties of concentrated solutions of linear and star branched polystyrenes, *Rheol. Acta*, 20, 163–178, 1981.

Yaws, C.L., X. Lin, and L. Bu, Calculate viscosities for 355 liquids, *Chem. Eng.*, 101, 119–126, April 1994.

4 Fluid Statics

"Archimedes' finding that the crown was of gold was a discovery; but he invented the method of determining the density of solids. Indeed, discoverers must generally be inventors; though inventors are not necessarily discoverers."

—William Ramsay, 1852–1916, Chemist

I. STRESS AND PRESSURE

The forces that exist within a fluid at any point may arise from various sources. These include gravity, or the "weight" of the fluid, an imposed static pressure from an external source, an external driving force such as a pump or compressor, and the internal resistance to relative motion between fluid elements, or inertial effects resulting from the local velocity and the mass flow rate of the fluid (e.g., the transport or rate of change of momentum).

Any or all of these forces may result in local stresses within the fluid. "Stress" can be thought of as a (local) "concentration of force," or the local force per unit area that acts upon an infinitesimal volume of the fluid. Now both force and area are vectors, the direction of the area being defined by the normal vector that points outward relative to the fluid volume bounded by the surface. Thus, each stress component has a magnitude and *two* directions associated with it, that is, the direction of the force that acts on the fluid element, F_j, and the orientation of the surface of the element upon which the force acts, A_i. These are the characteristics of a "second-order tensor" or "dyad." If the direction in which the local force acts is designated by subscript j (e.g., $j = x, y,$ or z in Cartesian coordinates) and the orientation (i.e., the *normal* to the surface) of the local area element upon which it acts is designated by the subscript i, then the corresponding stress component (σ_{ij}) is given by

$$\sigma_{ij} = \frac{F_j}{A_i} \quad \text{where } i, j, k = 1, 2 \text{ or } 3 \text{ (e.g., } x, y, \text{ or } z) \tag{4.1}$$

Note that since i and j each represent any of three possible directions, there are a total of nine possible components of the stress tensor (at any given point in a fluid). However, it can readily be shown that the stress tensor is symmetric (i.e., the ij components are the same as the ji components for $i \neq j$) so there are at most six independent stress components. Of these six, three are normal components ($i = j$) and three are tangential ($i \neq j$) components of the stress tensor. Because of the various origins of these forces, as mentioned earlier, there are different "types" of stresses. For example, the only stress that can exist in a fluid at rest is pressure, which can result from gravity (e.g., hydrostatic head) or various other static forces acting on the fluid. Although pressure is a stress (e.g., a force per unit area), it is isotropic, that is, the pressure at a point acts uniformly in all directions, normal to any local surface at a given point in the fluid. Such a stress has no directional character and is thus a *scalar*. (Any *isotropic* tensor is, by definition, a *scalar*, because it has magnitude only and no direction.) However, the stress components that arise from the fluid motion do have directional characteristics, which are determined by the relative motion (or rather by the local velocity gradients) present in the fluid. These stresses are associated with the local resistance to motion due to viscous or inertial properties and are anisotropic in nature, that is, their components depend upon direction at any given point in the fluid. These anisotropic stress components are designated by τ_{ij}, where the i and j have the same significance as in Equation 4.1.

Thus, the total stress, σ_{ij}, at any point within a fluid is composed of both the isotropic pressure and anisotropic stress components, as follows:

$$\sigma_{ij} = -P\delta_{ij} + \tau_{ij} \tag{4.2}$$

where P is the (isotropic) pressure. By convention, tensile stresses are considered positive and compressive stresses are negative, so that pressure is a "negative" stress because it is always compressive, whereas tensile stresses are positive (i.e., a positive F_j acting on a positive A_i or a negative F_j on a negative A_i). The term δ_{ij} in Equation 4.2 is a "unit tensor" (or Kronecker delta), which has a value of zero if $i \neq j$, and unity if $i = j$. This is required, because the isotropic pressure acts only in the normal direction (e.g., $i = j$) and has only one component (i.e., it is a scalar). As mentioned earlier, the anisotropic shear stress components τ_{ij} in a fluid are associated with relative motion within the fluid and are therefore zero in any fluid at rest. It follows that the only stress that can exist in a fluid at rest or in a state of uniform motion (in which there are no velocity gradients present in the fluid) is pressure. This is a major distinction between a fluid and a solid, since solids can support a shear stress in a state of rest. It is this situation with which we will be concerned in this chapter.

In Cartesian coordinates, the components of the stress tensor are

$$\begin{pmatrix} \sigma_{xx} & \sigma_{xy} & \sigma_{xz} \\ \sigma_{yx} & \sigma_{yy} & \sigma_{yz} \\ \sigma_{zx} & \sigma_{zy} & \sigma_{zz} \end{pmatrix} = \begin{pmatrix} -P + \tau_{xx} & \tau_{xy} & \tau_{xz} \\ \tau_{xy} & -P + \tau_{yy} & \tau_{yz} \\ \tau_{xz} & \tau_{yz} & -P + \tau_{zz} \end{pmatrix} \tag{4.3}$$

Example 4.1 Flow in a Vertical Slit

Consider a Newtonian liquid contained in a slit between two vertical parallel planes (see Figure E4.1). For case (a), the slit is closed at the bottom so that the fluid cannot flow out. For case (b), the slit is open at the bottom so that the liquid can flow out. If the coordinate system is as shown in the diagram, show the nonzero components of the stress tensor for both case (a) and (b).

Solution:

Case (a): As there is no motion, then all shear stresses are zero, so

$$\sigma_{ij} = \begin{pmatrix} -P & 0 & 0 \\ 0 & -P & 0 \\ 0 & 0 & -P \end{pmatrix} + (0)$$

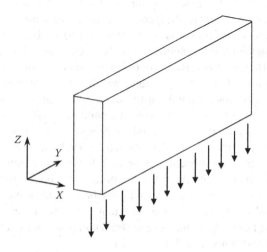

FIGURE E4.1 Flow in a vertical slit.

Case (b): As there is motion only in the $-z$ direction and the z component of velocity varies only in the x direction (i.e., the fluid sticks to the walls, so the velocity must be zero at the walls and nonzero away from the walls), then the shear stress components must be $\tau_{xz} = \tau_{zx}$:

$$\sigma_{ij} = \begin{pmatrix} -P & 0 & 0 \\ 0 & -P & 0 \\ 0 & 0 & -P \end{pmatrix} + \begin{pmatrix} 0 & 0 & \tau_{xz} \\ 0 & 0 & 0 \\ \tau_{zx} & 0 & 0 \end{pmatrix}$$

Note that $\tau_{xz} = \tau_{zx}$ due to symmetry of the stress tensor.

"Students often feel under stress and pressure when trying to understand pressure and stress."

II. THE BASIC EQUATION OF FLUID STATICS

Consider a cylindrical region of arbitrary size and shape within a fluid, as shown in Figure 4.1. We will apply a momentum balance to a "slice" of the fluid that has a z area A_z, a thickness Δz and is located a vertical distance z above some horizontal reference plane. The density of the fluid in the slice is ρ, and the force of gravity (g) acts in the $-z$ direction. A momentum balance on a "closed" system (e.g., the slice) is equivalent to Newton's second law of motion, that is,

$$\sum F_z = ma_z \tag{4.4}$$

Because this is a vector equation, we apply it to the z vector components. ΣF_z is the sum of all of the forces acting *on the system* (the "slice") in the z direction, m is the mass of the system, and a_z is the acceleration in the z direction. Since the fluid is not moving, $a_z = 0$, and the momentum balance reduces to a force balance. The z forces acting on the system include the ($-$) pressure on the bottom (at z), times the ($-$) z area, the ($-$) pressure on the top (at $z + \Delta z$), times the ($+$) z area, and the z component of gravity, that is, the "weight" of the fluid, ($-\rho g A_z \Delta z$). The first force is positive, and the latter two are negative because they act in the ($-z$) direction. The momentum (force) balance thus becomes

$$(A_z P)_z - (A_z P)_{z+\Delta z} - \rho g A_z \Delta z = 0 \tag{4.5}$$

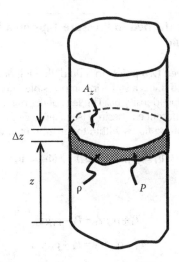

FIGURE 4.1 Arbitrary region within a fluid.

If we divide through by $A_z \Delta z$, then take the limit as the thickness of the slice shrinks to zero ($\Delta z \to 0$), the result is

$$\frac{dP}{dz} = -\rho g \qquad (4.6)$$

which is the *basic equation of fluid statics*. This equation states that at any point within a given fluid the pressure decreases as the elevation (z) increases, at a local rate that is equal to the product of the fluid density and the gravitational acceleration at that point. This equation is valid at all points within any given *static* fluid regardless of the nature of the fluid (irrespective of whether the fluid is Newtonian or non-Newtonian, compressible or incompressible). We shall now show how the equation can be applied to various special situations.

A. CONSTANT DENSITY FLUIDS

If the density (ρ) is constant, the fluid is referred to as "isochoric" (i.e., a given mass occupies a *constant volume*), although the somewhat more restrictive term "incompressible" is commonly used for this property (liquids are considered to be incompressible or isochoric fluids under most conditions). If gravity (g) is also constant, the only variables in Equation 4.6 are pressure and elevation. This can then be integrated between any two points (1 and 2) in a given fluid to give

$$P_1 - P_2 = \rho g(z_2 - z_1) \qquad (4.7)$$

which can also be written as

$$\Phi_1 = \Phi_2 = \text{Constant} \qquad (4.8)$$

where

$$\Phi = P + \rho g z \qquad (4.9)$$

This says that the sum of the local pressure (P) and static head ($\rho g z$), that is, the *potential* (Φ), is constant at all points within a given isochoric (incompressible) non-flowing fluid. This is an important result for such fluids, and it can be applied directly to determine how the pressure varies with elevation in a static liquid, as illustrated by the following example.

Example 4.2 Manometer Attached to Pressure Taps on a Pipe Carrying a Flowing Fluid

The pressure difference between two points in a fluid (flowing or static) can be measured using a manometer. The manometer contains a static incompressible liquid (density ρ_m) that is immiscible with the fluid flowing in the pipe (density ρ_f). The legs of the manometer are connected to taps on the pipe where the pressure difference is desired (see Figure E4.2). By applying Equation 4.8 to any two points within either one of the fluids within the manometer, we see that

$$(\Phi_1 = \Phi_3, \Phi_2 = \Phi_4)_f, \quad (\Phi_3 = \Phi_4)_m \qquad (4.10)$$

or

$$P_1 + \rho_f g z_1 = P_3 + \rho_f g z_3$$
$$P_3 + \rho_m g z_3 = P_4 + \rho_m g z_4 \qquad (4.11)$$
$$P_4 + \rho_f g z_4 = P_2 + \rho_f g z_2$$

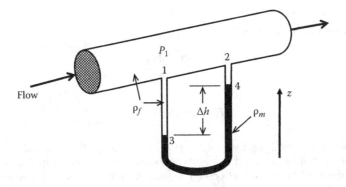

FIGURE E4.2   Manometer.

When these three equations are added, P_3 and P_4 cancel out. The remaining terms can be collected to give

$$\Delta\Phi = -\Delta\rho g \Delta h \qquad (4.12)$$

where $\Phi = P + \rho g z$ and $\Delta\Phi = \Phi_2 - \Phi_1$, $\Delta\rho = \rho_m - \rho_f$, $\Delta h = z_4 - z_3$. Equation 4.12 is the *basic manometer equation* and can be applied to a manometer in any orientation. Note that the manometer reading (Δh) is a direct measure of the *potential difference* ($\Phi_2 - \Phi_1$), which is identical to the pressure difference ($P_2 - P_1$) only if the pipe is horizontal (i.e., $z_2 = z_1$). It should be noted that these static fluid equations, Equations 4.10 through 4.12, apply only within the manometer but not within the pipe, as the fluid in the pipe is not static.

Example 4.3 Barometer

A straightforward and most common application of Equation 4.8 is seen in the everyday mercury barometer used to measure the atmospheric pressure. In its basic form, a glass tube filled with mercury is inverted so that its open end is well submerged in a mercury reservoir as shown in Figure E4.3. Determine the relation between the height of the mercury in the barometer (z_1) and the local atmospheric pressure.

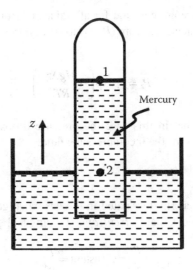

FIGURE E4.3   Barometer.

Solution:

Applying Equation 4.8 to points 1 and 2

$$\Phi_1 = \Phi_2,$$

or

$$P_1 + \rho_m g z_1 = P_2 + \rho_m g z_2$$

Since the mercury at point 2 is at the same level in a common (static) fluid both inside and outside the barometer and is in contact with the open atmosphere, $P_2 = 1$ atm $= 101,325$ Pa. Similarly, the space above the level of mercury at point 1 is filled with mercury vapor. Therefore, pressure at point 1 is equal to the vapor pressure of mercury, which is negligibly small, for example, at 300 K, the vapor pressure of mercury is about 0.3 Pa and hence $P_1 \sim 0$. The density of mercury at 20°C is 13,500 kg/m³, so that substituting these values into the above equation gives

$$0_1 + (13,500\,\text{kg/m}^3)(9.81\,\text{m/s}^2)(z_1\,\text{m}) = (1\,\text{atm})(101,325\,\text{Pa/atm})$$

This yields $\boxed{z_1 = 0.765\text{m, or } 765\text{ mm, or } 30 \text{ in.,}}$ which is the value reported in most weather reports in newspapers and TV.

B.   IDEAL GAS: ISOTHERMAL

If the fluid can be described by the ideal gas law (e.g., air under normal atmospheric conditions follows this law quite well), then

$$\rho = \frac{PM}{RT} \tag{4.13}$$

and Equation 4.6 becomes

$$\frac{dP}{dz} = -\frac{PMg}{RT} \tag{4.14}$$

Now if gravity and temperature are constant for all values of z of interest (i.e., isothermal conditions), Equation 4.14 can be integrated from (P_1, z_1) to (P_2, z_2) to give the pressure as a function of elevation:

$$P_2 = P_1 \exp\left(-\frac{Mg\Delta z}{RT}\right) \tag{4.15}$$

where $\Delta z = z_2 - z_1$. Notice that in this case the pressure drops exponentially as the elevation increases, instead of linearly as for the incompressible fluid.

C.   IDEAL GAS: ISENTROPIC

If there is no heat transfer or energy dissipated in the gas when going from state 1 to state 2, the process is adiabatic and reversible, that is, *isentropic*. For an ideal gas under these conditions,

$$\frac{P}{\rho^k} = \text{Constant} = \frac{P_1}{\rho_1^k} \tag{4.16}$$

where $k = c_p/c_v$ is the specific heat ratio for the gas (for an ideal gas, $c_p = c_v + R/M$). If the density is eliminated from Equations 4.16 and 4.13, the result is

$$\frac{T}{T_1} = \left(\frac{P}{P_1} \right)^{(k-1)/k}$$

(4.17)

which relates the temperature and pressure at any two points in an isentropic ideal gas. If Equation 4.17 is used to eliminate T from Equation 4.14, the latter can be integrated to give the pressure as a function of elevation:

$$P_2 = P_1 \left[1 - \left(\frac{k-1}{k} \right) \frac{gM\Delta z}{RT_1} \right]^{k/(k-1)}$$

(4.18)

which is a nonlinear relationship between pressure and elevation. Equation 4.15 can be used to eliminate P_2/P_1 from this equation to give an expression for the temperature as a function of elevation under isentropic conditions:

$$T_2 = T_1 \left[1 - \left(\frac{k-1}{k} \right) \frac{gM\Delta z}{RT_1} \right]$$

(4.19)

This indicates that the temperature drops linearly as the elevation increases.

D. THE STANDARD ATMOSPHERE

Neither Equation 4.15 nor Equation 4.18 would be expected to provide a very good representation of the pressure and temperature in the real atmosphere, which is neither isothermal nor isentropic. Thus, we must resort to the use of observations (i.e., empiricism) to describe the real atmosphere. In fact, atmospheric conditions vary considerably from time to time and from place to place over the earth. However, a reasonable representation of atmospheric conditions "averaged" over the year and over the earth based on observations results in the following:

$$\text{For} \quad 0 < z < 11\,\text{km:} \quad \frac{dT}{dz} = -6.5°\text{C/km} = -G$$

(4.20)

$$\text{For} \quad z > 11\,\text{km:} \quad T = -56.5°\text{C} \approx \text{const.}$$

where the average temperature at ground level ($z = 0$) is assumed to be 15°C (288 K). These equations describe what is known as the "standard atmosphere" which represents an average state over the world and throughout the year. Using Equation 4.20 for the temperature as a function of elevation and incorporating this into Equation 4.14 gives

$$\frac{dP}{dz} = -\frac{PMg}{R(T_o - Gz)}$$

(4.21)

where
$T_o = 288$ K
$G = 6.5°\text{C/km}$

Integrating Equation 4.21, assuming that g is constant, gives the pressure as a function of elevation:

$$P_2 = P_1 \left[1 - \frac{G\Delta z}{T_o - Gz_1} \right]^{Mg/RG}$$

(4.22)

This applies for $0 < z < 11$ km.

III. MOVING SYSTEMS

We have stated that the only stress that can exist in a fluid at rest is pressure, because the shear stresses (which resist motion) are zero when the fluid is at rest. This also applies to fluids in motion provided there is no *relative* motion within the fluid (the shear stresses are determined by the velocity *gradients*, e.g., the shear rate). However, if the motion involves a uniform acceleration, this can contribute an additional component to the pressure, as illustrated by the examples in this section.

A. VERTICAL ACCELERATION

Consider the vertical column of fluid illustrated in Figure 4.1, but now imagine it to be on an elevator that is accelerating upward with an acceleration of a_z, as illustrated in Figure 4.2. Application of the momentum balance to the "slice" of fluid, as before, gives

$$\sum F_z = ma_z \tag{4.23}$$

which is the same as Equation 4.3, except that now $a_z \neq 0$. The same procedure that led to Equation 4.6 now gives

$$\frac{dP}{dz} = -\rho(g + a_z) \tag{4.24}$$

which shows that the effect of a superimposed upward vertical acceleration is equivalent to increasing the gravitational acceleration by an amount a_z (which is why you feel "heavier" on a rapidly accelerating elevator). In fact, this result may be generalized to any direction, that is, an acceleration in the i direction will result in a pressure gradient within the fluid in the $-i$ direction of magnitude ρa_i:

$$\frac{\partial P}{\partial x_i} = -\rho a_i \tag{4.25}$$

Two applications of this equation are illustrated in the following sections.

B. HORIZONTALLY ACCELERATING FREE SURFACE

Consider a pool of water in the bed of your pickup truck. If you accelerate from rest, the water will slosh toward the rear, and you want to know how fast you can accelerate (a_x) without spilling the

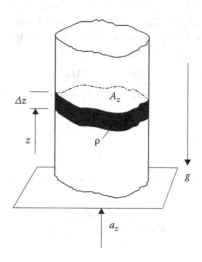

FIGURE 4.2 Vertically accelerating column of fluid.

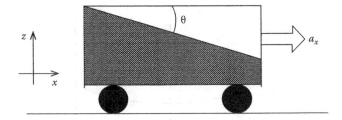

FIGURE 4.3 Horizontally accelerating tank.

water out of the back of the truck (see Figure 4.3). That is, you must determine the slope (tan θ) of the water surface as a function of the rate of acceleration (a_x). Now at any point within the liquid there is a vertical pressure gradient due to gravity (Equation 4.6) and a horizontal pressure gradient due to the acceleration a_x (Equation 4.25). Thus, at any location within the liquid, the total differential pressure dP between two points separated by a small distance dx in the horizontal direction and dz in the vertical direction is given by

$$dP = \frac{\partial P}{\partial x} dx + \frac{\partial P}{\partial z} dz$$

$$= -\rho a_x \, dx - \rho g \, dz \tag{4.26}$$

Because the surface of the water is open to the atmosphere where P = constant (1 atm), one can apply Equation 4.26 on the free surface as

$$(dP)_s = 0 = -\rho g (dz)_s - \rho a_x (dx)_s \tag{4.27}$$

or

$$\boxed{\left(\frac{dz}{dx}\right)_s = -\frac{a_x}{g} = \tan \theta} \tag{4.28}$$

which is the slope of the surface and is seen to be independent of fluid properties. Knowledge of the initial position of the surface plus the surface slope determines the elevation of the water level at the rear of the truck bed and hence whether or not the water will spill out.

C. ROTATING FLUID

Consider an open bucket of water resting on a turntable that is rotating at an angular velocity ω (see Figure 4.4). The (inward) radial acceleration due to the rotation is $\omega^2 r$, which results in a corresponding radial pressure gradient at all points in the fluid, in addition to the vertical pressure gradient due to gravity. Thus, the pressure differential between any two points within the fluid separated by dr and dz is

$$dP = \left(\frac{\partial P}{\partial z}\right) dz + \left(\frac{\partial P}{\partial r}\right) dr = \rho(-g \, dz + \omega^2 r \, dr) \tag{4.29}$$

Just as for the accelerating tank, the shape of the free surface can be determined from the fact that the pressure is constant at the free surface, that is,

$$(dP)_s = 0 = -g(dz)_s + \omega^2 r (dr)_s \tag{4.30}$$

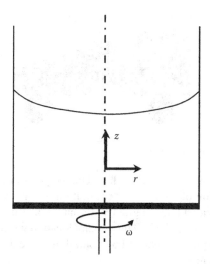

FIGURE 4.4   Rotating fluid.

This can be integrated to give an equation for the shape of the surface:

$$z = z_o + \frac{\omega^2 r^2}{2g}$$

(4.31)

which shows that the shape of the rotating surface is parabolic.

IV.   BUOYANCY

As a consequence of Archimedes' principle, the buoyant force exerted on a submerged body is equal to the weight of the displaced fluid, and it acts in a direction opposite to the gravity vector. Thus, the "effective net weight" of a submerged body is its actual weight (in air) less the weight of an equal volume of the fluid. The result is equivalent to replacing the density of the body (ρ_s) in the expression for its weight ($\rho_s g \tilde{V}_s$, where $\tilde{V}_s$ is the volume of the solid body) by the difference between the density of the body and that of the fluid (i.e., $\Delta\rho g \tilde{V}_s$, where $\Delta\rho = \rho_s - \rho_f$).

This also applies to a body submerged in a fluid that is subject to any acceleration. For example, a solid particle of volume $\tilde{V}_s$ submerged in a fluid within a centrifuge at a point r where the angular velocity is ω is subjected to a net radial force equal to $\Delta\rho\omega^2 r \tilde{V}_s$. Thus, the effect of buoyancy is to effectively reduce the density of the body by an amount equal to the density of the surrounding fluid.

V.   STATIC FORCES ON SOLID BOUNDARIES

The force exerted on a solid boundary by a static pressure is given by

$$\vec{F} = \int_A P \, d\vec{A}$$

(4.32)

Note that both force and area are vectors, whereas pressure is a scalar. Hence, the directional character of the force is determined by the orientation of the surface on which the pressure acts.

That is, the component of force acting in a given direction on a surface is the integral of the pressure over the *projected component* area of the surface normal to the direction of the force vector. The surface vector that is normal to the surface is parallel to the direction of the force (recall that pressure is a *negative* isotropic stress and the *outward* normal to the [fluid] system boundary represents a positive area). Also, from Newton's third law ("action equals reaction"), the force exerted *on* the fluid system boundary is of opposite sign to the force exerted *by* the fluid system on the solid boundary.

Example 4.4

Consider the force within the wall of a pipe resulting from the pressure of the fluid inside the pipe, as illustrated in Figure E4.4.

The pressure P acts equally in all directions on the inside wall of the pipe. The resulting force exerted within the pipe wall normal to a vertical (or any other orientation) plane through the pipe axis is simply the product of the pressure and the projected area of the inside pipe wall on this plane, for example, $F_x = PA_x = 2PRL$. This force acts to pull the metal in the wall apart and is resisted by the internal stress within the metal holding it together. This is the effective *working stress*, σ, of the particular material of which the pipe is made, acting on the cross section of pipe wall, or $2\sigma tL$. If we assume a thin-walled pipe (i.e., we neglect the radial variation of the stress from point to point within the wall), a force balance between the "disruptive" pressure force and the "restorative" force due to the internal stress in the metal gives

$$2PRL = 2\sigma tL \tag{4.33}$$

or

$$\frac{t}{R} \cong \frac{P}{\sigma} \tag{4.34}$$

This relation determines the pipe wall thickness (t) required to withstand a fluid pressure P in a pipe of radius R made of a material with a working stress σ. The dimensionless pipe wall thickness (times 1000) is known as the *Schedule number* of the pipe:

$$\text{Schedule No.} = \frac{1000t}{R} = \frac{1000P}{\sigma} \tag{4.35}$$

This expression is only approximate, as it does not make any allowance for the effects of such things as pipe threads, corrosion, or wall damage. To compensate for these factors, an additional allowance is made for the wall thickness in the working definition of the "schedule thickness," t_s:

$$\text{Schedule No.} = \frac{1000P}{\sigma} = \frac{1750t_s - 200}{D_o} \tag{4.36}$$

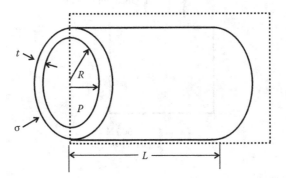

FIGURE E4.4 Fluid pressure inside a pipe.

where both t_s and D_o (the pipe outside diameter) are measured in inches, and the factor of 200 allows for damage, threads, etc., on the pipe wall. This relation between schedule number and pipe dimensions can be compared with the actual dimensions of commercial pipes for various schedule pipe sizes, as tabulated in Appendix F.

Example 4.5

A composite block made up of concrete and iron weighs 500 N in air and only 280 N in fresh water (density 1000 kg/m³). Calculate the specific gravity of the block.

Solution:

For the given data, one can draw the free body diagram with various forces acting on the block as shown in Figure E4.5. A force balance in the vertical direction yields

$$F_B + 280\,\text{N} = 500\,\text{N}$$

or

$$F_B = 500 - 280 = 220\,\text{N}$$

By definition, the buoyancy force is given by

$$F_B = \tilde{V}_s \rho_w g = \tilde{V}_s(\text{m}^3)(1000\,\text{kg/m}^3)(9.81\,\text{m/s}^2)$$

$$\therefore \tilde{V}_s = \frac{220\,\text{N}}{(9.81\,\text{m/s}^2)(1000\,\text{kg/m}^3)} = 0.0224\,\text{m}^3$$

$$\therefore \text{Density of the block} = \frac{M_s}{\tilde{V}_s} = \frac{500\,\text{N}}{\left(9.81\,\text{m/s}^2\right)\left(0.0224\,\text{m}^3\right)}$$

$$= 2270\,\text{kg/m}^3$$

Specific gravity of the block is 2.27.

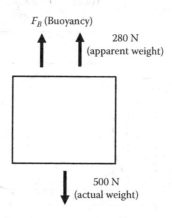

FIGURE E4.5   Submerged block.

Example 4.6

Calculate the pressure at the bottom of the tank shown in Figure E4.6. Also, calculate the magnitude and direction of the force (per unit width) exerted on the door at the bottom, with dimensions A (height) and B (width).

Solution:

The pressure is calculated at the bottom of the tank by successive application of Equation 4.8 across points 1, 2, 3, that is,

$$\phi_1 = \phi_2, \quad \phi_2 = \phi_3$$

In terms of pressure and density,

$$P_1 + \rho_{oil}\, gz_1 = P_2 + \rho_{oil}\, gz_2 \quad \text{solve for } P_2:$$

$$P_2 = P_1 + \rho_{oil} g\,(z_1 - z_2) = 200 \times 10^3 \text{ Pa} + (800 \text{ kg/m}^3)(9.81 \text{ m/s}^2)(1 \text{ m}) = 208 \times 10^3 \text{ Pa}$$

Now using $\phi_2 = \phi_3$,

$$P_2 + \rho_w\, gz_2 = P_3 + \rho_w\, gz_3 \quad \text{solve for } P_3:$$

$$P_3 = P_2 + \rho_w g\,(z_2 - z_3) = 208{,}000\text{Pa} + (1{,}000 \text{ kg/m}^3)(9.81 \text{ m/s}^2)(2 \text{ m}) = 227 \times 10^3\text{Pa}$$

which is the pressure at the bottom of the tank.

   In order to calculate the force on the door due to the pressure, Equation 4.32 can be applied to a differential area $B\,dz$, where B is the width of the door, oriented in the $-x$ *direction*. This force has only one component, in the x direction:

$$F_x = \int_{AB} PB\,dz$$

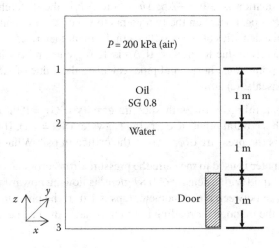

FIGURE E4.6 Tank Containing oil and water.

The pressure at any height z acting on the door is

$$P_2 = P_3 - \rho_w g z$$

$$F_x = \int_{z=0}^{z=1m} (P_3 - \rho_w g z) B \, dz$$

or

$$\frac{F_x}{B} = \left[P_3 z - \rho_w g \frac{z^2}{2} \right]_{z=0}^{z=1m}$$

$$= (227 \times 10^3 \, \text{Pa})(1\text{m}) - (1000 \, \text{kg/m}^3)(9.81 \text{m/s}^2)(1\text{m})^2/2 = \boxed{222 \times 10^3 \, \text{N/m}}$$

This force acts in the positive x direction on the door with dimensions AB per unit width (B).

SUMMARY

The key points covered in this chapter include

- Pressure/stress in static and uniformly moving systems
- The basic equation of fluid statics
- The manometer equation
- Pressure distribution in static gases
- Pressure distribution in uniformly rotating and accelerating systems
- Static forces on solid boundaries—pipe wall stress

PROBLEMS

STATICS

1. The manometer equation is $\Delta\Phi = -\Delta\rho g \Delta h$, where $\Delta\Phi$ is the difference in the total pressure plus static head ($P + \rho g z$) between the two points to which the manometer is connected, $\Delta\rho$ is the difference in the densities of the two fluids in the manometer, Δh is the manometer reading, and g is the acceleration due to gravity. If $\Delta\rho$ is 12.6 g/cm^3 and Δh is 6 in. for a manometer connected to two points on a horizontal pipe, calculate the value of ΔP in the following units: (a) dyn/cm^2; (b) pascals; (c) atm.

2. A manometer containing an oil with specific gravity (SG) = 0.92 is connected across an orifice plate in a horizontal pipeline carrying seawater (SG = 1.1). If the manometer reading is 16.8 cm, what is the pressure drop across the orifice in psi? What is it in inches of water?

3. A mercury manometer is used to measure the pressure drop across an orifice that is mounted in a vertical pipe. A liquid with a density of 0.87 g/cm^3 is flowing upward through the pipe and the orifice. The distance between the manometer taps is 1 ft. If the pressure in the pipe at the upper tap is 30 psig and the manometer reading is 15 cm, what is the pressure in the pipe at the lower manometer tap in psig?

4. A mercury manometer is connected between two points in a piping system that contains water. The downstream tap is 6 ft higher than the upstream tap, and the manometer reading is 16 in.

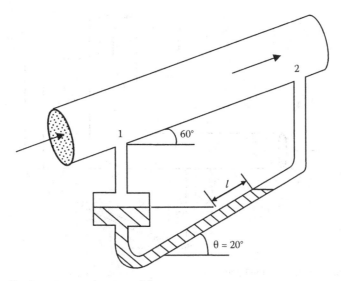

FIGURE P4.5   Inclined manometer (not to scale).

If a pressure gage in the pipe at the upstream tap reads 40 psia, what would a pressure gage at the downstream tap read in (a) psia, (b) dyn/cm^2, (c) Pa, (d) kg_f/m^2?

5.   An inclined tube manometer with a reservoir is used to measure the pressure gradient in a large pipe carrying oil (SG = 0.91) (see Figure P4.5). The pipe is inclined at an angle of 60° to the horizontal, and flow is uphill. The manometer tube is inclined at an angle of 20° to the horizontal, and the pressure taps on the pipe are 5 in. apart. The manometer reservoir diameter is eight times as large as the manometer tube diameter, and the manometer fluid is water. If the manometer reading (*l*) is 3 in. and the displacement of the interface in the reservoir is neglected, what is the pressure drop in the pipe in (a) psi, (b) Pa, (c) in. H_2O? What is the percentage error introduced by neglecting the change in elevation of the interface in the reservoir?

6.   Water is flowing downhill in a pipe that is inclined 30° to the horizontal. A mercury manometer is attached to pressure taps 5 cm apart on the pipe. The interface in the downstream manometer leg is 2 cm higher than the interface in the upstream leg. What is the pressure gradient ($\Delta P/L$) in the pipe in (a) Pa/m, (b) dyn/cm^3, (c) in. H_2O/ft, (d) psi/mi?

7.   Repeat Problem 6 for the case in which the water in the pipe is flowing uphill instead of downhill, all other conditions remaining the same.

8.   Two horizontal pipelines are parallel, with one carrying salt water (ρ = 1.988 slugs/ft³) and the other carrying fresh water (ρ = 1.937 slugs/ft³). An inverted manometer using linseed oil (ρ = 1.828 slugs/ft³) as the manometer fluid is connected between the two pipelines. The interface between the oil and the fresh water in the manometer is 38 in. above the centerline of the freshwater pipeline, and the oil/salt water interface in the manometer is 20 in. above the centerline of the salt water pipeline. If the manometer reading is 8 in., determine the difference in the pressures between the pipelines in (a) Pa and in (b) psi.

9.   Two identical tanks are 3 ft in diameter and 3 ft high, and they are both vented to the atmosphere. The top of tank B is level with the bottom of tank A, and they are connected by a line from the bottom of A to the top of B with a valve in it. Initially A is full of water, and B is empty. The valve is opened for a short time, letting some of the water drain into B. An inverted manometer having an oil with SG = 0.7 is connected between taps on the bottom of each tank. The manometer reading is 6 in., and the oil/water interface in the leg connected to tank A is higher. What is the water level in each of the tanks?

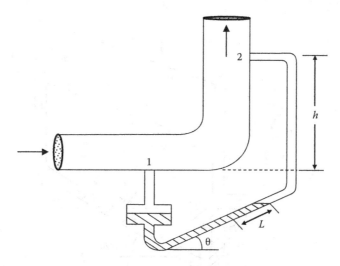

FIGURE P4.10   Manometer on pipe bend.

10. An inclined tube manometer is used to measure the pressure drop in an elbow through which water is flowing (see Figure P4.10). The manometer fluid is an oil with SG = 1.15. The distance L is the distance along the inclined tube that the interface has moved from its equilibrium (no pressure differential) position. If $h = 6$ in., $L = 3$ in., $\theta = 30°$, the reservoir diameter is 2 in., and the tubing diameter is 0.25 in., calculate the pressure drop ($P_1 - P_2$) in (a) atm; (b) Pa; (c) cmH$_2$O, and; (d) dyn/cm^2. What would be the percentage error in pressure difference as read by the manometer if the change in level in the reservoir is neglected?

11. The three-fluid manometer illustrated in Figure P4.11 is used to measure a very small pressure difference ($P_1 - P_2$). The cross-sectional area of each of the reservoirs is A and that of the manometer legs is a. The three fluids have densities ρ_a, ρ_b, and ρ_c, and the

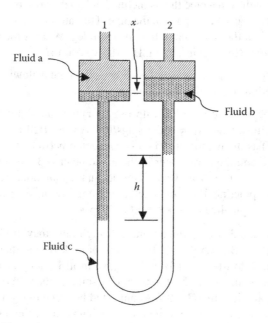

FIGURE P4.11   Three fluid manometer.

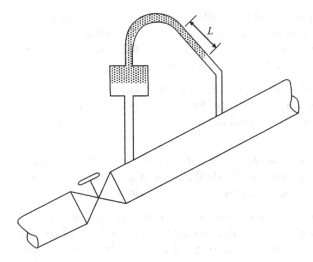

FIGURE P4.13 Manometer on inclined tube.

difference in elevation of the interfaces in the reservoir is x. Derive the equation that relates the manometer reading h to the pressure difference $(P_1 - P_2)$. How would the relation be simplified if $A \gg a$?

12. A tank that is vented to the atmosphere contains a liquid with a density of 0.9 g/cm^3. A dip tube inserted into the top of the tank extends to a point 1 ft from the bottom of the tank. Air is bubbled slowly through the dip tube, and the air pressure in the tube is measured with a mercury (SG = 13.6) manometer. One leg of the manometer is connected to the air line feeding the dip tube, and the other leg is open to the atmosphere. If the manometer reading is 5 in., what is the depth of the liquid in the tank?

13. An inclined manometer is used to measure the pressure drop between two taps on a pipe carrying water, as shown in Figure P4.13. The manometer fluid is an oil with SG = 0.92, and the manometer reading (L) is 8 in. The manometer reservoir is 4 in. in diameter, the tubing is 1/4 in. in diameter, and the manometer tube is inclined at an angle of 30° to the horizontal. The pipe is inclined at 20° to the horizontal, and the pressure taps are 40 in. apart.
    (a) What is the pressure difference between the two pipe taps that would be indicated by the difference in readings of two pressure gages attached to the taps, in (1) psi, (2) Pa, and (3) in. H$_2$O?
    (b) Which way is the water flowing?
    (c) What would the manometer reading be if the valve were closed?

14. The pressure gradient required to force water through a straight horizontal 1/4 in. ID tube at a rate of 2 gpm is 1.2 psi/ft. Consider this same tubing coiled in an expanding helix with a vertical axis. Water enters the bottom of the coil and flows upward at a rate of 2 gpm. A mercury manometer is connected between two pressure taps on the coil, one near the bottom where the coil radius is 6 in., and the other near the top where the coil radius is 12 in. The taps are 2 ft apart in the vertical direction, and there is a total of 5 ft of tubing between the two taps. Determine the manometer reading, in cm.

15. It is possible to achieve a weightless condition for a limited time in an airplane by flying in a circular arc above the earth (like a rainbow). If the plane flies at 650 mph, what should the radius of the flight path be (in miles) to achieve weightlessness?

16. Water is flowing in a horizontal pipe bend at a velocity of 10 ft/s. The radius of curvature of the inside of the bend is 4 in., and the pipe ID is 2 in. A mercury manometer is connected to taps

located radially opposite to each other on the inside and outside of the bend. Assuming that the water velocity is uniform over the pipe cross section, what would be the manometer reading in centimeters? What would it be if the water velocity were 5 ft/s? Convert the manometer reading to equivalent pressure difference in psi and Pa.

17. Calculate the atmospheric pressure at an elevation of 3000 m, assuming (a) air is incompressible, at a temperature of 59°F; (b) air is isothermal at 59°F and an ideal gas; (c) the pressure distribution follows the model of the standard atmosphere, with a temperature of 59°F at the surface of the earth.

18. One pound mass of air (MW = 29) at sea level and 70°F is contained in a balloon, which is then carried to an elevation of 10,000 ft in the atmosphere. If the balloon offers no resistance to expansion of the gas, what is its volume at this elevation?

19. A gas well contains hydrocarbon gases with an average molecular weight of 24, which can be assumed to be an ideal gas with a specific heat ratio of 1.3. The pressure and temperature at the top of the well are 250 psig and 70°F, respectively. The gas is being produced at a slow rate, so that the conditions in the well can be considered to be isentropic.
    (a) What are the pressure and temperature at a depth of 10,000 ft?
    (b) What would the pressure be at this depth if the gas were assumed to be isothermal?
    (c) What would the pressure be at this depth if the gas were assumed to be incompressible?

20. The adiabatic atmosphere obeys the equation

$$P/\rho^k = \text{Constant}$$

where
    k is a constant
    ρ is density

If the temperature decreases 0.3°C for every 100 ft increase in altitude, what is the value of k? [*Note:* Air is an ideal gas; $g = 32.2$ ft/s^2; $R = 1544$ ft lb$_f$/(°R lbmol).

21. Using the actual dimensions of commercial steel pipe from Appendix F, plot the pipe wall thickness versus the pipe diameter for both Schedule 40 and Schedule 80 pipes, and fit the plot with a straight line by linear regression analysis. Rearrange your equation for the line in a form consistent with the given equation for the schedule number as a function of the wall thickness and diameter:

$$\text{Schedule No} = (1750t_s - 200)/D$$

and use the results of the regression to calculate values corresponding to the parameters 1750 and 200 in this equation. Do this using (for D) (a) the nominal pipe diameter and (b) the outside pipe diameter. Explain any discrepancies or differences in the numerical values determined from the data fit compared to those in the equation.

22. The "yield stress" for carbon steel is 35,000 psi, and the "working stress" is one-half of this value. What schedule number would you recommend for a pipe carrying ethylene at a pressure of 2500 psi if the pipeline design calls for a pipe of 2 in. ID? Give the dimensions of the pipe that you would recommend. What would be a safe maximum pressure to recommend for this pipe?

23. Consider a 90° elbow in a 2 in. pipe (all of which is in the horizontal plane). A pipe tap is drilled through the wall of the elbow on the inside curve of the elbow, and another through the outer wall of the elbow directly across from the inside tap. The radius of curvature of the inside of the bend is 2 in., and that of the outside of the bend is 4 in. The pipe is carrying water, and a manometer containing an immiscible oil with SG of 0.90 is connected across the two taps on the elbow. If the reading of the manometer is 7 in., what is the average velocity of the water in the pipe, assuming that the flow is uniform across the pipe inside the elbow?

24. A pipe carrying water is inclined at an angle of 45° to the horizontal. A manometer containing a fluid with SG of 1.2 is attached to taps on the pipe, which are 1 ft apart. If the liquid interface in the manometer leg that is attached to the lower tap is 3 in. below the interface in the other leg, what is the pressure gradient in the pipe ($\Delta P/L$), in units of (a) psi/ft and (b) Pa/m? Which direction is the water flowing?

25. A tank contains a liquid of unknown density (see Figure P4.25). Two dip tubes are inserted into the tank, each to a different level in the tank, through which air is bubbled very slowly through the liquid. A manometer is used to measure the difference in pressure between the two dip tubes. If the difference in level of the ends of the dip tubes (H) is 1 ft, and the manometer reads 1.5 ft (h) with water as the manometer fluid, what is the density of the liquid in the tank?

26. The tank shown in Figure P4.26 has a partition that separates two immiscible liquids. Most of the tank contains water, and oil is floating above the water on the right of the partition. The height of the water in the standpipe (h) is 10 cm, and the interface between the oil and water is 20 cm below the top of the tank and 25 cm above the bottom of the tank. If the specific gravity of the oil is 0.82, what is the height of the oil in the standpipe (H)?

27. A manometer that is open to the atmosphere contains water, with a layer of oil floating on the water in one leg (see Figure P4.27). If the level of the water in the left leg is 1 cm above the center of the leg, the interface between the water and oil is 1 cm below the center in the right leg, and the oil layer on the right extends 2 cm above the center, what is the density of the oil?

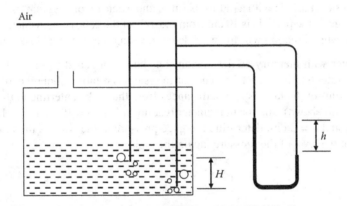

FIGURE P4.25 Manometer measuring density of liquid.

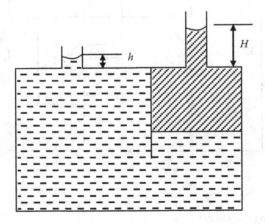

FIGURE P4.26 Immiscible fluids in tank.

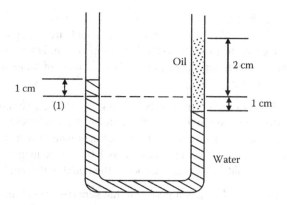

FIGURE P4.27   Density of oil in two-fluid manometer.

28. An open cylindrical drum, with a diameter of 2 ft and a length of 4 ft, is turned upside down in the atmosphere and then submerged in a liquid so that it floats partially submerged upside down, with air trapped inside. If the drum weighs 150 lb_f, and it floats with 1 ft extending above the surface of the liquid, what is the density of the liquid? How much additional weight must be added to the drum to make it sink to the point where it floats just level with the liquid?

29. A solid spherical particle with a radius of 1 mm and a density of 1.3 g/cm^3 is immersed in water in a centrifuge. If the particle is 10 cm from the axis of the centrifuge, which is rotating at a rate of 100 rpm, what direction will the particle be traveling relative to a horizontal plane?

30. A manometer with mercury as the manometer fluid is attached to the wall of a closed tank containing water (see Figure P4.30). The entire system is rotating about the axis of the tank at N rpm. The radius of the tank is r_1, the distances from the tank centerline to the manometer legs are r_2 and r_3 (as shown), and the manometer reading is h. If $N = 30$ rpm, $r_1 = 12$ cm, $r_2 = 15$ cm, $r_3 = 18$ cm, and $h = 2$ cm, determine the gage pressure at the wall of the tank and also at the centerline at the level of the pressure tap on the tank.

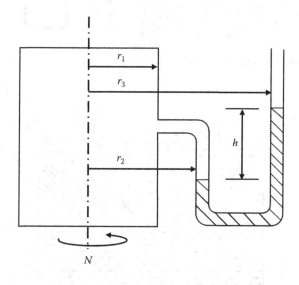

FIGURE P4.30   Revolving manometer.

31. With reference to Figure P4.30, the manometer contains water as the manometer fluid and is attached to a tank that is empty and open to the atmosphere. When the tank is stationary, the water level is the same in both legs of the manometer. If the entire system is rotated about the centerline of the tank at a rate of N (rpm):
    (a) What happens to the water levels in the legs of the manometer?
    (b) Derive an equation for the difference in elevation of the levels (h) in the legs of the manometer as a function of known quantities.

32. You want to measure the specific gravity of a liquid. To do this, you first weigh a beaker of the liquid on a scale (W_{Lo}). You then attach a string to a solid body that is heavier than the liquid and while holding the string you immerse the solid body in the liquid and measure the weight of the beaker containing the liquid with the solid submerged (W_{Ls}). You then repeat the same procedure using the same weight but with water instead of the "unknown" liquid. The corresponding weight of the water without the weight submerged is W_{Wo} and with the solid submerged is W_{Ws}. Show how the specific gravity of the "unknown" liquid can be determined from these four weights, and show that the result is independent of the size, shape, or weight of the solid body used (provided, of course, that it is heavier than the liquids and is large enough that the difference in the weights can be measured precisely).

33. A vertical U-tube manometer is open to the atmosphere and contains a liquid that has a SG of 0.87 and a vapor pressure of 450 mmHg at the operating temperature. The vertical tubes are 4 in. apart, and the level of the liquid in the tubes is 6 in. above the bottom of the manometer. The manometer is then rotated about a vertical axis through its centerline. Determine what the rotation rate would have to be (in rpm) for the liquid to start to boil.

34. A spherical particle with SG = 2.5 and a diameter of 2 mm is immersed in water in a cylindrical centrifuge with a diameter of 20 cm. If the particle is initially 8 cm above the bottom of the centrifuge and 1 cm from the centerline, what is the speed of the centrifuge (in rpm) if this particle strikes the wall of the centrifuge just before it hits the bottom?

NOTATION

A	Area, $[L^2]$
A_i	Bounding area component with outward normal in the i direction, $[L^2]$
a_z	Acceleration in the z direction, $[L/t^2]$
c_p	Specific heat at constant pressure $\{FL/MT = [L^2/Mt^2]$
c_v	Specific heat at constant volume $\{FL/MT = [L^2/Mt^2]$
D_o	Pipe diameter (outer), $[L]$
F_j	Force in the j direction, $[F = ML/t^2]$
G	Atmospheric temperature gradient ($= 6.5°C/km$), $[T/L]$
g	Acceleration due to gravity, $[L/t^2]$
h	Vertical displacement of manometer interface, $[L]$
k	Isentropic exponent ($= c_p/c_v$ for ideal gas), $[—]$
M	Molecular weight, $[M/mol]$
P	Pressure, $[F/L^2 = M/Lt^2]$
R	Gas constant, $[FL/(mol\ T) = M\ L^2\ (mole\ t^2\ T)]$
r	Radial direction, $[L]$
T	Temperature, $[T]$
t	Pipe thickness, $[L]$
$\tilde{V}$	Volume, $[L^3]$
z	Vertical direction, measured upward, $[L]$

GREEK

$\Delta()$	Difference between two values, $[= ()_2 - ()_1)]$
ρ	Density, $[M/L^3]$
Φ	Potential $= P + \rho gz$, $[F/L^2 = M/(L\ t^2)]$
σ	Working stress of metal, $[F/L^2 = M/(L\ t^2)]$
σ_{ij}	Total stress component, force in j direction on i surface, $[F/L^2 = M/Lt^2]$
ω	Angular velocity, $[1/t]$

SUBSCRIPTS

1	Reference point 1
2	Reference point 2
i, j, k, x, y, z	Coordinate directions
i	Inner
o	Outer

5 Conservation Principles

"As simple as possible, but no simpler."

—**Albert Einstein, 1879–1955, Physicist**

I. THE SYSTEM

As discussed in Chapter 1, the basic principles that apply to the analysis and solution of flow problems include the conservation of mass, energy, and momentum, in addition to appropriate transport relations for these conserved quantities. For flow problems, these conservation laws are applied to a *system*, which is defined as any clearly specified region or volume of fluid with either macroscopic or microscopic dimensions (this is also sometimes referred to as a "control volume"), as illustrated in Figure 5.1. The material (fluid) within the system is assumed to be a *continuum*, with physical properties that can be defined at all points within the system. This means that only dimensions which are very large compared to the molecular or particulate structure of the material are considered. The general conservation law is

$$\begin{Bmatrix} \text{Rate of } X \\ \text{into the system} \end{Bmatrix} - \begin{Bmatrix} \text{Rate of } X \\ \text{out of the system} \end{Bmatrix} = \begin{Bmatrix} \text{Rate of accumulation} \\ \text{of } X \text{ in the system} \end{Bmatrix}$$

where X is the conserved quantity, that is, mass, energy, or momentum. In the case of momentum, because a "rate of momentum" is equivalent to a force (by Newton's second law), the "rate in" term must also include any (net) forces acting on the system. It is emphasized that the *system* is *not* the "containing vessel" *per se* (e.g., a pipe, tank, or pump) but is the *fluid* contained within the designated boundary. We will show how this generic expression is applied for each of the conserved quantities.

II. CONSERVATION OF MASS

A. MACROSCOPIC MASS BALANCE

For a given system (e.g., Figure 5.1), each entering stream (i) will carry mass into the system (at rate $\dot{m}_i$) and each exiting stream (o) carries mass out of the system (at rate $\dot{m}_o$). Hence, the conservation of mass, or "continuity," equation for the system is

$$\sum_{in} \dot{m}_i - \sum_{out} \dot{m}_o = \frac{dm_s}{dt} \tag{5.1}$$

where m_s is the mass of the system. For each stream

$$\dot{m} = \int_A d\dot{m} = \int_A \rho \vec{v} \cdot d\vec{A} = \rho \vec{V} \cdot \vec{A} \tag{5.2}$$

where
  $\vec{v}$ is the local velocity through area $d\vec{A}$
  $\vec{V}$ is the average velocity through the total cross section $\vec{A}$

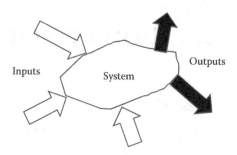

FIGURE 5.1 A system with inputs and outputs.

That is, the total mass flow rate through a given area for any stream is the integrated value of the local mass flow rate over that area. Note that mass flow rate is a scalar, whereas velocity and area are vectors. The scalar (or dot) product of the velocity and area vectors gives the volumetric flow rate which, when multiplied by the density, gives the mass flow rate. The "direction" or orientation of the area is that of the unit vector that is normal to the area and pointing outward from the system. The corresponding definition of the average velocity through the conduit is

$$\vec{V} = \frac{1}{\vec{A}} \int \vec{v} \cdot d\vec{A} = \frac{Q}{\vec{A}} \tag{5.3}$$

where $Q = \dot{m}/\rho$ is the volumetric flow rate and the area A is the projected component of $\vec{A}$ that is parallel to $\vec{V}$, that is, the component of $\vec{A}$ whose normal is in the same direction as $\vec{V}$.

For a system at steady state, there will be no accumulation of mass in the system, and Equation 5.1 reduces to

$$\sum_{in} \dot{m}_i = \sum_{out} \dot{m}_o \tag{5.4}$$

or

$$\sum_{in} (\rho \vec{V} \cdot \vec{A})_i = \sum_{out} (\rho \vec{V} \cdot \vec{A})_o \tag{5.5}$$

Example 5.1

Water is flowing at a velocity of 7 ft/s in both 1 in. and 2 in. ID pipes, which are joined together and fed into a 3 in. ID pipe, as shown in Figure E5.1. Determine the water velocity in the 3 in. pipe.

Solution:

The system is the fluid within the pipe branch, which is at a steady state so that Equation 5.5 applies:

$$(\rho VA)_1 + (\rho VA)_2 = (\rho VA)_3$$

For a constant density fluid, this may be solved for V_3:

$$V_3 = V_1 \frac{A_1}{A_3} + V_2 \frac{A_2}{A_3}$$

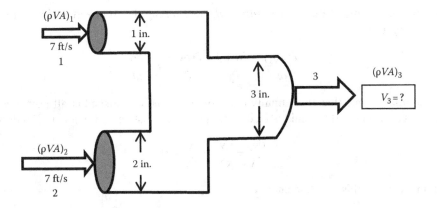

FIGURE E5.1 Application of continuity equation.

Since $A = \pi D^2/4$, this gives

$$V_3 = 7\,\text{ft/s}\left(\frac{1}{9} + \frac{4}{9}\right) = 3.89\,\text{ft/s}$$

(*Note:* The number of significant figures in this problem is indeterminate, so we report the answer to three figures, which is the maximum allowed unless more can be justified.)

B. MICROSCOPIC MASS BALANCE

The conservation of mass can be applied to an arbitrarily small fluid element (see Figure 5.2) to derive the "microscopic continuity" equation, which must be satisfied at all points within any continuous fluid. This can be done by considering an arbitrary (cubical) differential element of dimensions dx, dy, dz, with mass flow components into or out of each surface, for example,

$$\dot{m}_{out} - \dot{m}_{in} = dy\,dz\left[\left(\rho v_x\right)_{x+dx} - \left(\rho v_x\right)_x\right]$$
$$+ dx\,dz\left[\left(\rho v_y\right)_{y+dy} - \left(\rho v_y\right)_y\right]$$
$$+ dx\,dy\left[\left(\rho v_z\right)_{z+dz} - \left(\rho v_z\right)_z\right]$$
$$= -\frac{\partial \rho}{\partial t}\,dx\,dy\,dz \qquad (5.6)$$

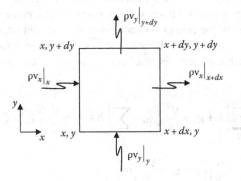

FIGURE 5.2 Two-dimensional microscopic element.

Dividing by the volume of the element ($dx\ dy\ dz$) and taking the limit as the size of the element shrinks to zero gives

$$\frac{\partial(\rho v_x)}{\partial x} + \frac{\partial(\rho v_y)}{\partial y} + \frac{\partial(\rho v_z)}{\partial z} = -\frac{\partial \rho}{\partial t} \tag{5.7}$$

This is the microscopic (local) continuity equation and must be satisfied at all points within any flowing fluid continuum. If the fluid is incompressible (i.e., constant ρ), Equation 5.7 reduces to

$$\frac{\partial v_x}{\partial x} + \frac{\partial v_y}{\partial y} + \frac{\partial v_z}{\partial z} = 0 \tag{5.8}$$

We will make use of this equation later on.

III. CONSERVATION OF ENERGY

Energy can take a wide variety of forms, such as internal (thermal), mechanical, work, kinetic, potential, surface, electrostatic, electromagnetic, and nuclear energy. Also, for nuclear reactions or velocities approaching the speed of light, the interconversion of mass and energy can be significant. However, we will not be concerned with situations involving nuclear reactions or velocities near that of light, and some other possible forms of energy (such as surface tension) may be negligible as well. Our purpose will be adequately served if we consider only internal (thermal), kinetic, potential (due to gravity and/or pressure), mechanical (work), and thermal (heat) forms of energy. For the system illustrated in Figure 5.1, a unit mass of fluid in each inlet and outlet stream may contain a certain amount of internal energy (u) by virtue of its temperature, kinetic energy ($v^2/2$) by virtue of its velocity, potential energy (gz) due to its position in a (gravitational) potential field, and "pressure (PV/m) or (P/ρ)" energy. The "pressure energy" is sometimes called the "flow work," because it is associated with the amount of work or energy required to "inject" a unit mass of fluid into the system or "eject" it out of the system at the appropriate pressure. In addition, energy can cross the boundaries of the system other than with the flow streams in the form of heat (Q), resulting from a temperature difference, and "shaft work" (W). Shaft work is so named because it is normally associated with work transmitted to or from the system by a shaft, such as that of a pump, compressor, mixer, or turbine.

The sign convention for heat (Q) and work (W) is arbitrary and consequently varies from one reference to another. Heat is usually taken to be positive when it is added to the system, so it would seem to be consistent to use this same convention for work (which is the convention in most "scientific" references). However, engineers, being "pragmatic," use a sign convention that is directly associated with "value." That is, if work can be extracted *from* the system (e.g., to drive a turbine), then it is *positive*, because a positive asset can be sold to produce revenue. However, if work must be put *into* the system (such as by a pump or a mixer), then it is *negative*, because the energy must be purchased to drive the pump (a negative asset). This convention is also more consistent with the "driving force" interpretation of the terms in the energy balance, as will be shown later.

With this introduction, we can write the rate form of the conservation of energy equation for any system as follows, in which each term in brackets (...) represents a form of energy per unit of mass of fluid:

$$\sum_{in}\left(h + gz + \frac{V^2}{2}\right)_i \dot{m}_i - \sum_{out}\left(h + gz + \frac{V^2}{2}\right)_o \dot{m}_o + \dot{Q} - \dot{W}$$

$$= \frac{d}{dt}\left[\left(u + gz + \frac{V^2}{2}\right)m\right]_{sys} \tag{5.9}$$

Here, $h = u + P/\rho$ is the *enthalpy* per unit mass of fluid. Note that the inlet and exit streams include enthalpy (i.e., internal energy, u, plus flow work, P/ρ), whereas the "system energy" includes only the internal energy but no P/ρ flow work (for obvious reasons). If there is only one inlet stream and one exit stream and the system is at steady state (i.e., $\dot{m}_i = \dot{m}_o = \dot{m}$), the energy balance becomes

$$\Delta h + g\Delta z + \frac{1}{2}\Delta V^2 = q - w \qquad (5.10)$$

where

$\Delta(\ldots) = (\text{"out"}) - (\text{"in"})$, and

$q = \dot{Q}/\dot{m}$

$w = \dot{W}/\dot{m}$ represents the heat added to the system and work done by the system, respectively, per unit mass of fluid

The system may also be composed of the fluid between any two points along a streamline (e.g., a "stream tube") within a flow field. Specifically, if these two points are only an infinitesimal distance apart, the result is the differential form of the energy balance:

$$dh + g\,dz + V\,dV = \delta q - \delta w \qquad (5.11)$$

where $dh = du + d(P/\rho)$. The "$d(\ )$" notation represents a total or "exact" differential and applies to those quantities that are determined only by the state (T, P) of the system and are thus "point" properties. The "$\delta(\ )$" notation represents quantities which are inexact differentials and depend upon the path taken from one point to another.

Note that the energy balance contains several different forms of energy, which may be generally classified as either *mechanical* energy, associated with motion or position, or *thermal* energy, associated with temperature. Mechanical energy is "useful," in that it can be converted directly into useful work, and includes potential energy, kinetic energy, "flow work," and shaft work. The thermal energy terms, that is, internal energy and heat, are not directly available to do useful work unless they are transformed into mechanical energy, in which case it is the mechanical energy that does the useful work.

In fact, the total amount of energy represented by a relatively small temperature change is equivalent to a relatively large amount of "mechanical energy." For example, 1 Btu of thermal energy is equivalent to 778 ft lb$_f$ of mechanical energy. This means that the amount of energy required to raise the temperature of lb$_m$ of water by 1°F (the definition of the Btu) is equivalent to the amount of energy required to raise the elevation of that same pound of water by 778 ft (e.g., an 80 story building!). Thus, for systems that involve significant temperature changes, the mechanical energy terms (e.g., pressure, potential and kinetic energy, and work) may be negligible compared with the thermal energy terms (e.g., heat transfer, internal energy). In such cases, the energy balance equation reduces to a "heat balance," that is, $\Delta h = q$. However, the reader should be warned that "heat" is not a conserved quantity and that the inherent assumption that the other forms of energy are negligible when a "heat balance" is being written should always be confirmed.

Before proceeding further, we will take a closer look at the significance of enthalpy and internal energy, because these cannot be measured directly but are determined indirectly by measuring other properties such as temperature and pressure.

A. INTERNAL ENERGY

An infinitesimal change in internal energy is an exact differential and is a unique function of temperature and pressure (for a given composition). Since the density of a given material is also uniquely determined by temperature and pressure (e.g., by an equation of state for the material), the internal energy may be expressed as a function of any two of the three variables T, P, or ρ (or $v = 1/\rho$). Hence, we may write

$$du = \left(\frac{\partial u}{\partial T}\right)_v dT + \left(\frac{\partial u}{\partial v}\right)_T dv \tag{5.12}$$

By making use of classical thermodynamic identities, this is found to be equivalent to

$$du = c_v\, dT + \left[T\left(\frac{\partial P}{\partial T}\right)_v - P\right] dv \tag{5.13}$$

where

$$c_v = \left(\frac{\partial u}{\partial T}\right)_v \tag{5.14}$$

is the specific heat at constant volume (e.g., constant density). We will now consider several special cases for various materials.

1. Ideal Gas

For an ideal gas

$$\rho = \frac{PM}{RT} \quad \text{so that} \quad T\left(\frac{\partial P}{\partial T}\right)_v = P \tag{5.15}$$

Thus, Equation 5.13 reduces to

$$du = c_v\, dT \quad \text{or} \quad \Delta u = \int_{T_1}^{T_2} c_v\, dT = \bar{c}_v(T_2 - T_1) \tag{5.16}$$

where $\bar{c}_v$ is the average constant volume heat capacity between the two temperatures T_1 and T_2. This shows that the internal energy for an ideal gas is a function of temperature only.

2. Nonideal Gas

For a nonideal gas, Equation 5.15 is not valid, so that

$$T\left(\frac{\partial P}{\partial T}\right)_v \neq P \tag{5.17}$$

Consequently, the term in brackets in Equation 5.13 does not cancel out as it did for the ideal gas, which means that, for a non-ideal gas,

$$\Delta u = fn(T, P) \tag{5.18}$$

The form of the implied function, $fn(T, P)$, may be analytical if the material is described by a nonideal equation of state or it could be empirical, such as for steam, for which the properties are expressed as data tabulated in steam tables.

3. Solids and Liquids

For solids and liquids, $\rho \approx$ constant (or $dv = 0$), so

$$du = c_v \, dT \quad \text{or} \quad \Delta u = \int c_v T = \bar{c}_v (T_2 - T_1) \tag{5.19}$$

This shows that the internal energy depends upon temperature only (just as for the ideal gas, but for an entirely different reason).

B. ENTHALPY

The enthalpy can be expressed as a function of temperature and pressure:

$$dh = \left(\frac{\partial h}{\partial T} \right)_P dT + \left(\frac{\partial h}{\partial P} \right)_T dP \tag{5.20}$$

which, from thermodynamic identities, is equivalent to

$$dh = c_p \, dT + \left[v - T \left(\frac{\partial v}{\partial T} \right)_P \right] dP \tag{5.21}$$

where

$$c_p = \left(\frac{\partial h}{\partial T} \right)_P \tag{5.22}$$

is the specific heat of the material at constant pressure. We again consider some special cases.

1. Ideal Gas

For an ideal gas

$$T \left(\frac{\partial v}{\partial T} \right)_P = v \quad \text{and} \quad c_p = c_v + \frac{R}{M} \tag{5.23}$$

Thus, Equation 5.21 for the enthalpy becomes

$$dh = c_p \, dT \quad \text{or} \quad \Delta h = \int_{T_1}^{T_2} c_p \, dT = \bar{c}_p (T_2 - T_1) \tag{5.24}$$

which shows that the enthalpy for an ideal gas is a function of temperature only (as is the internal energy).

2. Nonideal Gas

For a nonideal gas

$$T \left(\frac{\partial v}{\partial T} \right)_P \neq v \quad \text{so that} \quad \Delta h = fn(T, P) \tag{5.25}$$

which, like Δu, may be either an analytical or an empirical function. All gases follow the ideal gas law under appropriate conditions (i.e., far enough from the critical point) and become more nonideal as the critical point is approached. That is, under conditions that are sufficiently far from the critical point that the enthalpy at constant temperature is essentially independent of pressure, the gas should be adequately described by the ideal gas law. This can be confirmed by inspecting the thermodynamic plots for various materials shown in Appendix D.

3. Solids and Liquids

For solids and liquids, $v = 1/\rho \approx$ constant, so that $(\partial v/\partial T)_P = 0$ and $c_p \approx c_v$. Therefore, Equation 5.21 reduces to

$$dh = c_p\, dT + v\, dP \qquad (5.26)$$

or

$$\Delta h = \int_{T_1}^{T_2} c_p\, dT + \int_{P_1}^{P_2} \frac{dP}{\rho} = \bar{c}_p(T_2 - T_1) + \frac{P_2 - P_1}{\rho} \qquad (5.27)$$

This shows that for solids and liquids, the enthalpy depends upon both temperature and pressure. This is in contrast to the internal energy, which depends upon the temperature only. Note that for solids and liquids, $c_p \approx c_v$.

The thermodynamic properties of a number of compounds are shown in Appendix D as pressure–enthalpy diagrams with lines of constant temperature, entropy, and specific volume. The vapor, liquid, and two-phase regions are clearly evident on these plots. The conditions under which each compound may exhibit ideal gas properties are identified by the region on the plot where the enthalpy is independent of pressure at a given temperature (i.e., the lower the pressure and the higher the temperature relative to the critical point, the more nearly the properties can be described by the ideal gas law).

IV. IRREVERSIBLE EFFECTS

We have noted that if there is a significant change in temperature, the thermal energy terms (i.e., q and u) may represent much more energy than the mechanical terms (i.e., pressure, potential and kinetic energy, and work). On the other hand, if the temperature difference between the system and its surroundings is very small, the only source of "heat" (thermal energy) is the internal (irreversible) dissipation of mechanical energy into thermal energy or "friction." The origin of this "friction loss" is the irreversible work required to overcome intermolecular forces when the fluid is in motion, that is, the attractive forces between the "fluid elements," under dynamic (nonequilibrium) conditions. This can be quantified as follows.

For a system at equilibrium (i.e., in a reversible or "static" state), thermodynamics tells us that

$$du = T\, ds - Pd\!\left(\frac{1}{\rho}\right) \quad \text{and} \quad T\, ds = \delta q \qquad (5.28)$$

That is, the total increase in entropy (which is a measure of "disorder" or "irreversibility") comes from heat transferred across the system boundary (δq). However, a flowing fluid is in a "dynamic" or irreversible state, which consumes "useful" energy (e.g., mechanical energy) and transforms it into "non-useful" thermal energy (e.g., entropy). Because entropy is proportional to the degree of departure from the most stable (equilibrium) conditions, this means that the further the system is from equilibrium (i.e., the faster the flow), the greater the increase in entropy, so for a dynamic (flow) system

$$T\, ds > \delta q, \quad \text{or} \quad T\, ds = \delta q + \delta e_f \qquad (5.29)$$

that is,

$$du = \delta q + \delta e_f - P d\left(\frac{1}{\rho}\right) \qquad (5.30)$$

where δe_f represents the "irreversible energy" associated with the departure of the system from equilibrium, which is extracted from mechanical energy and transformed (or "dissipated") into thermal energy. The farther the system is from equilibrium (e.g., the faster the motion), the greater is this irreversible energy. The origin of this energy (or "extra entropy") is the mechanical energy that drives the system and is thus converted to e_f. This energy ultimately appears as an increase in the temperature of the system (du), heat transferred from the system (δq), and/or expansion energy [$Pd(1/\rho)$] (if the fluid is compressible). This mechanism of transfer of useful mechanical energy to low-grade (non-useful) thermal energy is referred to as "energy dissipation." Although e_f is often referred to as the "friction loss," it is evident that this energy is not really lost but is transformed (dissipated) from useful high-level mechanical energy to non-useful low-grade thermal energy. It should be clear that e_f must always be positive because energy can be transformed spontaneously only from a higher state (mechanical) to a lower state (thermal) and not in the reverse direction, as a consequence of the second law of thermodynamics.

When Equation 5.30 is introduced into the definition of enthalpy, we get

$$dh = du + d\left(\frac{P}{\rho}\right) = \delta q + \delta e_f + \frac{dP}{\rho} \qquad (5.31)$$

Substituting this for the enthalpy in the differential energy balance, Equation 5.11, gives

$$\frac{dP}{\rho} + g\,dz + V\,dV + \delta w + \delta e_f = 0 \qquad (5.32)$$

This can be integrated along a streamline from the inlet to the outlet of the system to give

$$\int_{P_i}^{P_o} \frac{dP}{\rho} + g(z_o - z_i) + \frac{1}{2}(V_o^2 - V_i^2) + e_f + w = 0 \qquad (5.33)$$

where, from Equation 5.30,

$$e_f = (u_o - u_i) - q + \int_{P_i}^{P_o} P d\left(\frac{1}{\rho}\right) \qquad (5.34)$$

Equations 5.33 and 5.34 are simply rearrangements of the steady-state total energy balance (Equation 5.10) but in much more useful form. Without the friction loss (e_f) term (which includes all of the thermal energy effects), Equation 5.33 represents a *mechanical energy balance* (although mechanical energy is not a conserved quantity). Equation 5.33 is the *engineering Bernoulli equation* or simply the *Bernoulli equation*. Along with Equation 5.34, it accounts for all of the possible thermal and mechanical energy effects of concern to us and is the form of the energy balance that is most convenient when mechanical energy dominates and thermal effects are minor. It should be stressed that the first three terms in Equation 5.32 are *point* functions, that is, they depend only on conditions at the inlet and outlet of the system, whereas the w and e_f terms are *path* functions, which depend on what is happening to the system *between* the inlet and outlet points (i.e., these are rate dependent and can be determined either empirically or from an appropriate rate or transport model, as will be shown later).

If the fluid is incompressible (constant density), Equation 5.33 can be written as

$$\frac{\Delta \Phi}{\rho} + \frac{1}{2}\Delta(V^2) + e_f + w = 0 \tag{5.35}$$

where $\Phi = P + \rho gz$ is the *potential*. For a fluid at rest, $e_f = V = w = 0$, and Equation 5.35 reduces to the basic equation of fluid statics for an incompressible fluid (i.e., $\Phi = $ constant), Equation 4.8. For any static fluid, Equation 5.35 reduces to the more general basic equation of fluid statics, Equation 4.6. For gases, if the pressure change is such that the density does not change more than about 30%, the incompressible equation can be applied with reasonable accuracy by assuming the density to be constant at a value equal to the average density in the system (a more general consideration of compressible fluids is given in Chapter 9).

Note that if each term of Equation 5.35 is divided by g, then all terms will have the dimension of length. The result is called the "*head*" form of the Bernoulli equation, and each term then represents the amount of energy in an equivalent static column of the system fluid. For example, the pressure term becomes the "pressure head, $(-\Delta P/\rho g = H_P)$," the potential energy term becomes the "static head, $(-\Delta z = H_z)$," the kinetic energy term becomes the "velocity head, $(\Delta V^2/2g = H_v)$," the friction loss becomes the "head loss, $(e_f/g = H_f)$," and the work term is, typically, the "pump head (work), $(-w/g = H_w)$."

A. KINETIC ENERGY CORRECTION

In the foregoing equations, we assumed that the fluid velocity (V) at a given point in the system (e.g., in a pipe or tube) is the same for all fluid elements at a given cross section in the flow stream. However, this is not true in conduits, because the fluid velocity is zero at a stationary boundary or wall and increases with the distance from the wall. The total rate at which kinetic energy is transported by a fluid element moving with local velocity $\vec{v}$ at a mass flow rate $d\dot{m}$ through a differential area $d\vec{A}$ is ($\vec{v}^2 d\dot{m}/2$) where $d\dot{m} = \rho \vec{v} \cdot d\vec{A}$. Therefore, since the density does not vary over the cross section, the total rate of transport of kinetic energy through the cross section A is

$$\int \frac{v^2}{2}\, d\dot{m} = \frac{\rho}{2} \int v^3\, dA \tag{5.36}$$

If the fluid velocity is uniform over the cross section at a value equal to the average velocity V (i.e., "plug flow"), then the rate at which kinetic energy is transported would be

$$\frac{1}{2}\rho V^3 A \tag{5.37}$$

Therefore, a *kinetic energy correction factor*, α, can be defined as the ratio of the true rate of kinetic energy transport, relative to that which would occur if the fluid velocity is everywhere equal to the average (plug flow) velocity, for example,

$$\alpha = \frac{\text{True KE transport rate}}{\text{Plug flow KE transport rate}} = \frac{1}{A}\int_A \left(\frac{v}{V}\right)^3 dA \tag{5.38}$$

The Bernoulli equation should therefore include this kinetic energy correction factor, that is,

$$\frac{\Delta\Phi}{\rho} + \frac{1}{2}\Delta(\alpha V^2) + e_f + w = 0 \tag{5.39}$$

As will be shown later, the velocity profile for a Newtonian fluid in fully developed laminar flow in a circular tube is parabolic. When this profile is introduced into Equation 5.38, the result is $\alpha = 2$. For highly turbulent flow, the profile is much flatter and $\alpha \approx 1.06$, although for practical applications it is usually assumed that $\alpha = 1$ for turbulent flow in a tube.

Example 5.2 Kinetic Energy Correction Factor for Laminar Flow of a Newtonian Fluid

We will show later that the velocity profile for the laminar flow of a Newtonian fluid in fully developed flow in a circular tube of radius R is parabolic. Because the velocity is zero at the wall of the tube and maximum in the center, the equation for the profile is

$$v(r) = V_{max}\left(1 - \frac{r^2}{R^2}\right)$$

This can be used to calculate the kinetic energy correction factor from Equation 5.38 as follows. First, we must calculate the average velocity, V, using Equation 5.3 and substitute $x = r/R$:

$$V = \frac{1}{\pi R^2}\int_0^R v 2\pi r\, dr$$

$$= 2V_{max}\int_0^1 (1 - x^2)x\, dx = \frac{V_{max}}{2}$$

This shows that the average velocity is simply one-half of the maximum (centerline) velocity. Thus, replacing V in Equation 5.38 by $V_{max}/2$ and then integrating the cube of the parabolic velocity profile over the tube cross section gives $\alpha = 2$. (The details of the manipulation are left as an exercise for the reader.)

Example 5.3 Diffuser

A diffuser is a section in a conduit over which the flow area increases gradually from upstream to downstream, as illustrated in Figure E5.2. If the inlet and outlet areas (A_1 and A_2) are known, and the upstream pressure and velocity (P_1 and V_1) are given, we would like to find the downstream pressure and velocity (P_2 and V_2). If the fluid is incompressible, the continuity equation gives V_2:

$$(\rho V A)_1 = (\rho V A)_2 \quad \text{or} \quad V_2 = V_1\frac{A_1}{A_2}$$

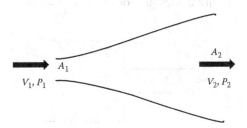

FIGURE E5.2 Diffuser.

The pressure P_2 is determined from the Bernoulli equation. If the diffuser is horizontal, there is no work done between the inlet and outlet, and the friction loss is small (which is a good assumption for a well-designed diffuser), utilizing the continuity (conservation of mass) equation gives

$$\rho V_1 A_1 = \rho V_2 A_2$$

Assuming $\alpha_1 = \alpha_2$, the Bernoulli equation gives

$$P_2 = P_1 + \frac{\rho}{2}\left(V_1^2 - V_2^2\right) - \rho e_f \cong P_1 + \frac{\rho V_1^2}{2}\left(1 - \frac{A_1^2}{A_2^2}\right)$$

Because $A_1 < A_2$ and the losses are small, this shows that $P_2 > P_1$, that is, the pressure increases downstream as the velocity decreases. This occurs because the decrease in kinetic energy is transformed into an increase in "pressure energy." Such a diffuser is said to have a "high-pressure recovery."

Example 5.4 Sudden Expansion

We now consider an incompressible fluid flowing from a small conduit through a sudden expansion into a larger conduit, as illustrated in Figure E5.3. The objective, as in the previous example, is to determine the exit pressure and velocity (P_2 and V_2), given the upstream conditions and the dimensions of the ducts. Note that the conditions are all identical to those of the previous diffuser example, so the continuity and Bernoulli equations are also identical. The major difference is that the friction loss (a "path" variable) is not small as it is for the diffuser. Because of inertia, the fluid cannot follow the sudden 90° change in direction of the boundary, so considerable secondary flow is generated after the fluid leaves the small duct and before it can expand to fill the large duct, resulting in much greater friction loss. The equation for P_2 is the same as before, except that now the friction loss is relatively large (methods for evaluating this will be given later):

$$P_2 = P_1 + \frac{\rho V_1^2}{2}\left(1 - \frac{A_1^2}{A_2^2}\right) - \rho e_f$$

The "pressure recovery" is reduced by the friction loss in the eddies which is relatively high for the sudden expansion. The pressure recovery is therefore relatively low and may be negative.

Example 5.5 The Torricelli Problem

Consider an open vessel with diameter D_1 containing a fluid at a depth h, that is draining out of a hole of diameter D_2 in the bottom of the tank. We would like to determine the velocity of the fluid flowing out of the hole in the bottom. As a first approximation, we neglect the friction loss in the tank and through the hole. Point 1 is taken at the free surface of the fluid in the tank, and point 2 is taken at the exit from the hole, since the pressure is known to be atmospheric at both points (Figure E5.4). The velocity in the tank is related to that through the hole by the continuity equation, with $\rho_2 = \rho_1$:

$$(\rho VA)_1 = (\rho VA)_2 \quad \text{or} \quad V_1 = V_2 \frac{A_2}{A_1} = V_2 \beta^2$$

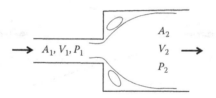

FIGURE E5.3   Sudden expansion.

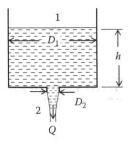

FIGURE E5.4 Draining tank—the Torricelli problem.

where $\beta = D_2/D_1$. The Bernoulli equation for an incompressible fluid between points 1 and 2 is

$$\frac{P_2 - P_1}{\rho} + g(z_2 - z_1) + \frac{1}{2}\left(\alpha_2 V_2^2 - \alpha_1 V_1^2\right) + w + e_f = 0$$

Because points 1 and 2 are both at atmospheric pressure, $P_2 = P_1$. We assume that $w = 0$, $\alpha = 1$, and we neglect friction, so $e_f = 0$ (actually a poor assumption in many cases). Setting $(z_2 - z_1) = -h$, eliminating V_1 from these two equations, and solving for V_2 gives

$$V_2 = \left(\frac{2gh}{1 - \beta^4}\right)^{1/2}$$

This is known as the *Torricelli equation*. We now consider what happens as the hole gets larger. Specifically, as $D_2 \rightarrow D_1$ (i.e., as $\beta \rightarrow 1$), the equation says that $V_2 \rightarrow \infty$! This is obviously an unrealistic limit, so there must be something wrong. Of course, our assumption that friction is negligible may be valid at low velocities, but as the velocity increases it becomes less valid and is obviously invalid long before the limiting condition is reached.

Upon examining the equation for V_2, we see that it is independent of the properties of the fluid in the tank. We might suspect that this is not accurate, because if the tank were to be filled with CO_2 we intuitively expect that it would drain more slowly than if it were filled with water. So, what is wrong? In this case, it is our assumption that $P_2 = P_1$. Of course, the pressure is atmospheric at both points 1 and 2, but we have neglected the static head of air between these points, which is the actual difference in the pressure. This results in a buoyant force due to the air and can have a significant effect on the drainage of CO_2 although it will be negligible for water. Thus, if we account for the static head of air, that is, $P_2 - P_1 = \rho_a gh$, in the Bernoulli equation and then solve for V_2, we get

$$V_2 = \left(\frac{2gh(1 - \rho_a/\rho)}{1 - \beta^4}\right)^{1/2}$$

where ρ is the density of the fluid in the tank. This also shows that as $\rho \rightarrow \rho_a$, the velocity goes to zero, as we would expect.

These examples illustrate the importance of knowing what can and cannot be neglected in a given problem (i.e., the "baby vs. the bathwater"), and the necessity for matching the appropriate assumptions to the specific problem conditions in order to arrive at a valid solution. They also illustrate the importance of understanding what is happening within the system in addition to knowing the inlet and outlet conditions.

V. CONSERVATION OF LINEAR MOMENTUM

A macroscopic momentum balance for a flow system must include all equivalent forms of momentum. In addition to the rate of linear momentum convected into and out of the system by the entering and leaving streams, the sum of all the forces that act *on* the system (the system being defined as a specified volume of *fluid*) must be included. This follows from Newton's second law, which provides an equivalence between force and the rate of change (and rate of transport) of linear momentum. The resulting macroscopic conservation of linear momentum thus becomes

$$\sum_{on\ system} \vec{F} + \sum_{in} (\dot{m}\vec{V})_i - \sum_{out} (\dot{m}\vec{V})_o = \frac{d}{dt}(m\vec{V})_{sys} \qquad (5.40)$$

Note that because momentum is a vector, this equation represents three component equations, one for each direction in three-dimensional space. If there is only one "in" (entering) and one "out" (leaving) stream, and the system is also at steady state, then $\dot{m}_i = \dot{m}_o = \dot{m}$ and the momentum balance becomes

$$\sum_{on\ system} \vec{F} = \dot{m}\left(\sum_{out} \vec{V}_o\right) - \left(\sum_{in} \vec{V}_i\right) \qquad (5.41)$$

Note that the vector (directional) character of the "convected" momentum terms (i.e., $\dot{m}\vec{V}$) is that of the velocity, because $\dot{m}$ is a scalar (i.e., $\dot{m} = \rho \vec{V} \cdot \vec{A}$ is a scalar product).

A. ONE-DIMENSIONAL FLOW IN A TUBE

We will apply the steady-state momentum balance to a fluid in plug flow (i.e., uniform velocity across the cross section) in a tube, as illustrated in Figure 5.3. (The "stream tube" may be bounded by either solid or imaginary boundaries; the only condition is that no fluid crosses the boundaries other than that through the "inlet" and "outlet" planes.)

The shape of the cross section does not have to be circular; it can be any shape. The fluid element in the "slice" of thickness dx is our system, and the momentum balance equation on this system is

$$\sum_{on\ fluid} F_x + \dot{m}V_x - \dot{m}(V_x + dV_x) = \sum_{on\ fluid} F_x - \dot{m}\,dV_x$$

$$= \frac{d}{dt}(\rho V_x A dx) = 0 \qquad (5.42)$$

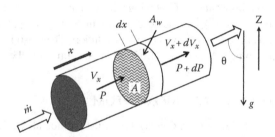

FIGURE 5.3 Momentum balance on a "slice" in a stream tube.

The forces acting on the fluid result from pressure (dF_P), gravity (dF_g), wall drag (dF_w), and external "shaft" work, that is, pump work ($\delta W = -F_{ext}\, dx$, not shown in Figure 5.3):

$$\sum_{on\ fluid} F_x = dF_p + dF_g + F_{ext} + dF_w \tag{5.43}$$

where

$$dF_p = A_x[P - (P + dP)] = -A_x\, dP$$

$$dF_g = \rho g_x A_x\, dx = -\rho g A_x\, dx \cos\theta = -\rho g A_x\, dz$$

$$dF_w = -\tau_w\, dA_w = -\tau_w W_p\, dx$$

$$-\delta W = F_{ext}\, dx \quad \text{or} \quad F_{ext} = -\frac{\delta W}{dx}$$

Here, τ_w is the stress exerted *by* the fluid *on* the wall (the reaction to the stress exerted *on* the fluid *by* the wall), and W_p is the perimeter of the cross section that is wetted by the fluid (the "wetted perimeter"). After substituting the expressions for the forces from Equation 5.43 into the momentum balance equation, Equation 5.42, and dividing the result by $-\rho A$, where $A = A_x$, the result is

$$\frac{dP}{\rho} + g\, dz + \frac{\tau_w W_p}{\rho A}\, dx + \delta w + V\, dV = -dV/dt\, dx \tag{5.44}$$

where $\delta w = \delta W/(\rho A\, dx)$ is the work done by the system per unit mass of fluid. Integrating this expression from the inlet (*i*) to the outlet (*o*) and assuming steady state gives

$$\int_{P_i}^{P_o} \frac{dP}{\rho} + g(z_o - z_i) + \frac{1}{2}\left(V_o^2 - V_i^2\right) + \int_L \frac{\tau_w W_p}{\rho A}\, dx + w = 0 \tag{5.45}$$

Comparing this with the Bernoulli equation (Equation 5.33) shows that they are identical provided

$$e_f = \int_L \left(\frac{\tau_w W_p}{\rho A}\right) dx \tag{5.46}$$

or, for steady flow in a uniform conduit,

$$e_f \approx \frac{\tau_w W_p L}{\rho A} = \frac{\tau_w}{\rho}\left(\frac{4L}{D_h}\right) \tag{5.47}$$

where

$$D_h = 4\frac{A}{W_p} \tag{5.48}$$

is called the *hydraulic diameter.* Note that this result applies to a conduit of any cross-sectional shape. For a circular tube, for example, D_h is identical to the tube diameter D.

We see that there are several ways of interpreting the term e_f. From the Bernoulli equation, it represents the "lost" (i.e., dissipated) energy associated with irreversible effects. From the momentum balance, e_f is also seen to be directly related to the stress between the fluid and the tube wall (τ_w), that is, it can be interpreted as the work required to overcome the resistance to flow in the conduit. Both of these interpretations are correct and are equivalent.

Although the energy and momentum balances lead to equivalent results for this special case of one-dimensional fully developed flow in a straight uniform tube, this is an exception and not the rule. In general, because momentum is a vector, the momentum balance gives additional information concerning the forces exerted on and/or by the fluid in the system through the boundaries, which is not given by the energy balance or Bernoulli equation. This will be illustrated shortly.

B.   THE LOSS COEFFICIENT

Looking at the Bernoulli equation, we see that the friction loss (e_f) can be made dimensionless by dividing it by the kinetic energy per unit mass of fluid. The result is the dimensionless *loss coefficient, K_f*:

$$K_f = \frac{e_f}{V^2/2} \tag{5.49}$$

A loss coefficient may be defined for any element that offers resistance to flow (i.e., in which energy is dissipated), such as a length of conduit, a valve, a pipe fitting, a bend, a contraction, or an expansion. The total friction loss can thus be expressed in terms of the sum of the losses in each element, that is, $e_f = \sum_i^o (K_{fi} V_i^2/2)$. This will be elaborated in Chapter 6.

As can be determined from Equations 5.47 and 5.49, the pipe wall stress can also be made dimensionless by dividing by the kinetic energy per unit volume of fluid. The result is known as the pipe *Fanning friction factor, f*:

$$f = \frac{\tau_w}{\rho V^2/2} \tag{5.50}$$

Although $\rho V^2/2$ represents kinetic energy per unit volume, ρV^2 is also the magnitude of the *flux of momentum* carried by the fluid along the conduit. The latter interpretation is more logical in Equation 5.50, because τ_w is also a flux of momentum from the fluid to the tube wall. This interpretation renders the factor of ½ arbitrary, although it is related to the kinetic energy of the fluid. Other definitions of the pipe friction factor are also in use, that are some multiple of two times the Fanning friction factor. For example, the *Darcy friction factor*, which is equal to $4f$, is used frequently by mechanical and civil engineers. Thus, it is important to know which definition is used or implied when data or charts for friction factors are used.

Because the friction loss and wall stress are related by Equation 5.47, the loss coefficient for pipe flow is related to the pipe Fanning friction factor as follows:

$$K_{f\,pipe} = \frac{4fL}{D_h} \tag{5.51}$$

Example 5.6   Friction Loss in a Sudden Expansion

Figure E5.5 shows the flow in a sudden expansion from a small conduit to a larger one. We assume that the conditions upstream of the expansion (point 1) are known, as well as the areas A_1 and A_2. We desire to find the velocity and pressure downstream of the expansion (V_2 and P_2), and the loss coefficient, K_f. As before, V_2 is determined from the mass balance (continuity equation) applied to

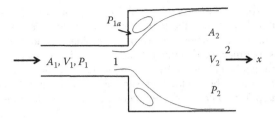

FIGURE E5.5   Loss coefficient for a sudden expansion.

the system (which is all of the fluid in the conduit between points 1 and 2). Assuming a constant density fluid

$$V_2 = V_1 \frac{A_1}{A_2}$$

For plug flow, the Bernoulli equation for this system is

$$\frac{P_2 - P_1}{\rho} + \frac{1}{2}\left(V_2^2 - V_1^2\right) + e_f = 0$$

which contains two unknowns, P_2 and e_f. So, we need another equation, which is the steady-state x component of the momentum balance:

$$\sum F_x = \dot{m}(V_{2x} - V_{1x})$$

where $V_{1x} = V_1$ and $V_{2x} = V_2$, because all velocities of interest are in the x direction. Accounting for all the forces that can act on the system through each section of the boundary, this becomes

$$P_1 A_1 + P_{1a}(A_2 - A_1) - P_2 A_2 + F_{wall} = \rho V_1 A_1 (V_2 - V_1)$$

where
  P_{1a} is the pressure on the left-hand boundary of the system (i.e., the "washer shaped" surface)
  F_{wall} is the force due to the drag of the wall on the fluid at the horizontal boundary of the system

The fluid pressure cannot change discontinuously, so $P_{1a} \approx P_1$. Also, because the contact area with the wall bounding the system is relatively small, we can neglect F_{wall} with no serious consequences. The result is

$$(P_1 - P_2)A_2 = \rho V_1^2 A_1\left(\frac{A_1}{A_2} - 1\right)$$

This can be solved for $P_2 - P_1$ which, when inserted into the Bernoulli equation, allows us to solve for e_f:

$$e_f = \frac{V_1^2}{2}\left(1 - \frac{A_1}{A_2}\right)^2 = \frac{K_f V_1^2}{2}$$

Thus,

$$\boxed{K_f = \left(1 - \frac{A_1}{A_2}\right)^2 = (1 - \beta^2)^2 \quad \text{where } \beta = \frac{D_1}{D_2}}$$

The loss coefficient is seen to be a function only of the geometry of the system (note that the assumption of plug flow implies that the flow is highly turbulent). For most systems (i.e., flow in valves, fittings, etc.), the loss coefficient cannot be determined accurately from simple theoretical concepts such as this, but must be determined experimentally. For example, the friction loss in a sudden contraction cannot be calculated by this simple method due to the occurrence of the *vena contracta* just downstream of the contraction (see Table 7.5 and the discussion in Section IV of Chapter 10). For a sharp 90° contraction, the contraction loss coefficient is given by

$$K_f = 0.5(1 - \beta^2)$$

where β is the ratio of the small to the large tube diameters.

Example 5.7 Flange Forces on a Pipe Bend

Consider an incompressible fluid flowing through a pipe bend, as illustrated in Figure E5.6. We would like to determine the forces in the bolts in the flanges that hold the bend in the pipe, knowing the geometry of the bend, the flow rate through the bend, and the exit pressure (P_2) from the bend. The system is the fluid within the pipe bend, and a steady-state "x momentum" balance on this system is

$$\sum (F_x)_{on\ sys} = \dot{m}(V_{2x} - V_{1x})$$

Various factors contribute to the forces on the left-hand side of this equation, that is,

$$\sum (F_x)_{on\ sys} = P_1 A_{1x} + P_2 A_{2x} - (F_x)_{on\ wall\ by\ fluid}$$

$$= P_1 A_1 - P_2 A_2 \cos\theta - (F_x)_{on\ bolts\ by\ fluid}$$

Since "action = reaction," the force exerted *on* the fluid *by* the solid wall is the negative of the force exerted *by* the fluid *on* the wall. This force is then transmitted to the bolts and supports holding the bend in place. The sign of the force resulting from the pressure acting on the inlet and outlet areas is intuitive, because pressure acts on any system boundary from the outside, that is, since the pressure at any point acts equally in all directions, the pressure on the left-hand boundary acts to the right on the system and vice versa. This is also consistent with previous definitions, because the sign of a surface element corresponds to the direction of the normal vector that points *outward* from the bounded volume, and pressure is a compressive (negative) stress. Thus, $P_1 A_{x1}$ is (+) since it is a negative stress acting on a negative area, and $P_2 A_{x2}$ is (−) because it is a negative stress acting on a positive area. These signs have been accounted for intuitively in the equation.

The right-hand side of the momentum balance reduces to

$$\dot{m}(V_{2x} - V_{1x}) = \dot{m}(V_2 \cos\theta - V_1)$$

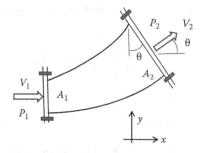

FIGURE E5.6   Flange forces in a pipe bend.

Equating the expressions for the RHS and LHS of the momentum equation and solving for $(F_x)_{on\ wall}$ gives

$$\boxed{(F_x)_{on\ wall} = (F_x)_{on\ bolts} = P_1 A_1 - P_2 A_2 \cos\theta - \dot{m}(V_2 \cos\theta - V_1)}$$

Similarly, the "y momentum" balance is

$$(F_y)_{on\ sys} = \dot{m}(V_{2y} - V_{1y})$$

which becomes

$$\boxed{(F_y)_{on\ wall} = (F_y)_{on\ bolts} = -P_2 A_2 \sin\theta - \dot{m}V_2 \sin\theta}$$

This assumes that the x–y plane is horizontal. If the y direction is vertical, the total weight of the bend, including the fluid inside, could be included as an additional (negative y) force component due to gravity. The magnitude and direction of the net force are

$$\boxed{\bar{F} = \sqrt{F_x^2 + F_y^2}, \qquad \varphi = \tan^{-1}\left(\frac{F_y}{F_x}\right)}$$

where φ is the direction of the net force vector measured counterclockwise from the $+x$ direction. Note that either P_1 or P_2 must be known, and the other is determined by the energy balance (Bernoulli equation), assuming that the loss coefficient, K_f, is known:

$$\frac{P_2 - P_1}{\rho} + g(z_2 - z_1) + \frac{1}{2}\left(V_2^2 - V_1^2\right) + e_f = 0 \quad \text{where } e_f = \frac{1}{2} K_f V_1^2$$

Methods for evaluating the loss coefficient K_f will be discussed in Chapter 6.

It should be noted that in evaluating the forces acting on the system, the effect of the external pressure transmitted through the boundaries to the system from the surrounding atmosphere was not included. Although this pressure does result in forces that act on the system, these forces all cancel out, so that the pressure that appears in the momentum balance equation is the *net* pressure in excess of atmospheric, that is, the *gage pressure*.

C. CONSERVATION OF ANGULAR MOMENTUM

In addition to linear momentum, angular momentum (or the moment of momentum) may be conserved. For a fixed mass (m) moving in the x direction with a velocity of V_x, the linear x momentum (M_x) is mV_x. Likewise, a mass (m) rotating counterclockwise about a center of rotation at an angular velocity $\omega = d\theta/dt$ has an angular momentum (L_θ) equal to $mV_\theta R = m\omega R^2$, where R is the distance from the center of rotation to the center of the mass m. Note that the angular momentum has dimensions of length times momentum and is therefore also referred to as the "moment of momentum." If the mass is not a point mass but a rigid distributed mass (M) rotating at a uniform angular velocity, the total angular momentum is given by

$$L_\theta = \int_M \omega r^2\, dm = \omega \int_M r^2\, dm = \omega I \tag{5.52}$$

where I is the moment of inertia of the body with respect to the center of rotation.

For a given mass, the conservation of linear momentum is equivalent to Newton's second law:

$$\sum \vec{F} = m\vec{a} = \frac{d(m\vec{V})}{dt} = m\frac{d\vec{V}}{dt} \tag{5.53}$$

The corresponding expression for the conservation of angular momentum is

$$\sum \Gamma_\theta = \sum F_\theta R = \frac{d(I\omega)}{dt} = I\frac{d\omega}{dt} = I\alpha \tag{5.54}$$

where
Γ_θ is the moment (torque) acting on the system
$d\omega/dt = \alpha$ is the angular acceleration

Note the similarity between Equations 5.53 and 5.54. For a flow system, streams with curved streamlines may carry angular momentum into and/or out of the system by convection. To account for this, the general macroscopic angular momentum balance applies:

$$\sum_{in} (\dot{m}RV_\theta)_i - \sum_{out} (\dot{m}RV_\theta)_o + \sum \Gamma_\theta = \frac{d(I\omega)}{dt} = I\alpha \tag{5.55}$$

For a steady-state system, with only one inlet and one outlet stream, this becomes

$$\sum \Gamma_\theta = \dot{m}\,[(RV_\theta)_o - (RV_\theta)_i] = \dot{m}[(R^2\omega)_o - (R^2\omega)_i] \tag{5.56}$$

This is also known as the *Euler turbine equation*, because it applies directly to turbines and all rotating fluid machinery. We will find it useful later in the analysis of the performance of centrifugal pumps.

D.   MOVING BOUNDARY SYSTEMS AND RELATIVE MOTION

We sometimes encounter a system that is in contact with a moving boundary, such that the fluid that comprises the system is carried along with the boundary while streams carrying momentum and/ or energy may flow into and/or out of the system. Examples of this include the flow impinging on a turbine blade (with the system being the fluid in contact with the moving blade) and the flow of exhaust gases from a moving rocket motor. In such cases, we often have direct information concerning the velocity of the fluid relative to the moving boundary (i.e., relative to the system), V_r, and so we must also consider the velocity of the system, V_s, to determine the absolute velocity of the fluid that is required for the conservation equations.

For example, consider a system that is moving in the x direction with a velocity of V_s, a fluid stream entering the system with a velocity in the x direction relative to the system of V_{ri}, and a stream which leaves the system with a velocity of V_{ro} relative to the system. The absolute stream velocity in the x direction V_x is related to the relative velocity V_{rx} and the system velocity V_{sx} by

$$V_x = V_{sx} + V_{rx} \tag{5.57}$$

and the linear momentum balance equation becomes

$$\sum \vec{F} = \dot{m}_o\vec{V}_o - \dot{m}_i\vec{V}_i + \frac{d(m\vec{V}_s)}{dt}$$

$$= \dot{m}_o(\vec{V}_{ro} + \vec{V}_s) - \dot{m}_i(\vec{V}_{ri} + \vec{V}_s) + \frac{d(m\vec{V}_s)}{dt} \tag{5.58}$$

Example 5.8 Turbine Blade

Consider a fluid stream impinging on a turbine blade that is moving with a velocity V_s as shown in Figure E5.7. We would like to know what the velocity of the impinging stream should be in order to transfer the maximum amount of energy to the blade.

The system is the fluid in contact with the blade, which is moving at a velocity V_s. The impinging stream velocity is V_i and the stream leaves the blade at velocity V_o. Since $V_o = V_{ro} + V_s$ and $V_i = V_{ri} + V_s$, the system velocity cancels out of the momentum equation:

$$F_x = \dot{m}(V_o - V_i) = \dot{m}(V_{ro} - V_{ri})$$

If the friction loss is negligible, the energy balance (Bernoulli equation) becomes

$$w = \frac{1}{2}\left(V_i^2 - V_o^2\right)$$

which shows that the maximum energy or work transferred from the fluid to the blade occurs when $V_o = 0$ or $V_{ro} = -V_s$. Now from continuity at steady state, recognizing that V_i and V_o are of opposite sign

$$|V_i| = |V_o| \quad \text{or} \quad V_{ri} = -V_{ro}$$

that is,

$$(V_i - V_s) = -(V_o - V_s)$$

Rearranging this for V_s gives

$$V_s = \frac{1}{2}(V_i + V_o)$$

Since the maximum energy is transferred when $V_o = 0$, this reduces to

$$V_s = \frac{1}{2}V_i$$

That is, the maximum efficiency for energy transfer from the fluid to the blade occurs when the velocity of the impinging fluid is twice that of the moving blade (Figure E5.7).

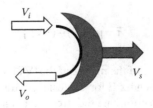

FIGURE E5.7 Turbine blade.

E. MICROSCOPIC MOMENTUM BALANCE

The conservation of momentum principle can be applied to a system composed of the fluid within an arbitrarily small (differential) cubical volume within any flow field. This is done by accounting for convection of mass and momentum through all six surfaces of the cube, all possible stress components acting on each of the six surfaces, and any body forces (e.g., gravity) acting on the mass as a whole. Dividing the result by the volume of the cube and taking the limit as the volume shrinks to zero results in a general microscopic form of the momentum equation that is valid at all points within any fluid. This is done in a manner similar to the earlier derivation of the microscopic mass balance (continuity) equation, Equation 5.7, for each of the three vector components of momentum (see, e.g., Darby, 1976, or many other references). The result can be expressed in general vector notation as

$$\rho\left(\frac{\partial \vec{v}}{\partial t} + \vec{v} \cdot \vec{\nabla}\vec{v}\right) = -\vec{\nabla}P + \vec{\nabla} \cdot \vec{\vec{\tau}} + \rho\vec{g} \tag{5.59}$$

The three components of this momentum equation, expressed in Cartesian, cylindrical, and spherical coordinates, are given in detail in Appendix E. Note that Equation 5.59 is simply a microscopic ("local") expression of the conservation of momentum, for example, Equation 5.40, and it applies locally at all points in flowing continuous media.

Note that there are 11 dependent variables, or "unknowns" in these equations (3 v_i's, 6 τ_{ij}'s, P, and ρ), all of which may depend on space and time. (For an incompressible fluid, ρ is constant so there are only 10 "unknowns.") There are four conservation equations involving these unknowns: the three momentum equations plus the conservation of mass or continuity equation, also given in component form in Appendix E. This means that we still need six more equations (seven, if the fluid is compressible, as the density is also unknown in this case). These additional equations are the "constitutive" equations that relate the local stress components to the rate of deformation of the particular fluid in laminar flow (as determined by the "constitution" or structure of the material), or equations for the local turbulent stress components (the "Reynolds stresses"; see Chapter 6). These equations describe the deformation or flow properties of the specific fluid of interest and relate the six stress components (τ_{ij}) to the deformation rate (i.e., the velocity gradient components). Note there are only six independent components of the shear stress tensor (τ_{ij}) because it is symmetrical, i.e., $\tau_{ij} = \tau_{ji}$, which is a result of the conservation of angular momentum. For a compressible fluid, the density is related to the pressure through an appropriate "constitutive" equation of state. When the equations for the six τ_{ij} components are coupled with the four conservation equations, the result is a set of differential equations for the 10 (or 11) unknowns that can be solved (in principle) with appropriate boundary conditions for the velocity components and pressure as a function of time and space for a given flow configuration. In laminar flows, the constitutive equation gives the stress components as a unique function of the velocity gradient components. For example, the constitutive equation for a Newtonian fluid, generalized from the one-dimensional form (i.e., $\tau = \mu\dot{\gamma}$), is

$$\vec{\vec{\tau}} = \mu[(\vec{\nabla}\vec{v}) + (\vec{\nabla}\vec{v})^t] \tag{5.60}$$

where $(\vec{\nabla}\vec{v})^t$ represents the transpose of the matrix of the $(\vec{\nabla}\vec{v})$ components. The component forms of this equation are also given in Appendix E for Cartesian, cylindrical, and spherical coordinate systems. If these equations are used to eliminate the stress components from the momentum equations, the result is called the *Navier–Stokes equations*, which apply to the laminar flow of any Newtonian fluid in any system and are the starting point for the detailed solution of many fluid flow problems.

Similar equations can be developed for non-Newtonian fluids, based upon the appropriate rheological (constitutive) model for the fluid (similar to Equation 5.60). For turbulent flows, additional equations are required to describe the momentum transported by the fluctuating ("eddy") components of the flow (see Chapter 6). However, the number of flow problems for which closed form analytical solutions are possible is rather limited, so numerical (computational) techniques are required for many problems of practical interest. These procedures are beyond the scope of this book, but we will illustrate the application of the continuity and momentum equations to the solution of an example problem.

Example 5.9 Flow Down an Inclined Plane

Consider the steady laminar flow of a thin layer or film of a liquid down a flat plate that is inclined at an angle θ to the vertical, as illustrated in Figure E5.8. The width of the plate is W (normal to the plane of the figure). For convenience, we take the x coordinate direction parallel to the plate in the flow direction, and the y-axis normal to the x-axis, measured from the fluid surface toward the plate. Flow is only in the x direction (parallel to the plane surface), and the velocity varies only in the y direction (normal to the surface). These prescribed conditions, along with the properties of the fluid, constitute the definition of the problem to be solved. The objective is to determine the film thickness, δ, as a function of the flow rate per unit width of plate (Q/W), the fluid properties (μ, ρ), and other parameters in the problem. Since in this case, $v_y = v_z = 0$ and the density is constant, the microscopic mass balance (continuity equation) from Appendix E reduces to

$$\frac{\partial \rho}{\partial t} + \frac{\partial}{\partial x}(\rho v_x) + \frac{\partial}{\partial y}(\rho v_y) + \frac{\partial}{\partial z}(\rho v_z) = 0$$

$$\text{or } \frac{\partial v_x}{\partial x} = 0$$

This tells us that the velocity v_x must be independent of x. Hence, the only independent variable is y. Considering the x component of the momentum equation (see Appendix E), and discarding all y and z velocity and stress components and all derivatives except those with respect to the y direction, the result is

$$\rho\left(\frac{\partial v_x}{\partial t} + v_x\frac{\partial v_x}{\partial x} + v_y\frac{\partial v_x}{\partial y} + v_z\frac{\partial v_x}{\partial z}\right)$$

$$= -\frac{\partial P}{\partial x} + \left(\frac{\partial \tau_{xx}}{\partial x} + \frac{\partial \tau_{yx}}{\partial y} + \frac{\partial \tau_{zx}}{\partial z}\right) + \rho g_x$$

$$\text{or } 0 = \frac{\partial \tau_{yx}}{\partial y} + \rho g \cos\theta$$

The pressure gradient term has been discarded, because the system is open to the atmosphere and thus the pressure is constant (or, at most, hydrostatic) everywhere (the proof of this is left as an exercise for the reader). This equation can be integrated to give the shear stress distribution in the film:

$$\tau_{yx} = -\rho g y \cos\theta$$

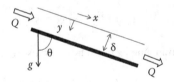

FIGURE E5.8   Flow down an inclined plane.

where the constant of integration is zero, because the shear stress is zero (negligible) at the free surface of the film ($y = 0$). Note that this result so far is valid for any fluid (Newtonian or non-Newtonian) under any flow conditions (laminar or turbulent), because it is simply a statement of the conservation of mass and momentum. If the fluid is Newtonian and the flow is laminar, the shear stress is

$$\tau_{yx} = \mu \frac{dv_x}{dy}$$

Eliminating the stress between the last two equations gives a differential equation for $v_x(y)$ that can be integrated to give the velocity distribution

$$v_x = \frac{\rho g \delta^2 \cos\theta}{2\mu} \left(1 - \frac{y^2}{\delta^2} \right)$$

where the boundary condition that $v_x = 0$ at $y = \delta$ (the wall) has been used to evaluate the constant of integration.

The volumetric flow rate can now be determined from

$$Q = W \int_0^\delta v_x \, dy = \frac{W\rho g \delta^3 \cos\theta}{3\mu}$$

The film thickness (δ) is seen to be proportional to the cube root of the fluid viscosity and the flow rate. The shear stress exerted on the plate is

$$\tau_w = (-\tau_{yx})_{y=\delta} = \left(-\mu \frac{dv_x}{dy} \right)_{y=\delta} = \rho g \delta \cos\theta$$

which is just the component of the weight of the fluid on the plate acting parallel to the plate.

It is also informative to express these results in dimensionless form, that is, in terms of appropriate dimensionless groups. Because this is a "noncircular conduit," the appropriate flow "length" parameter is the hydraulic diameter, defined by Equation 5.48:

$$D_h = 4 \frac{A}{W_p} = \frac{4W\delta}{W} = 4\delta$$

where A is the cross-sectional area of the fluid stream perpendicular to the flow direction and the "wetted perimeter" is simply the width of the plate.

The appropriate form for the Reynolds number is thus

$$N_{Re} = \frac{D_h V \rho}{\mu} = \frac{4\delta V \rho}{\mu} = \frac{4\rho Q/W}{\mu}$$

because $V = Q/A = Q/(W\delta)$. The wall stress can also be expressed in terms of the (Fanning) friction factor (Equation 5.50):

$$\tau_w = f \frac{\rho V^2}{2} = \rho g \delta \cos\theta$$

Substituting $V = Q/W\delta$ and eliminating $\rho g \cos\theta$ from the expression for Q gives

$$f \frac{\rho}{2} \left(\frac{Q}{W\delta} \right)^2 = \left(\frac{Q}{W} \right) \frac{3\mu}{\delta^2}$$

or

$$f = \frac{24}{N_{Re}}$$

that is, $fN_{Re} = 24 = $ constant. This can be compared with the results of the dimensional analysis for the laminar flow of a Newtonian fluid in a pipe (Section V of Chapter 2), for which we deduced that $fN_{Re} = $ constant. In this case, we have determined the value of the constant analytically using first principles rather than by experiment.

The foregoing procedure can be used to solve a variety of steady, fully developed laminar flow problems, such as flow in a tube, in a concentric annulus, or in a slit between parallel walls, for Newtonian or non-Newtonian fluids. However, if the flow is turbulent, the turbulent eddies transport momentum in three dimensions within the flow field. These contribute additional momentum flux components to the shear stress terms in the momentum equation that must be accounted for. The resulting equations cannot be solved exactly for such flows, and empirical and semiempirical methods for treating turbulent flows will be discussed in Chapter 6.

SUMMARY

The key concepts that should be retained from this chapter include:

- The concept of a "system," comprising a defined volume of fluid on which a balance of conserved quantities can be written.
- The difference between a macroscopic and microscopic balance and the application of each.
- The similarities and differences between the balances of mass, energy, and momentum and the general equations for each.
- The similarities and differences between internal energy and enthalpy for ideal gases and nonideal gases and solids and liquids.
- Irreversible effects in the general Bernoulli equation.
- Conservation of energy and momentum in a straight uniform circular pipe at steady state.
- Definitions of the loss coefficient and friction factor.
- Application of the conservation of momentum to multidimensional systems.
- The conservation of angular momentum.

PROBLEMS

CONSERVATION OF MASS AND ENERGY

1. Water is flowing into the top of a tank at a rate of 200 gpm. The tank is 18 in. in diameter and has a 3 in. diameter hole in the bottom, through which the water flows out. If the inflow rate is adjusted to match the outflow rate, what will the height of the water be in the tank if friction is negligible?

2. A vacuum pump operates at a constant volumetric flow rate of 10 L/min, evaluated at the pump inlet conditions. How long will it take to pump down a 100 L tank containing air from 1 to 0.01 atm, assuming that the temperature is constant?

3. Air is flowing at a constant mass flow rate into a tank that has a volume of 3 ft³. The temperature of both the tank and the air is constant at 70°F. If the pressure in the tank is observed to increase at a rate of 5 psi/min, what is the mass flow rate of air into the tank?

4. A tank contains water initially at a depth of 3 ft. The water flows out of a hole in the bottom of the tank, and air at a constant pressure of 10 psig is admitted to the top of the tank. If the water

flow rate is directly proportional to the square root of the gage pressure inside the bottom of the tank, derive expressions for the water mass flow rate and air mass flow rate as a function of time. Be sure to define all symbols you use in your equations.

5. The flow rate of a hot coal/oil slurry in a pipeline is measured by injecting a small side stream of cool oil and measuring the resulting temperature change downstream in the pipeline. The slurry is initially at 300°F and has a density of 1.2 g/cm³ and a specific heat of 0.7 Btu/(lb$_m$ °F). With no side stream injected, the temperature downstream of the mixing point is 298°F. With a side stream at 60°F and a flow rate of 1 lb$_m$/s, the temperature at this point is 295°F. The side stream has a density of 0.8 g/cm³ and a c_p of 0.6 Btu/(lb$_m$ °F). What is the mass flow rate of the slurry?

6. A gas enters a horizontal 3 in. schedule 40 pipe at a constant rate of 0.5 lb$_m$/s, with a temperature of 70°F and a pressure of 1.15 atm. The pipe is wrapped with a 20 kW heating coil, covered with a thick layer of insulation. At the point where the gas is discharged, the pressure is 1.05 atm. What is the gas temperature at the discharge point, assuming it to be ideal with a MW of 29 and a c_p of 0.24 Btu/(lb$_m$ °F)?

7. Water is flowing into the top of an open cylindrical tank (diameter D) at a volume flow rate of Q_i and out of a hole in the bottom at a rate of Q_o. The tank is made of wood that is very porous, and the water is leaking out through the wall uniformly at a rate of q per unit of wetted surface area. The initial depth of water in the tank is z_1. Derive an equation for the depth of water in the tank as a function of time. If $Q_i = 10$ gpm, $Q_o = 5$ gpm, $D = 5$ ft, $q = 0.1$ gpm/ft², and $z_1 = 3$ ft, is the level in the tank rising or falling?

8. Air is flowing steadily through a horizontal tube at a constant temperature of 32°C and a mass flow rate of 1 kg/s. At one point upstream where the tube diameter is 50 mm the pressure is 345 kPa. At another point downstream, the diameter is 75 mm and the pressure is 359 kPa. What is the value of the friction loss (e_f) between these two points? [$c_p = 1005$ J/(kg K)].

9. Steam is flowing through a horizontal nozzle. At the inlet the velocity is 1000 ft/s and the enthalpy is 1320 Btu/lb$_m$. At the outlet the enthalpy is 1200 Btu/lb$_m$. If heat is lost through the nozzle at a rate of 5 Btu/lb$_m$ of steam, what is the outlet velocity?

10. Oil is being pumped from a large storage tank, where the temperature is 70°F, through a 6 in. ID pipeline. The oil level in the tank is 30 ft above the pipe exit. If a 25 hp pump is required to pump the oil at a rate of 600 gpm through the pipeline, what would the temperature of the oil at the exit be, if no heat is transferred across the pipe wall? State any assumptions that you make. Oil properties: SG = 0.92, $\mu = 35$ cp, $c_p = 0.5$ Btu/(lb$_m$ °F).

11. Ethylene enters a 1 in. schedule 80 pipe at 170°F and 100 psia and a velocity of 10 ft/s. At a point somewhere downstream, the temperature has dropped to 140°F and the pressure to 15 psia. Calculate the velocity at the downstream conditions and the Reynolds number at both the upstream and downstream conditions.

12. Number 3 fuel oil (30° API) is transferred from a storage tank at 60°F to a feed tank in a power plant at a rate of 2000 bbl/day. Both tanks are open to the atmosphere and are connected by a pipeline containing 1200 ft equivalent length of 1½ in. sch 40 steel pipe and fittings. The level in the feed tank is 20 ft higher than that in the storage tank, and the transfer pump is 60% efficient. The Fanning friction factor is given by

$$f = \frac{0.0791}{N_{Re}^{1/4}}.$$

(a) What horsepower motor is required to drive the pump?
(b) If the specific heat of the oil is 0.5 Btu/(lb$_m$ °F) and the pump and transfer line are perfectly insulated, what is the temperature of the oil entering the feed tank?

13. Oil with a viscosity of 35 cP, SG of 0.9, and a specific heat of 0.5 Btu/(lb$_m$ °F) is flowing through a straight pipe at a rate of 100 gpm. The pipe is 1 in. sch 40, 100 ft long, and the Fanning friction factor is given by $f = 0.0791/N_{Re}^{1/4}$. If the temperature of the oil entering the pipe is 150°F, determine
    (a) The Reynolds number
    (b) The pressure drop in the pipe, assuming that it is horizontal
    (c) The temperature of the oil at the end of the pipe, assuming the pipe to be perfectly insulated
    (d) The rate at which heat must be removed from the oil (in Btu/h) to maintain it at a constant temperature if there is no insulation on the pipe

14. Water is pumped at a rate of 90 gpm by a centrifugal pump driven by a 10 hp motor. The water enters the pump through a 3 in. sch 40 pipe at 60°F and 10 psig and leaves through a 2 in. sch 40 pipe at 100 psig. If the water gains 0.1 Btu/lb$_m$ while passing through the pump, what is the water temperature leaving the pump?

15. A pump driven by a 7.5 hp motor takes water in at 75°F and 5 psig and discharges it at 60 psig at a flow rate of 600 lb$_m$/min. If no heat is transferred to or from the water while it is in the pump, what will the temperature of the water be leaving the pump?

16. A high-pressure pump takes water in at 70°F, 1 atm, through a 1 in. ID suction line and discharges it at 1000 psig through a 1/8 in. ID line. The pump is driven by a 20 hp motor and is 65% efficient. If the flow rate is 500 g/s and the temperature of the discharge is 73°F, how much heat is transferred between the pump casing and the water per pound of water? Does the heat go into or out of the water?

THE BERNOULLI EQUATION

17. Water is contained in two closed tanks (A and B) which are connected by a pipe. The pressure in tank A is 5 psig and that in tank B is 20 psig, and the water level in tank A is 40 ft above that in tank B. Which direction does the water flow?

18. A pump that is driven by a 7.5 hp motor takes water in at 75°F and 5 psig and discharges it at 60 psig at a flow rate of 600 lb$_m$/min. If no heat is transferred between the water in the pump and the surroundings, what will be the temperature of the water leaving the pump?

19. A 90% efficient pump driven by a 50 hp motor is used to transfer water at 70°F from a cooling pond to a heat exchanger through a 6 in. sch 40 pipeline. The heat exchanger is located 25 ft above the level of the cooling pond, and the water pressure at the discharge end of the pipeline is 40 psig. With all valves in the line wide open, the water flow rate is 650 gpm. What is the rate of energy dissipation (friction loss) in the pipeline in kilowatts (kW)?

20. A pump takes water from the bottom of a large tank where the pressure is 50 psig and delivers it through a hose to a nozzle that is 50 ft above the bottom of the tank, at a rate of 100 lb$_m$/s. The water exits the nozzle into the atmosphere at a velocity of 70 ft/s. If a 10 hp motor is required to drive the pump which is 75% efficient, find
    (a) The friction loss in the pump
    (b) The friction loss in the rest of the system
    Express your answer in units of ft lb$_f$/lb$_m$ and Nm/kg.

21. You have purchased a centrifugal pump to transport water at a maximum rate of 1000 gpm from one reservoir to another through an 8 in. sch 40 pipeline. The total pressure drop through the pipeline is 50 psi. If the pump has an efficiency of 65% at maximum flow conditions and there is no heat transferred across the pipe wall or the pump casing, calculate
    (a) The temperature change of the water through the pump
    (b) The horsepower of the motor that would be required to drive the pump

22. The hydraulic turbines at Boulder dam power plant are rated at 86,000 kW when water is supplied at a rate of 66.3 m³/s. The water enters at a head of 145 m at 20°C and leaves through a 6 m diameter duct.
    (a) Determine the efficiency of the turbines.
    (b) What would be the rating of these turbines if the dam power plant was on Jupiter ($g = 26$ m/s²)?

23. Water is draining from an open conical funnel at the same rate at which it is entering the top. The diameter of the funnel is 1 cm at the top and is 0.5 cm at the bottom, and it is 5 cm high. The friction loss in the funnel per unit mass of fluid is given by $0.4V^2$, where V is the velocity leaving the funnel. What is (a) the volumetric flow rate of the water and (b) the value of the Reynolds number entering and leaving the funnel?

24. Water is being transferred by a pump between two open tanks (from A to B) at a rate of 100 gpm. The pump receives the water from the bottom of tank A through a 3 in. sch 40 pipe and discharges it into the top of tank B through a 2 in. sch 40 pipe. The point of discharge into B is 75 ft higher than the surface of the water in A. The friction loss in the piping system is 8 psi, and both tanks are 50 ft in diameter. What is the head (in ft) which must be delivered by the pump to move the water at the desired rate? If the pump is 70% efficient, what horsepower motor is required to drive the pump?

25. A 4 in. diameter open can has a 1/4 in. diameter hole in the bottom. The can is immersed bottom down in a pool of water to a point where the bottom is 6 in. below the water surface and is held there while the water flows through the hole into the can. How long will it take for the water in the can to rise to the same level as that outside the can? Neglect friction, and assume a "pseudo steady state," that is, time changes are so slow that at any instant the steady-state Bernoulli equation applies.

26. Carbon tetrachloride (SG = 1.6) is pumped at a rate of 2 gpm through a pipe that is inclined upward at an angle of 30°. An inclined tube manometer (with a 10° angle of inclination) using mercury as the manometer fluid (SG = 13.6) is connected between two taps on the pipe that are 2 ft apart. The manometer reading is 6 in. If no heat is lost through the tube wall, what is the temperature rise of the CCl_4 over a 100 ft length of the tube?

27. A pump that is taking water at 50°F from an open tank at a rate of 500 gpm is located directly over the tank. The suction line entering the pump is a nominal 6 in. sch 40 straight pipe 10 ft long and extends 6 ft below the surface of the water in the tank. If friction in the suction line is neglected, what is the pressure at the pump inlet (in psi)?

28. A pump is transferring water from tank A to tank B, both of which are open to the atmosphere, at a rate of 200 gpm. The surface of the water in tank A is 10 ft above ground level, and that in tank B is 45 ft above ground level. The pump is located at ground level, and the discharge line that enters tank B is 50 ft above ground level at its highest point. All piping is 2 in. ID, and the tanks are 20 ft in diameter. If friction is neglected, what would be the required pump head rating for this application (in ft), and what size motor (horsepower) would be needed to drive the pump if it is 60% efficient? (Assume the temperature is constant at 77°F.)

29. A surface effect (air cushion) vehicle measures 10 ft × 20 ft and weighs 6000 lb$_f$. The air is supplied by a blower mounted on top of the vehicle, which must supply sufficient power to lift the vehicle 1 in. off the ground. Calculate the required blower capacity in standard cubic feet per minute (scfm), and the horsepower of the motor required to drive the blower if it is 80% efficient. Neglect friction, and assume that the air is an ideal gas at 80°F with properties evaluated at an average pressure.

30. The air cushion car in Problem 29 is equipped with a 2 hp blower that is 70% efficient.
    (a) What would be the clearance between the skirt of the car and the ground?
    (b) What is the air flow rate in scfm?

31. An ejector pump operates by injecting a high-speed fluid stream into a slower stream to increase its pressure. Consider water flowing at a rate of 50 gpm through a 90° elbow in a 2 in. ID pipe. A stream of water is injected at a rate of 10 gpm through a 1/2 in. ID pipe through the center of the elbow in a direction parallel to the downstream flow in the larger pipe. If both streams are at 70°F, determine the increase in pressure in the larger pipe at the point where the two streams mix.

32. A large tank containing water has a 51 mm diameter hole in the bottom. When the depth of the water is 15 m above the hole, the flow rate through the hole is found to be 0.0324 m³/s. What is the head loss due to friction in the hole?

33. Water at 68°F is pumped through a 1000 ft length of 6 in. sch 40 pipe. The discharge end of the pipe is 100 ft above the suction end. The pump is 90% efficient, and it is driven by a 25 hp motor. If the friction loss in the pipe is 70 ft lb$_f$/lb$_m$, what is the flow rate through the pipe in gpm? ($P_{in} = P_{out} = 1$ atm.)

34. You want to siphon water out of a large tank using a 5/8 in. ID hose. The highest point of the hose is 10 ft above the water surface in the tank, and the hose exit outside the tank is 5 ft below the inside surface level. If friction is neglected, (a) what would be the flow rate through the hose (in gpm), and (b) what is the minimum pressure in the hose (in psi)?

35. It is desired to siphon a volatile liquid out of a deep open tank. If the liquid has a vapor pressure of 200 mmHg and a density of 45 lb$_m$/ft³ and the surface of the liquid is 30 ft below the top of the tank, is it possible to siphon the liquid? If so, what would the velocity be through a frictionless siphon, 1/2 in. in diameter, if the exit of the siphon tube is 3 ft below the level in the tank?

36. The propeller of a speedboat is 1 ft in diameter and 1 ft below the surface of the water. At what speed (rpm) will cavitation occur? The vapor pressure of the water is 18.65 mmHg at 70°F.

37. A conical funnel is full of liquid. The diameter of the top (mouth) is D_1, and that of the bottom (spout) is D_2 (where $D_2 \ll D_1$), and the depth of the fluid above the bottom is H_o. Derive an expression for the time required for the fluid to drain by gravity to a level of $H_o/2$, assuming frictionless flow.

38. An open cylindrical tank of diameter D contains a liquid of density ρ at a depth H. The liquid drains through a hole of diameter d in the bottom of the tank. The velocity of the liquid through the hole is $C\sqrt{h}$, where h is the depth of the liquid at any time t. Derive an equation for the time required for 90% of the liquid to drain out of the tank.

39. An open cylindrical tank, that is 2 ft in diameter and 4 ft high is full of water. If the tank has a 2 in. diameter hole in the bottom, how long will it take for half of the water to drain out, if friction is neglected?

40. A large tank has a 5.1 mm diameter hole in the bottom. When the depth of liquid in the tank is 1.5 m above the hole, the flow rate through the hole is found to be 324 cm³/s. What is the head loss due to friction in the hole (in ft)?

41. A window is left slightly open while the air conditioning system is running. The air conditioning blower develops a pressure of 2 in. H$_2$O (gage) inside the house, and the window opening measures 1/8 in. × 20 in. Neglecting friction, what is the flow rate of air through the opening in scfm (ft³/min at 60°F, 1 atm)? How much horsepower is required to move this air?

42. Water at 68°F is pumped through a 1000 ft length of 6 in. sch 40 pipe. The discharge end of the pipe is 100 ft above the suction end. The pump is 90% efficient and is driven by a 25 hp motor. If the friction loss in the pipe is 70 ft lb$_f$/lb$_m$, what is the flow rate through the pipe (in gpm)?

43. The plumbing in your house is 3/4 in. sch 40 galvanized pipe, and it is connected to an 8 in. sch 80 water main in which the pressure is 15 psig. When you turn on a faucet in your bathroom (which is 12 ft higher than the water main), the water flows out at a rate of 20 gpm.
    (a) How much energy is lost due to friction in the plumbing?
    (b) If the water temperature in the water main is 60°F, and the pipes are well insulated, what would the temperature of the water be leaving the faucet?
    (c) If there were no friction loss in the plumbing, what would the flow rate be (in gpm)?

44. A 60% efficient pump driven by a 10 hp motor is used to transfer bunker C fuel oil from a storage tank to a boiler through a well-insulated line. The pressure in the tank is 1 atm, and the temperature is 100°F. The pressure at the burner in the boiler is 100 psig, and it is 100 ft above the level in the tank. If the temperature of the oil entering the burner is 102°F, what is the oil flow rate, in gpm? (Oil properties: SG = 0.8, $c_p = 0.5$ Btu/(lb$_m$ °F))

FLUID FORCES, MOMENTUM TRANSFER

45. You have probably noticed that when you turn on the garden hose, it will whip about uncontrollably if it is not restrained. This is because of the unbalanced forces developed by the change of momentum in the tube. If a 1/2 in. ID hose carries water at a rate of 50 gpm, and the open end of the hose is bent at an angle of 30° to the rest of the hose, calculate the components of the force (magnitude and direction) exerted by the water on the bend in the hose. Assume that the loss coefficient in the hose is 0.25.

46. Repeat Problem 45 for the case in which a nozzle is attached to the end of the hose and the water exits the nozzle through a 1/4 in. opening. The loss coefficient for the nozzle is 0.3 based on the velocity through the nozzle.

47. You are watering your garden with a hose that has a 3/4 in. ID, and the water is flowing at a rate of 10 gpm. A nozzle attached to the end of the hose has an ID of 1/4 in. The loss coefficient for the nozzle is 20 based on the velocity in the hose. Determine the force (magnitude and direction) that you must apply to the nozzle in order to deflect the free end of the hose (nozzle) by an angle of 30° relative to the straight hose.

48. A 4 in. ID fire hose discharges water at a rate of 1500 gpm through a nozzle that has a 2 in. ID exit. The nozzle is conical and converges through a total included angle of 30°. What is the total force transmitted to the bolts in the flange where the nozzle is attached to the hose? Assume the loss coefficient in the nozzle is 3.0 based on the velocity in the hose.

49. A 90° horizontal reducing bend has an inlet diameter of 4 in. and an outlet diameter of 2 in. If water enters the bend at a pressure of 40 psig and a flow rate of 500 gpm, calculate the force (net magnitude and direction) exerted on the supports that hold the bend in place. The loss coefficient for the bend may be assumed to be 0.75 based on the highest velocity in the bend.

50. A fireman is holding the nozzle of a fire hose that he is using to put out a fire. The hose is 3 in. in diameter, and the nozzle is 1 in. in diameter. The water flow rate is 200 gpm, and the loss coefficient for the nozzle is 0.25 (based on the exit velocity). How much force must the fireman use to restrain the nozzle? Must he push or pull on the nozzle to apply the force? What is the pressure at the end of the hose where the water enters the nozzle?

51. Water flows through a 30° pipe bend at a rate of 200 gpm. The diameter of the entrance to the bend is 2.5 in. and that of the exit is 3 in. The pressure in the pipe is 30 psig, and the pressure drop in the bend is negligible. What is the total force (magnitude and direction) exerted by the fluid on the pipe bend?

52. A nozzle with a 1 in. ID outlet is attached to a 3 in. ID fire hose. Water pressure inside the hose is 100 psig and the flow rate is 100 gpm. Calculate the force (magnitude and direction) required to hold the nozzle at an angle of 45° relative to the axis of the hose. (Neglect friction in the nozzle).

53. Water flows through a 45° expansion pipe bend at a rate of 200 gpm, exiting into the atmosphere. The inlet to the bend is 2 in. ID, the exit is 3 in. ID, and the loss coefficient for the bend is 0.3 based on the inlet velocity. Calculate the force (magnitude and direction) exerted by the fluid on the bend relative to the direction of the entering stream.

54. A patrol boat is powered by a water jet engine, which takes water in at the bow through a 1 ft diameter duct and pumps it out the stern through a 3 in. diameter exhaust jet. If the water is pumped at a rate of 5000 gpm, determine
    (a) The thrust rating of the engine
    (b) The maximum speed of the boat, if the drag coefficient is 0.5 based on an underwater area of 600 ft^2
    (c) The horsepower required to operate the motor (neglecting friction in the motor, pump, and ducts)

55. A patrol boat is powered by a water jet pump engine. The engine takes water in through a 3 ft diameter duct in the bow and discharges it through a 1 ft diameter duct in the stern. The drag coefficient of the boat has a value of 0.1 based on a total underwater area of 1500 ft^2. Calculate the pump capacity in gpm and the engine horsepower required to achieve a speed of 35 mph, neglecting friction in the pump and ducts.

56. Water is flowing through a 45° pipe bend at a rate of 200 gpm and exits into the atmosphere. The inlet to the bend is 1½ in. inside diameter, and the exit is 1 in. in diameter. The friction loss in the bend can be characterized by a loss coefficient of 0.3 (based on the inlet velocity). Calculate the net force (magnitude and direction) transmitted to the flange holding the pipe section in place.

57. The arms of a lawn sprinkler are 8 in. long and 3/8 in. ID. Nozzles at the end of each arm direct the water in a direction that is 45° from the arms. If the total flow rate is 10 gpm, determine:
    (a) The moment developed by the sprinkler if it is held stationary and not allowed to rotate.
    (b) The angular velocity (in rpm) of the sprinkler if there is no friction in the bearings.
    (c) The trajectory of the water from the end of the rotating sprinkler (i.e., the radial and angular velocity components).

58. A water sprinkler contains two 1/4 in. ID jets at the ends of a rotating hollow (3/8 in. ID) tube, which direct the water 90° to the axis of the tube. If the water leaves at 20 ft/s, what torque would be necessary to hold the sprinkler in place?

59. An open container 8 in. high with an inside diameter of 4 in. weighs 5 lb$_f$ when empty. The container is placed on a scale and water flows into the top of the container through a 1 in. diameter tube at a rate of 40 gpm. The water flows horizontally out into the atmosphere through two 1/2 in. holes on opposite sides of the container. Under steady conditions, the height of the water in the tank is 7 in.

    (a) Determine the reading on the scale.
    (b) Determine how far the holes in the sides of the container should be from the bottom so that the level in the container will be constant.

60. A boat is tied to a dock by a line from the stern of the boat to the dock. A pump inside the boat takes water in through the bow and discharges it out the stern at a rate of 3 ft^3/s through a pipe running through the hull. The pipe inside area is 0.25 ft^2 at the bow and 0.15 ft^2 at the stern. Calculate the tension on the line, assuming inlet and outlet pressures are equal.

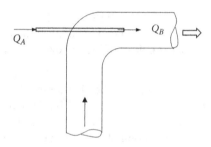

FIGURE P5.61   Ejector pump.

61. A jet ejector pump is shown in Figure P5.61. A high-speed stream (Q_A) is injected at a rate of 50 gpm through a small tube 1 in. in diameter, into a stream (Q_B) in a larger, 3 in. diameter, tube. The energy and momentum are transferred from the small stream to the larger stream, which increases the pressure in the pump. The fluids come in contact at the end of the small tube and become perfectly mixed a short distance downstream (the flow is turbulent). The energy dissipated in the system is significant, but the wall force between the end of the small tube and the point where mixing is complete can be neglected. If both streams are water at 60°F, and $Q_B = 100$ gpm, calculate the pressure rise in the pump.

62. Figure P5.62 illustrates two relief valves. The valve disk is designed to lift when the upstream pressure in the vessel (P_1) reaches the valve set pressure. Valve A has a disk that diverts the fluid leaving the valve by 90° (i.e., to the horizontal direction), whereas the disk in valve B diverts the fluid to a direction that is 60° downward from the horizontal. The diameter of the valve nozzle is 3 in., and the clearance between the end of the nozzle and the disk is 1 in., for both valves. If the fluid is water at 200°F, $P_1 = 100$ psig, and the discharge pressure is atmospheric, determine the force exerted on the disk for both cases A and B. The loss coefficient for the valve in both cases is 2.4 based on the velocity in the nozzle.

63. A relief valve is mounted on top of a large vessel containing hot water. The inlet diameter to the valve is 4 in., and the outlet diameter is 6 in. The valve is set to open when the pressure in the vessel reaches 100 psig, which happens when the water is at 200°F. The liquid flows through the open valve and exits to the atmosphere on the side of the valve, 90° from the entering direction. The loss coefficient for the valve has a value of 5, based on the exit velocity from the valve.
    (a) Determine the net force (magnitude and direction) acting on the valve.
    (b) You want to attach a cable to the valve to brace it such that the tensile force in the cable balances the net force on the valve. Show exactly where you would attach the cable at both ends.

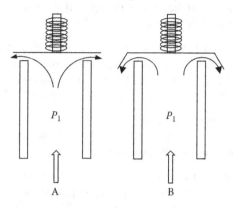

FIGURE P5.62   Relief valve disk geometry.

64. A relief valve is installed on the bottom of a pressure vessel. The entrance to the valve is 4.5 in. diameter and the exit (which discharges in the horizontal direction, 90° from the entrance) is 5 in. diameter. The loss coefficient for the valve is 4.5 based on the inlet velocity. The fluid in the tank is a liquid, with a density of 0.8 g/cm³. If the valve opens when the pressure at the valve reaches 150 psig, determine
    (a) The mass flow rate through the valve, in lb_m/s.
    (b) The net force (magnitude and direction) exerted on the valve.
    (c) Determine the location (orientation) of a cable that is to be attached to the valve to balance the net force (note that a cable can support only a tensile force).

65. An emergency relief valve is installed on a reactor to relieve excess pressure in case of a runaway reaction. The lines upstream and downstream of the valve are 6 in. sch 40 pipe. The valve is designed to open when the tank pressure reaches 100 psig, and the vent exhausts to the atmosphere at 90° to the direction entering the valve. The fluid can be assumed to be incompressible, with an SG of 0.95, a viscosity of 3.5 cP, and a specific heat of 0.5 Btu/lb_m °F. If the sum of the loss coefficients for the valve and the vent line is 6.5, determine
    (a) The mass flow rate of the fluid through the valve in lb_m/s and the value of the Reynolds number in the pipe when the valve opens.
    (b) The rise in temperature of the fluid from the tank to the vent exit, if the heat transferred through the walls of the system is negligible.
    (c) The force exerted on the valve supports by the fluid flowing through the system. If you could install only one support cable to balance this force, show where you would put it.

66. Consider the "tank on wheels" shown in Figure P5.66. Water is draining out of a hole in the side of the open tank, at a rate of 10 gpm. If the tank diameter is 2 ft and the diameter of the hole is 2 in., determine the magnitude and direction of the force transmitted from the water to the tank.

67. The tank in Problem 66 is 6 in. in diameter and contains water at a depth of 3 ft. On the side of the tank near the bottom is a 1.5 in. ID outlet to which is attached a ball valve, which has a loss coefficient of 1.2. When the valve is opened, the water flows out in a horizontal stream. Calculate
    (a) The flow rate of the water (in gpm).
    (b) The thrust exerted on the tank by the escaping water, and the direction that the tank will move. If the diameter of the outlet and valve are increased, will the thrust on the tank increase or decrease? Why?

LAMINAR FLOW

68. Use the microscopic equations of motion in Appendix E as a starting point to derive a relationship between the volumetric flow rate and the pressure gradient for a Newtonian fluid in a pipe that is valid for any orientation of the pipe axis. (*Hint*: the critical starting point requires that you identify which velocity and velocity gradients are nonzero, and hence the corresponding nonzero stress components, for this problem. This allows you to tailor the differential

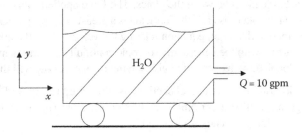

FIGURE P5.66   Water jet force on tank.

equations to suit the problem, and the resulting equations can be integrated, with appropriate boundary conditions, to get the answer.)

69. A viscous molten polymer is pumped through a thin slit between two flat surfaces. The slit has a depth H, width W, and length L, and is inclined upward at an angle θ to the horizontal ($H \ll W$). The flow is laminar, and the polymer is non-Newtonian, with properties that can be represented by the power law model.
    (a) Derive an equation relating the volume flow rate of the polymer (Q) to the applied pressure difference along the slit, the slit dimensions, and the fluid properties.
    (b) Using the definition of the Fanning friction factor (f), solve your equation for f in terms of the remaining quantities. The corresponding solution for a Newtonian fluid can be written as $f = 24/N_{Re}$. Use your solution to obtain an equivalent expression for the power law Reynolds number (i.e., $N_{Re, pl} = 24/f$). Use the hydraulic diameter as the length scale in the Reynolds number.
    (*Note*: It is easiest to take the origin of your coordinates at the center of the slit, then calculate the flow rate for one-half the slit and double this to get the answer. Why is this the easiest way?)

70. Acrylic latex paint can be described as a Bingham plastic with a yield stress of 200 dyn/cm^2, a limiting viscosity of 50 cP, and a density of 0.95 g/cm^3.
    (a) What is the maximum thickness at which a film of this paint could be spread on a vertical wall without running?
    (b) If the power law model were used to describe this paint, such that the apparent viscosity predicted by both the power law and Bingham plastic models is the same at shear rates of 1 and 100 s^{-1}, what would the flow rate of the film be if it has the thickness predicted in (a)?

71. A vertical belt is moving upward continuously through a liquid bath at a velocity V. A film of the liquid adheres to the belt, which tends to drain downward due to gravity. The equilibrium thickness of the film is determined by the steady-state condition at which the downward drainage velocity of the surface of the film is exactly equal to the upward velocity of the belt. Derive an equation for the film thickness if the fluid is
    (a) Newtonian
    (b) A Bingham plastic

72. Water at 70°F flows upward through a vertical tube and overflows over the top and down the outside wall. The OD of the tube is 4 in., and the water flow rate is 1 gpm. Determine the thickness of the film. Is the flow laminar or turbulent?

73. For laminar flow of a Newtonian fluid in a tube
    (a) Show that the average velocity over the cross section is half of the maximum velocity in the tube.
    (b) Derive the kinetic energy correction factor for laminar flow of a Newtonian fluid in a tube (i.e., $\alpha = 2$).

74. A slider bearing can be described as one plate moving with a velocity V parallel to a stationary plate, with a viscous lubricant in between the plates. The force applied to the moving plate is F, and the distance between the plates is H. If the lubricant is a grease with properties that can be described by the power law model, derive an equation relating the velocity V to the applied force F and the gap clearance H, starting with the general microscopic continuity and momentum equations. If the area of the plate is doubled, with everything else staying the same, how will the velocity V change?

75. Consider a fluid flowing in a conical section, as illustrated in Figure P5.75. The mass flow rate is the same going in (through point 1) as it is coming out (point 2), but the velocity changes because the area changes. They are related by

$$(\rho V A)_1 = (\rho V A)_2$$

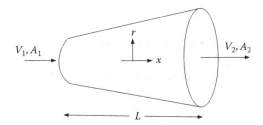

FIGURE P5.75   Flow in a cone.

where ρ is the fluid density (assumed to be constant here). Because the velocity changes, the transport of momentum will be different going in than going out, which results in a net force in the fluid. (a) Derive an expression for the magnitude of this force associated with the change in momentum. (b) Which direction will the force transmitted from the fluid to the cone act?

NOTATION

A	Cross-sectional area, $[L^2]$	
A_w	Area of wall, $[L^2]$	
c_v	Specific heat at constant volume, $[FL/MT = L^2/Mt^2]$	
c_p	Specific heat at constant pressure, $[FL/MT = L^2/Mt^2]$	
d	Diameter, $[L]$	
D	Diameter, $[L]$	
D_h	Hydraulic diameter, $[L]$	
f	Fanning friction factor, $[-]$	
e_f	Energy dissipated per unit mass of fluid, $[FL/M = L^2/t^2]$	
F_x	Force component in x direction, $[F = ML/t^2]$	
g	Acceleration due to gravity, $[L/t^2]$	
h	Enthalpy per unit mass, $[FL/M = L^2/t^2]$	
H_f	Friction (loss) head, $[L]$	
H_p	Pressure head, $[L]$	
H_v	Velocity head, $[L]$	
H_w	Work (pump) head, $[L]$	
H_z	Static head, $[L]$	
I	Moment of inertia, $[FLt^2 = ML^2]$	
K_f	Loss coefficient, $[-]$	
L_θ	Angular momentum in the θ direction, $[ML^2/t]$	
M	Molecular weight, $[M/mol]$	
m	Mass, $[M]$	
$\dot{m}$	Mass flow rate, $[M/t]$	
P	Pressure, $[F/L^2 = M/Lt^2]$	
Q	Volumetric flow rate, $[L^3/t]$	
$\dot{Q}$	Rate of heat transfer into the system, $[FL/t = ML^2/t^3]$	
q	Heat transferred into the system per unit mass of fluid, $[FL/M = L^2/t^2]$	
R	Gas constant, $[FL/(mol\ T) = ML^2\ (mol\ t^2T)]$	
r	Radial coordinate $[L]$	
s	Entropy per unit mass, $[FL/M = L^2/t^2]$	
T	Temperature, $[T]$	
t	Time, $[t]$	
u	Internal energy per unit mass, $[FL/M = L^2/t^2]$	
V	Spatial average velocity, $[L/t]$	

v   Local velocity, [L/t]
W Width of plate, [L]
$\dot{W}$ Rate of work done by fluid system, $[FL/t = ML^2/t^3]$
W_p Wetted perimeter, [L]
w   Work done by fluid system per unit mass of fluid, $[FL/M = L^2/t^2]$

GREEK

α Kinetic energy correction factor, [—]
β d/D, where $d < D$, [—]
Γ Moment or torque, $[FL = ML^2/t^2]$
$\Delta()$ $()_2 - ()_1$
$\vec{\nabla}$ Gradient vector operator, [1/L]
δ Film thickness, [L]
μ Viscosity (constant), [M/Lt]
v   Specific volume, $[L^3/M]$
Φ Potential $= P + \rho gz$, $[F/L^2 = M/Lt^2]$
ρ Density, $[M/L^3]$
τ_{yx} Shear stress component, force in x direction on y area component, $[F/L^2 = M/Lt^2]$
$\ddot{\tau}$ Shear stress tensor, $[F/L^2 = M/(Lt^2)]$
τ_w Stress exerted on wall by fluid, $[F/L^2 = M/Lt^2]$
ω Angular velocity, [1/t]

SUBSCRIPTS

1   Reference point 1
2   Reference point 2
i   Input
o   Output
s   System
x, y, z   Coordinate directions, [L]

REFERENCE

Darby, R., *Viscoelastic Fluids*, Marcel Dekker, New York, 1976.

6 Pipe Flow

> "*Simple laws can very well describe complex structures. The miracle is not the complexity of our world, but the simplicity of the equations describing that complexity.*"
>
> **—Sander Bais, b. 1945, Theoretical Physicist**

I. FLOW REGIMES

In 1883, Osborne Reynolds conducted a classical experiment, illustrated in Figure 6.1, in which he measured the pressure drop, ΔP, as a function of flow rate, Q, for water in a tube. He found that, at low flow rates, the pressure drop was directly proportional to the flow rate, but as the flow rate was increased, a point was reached where the relation was no longer linear and the "noise" or scatter in the data increased considerably. At still higher flow rates, the data became more reproducible, but the relationship between the pressure drop and the flow rate became almost quadratic instead of linear.

To investigate this phenomenon further, Reynolds introduced a trace of dye into the flow to observe what was happening. At the low flow rates where the linear relationship (ΔP vs. Q) was observed, the dye was seen to remain a coherent, rather smooth thread throughout most of the tube except for a little dispersion due to molecular diffusion. However, where the data scatter occurred, the dye trace was seen to be rather unstable, and it broke up after a short distance. At still higher flow rates, where the quadratic relationship was observed, the dye dispersed almost immediately into a uniform "cloud" throughout the tube. The stable flow observed initially was called *laminar* because it was observed that the fluid elements moved in smooth layers or "lamella" relative to each other with no mixing. The unstable flow pattern, characterized by a high degree of mixing between the fluid elements, was called *turbulent*. Although the transition from laminar to turbulent flow occurs rather abruptly, there is nevertheless a *transition region* where the flow is unstable and not very reproducible.

Careful study of various fluids in tubes of different sizes has indicated that laminar flow in a tube persists up to a point where the value of the Reynolds number ($N_{Re} = DV\rho/\mu$) is about 2000, and turbulent flow occurs when N_{Re} is greater than about 4000, with a transition region in between. Actually, unstable flow (turbulence) occurs when disturbances to the flow are amplified, whereas laminar flow occurs when these disturbances are damped out by the viscous forces. Because turbulent flow cannot occur unless there are disturbances, studies have been conducted on systems in which extreme care has been taken to eliminate any disturbances due to irregularities in the boundary surfaces, sudden changes in direction, vibrations, etc. Under these conditions, it has been possible to sustain laminar flow in a tube to a Reynolds number of the order of 100,000 or more. However, under all but the most unusual conditions, there are sufficient natural disturbances in all practical systems that the laminar flow ceases and turbulence begins in a pipe at a Reynolds number of about 2000.

The physical significance of the Reynolds number can be appreciated better if it is rearranged as follows:

$$N_{Re} = \frac{DV\rho}{\mu} = \frac{\rho V^2}{\mu V/D} \tag{6.1}$$

The numerator (ρV^2) is the flux of "inertial" momentum, carried by the fluid along the tube in the axial direction. The denominator ($\mu V/D$) is proportional to the viscous shear stress in the tube, which is equivalent to the "viscous" flux of momentum normal to the flow direction, that is, in the radial direction. Thus, the Reynolds number is the ratio of the inertial momentum flux in the flow direction

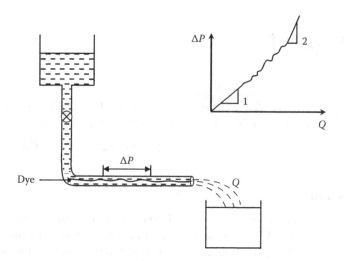

FIGURE 6.1 Reynolds' experiment.

to the viscous momentum flux in the transverse (radial) direction. Because viscous forces are a manifestation of intermolecular attractive forces, they are stabilizing, whereas inertial forces tend to pull the fluid elements apart and are therefore destabilizing. It is thus quite logical that stable (laminar) flow should occur at low Reynolds numbers where viscous forces dominate, whereas unstable (turbulent) flow occurs at high Reynolds numbers where inertial forces dominate. Also, laminar flows are dominated by viscosity and are independent of the fluid density, whereas fully turbulent flows are dominated by the fluid density and are independent of the fluid viscosity at high turbulence levels. For fluids flowing near solid boundaries (e.g., inside conduits or around solid bodies), viscous forces dominate in the immediate vicinity of the boundary, whereas for turbulent flows (high Reynolds numbers) inertial forces dominate in the region far from the boundary. We will consider both the laminar and turbulent flow of Newtonian and non-Newtonian fluids in pipes in this chapter.

II. GENERAL RELATIONS FOR PIPE FLOWS

For steady, uniform, fully developed flow in a pipe (or any conduit), the conservation of mass, energy, and momentum equations can be arranged in specific forms that are most useful for the analysis of such problems. There are several different approaches to this analysis, as described in the following. Each of these is a general expression that is valid for both Newtonian and non-Newtonian fluids in either laminar or turbulent flow.

A. ENERGY BALANCE

Consider a section of uniform cylindrical pipe of length L and radius R, inclined upward at an angle θ to the horizontal, as shown in Figure 6.2. The steady-state energy balance (or Bernoulli equation) applied to an incompressible fluid flowing in a uniform pipe can be written as

$$\frac{-\Delta\Phi}{\rho} = e_f = K_f \frac{V^2}{2} \tag{6.2}$$

where
  $\Phi = P + \rho g z$
  $K_f = 4fL/D$
  f is the Fanning friction factor

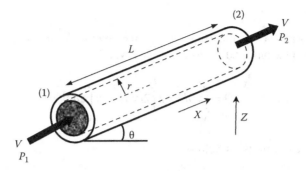

FIGURE 6.2 Steady flow in a uniform pipe.

B. MOMENTUM BALANCE

Another approach is to write a momentum balance on a cylindrical volume of fluid of radius r and length L, centered on the pipe centerline (see Figure 6.2) as follows:

$$\sum F_x = (P_1 - P_2)\pi r^2 - \pi r^2 L\rho g \sin\theta + 2\pi rL\tau_{rx} = 0 \tag{6.3}$$

where τ_{rx} is the tangential shear stress in the x direction acting on the r surface of the fluid system. Solving Equation 6.3 for τ_{rx} gives

$$\tau_{rx} = \frac{\Delta\Phi r}{2L} = -\tau_w \frac{r}{R} \tag{6.4}$$

where
$\Delta\Phi = \Delta P + \rho g L \sin\theta = \Delta P + \rho g\Delta z$
τ_w is the stress exerted *by* the fluid *on* the tube wall (i.e., $\tau_w = (-\tau_{rx})_{r=R}$)

Note that this can also be obtained directly by integrating the axial component of the microscopic momentum equation of motion in cylindrical coordinates (i.e., the z-component equation in Appendix E).

Equation 6.4 is equivalent to Equation 6.2, because

$$f = \frac{\tau_w}{\rho V^2/2} = \frac{K_f}{4L/D} = \frac{e_f}{(4L/D)(V^2/2)} \tag{6.5}$$

Note that from Equation 6.4 the shear stress is negative, that is, the fluid outside the cylindrical system of radius r is moving slower than the fluid inside the system and hence exerts a force in the $-x$ direction on the fluid in the system, which is bounded by the r surface. However, the stress at the wall (τ_w) is defined as the force exerted in the $+x$ direction *by* the fluid *on the wall* (which is positive). The fact that the energy balance equation (Bernoulli) and the momentum balance both give identical results for this particular problem is not a general result. In general, the momentum balance is a vector equation, but when applied to this 1-D problem the result is identical to the (scalar) Bernoulli equation. The momentum balance is most useful in determining forces on fluid systems involving multidirectional flows, such as flows through pipe bends and elbows. This is illustrated later on.

C. CONTINUITY

Continuity provides another relationship between the volumetric flow rate (Q) passing through a given cross section in the pipe and the local velocity (v_x), that is,

$$Q = \int_A v_x \, dA = \pi \int_o^R 2rv_x \, dr = \pi \int_A v_x dr^2 \tag{6.6}$$

This can be integrated by parts, as follows:

$$Q = \pi \int_A v_x \, dr^2 = -\pi \int_A r^2 \, dv_x = -\pi \int_0^R r^2 \frac{dv_x}{dr} \, dr \tag{6.7}$$

Thus, if the radial dependence of the velocity or shear rate (dv_x/dr) is known or can be found, the flow rate can be determined directly from Equation 6.7. Application of this principle to laminar flow is shown in the following section.

D. ENERGY DISSIPATION

A different, but related, approach to pipe flow that provides additional insight involves consideration of the rate at which energy is dissipated per unit volume of fluid. In general, the rate of energy (or power) expended in a system subjected to a force $\vec{F}$ and moving at a velocity $\vec{V}$ is simply $\vec{F} \cdot \vec{V}$. With reference to the "simple shear" deformation shown in Figure 3.1, the corresponding rate of energy dissipation per unit volume of fluid is $\vec{F} \cdot \vec{V}/(Ah) = \tau_{yx} dv_x/dy$. This can be generalized for any system as follows:

$$\dot{e}_f = e_f \dot{m} = e_f \rho Q = \int_{\text{Vol}} \bar{\bar{\tau}} : \nabla \tilde{v} \, d\tilde{V} \tag{6.8}$$

where
  $\bar{\bar{\tau}}$ is the shear stress tensor (defined in Equation 5.60)
  $\tilde{V}$ is the volume of the fluid in the pipe (a scalar)

The ":" operator in Equation (6.8) represents the scalar product of two dyads. Thus, integration of the local rate of energy dissipation throughout the entire flow volume, along with the Bernoulli equation, which relates the energy dissipated per unit mass (e_f) to the driving force ($\Delta\Phi$), can be used to determine the flow rate. All of the equations to this point are general because they apply to *any fluid (Newtonian or non-Newtonian) in any type of flow (laminar or turbulent) in steady, fully developed flow in a uniform cylindrical tube at any orientation*. The following section will illustrate the application of these relations to laminar flow in a pipe.

III. NEWTONIAN FLUIDS

A. LAMINAR FLOW

For a Newtonian fluid in laminar flow

$$\tau_{rx} = \mu \frac{dv_x}{dr} \quad \text{or} \quad \frac{dv_x}{dr} = \frac{\tau_{rx}}{\mu} \tag{6.9}$$

When the velocity gradient from Equation 6.9 is substituted into Equation 6.7 and Equation 6.4 is used to eliminate the shear stress, Equation 6.7 becomes

$$Q = -\pi \int_0^R r^2 \frac{dv_x}{dr} dr = -\frac{\pi}{\mu} \int_0^R r^2 \tau_{rx} \, dr = \frac{\pi \tau_w}{\mu R} \int_0^R r^3 \, dr \tag{6.10}$$

or

$$Q = \frac{\pi \tau_w R^3}{4\mu} = -\frac{\pi \nabla\Phi R^4}{8\mu L} = -\frac{\pi \nabla\Phi D^4}{128\mu L} \tag{6.11}$$

Equation 6.11 is known as the *Hagen–Poiseuille equation* for laminar flow of a Newtonian fluid in a tube.

This result can also be derived by equating the shear stress for a Newtonian fluid (Equation 6.9) to the expression obtained from the momentum balance for tube flow (Equation 6.4) and integrating (using the no-slip condition at the tube wall) to obtain the velocity profile:

$$v(r) = \frac{\tau_w R}{2\mu} \left(1 - \frac{r^2}{R^2} \right) \tag{6.12}$$

Inserting this into Equation 6.6 and integrating over the tube cross section gives Equation 6.11 for the volumetric flow rate.

Another approach is to use the Bernoulli equation (Equation 6.2) and Equation 6.8 for the friction loss term e_f. The integral in the latter equation is evaluated in a manner similar to that leading to Equation 6.10, as follows. Eliminating e_f between Equation 6.8 and the Bernoulli equation (Equation 6.2, i.e., $\rho e_f = -\nabla\Phi$) leads directly to

$$\dot{e}_f = \rho Q e_f = -\nabla\Phi Q = \int_{vol} \tau : \nabla \tilde{v} \, d\tilde{V} = L \int_0^R \tau \frac{dv}{dr} 2\pi r \, dr$$

$$= \frac{2\pi L}{\mu} \int_0^R \tau^2 r \, dr = \frac{2\pi \tau_w^2}{\mu R^2} \int_0^R r^3 \, dr = \frac{\pi L R^2 \tau_w^2}{2\mu} = \frac{\pi(-\nabla\Phi)^2 D^4}{128\mu L} \tag{6.13}$$

which is, again, the *Hagen–Poiseuille* Equation (Equation 6.11).

If the wall stress (τ_w) in Equation 6.11 is expressed in terms of the Fanning friction factor (i.e., $\tau_w = f\rho V^2/2$) and the result solved for f, the dimensionless form of the Hagen–Poiseuille equation follows:

$$f = \frac{4\pi D\mu}{Q\rho} = \frac{16\mu}{DV\rho} = \frac{16}{N_{Re}} \tag{6.14}$$

It may be recalled that application of dimensional analysis (Chapter 2) showed that the steady fully developed laminar flow of a Newtonian fluid in a cylindrical tube can be characterized by a single dimensionless group, which is seen from Equation 6.14 to be the product fN_{Re} (note that this group is independent of the fluid density, which cancels out). Since there is only one dimensionless group, it follows that this group must be the same (i.e., constant) for all such flows, regardless of the fluid viscosity or density, the size of the tube, the flow rate, etc. Although the magnitude of this constant could not be obtained from dimensional analysis, we have shown from basic principles that this value is 16, which is also in agreement with experimental observations. Equation 6.14 is valid for a Newtonian fluid with $N_{Re} < 2000$, as previously discussed.

It should be emphasized here that these results are applicable only to "fully developed" flow (i.e., far enough from the tube entrance so that the flow conditions are independent of the distance downstream). However, if the fluid enters a pipe with a uniform ("plug") velocity distribution, a minimum hydrodynamic entry length (L_e) is required for the parabolic velocity flow profile (Equation 6.12) to develop and the pressure gradient to become uniform. It can be shown that this (dimensionless) "hydrodynamic entry length" is approximately $L_e/D = N_{Re}/20$.

B. TURBULENT FLOW

As previously noted, if the Reynolds number in the tube is larger than about 2000, the flow will no longer be laminar. Because fluid elements in contact with a stationary solid boundary are also stationary (i.e., the fluid sticks to the wall), the velocity increases from zero at the boundary to a maximum value at some distance from the boundary. For uniform flow in a symmetrical duct, the maximum velocity occurs at the centerline of the duct. The region of flow over which the velocity varies with the distance from the boundary is called the *boundary layer* and is illustrated in Figure 6.3.

1. Boundary Layer

Because the fluid velocity at a solid boundary is zero, there will always be a region adjacent to the wall that is laminar. This is called the *laminar sublayer* and is designated δ_L in Figure 6.3 (in which the distances are not to scale, as the laminar sublayer is normally quite small). Note that for tube flow, if $N_{Re} < 2000$ the entire flow is laminar and $\delta_L = R$. The turbulent boundary layer (δ_T) includes the region in the vicinity of the wall in which the flow is turbulent and the velocity varies with the distance from the wall (y). Beyond this region, the fluid is almost completely mixed in what is often called the *turbulent core*, in which the velocity is independent of y. The transition from the laminar sublayer to the turbulent boundary layer is gradual, not abrupt, and the transition region is called the *buffer zone*.

2. Turbulent Momentum Flux

The velocity field in turbulent flow can be described by a local "mean" (or time-average) velocity, upon which is superimposed a time-dependent fluctuating component or "eddy." Even in "1-D" flow, in which the overall average velocity has only one directional component, that is, the *turbulent core* (as illustrated in Figure 6.3), the turbulent eddies have a 3D structure. Thus, for the flow illustrated in Figure 6.3, the local velocity components are

$$v_x(y,t) = \bar{v}_x(y) + v'_x(y,t)$$

$$v_y(y,t) = 0 + v'_y(y,t) \tag{6.15}$$

$$v_z(y,t) = 0 + v'_z(y,t)$$

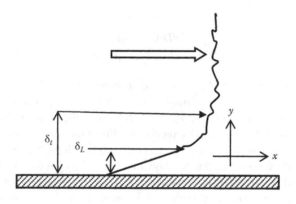

FIGURE 6.3 The boundary layer (not to scale).

The time-average velocity ($\bar{v}_x$) obviously has zero components in the y and z directions, but the eddy (fluctuating) velocity components are nonzero in all three directions. The time-average velocity is defined as

$$\bar{v}_x = \frac{1}{T} \int_0^T v_x \, dt \quad \text{so that} \quad \int_0^T v_x' \, dt = 0 \tag{6.16}$$

Now the eddies transport momentum and the corresponding momentum flux components are equivalent to (negative) fluctuating stress components, as follows:

$$
\begin{aligned}
\tau_{xx}' &= -\rho(v_x')^2 & \tau_{xy}' &= -\rho v_x' v_y' & \tau_{xz}' &= -\rho v_x' v_z' \\
\tau_{yx}' &= \tau_{xy}' & \tau_{yy}' &= -\rho(v_y')^2 & \tau_{yz}' &= -\rho v_y' v_z' \\
\tau_{zx}' &= \tau_{xz}' & \tau_{zy}' &= \tau_{yz}' & \tau_{zz}' &= -\rho(v_z')^2
\end{aligned}
\tag{6.17}
$$

These "turbulent momentum flux components" are also called *Reynolds stresses*. Thus, the total stress in a Newtonian fluid in turbulent flow is composed of both viscous and turbulent (Reynolds) stresses:

$$\tau_{ij} = \mu \left(\frac{\partial \bar{v}_i}{\partial x_j} + \frac{\partial \bar{v}_j}{\partial x_i} \right) - \rho v_i' v_j' \tag{6.18}$$

Although Equation 6.18 can be used to eliminate the stress components from the general microscopic equations of motion, a solution for the turbulent flow field still cannot be obtained unless some information about the spatial dependence and structure of the eddy velocities or turbulent (Reynolds) stresses is known. A classical (simplified) model for the turbulent stresses, attributed to Prandtl, is outlined in the following section.

3. Mixing Length Theory

Turbulent eddies (with fluctuating velocity components v_x', v_y', v_z') are continuously being generated, grow, and die out. During this process, there is an exchange of momentum between the eddies and the mean flow. Considering a 2D turbulent field in a smooth tube, Prandtl assumed (a gross approximation) that $v_x' \approx v_y'$ so that

$$\tau_{yx}' = -\rho v_x' v_y' \cong -\rho(v_x')^2 \tag{6.19}$$

He also assumed that each eddy moves a distance l (the "mixing length") during the time it takes to exchange its momentum with the mean flow, that is, the eddy "lifetime"

$$\frac{v_x'}{l} \cong \frac{d\bar{v}_x}{dy} \tag{6.20}$$

Using Equation 6.20 to eliminate the eddy velocity from Equation 6.19 gives

$$\tau_{yx}' = \mu_e \frac{d\bar{v}_x}{dy} \tag{6.21}$$

where

$$\mu_e = \rho l^2 \left| \frac{d\bar{v}_x}{dy} \right| \tag{6.22}$$

is called the *eddy viscosity*. Note that the eddy viscosity is *not* a fluid property; it is a function of the eddy characteristics (e.g., the mixing length or the degree of turbulence) and the mean velocity gradient. The only fluid property involved is the density because turbulent momentum transport is an inertial (i.e., mass dominated) effect. Since turbulence (and all motion) is zero at the wall, Prandtl further assumed that the mixing length should be proportional to the distance from the wall, that is,

$$l = \kappa y \tag{6.23}$$

Because these relations apply only in the vicinity of the wall, Prandtl also assumed that the eddy (Reynolds) stress must be of the same order as the wall stress, that is,

$$\tau'_{yx} \cong \tau_w = \rho \kappa^2 y^2 \left(\frac{d\bar{v}_x}{dy} \right)^2 \tag{6.24}$$

Integrating Equation 6.24 over the turbulent boundary layer (from y_1, the edge of the buffer layer, to y) gives

$$\bar{v}_x = \frac{1}{\kappa} \left(\frac{\tau_w}{\rho} \right)^{1/2} \ln y + C_1 \tag{6.25}$$

This equation is called the *von Karman equation* (or, sometimes, the "law of the wall") and can be written in the following dimensionless form:

$$v^+ = \frac{1}{\kappa} \ln y^+ + A \tag{6.26}$$

where

$$v^+ = \frac{\bar{v}_x}{v_*} = \frac{\bar{v}_x}{V} \sqrt{\frac{2}{f}}, \quad y^+ = \frac{y v_* \rho}{\mu} = \frac{y V \rho}{\mu} \sqrt{\frac{f}{2}} \tag{6.27}$$

The term

$$v_* = \sqrt{\frac{\tau_w}{\rho}} = V \sqrt{\frac{f}{2}} \tag{6.28}$$

is called the *friction velocity* because it is a wall stress parameter with dimensions of velocity. The parameters κ and A in the von Karman equation have been determined from experimental data on Newtonian fluids in smooth pipes to be $\kappa = 0.4$ and $A = 5.5$. Equation 6.26 applies only within the turbulent boundary layer (outside the buffer region), which has been found empirically to correspond to $y^+ \geq 26$. Within the laminar sublayer, the turbulent eddies are negligible, so that

$$\tau_{yx} \cong \tau_w = \mu \frac{d\bar{v}_x}{dy} \tag{6.29}$$

The corresponding dimensionless form of this equation is

$$\frac{dv^+}{dy^+} = 1 \tag{6.30}$$

or

$$v^+ = y^+ \tag{6.31}$$

Equation 6.31 applies to the laminar sublayer region in a Newtonian fluid, which has been found to correspond to $0 \leq y^+ \leq 5$. The intermediate region or "buffer zone" between the laminar sublayer and the turbulent boundary layer can be represented by the empirical equation

$$v^+ = -3.05 + 5.0 \ln y^+ \tag{6.32}$$

which applies for $5 < y^+ < 26$.

4. Friction Loss in Smooth Pipe

For a Newtonian fluid in a smooth pipe, these equations can be integrated over the pipe cross section to give the average fluid velocity, for example,

$$V = \frac{2}{R^2} \int_0^R \bar{v}_x r\, dr = 2v_* \int_0^1 v^+ (1-x)dx \tag{6.33}$$

where $x = y/R = 1 - r/R$ is the dimensionless distance from the wall. If the von Karman equation (Equation 6.26) for v^+ is introduced into this equation and the laminar sublayer and buffer zones are neglected, the integral can be evaluated and the result solved for $1/\sqrt{f}$ to give

$$\frac{1}{\sqrt{f}} = 4.1 \log(N_{Re}\sqrt{f}) - 0.60 \tag{6.34}$$

The constants in this equation were modified by Nikuradse from observed data taken in smooth pipe as follows:

$$\frac{1}{\sqrt{f}} = 4.0 \log(N_{Re}\sqrt{f}) - 0.40 \tag{6.35}$$

This is also known as the *von Karman–Nikuradse equation* and agrees well with observations for friction loss in smooth pipe over the range $5 \times 10^3 < N_{Re} < 5 \times 10^6$.

An alternative equation for smooth tubes was derived by Blasius based on observations that the mean velocity profile in the tube could be represented approximately by

$$\bar{v}_x = v_{max} \left(1 - \frac{r}{R}\right)^{1/7} \tag{6.36}$$

A corresponding expression for the friction factor can be obtained by writing this expression in dimensionless form and substituting the result into Equation 6.33. Evaluating the integral and solving for f gives the Blasius equation

$$f = \frac{0.0791}{N_{Re}^{1/4}} \tag{6.37}$$

This represents the friction factor for Newtonian fluids in smooth tubes quite well over a range of Reynolds numbers from about 5000 to 10^5. The Prandtl mixing length theory and the von Karman and Blasius equations are referred to as "semiempirical" models. That is, even though these models result from a process of logical reasoning, the results cannot be deduced solely from first principles because they require the introduction of certain parameters that can be evaluated only experimentally.

5. Friction Loss in Rough Tubes

All models for turbulent flows are semiempirical in nature, so it is necessary to rely upon empirical observations (e.g., data) for a quantitative description of friction loss in such flows. For Newtonian fluids in long tubes, we have shown from dimensional analysis that the friction factor should be a unique function of the Reynolds number and the relative roughness of the tube wall. This result has been used to correlate a wide range of measurements in tubes of a variety of sizes and roughness factors, with a variety of fluids and for wide range of flow rates in terms of a generalized plot of f versus N_{Re}, with ε/D as a parameter. This correlation, shown in Figure 6.4, is called a *Moody diagram*.

The laminar region (for $N_{Re} < 2000$) is represented by the theoretical Hagen–Poiseuille equation (Equation 6.14), which is plotted in Figure 6.4 as the $f = 16/N_{Re}$ line. In this region, the only fluid property that influences the friction loss (or energy dissipation) is the viscosity, as the density cancels out. Furthermore, the roughness has a negligible effect in laminar flow, as will be explained shortly. The "critical zone" is the range of transition from laminar to turbulent flow, which corresponds to values of N_{Re} from about 2000 to 4000. Data are not very reproducible in this region, where the friction factor depends strongly on both the Reynolds number and relative roughness. The region in the upper right of the diagram, where the relative roughness lines are horizontal, is called "complete turbulence," or "fully turbulent." In this region, the friction factor is independent of Reynolds number (i.e., independent of viscosity) and is a function only of the relative roughness. For turbulence in smooth tubes, the semiempirical Prandtl–von Karman/Nikuradse or Blasius models represent the friction factor quite well.

Whether a tube is hydraulically "smooth" or "rough" depends upon the size of the wall roughness elements relative to the thickness of the laminar sublayer. Because laminar flow is stable, if the flow perturbations due to the roughness elements lie entirely within the laminar wall region, the disturbances will be damped out and will not affect the rest of the flow field. However, if the roughness elements protrude through the laminar sublayer into the turbulent region, which is unstable, the disturbance will grow, thus enhancing the Reynolds stresses and consequently the energy dissipation or friction loss in the pipe. Because the thickness of the laminar sublayer decreases as the Reynolds number increases, a tube with a given roughness may be hydraulically smooth at a low Reynolds number but it may become hydraulically rough at a high Reynolds number.

6. Friction Loss in Rough Pipe

For rough tubes in turbulent flow ($N_{Re} > 4000$), the von Karman equation was modified empirically by Colebrook to include the effect of wall roughness, as follows:

$$\frac{1}{\sqrt{f}} = -4\log\left[\frac{\varepsilon/D}{3.7} + \frac{1.255}{N_{Re}\sqrt{f}}\right] \tag{6.38}$$

The term $N_{Re}\sqrt{f}$ is, by definition,

$$N_{Re}\sqrt{f} = \left(\frac{e_f D^3 \rho^2}{2L\mu^2}\right)^{1/2} \tag{6.39}$$

and is independent of velocity or flow rate. Thus, the dimensionless groups in the Colebrook equation are in a form that is convenient if the flow rate is to be found and the allowable friction loss (e.g., driving force), tube size, and fluid properties are known.

In the *fully turbulent* region, f is independent of N_{Re}, so that the Colebrook equation reduces to

$$f = \left(\frac{1}{4\log[3.7/(\varepsilon/D)]}\right)^2 \tag{6.40}$$

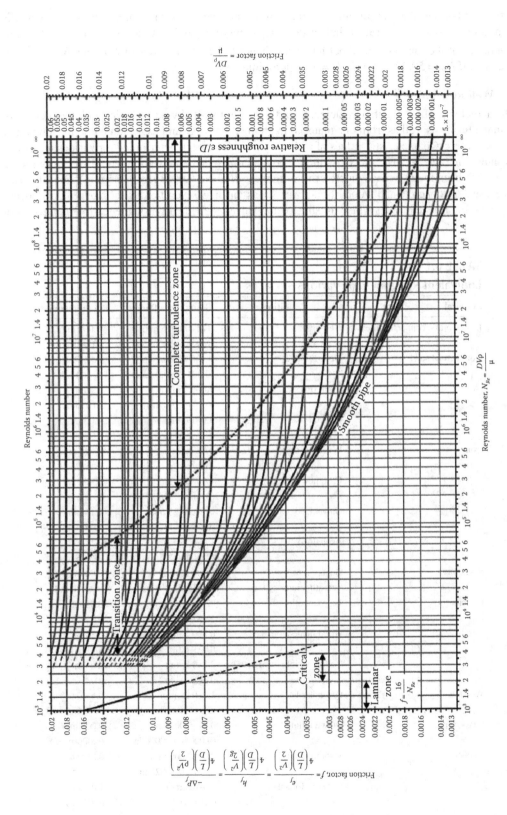

FIGURE 6.4 Moody diagram.

Just as for laminar flow, a minimum hydrodynamic entry length (L_e) is required for the flow profile to become fully developed in turbulent flow. This length depends on the exact nature of the flow conditions at the tube entrance, but has been shown to be on the order of $L_e/D = 0.623N_{Re}^{0.25}$. For example, if $N_{Re} = 50,000$, then $L_e/D = 10$ (approximately).

7. Wall Roughness

The actual size of the roughness elements on the conduit wall obviously varies with the method of manufacturing as well as from one material to another, with age and usage, and with the amount of dirt, scale, etc. Characteristic values of wall roughness have been determined for various materials, as shown in Table 6.1. The most common pipe material—clean, new commercial steel or wrought iron—has been found to have an effective roughness of about 0.0018 in. (0.045 mm). For other surfaces, such as concrete and wood, it may vary by as much as several orders of magnitude, depending upon the nature of the surface finish. Conduit surfaces artificially roughened by sand grains of

TABLE 6.1

Equivalent Roughness of Various Surfaces

Material	Condition	Roughness Range	Recommended
Drawn brass, copper, stainless	New	0.01–0.0015 mm (0.0004–0.00006 in.)	0.002 mm (0.00008 in.)
Commercial steel	New	0.1–0.02 mm (0.004–0.0008 in.)	0.045 mm (0.0018 in.)
	Light rust	1.0–0.15 mm (0.04–0.006 in.)	0.3 mm (0.015 in.)
	General rust	3.0–1.0 mm (0.1–0.04 in.)	2.0 mm (0.08 in.)
Iron	Wrought, new	0.045 mm (0.002 in.)	0.045 mm (0.002 in.)
	Cast, new	1.0–0.25 mm (0.04–0.01 in.)	0.30 mm (0.025 in.)
	Galvanized	0.15–0.025 mm (0.006–0.001 in.)	0.15 mm (0.006 in.)
	Asphalt coated	1.0–0.1 mm (0.04–0.004 in.)	0.15 mm (0.006 in.)
Sheet metal	Ducts Smooth joints	0.1–0.02 mm (0.004–0.0008 in.)	0.03 mm (0.0012 in.)
Concrete	Very smooth	0.18–0.025 mm (0.007–0.001 in.)	0.04 mm (0.0016 in.)
	Wood floated, brushed	0.8–0.2 mm (0.03–0.007 in.)	0.3 mm (0.012 in.)
	Rough, visible form marks	2.5–0.8 mm (0.1–0.03 in.)	2.0 mm (0.08 in.)
Wood	Stave, used	1.0–0.25 mm (0.035–0.01 in.)	0.5 mm (0.02 in.)
Glass or plastic	Drawn tubing	0.01–0.0015 mm (0.0004–0.00006 in.)	0.002 mm (0.00008 in.)
Rubber	Smooth tubing	0.07–0.006 mm (0.003–0.00025 in.)	0.01 mm (0.0004 in.)
	Wire reinforced	4.0–0.3 mm (0.15–0.01 in.)	1.0 mm (0.04 in.)

various sizes were studied initially by Nikuradse, and measurements of f and N_{Re} were plotted to establish the reference curves for various known values of ε/D for these surfaces, as shown on the Moody diagram. The equivalent roughness factors for other materials are determined from similar measurements in conduits made of the material, by plotting the data on the Moody diagram and comparing the results with the reference curves (or by using the Colebrook equation). For this reason, these roughness values are sometimes termed the *equivalent sand grain roughness*.

C. ALL FLOW REGIMES

The expressions for the friction factor in both laminar and turbulent flow were combined into a single expression by Churchill (1977) as follows:

$$f = 2\left[\left(\frac{8}{N_{Re}}\right)^{12} + \frac{1}{(A+B)^{3/2}}\right]^{1/12} \tag{6.41}$$

where

$$A = \left[2.457\ln\left(\frac{1}{(7/N_{Re})^{0.9} + (0.27\varepsilon/D)}\right)\right]^{16}$$

and

$$B = \left(\frac{37,530}{N_{Re}}\right)^{16}$$

Equation 6.41 adequately represents the Fanning friction factor over the entire range of Reynolds numbers, from laminar to fully turbulent, within the accuracy of the data used to construct the Moody diagram. It also gives a reasonable estimate for the intermediate or transition region between laminar and turbulent flow. Note that it is explicit in f.

IV. POWER LAW FLUIDS

Corresponding expressions for the friction loss in laminar and turbulent flow for non-Newtonian fluids in pipes for the two simplest (two-parameter) models—the power law and Bingham plastic—can be evaluated in a similar manner. The power law model is very popular for representing the viscosity of a wide variety of non-Newtonian fluids because of its simplicity and versatility. However, extreme care should be exercised in its application. Because the model *breaks down at both very low and very high shear rates*, any application involving extrapolation beyond the range of shear stress (or shear rate) represented by the data used to determine the model parameters can lead to misleading or erroneous results. Both laminar and turbulent pipe flow of highly loaded slurries of fine particles, for example, can often be adequately represented by either of these two models over an appreciable shear rate range, as shown by Darby et al. (1992).

A. LAMINAR FLOW

Because the shear stress and shear rate are negative in pipe flow, the appropriate form of the power law model for laminar pipe flow is

$$\tau_{rx} = m\dot{\gamma}_{rx}^n = -m\left(-\frac{dv_x}{dr}\right)^n \tag{6.42}$$

By equating the shear stress from Equations 6.42 and 6.4, solving for the velocity gradient, and introducing the result into Equation 6.7 (as was done for the Newtonian fluid), the flow rate is found to be

$$Q = \pi \left(\frac{\tau_w}{mR} \right)^{1/n} \int_0^R r^{2+1/n} \, dr = \pi \left(\frac{\tau_w}{mR} \right)^{1/n} \left(\frac{n}{3n+1} \right) R^{(3n+1)/n} \tag{6.43}$$

This is the power law equivalent of the Newtonian Hagen–Poiseuille equation, which it reduces to if $n = 1$. It can be written in dimensionless form by expressing the wall stress in terms of the friction factor using Equation 6.5, solving for f, and equating the result to $16/N_{Re}$ (i.e., the form of the Newtonian result). The result is an expression that is identical to the dimensionless Hagen–Poiseuille equation

$$fN_{Re,pl} = 16 \tag{6.44}$$

which defines the Reynolds number for a power law fluid as

$$N_{Re,pl} = \frac{8D^n V^{2-n} \rho}{m[2(3n+1)/n]^n} \tag{6.45}$$

It should be noted that a dimensional analysis of this problem results in one more dimensionless group than for the Newtonian fluid because there is one more fluid rheological property (e.g., m and n for the power law fluid vs. μ for the Newtonian fluid). However, the parameter n is itself dimensionless and thus constitutes the additional "dimensionless group," even though it is integrated into the Reynolds number as it has been defined here. Note also that because n is an empirical parameter and can take on any value, the units in expressions for power law fluids can be complex. Thus, the calculations are simplified if a scientific system of dimensions/units is used (e.g., SI or cgs), which avoids the necessity of introducing the conversion factor g_c. In fact, the evaluation of most dimensionless groups is usually simplified by the use of scientific units.

B.   TURBULENT FLOW

Dodge and Metzner (1959) modified the von Karman equation to apply to power law fluids, with the following result:

$$\frac{1}{\sqrt{f}} = \frac{4}{n^{0.75}} \log \left[N_{Re,pl} f^{(1-n)/2} \right] - \frac{0.4}{n^{1.2}} \tag{6.46}$$

Like the von Karman equation, this equation is implicit in f. Equation 6.46 can be applied to any non-Newtonian fluid if the parameter n is interpreted to be the point slope of the shear stress versus shear rate plot from (laminar) viscosity measurements, at the wall shear stress (or shear rate) corresponding to the conditions of interest in turbulent flow. However, it is not a simple matter to acquire the needed data over the appropriate range, or to solve the equation for f for a given flow rate and pipe diameter, in turbulent flow.

Note that there is no effect of pipe wall roughness in Equation 6.46, in contrast to the case for Newtonian fluids. There are insufficient data in the literature to provide a reliable estimate of the effect of roughness on friction loss for non-Newtonian fluids in turbulent flow. However, the evidence that does exist suggests that the roughness is not as significant for non-Newtonian fluids as for Newtonian fluids. This is partly due to the fact that the majority of non-Newtonian turbulent flows lie in the low Reynolds number range and partly due to the fact that the laminar boundary layer tends to be thicker for non-Newtonian fluids than for Newtonian fluids (i.e., the flows are generally in the "hydraulically smooth" range for common pipe materials).

C. ALL FLOW REGIMES

An expression that represents the friction factor for the power law fluid over the entire range of Reynolds numbers (laminar through turbulent) and encompasses Equations 6.44 and 6.46 has been given by Darby et al. (1992):

$$f = (1-\alpha)f_L + \frac{\alpha}{\left[f_T^{-8} + f_{Tr}^{-8}\right]^{1/8}} \tag{6.47}$$

where

$$f_L = \frac{16}{N_{Re,pl}} \tag{6.48}$$

$$f_{Tr} = 1.79 \times 10^{-4} \exp\left[-5.24\,n\right]N_{Re,pl}^{0.414+0.757n} \tag{6.49}$$

$$f_T = \frac{0.0682n^{-1/2}}{\left(N_{Re,pl}\right)^{1/(1.87+2.39n)}} \tag{6.50}$$

The parameter α is given by

$$\alpha = \frac{1}{1+4^{-\Delta}} \tag{6.51}$$

where

$$\Delta = N_{Re,pl} - N_{Re,plc} \tag{6.52}$$

and $N_{Re,plc}$ is the critical power law Reynolds number at which laminar flow ceases

$$N_{Re,plc} = 2100 + 875(1-n) \tag{6.53}$$

Equation 6.48 applies for $N_{Re,pl} < N_{Re,plc}$, Equation 6.49 applies for $N_{Re,plc} < N_{Re,pl} < 4000$, Equation 6.50 applies for $4000 < N_{Re,pl} < 10^5$, and all are encompassed by Equation 6.47 for all values of $N_{Re,pl}$.

V. BINGHAM PLASTICS

The Bingham plastic model usually provides a good representation for the viscosity of concentrated slurries, suspensions, emulsions, foams, etc. Such materials often exhibit a yield stress that must be exceeded before the material will flow at a significant rate. Other examples include paint, shaving cream, and mayonnaise. There are also many fluids, such as blood, that may have a yield stress that is not as pronounced.

It is recalled that a "plastic" is really two materials. At low stresses below the critical or yield stress (τ_o), the material behaves as a solid, whereas for stresses above the yield stress, the material behaves as a fluid. The Bingham model for this behavior is

$$\text{For } |\tau| < \tau_o: \quad \dot{\gamma} = 0$$

$$\text{For } |\tau| > \tau_o: \quad \tau = \pm\tau_o + \mu_\infty\dot{\gamma} \tag{6.54}$$

Because the shear stress and shear rate can be either positive or negative, the plus/minus sign in Equation 6.54 is "plus" in the former case, and "minus" in the latter. For pipe flow, since the shear stress and shear rate are both negative, the appropriate form of the model is

$$\text{For } |\tau_{rx}| < \tau_o: \quad \frac{dv_x}{dr} = 0$$

$$\text{For } |\tau_{rx}| > \tau_o: \quad \tau_{rx} = -\tau_o + \mu_\infty \frac{dv_x}{dr}$$

(6.55)

A. LAMINAR FLOW

Because the shear stress is always zero at the centerline in pipe flow and increases linearly with distance from the center toward the wall (Equation 6.4), there will be a finite distance from the center over which the stress is always less than the yield stress. In this region, the material has solid-like properties and does not yield but moves as a rigid plug. The radius of this plug (r_o) is, from Equation 6.4:

$$r_o = R \frac{\tau_o}{\tau_w}$$

(6.56)

Because the stress outside this plug region exceeds the yield stress, the material will deform or flow as a fluid between the plug and the wall. The flow rate must thus be determined by combining the flow rate of the "plug" with that of the "fluid" region:

$$Q = \int_A v_x \, dA = Q_{plug} + \pi \int_{r_o^2}^{R^2} v_x \, dr^2$$

(6.57)

Evaluating the integral by parts and noting that the Q_{plug} term cancels with $\pi r_o^2 V_{plug}$ from the lower limit, the result is

$$Q = -\pi \int_{r_o}^{R} r^2 \dot{\gamma} \, dr$$

(6.58)

When Equation 6.55 is used for the shear rate ($\dot{\gamma} = dv_x/dr$) in terms of the shear stress and Equation 6.4 is used for the shear stress as a function of r, the integral can be evaluated to give

$$Q = \frac{\pi R^3 \tau_w}{4\mu_\infty} \left[1 - \frac{4}{3} \left(\frac{\tau_o}{\tau_w} \right) + \frac{1}{3} \left(\frac{\tau_o}{\tau_w} \right)^4 \right]$$

(6.59)

This equation is known as the *Buckingham–Reiner equation*. It can be cast in dimensionless form and rearranged as follows:

$$f_L = \frac{16}{N_{Re}} \left[1 + \frac{1}{6} \frac{N_{He}}{N_{Re}} - \frac{1}{3} \frac{N_{He}^4}{f_L^3 N_{Re}^7} \right] \approx \frac{16}{N_{Re}} \left[1 + \frac{N_{He}}{8N_{Re}} \right]$$

(6.60)

where the Reynolds number is given by

$$N_{Re} = \frac{DV\rho}{\mu_\infty}$$

(6.61)

and

$$N_{He} = \frac{D^2 \rho \tau_o}{\mu_\infty^2} \tag{6.62}$$

is the *Hedstrom number*. Although the dimensionless Buckingham–Reiner equation is implicit in f, the approximate expression on the far right of Equation 6.60 follows from considerations of Equation 7.41 (Chapter 7) and is an excellent explicit approximation for the laminar friction factor. Note that the Bingham plastic model reduces to a Newtonian fluid if $\tau_o = 0 = N_{He}$. In this case, Equation 6.60 reduces to the Newtonian result, that is, $f = 16/N_{Re}$ (see Equation 6.14). Note that there are actually only two independent dimensionless groups in Equation 6.60 (consistent with the results of dimensional analysis for a fluid with two rheological properties, τ_o and μ_∞), which are the combined groups fN_{Re} and N_{He}/N_{Re}. The ratio N_{He}/N_{Re} is also called the *Bingham number*, $N_{Bi} = D\tau_o/\mu_\infty V$. The Buckingham–Reiner equation is implicit in f, so it must be solved by iteration for known values of N_{Re} and N_{He}. However, the approximate expression on the right of Equation 6.60 is explicit in f and gives an excellent approximation for f in almost all cases.

B. TURBULENT FLOW

For the Bingham plastic, there is no abrupt transition from laminar to turbulent flow as is observed for Newtonian fluids. Instead, there is a gradual deviation from purely laminar flow to fully turbulent flow. For turbulent flow, the friction factor can be represented by the empirical expression of Darby and Melson (1981) (as modified by Darby et al. (1992))

$$f_T = \frac{10^a}{N_{Re}^{0.193}} \tag{6.63}$$

where

$$a = -1.47 \left[1 + 0.146 \exp \left(-2.9 \times 10^{-5} N_{He} \right) \right] \tag{6.64}$$

C. ALL REYNOLDS NUMBERS

The friction factor for a Bingham plastic can be calculated for any Reynolds number, from laminar through turbulent, from the equation

$$f = \left(f_L^m + f_T^m \right)^{1/m} \tag{6.65}$$

where

$$m = 1.7 + \frac{40{,}000}{N_{Re}} \tag{6.66}$$

In Equation 6.65, f_T is given by Equation 6.63 and f_L is given by Equation 6.60.

VI. PIPE FLOW PROBLEMS

There are three typical problems encountered in pipe flows, depending upon what is known and what is to be found. These are the "unknown driving force," "unknown flow rate," and "unknown diameter" problems. We will outline here the procedure for the solution of each of these for both Newtonian and non-Newtonian (power law and Bingham plastic) fluids. A fourth problem, perhaps of even more practical interest for piping system design, is the "most economical diameter" problem that will be considered in Chapter 7.

We note first that the Bernoulli equation can be written in terms of the driving force DF as

$$DF = e_f + \frac{1}{2}\left(\alpha_2 v_2^2 - \alpha_1 v_1^2\right) \tag{6.67}$$

where

$$e_f = \left(\frac{4fL}{D}\right)\left(\frac{v^2}{2}\right) = \frac{32 fLQ^2}{\pi^2 D^5} \tag{6.68}$$

and

$$DF = -\left(\frac{\Delta\Phi}{\rho} + w\right) \tag{6.69}$$

DF represents the net energy input into the fluid per unit mass (or the net "driving force") and is the combination of static head, pressure difference, and pump work. When any of the terms in Equation 6.69 are negative, they represent a positive "driving force" (or energy input) for moving the fluid through the pipe. Positive terms represent forces resisting the flow, for example, an increase in elevation, pressure, etc., correspond to a negative driving force.

We will use the Bernoulli equation in the form of Equation 6.67 for analyzing pipe flows and we will use the total volumetric flow rate (Q) as the flow variable instead of the velocity, because this is the usual measure of capacity in a pipeline. For Newtonian fluids, the problem thus reduces to a relation between the three dimensionless variables:

$$N_{Re} = \frac{4Q\rho}{\pi D \mu}, \quad f = \frac{e_f \pi^2 D^5}{32LQ^2}, \quad \frac{\varepsilon}{D} \tag{6.70}$$

For a uniform pipe, the velocity is everywhere the same so the kinetic energy terms cancel out. In many other applications, the kinetic energy terms are often negligible or cancel out as well, although this should be verified for each situation.

A.   UNKNOWN DRIVING FORCE

For this problem, we want to know the net driving force (DF) that is required to move a given fluid (μ, ρ) at a specified rate (Q) through a specified pipe (D, L, ε). The Bernoulli equation in the form $DF = e_f$ applies.

1.   Newtonian Fluid

The "knowns" and "unknowns" in this case are as follows:

Given: Q, μ, ρ, D, L, ε *Find*: DF

All the relevant variables and parameters are uniquely related through the three dimensionless variables f, N_{Re}, and ε/D by the Moody diagram or the Churchill equation. Furthermore, the unknown ($DF = e_f$) appears in only one of these groups (f). The procedure is thus straightforward:

   1. Calculate the Reynolds number from Equation 6.70.
   2. Calculate ε/D.

3. Determine f from the Moody diagram or Churchill equation (Equation 6.41) (if $N_{Re} < 2000$, use $f = 16/N_{Re}$).
4. Calculate e_f (hence DF) from the Bernoulli equation, Equation 6.67.

From the resulting value of DF, the required pump head $(-w/g)$, can be determined, for example, from a knowledge of the upstream and downstream pressures and elevations using Equation 6.69.

2. Power Law Fluid

The equivalent problem statement is as follows:

Given: Q, m, n, ρ, D, L *Find*: *DF*

Note that we have an additional fluid property (m and n instead of μ), but we also assume that pipe roughness has a negligible effect, so that the total number of variables is the same. The corresponding dimensionless variables are f, $N_{Re,pl}$, and n (which are related by Equation 6.47), and the unknown ($DF = e_f$) appears in only one group (f). The procedure just followed for the Newtonian fluid can thus also be applied to a power law fluid if the appropriate equations are used, as follows:

1. Calculate the Reynolds number ($N_{Re,pl}$) using Equation 6.45 and the volumetric flow rate instead of the velocity, that is,

$$N_{Re,pl} = \frac{2^{7-3n}\rho Q^{2-n}}{m\pi^{2-n}D^{4-3n}}\left(\frac{n}{3n+1}\right)^n \tag{6.71}$$

2. Calculate f from Equation 6.47.
3. Calculate e_f (hence DF) from Equation 6.68.

3. Bingham Plastic

The problem statement is as follows:

Given: $Q, \mu_\infty, \tau_o, \rho, D,$ and L *Find*: *DF*

The number of variables is the same as in the foregoing problems; hence, the number of groups relating these variables is the same. For the Bingham plastic, these are f, N_{Re}, and N_{He}, which are related by Equation 6.65 (along with Equations 6.60 and 6.63). The unknown ($DF = e_f$) appears only in f, as before. The solution procedure is similar to that followed for Newtonian and power law fluids.

1. Calculate the Reynolds number:

$$N_{Re} = \frac{4Q\rho}{\pi D\mu_\infty} \tag{6.72}$$

2. Calculate the Hedstrom number:

$$N_{He} = \frac{D^2\rho\,\tau_o}{\mu_\infty^2} \tag{6.73}$$

3. Determine f from Equations 6.65, 6.63, and 6.60.
4. Calculate e_f, hence DF, from Equation 6.68.

Example 6.1 Slurry Draining from Tank

Find the static head required to provide a flow of 150 gpm (gallons/min) for a Bingham plastic slurry through a 3 in. ID pipe, 400 ft long (See Figure E6.1). The slurry properties are as follows: density 1.8 g/cm^3, yield stress 150 dyn/cm^2, and limiting viscosity of 85 cP.

Solution:

As fluid properties are often expressed in cgs units, it is convenient to use these units for all variables in the problem. Therefore, we obtain the following:

$$Q = (150 \text{ gpm})(63.1 \text{ cm}^3/\text{s and } 1/\text{gpm}) = 9470 \text{ cm}^3/\text{s}$$

$$D = 3 \text{ in.} = 7.62 \text{ cm}$$

$$L = 400 \text{ ft} = 12{,}300 \text{ cm}$$

$$\tau_o = 150 \text{ dyn/cm}^2 = 150 \text{ g/(cm s}^2)$$

$$\mu_\infty = 85 \text{ cP} = 0.85 \text{ P} = 0.85 \text{ dyn s/cm}^2 = 0.85 \text{ g/(cm s)}$$

$$\rho = 1.8 \text{ g/cm}^3$$

We must first define the system for this problem. We may assume that the slurry is transported from an elevated open vessel to which the pipe is attached and exits at ground level (see Figure E6.1). The *system* is therefore all the slurry between the surface in the tank, point 1 where $z = z_1$, and the exit from the pipe, point 2 where $z_2 = 0$. As the pressure difference is zero (both points 1 and 2 are at atmospheric pressure) and there is no pump, the only component of driving force is gravity, that is, $DF = -g\Delta z$. The Bernoulli equation thus reduces to

$$DF = -g\Delta z = e_f + \frac{1}{2}\left(\alpha_2 v_2^2 - \alpha_1 v_1^2\right)$$

where

$$e_f = \left(\frac{4fL}{D}\right)\left(\frac{v^2}{2}\right) = \frac{32fLQ^2}{\pi^2 D^5}$$

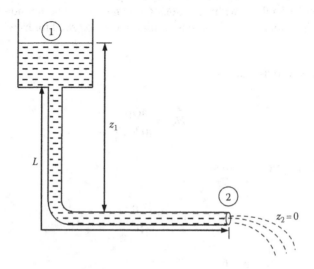

FIGURE E6.1 Slurry draining from a tank.

In order to get f, we must first calculate the Reynolds number

$$N_{Re} = \frac{4Q\rho}{\pi D\mu_\infty} = \frac{4(9470\,\text{cm}^3/\text{s})(1.8\,\text{g/cm}^3)}{\pi(7.62\,\text{cm})(0.85\,\text{g/cm s})} = 3351$$

and the Hedstrom number

$$N_{He} = \frac{D^2\rho\tau_o}{\mu_\infty^2} = \frac{(7.62\,\text{cm})^2(1.8\,\text{g/cm}^3)(150\,\text{g/cm s}^2)}{(0.85\,\text{g/cm s})^2} = 2420$$

Then we determine a from Equation 6.63

$$a = -1.47(1 + 0.146\exp[-2.9\times10^{-5}N_{He}])$$

$$= -1.47(1 + 0.146\exp[-2.9\times10^{-5}(2420)]) = -1.670$$

and f_L from Equation 6.60

$$f_L \approx \frac{16}{N_{Re}}\left[1 + \frac{N_{He}}{8N_{Re}}\right] = \frac{16}{3351}\left[1 + \frac{2420}{8(3351)}\right] = 0.00521$$

Then we obtain f_T from Equation 6.63

$$f_T = \frac{10^a}{N_{Re}^{0.193}} = \frac{10^{-1.670}}{3351^{0.193}} = 0.00446$$

m is found from Equation 6.66

$$m = 1.7 + \frac{40{,}000}{N_{Re}} = 1.7 + \frac{40{,}000}{3{,}351} = 13.64$$

And finally, we solve for f from Equation 6.65

$$f = \left(f_L^m + f_T^m\right)^{1/m} = (0.00521^{13.64} + 0.00446^{13.64})^{1/13.64} = 0.00525$$

Now, we can determine the friction loss from Equation 6.68

$$e_f = \frac{32fLQ^2}{\pi^2 D^5} = \frac{32(0.00525)(12{,}300\,\text{cm})(9{,}470\,\text{cm}^3/\text{s})^2}{\pi^2(7.62\,\text{cm})^5} = 7.31\times10^5\,\text{cm}^2/\text{s}^2$$

With regard to the kinetic energy terms in the Bernoulli equation, we may assume that the velocity in the tank is negligible compared to that in the pipe, that is, $v_1 \ll v_2$. We will also neglect the kinetic energy of the stream leaving the pipe, as well as the friction loss in the contraction from the tank to the pipe, and any fittings such as elbows (these items may not be always negligible, however, and methods for evaluating them will be given in Chapter 7).

We may now solve the Bernoulli equation for the unknown z_1:

$$z_1 = \frac{e_f}{g} + z_2 = \frac{7.31\times10^5\,(\text{cm}^2/\text{s}^2)}{980\,\text{cm/s}^2} + 0 = 746\,\text{cm} = 7.46\,\text{m}$$

However, this vertical elevation will provide the required flow rate through only 115.4 m of horizontal pipe (and 7.46 m of vertical pipe). In order to provide the desired flow rate through 123 m of horizontal pipe, the desired elevation can be obtained by a simple ratio because the friction loss is a linear function of pipe length:

$$z_{123} = z_{115}\left(\frac{123}{115.4}\right) = 7.46\,\text{m}\left(\frac{123}{115.4}\right) = 7.95\,\text{m}$$

B.   UNKNOWN FLOW RATE

In this case, the flow rate is to be determined when a given fluid is transported in a given pipe with a known net driving force (e.g., pump head, pressure head, and/or hydrostatic head). The same total variables are involved as before, and hence the dimensionless variables are the same and are related in the same way as for the unknown driving force problems. The main difference is that now the unknown (Q) appears in two of the dimensionless variables (f and N_{Re}), which requires a different solution strategy.

1.   Newtonian Fluid

The problem statement is as follows:

Given: $DF, D, L, \varepsilon, \mu, \rho$ *Find*: Q

The strategy is to redefine the relevant dimensionless variables by combining the original groups in such a way that the unknown variable appears in one group. For example, f and N_{Re} can be combined to eliminate the unknown (Q) as follows:

$$fN_{Re}^2 = \left[\frac{DF\pi^2 D^5}{32LQ^2}\right]\left[\frac{4Q\rho}{\pi D\mu}\right]^2 = \frac{DF\rho^2 D^3}{2L\mu^2} \tag{6.74}$$

Thus, if we work with the three dimensionless variables fN_{Re}^2, N_{Re}, and ε/D, the unknown (Q) appears in only N_{Re}, which then becomes the unknown (dimensionless) variable.

There are various approaches that we can take to solve this problem. Since the Reynolds number is unknown, an explicit solution is not possible using the established relations between the friction factor and Reynolds number (e.g., the Moody diagram or Churchill equation). We can, however, proceed by a trial-and-error method that requires an initial guess for an unknown variable, use the basic relations to solve for this variable, revise the guess accordingly, and repeat the process (iterating) until agreement between calculated and guessed values is achieved.

Note that in this context, either f or N_{Re} can be considered the unknown dimensionless variable because they both involve the unknown Q. As an aid in making the choice between these, a glance at the Moody diagram shows that the practical range of possible values of f is approximately one order of magnitude, whereas the corresponding possible range of N_{Re} values is over five orders of magnitude. Thus, the chances of our initial guess being close to the final answer are greatly enhanced if we choose to iterate on f instead of N_{Re}. Using this approach, the procedure is as follows:

1. A reasonable guess might be based on the assumption that the flow conditions are turbulent, for which the Colebrook equation, Equation 6.38, applies.
2. Calculate the value of fN_{Re}^2 from given values.
3. Calculate f using the Colebrook equation, Equation 6.38.
4. Calculate $N_{Re} = (fN_{Re}^2/f)^{1/2}$, using f from step 3.
5. Using the value of N_{Re} from step 4 and the known value of ε/D, determine f from the Moody diagram or Churchill equation (if $N_{Re} < 2000$, use $f = 16/N_{Re}$).
6. If this value of f does not agree with that from step 3, insert the value of f from step 5 into step 4 to get a revised value of N_{Re}.
7. Repeat steps 5 and 6 until f no longer changes.
8. Calculate $Q = \pi D\mu N_{Re}/4\rho$.

2. Power Law Fluid

The problem statement is as follows:

Given: DF, D, L, m, n, ρ   *Find*: Q

The simplest approach for this problem is also an iteration procedure, based on an assumed value of f:

1. A reasonable starting value for f is 0.005 based on a "dart throw" at the (equivalent) Moody diagram.
2. Calculate Q from Equation 6.68:

$$Q = \pi \left(\frac{D^5 DF}{32 fL} \right)^{1/2}. \tag{6.75}$$

3. Calculate the Reynolds number from Equation 6.71,

$$N_{Re,pl} = \left(\frac{2^{7-3n}}{m\pi^{2-n}} \right) \frac{\rho Q^{2-n}}{D^{4-3n}} \left(\frac{n}{3n+1} \right)^n. \tag{6.76}$$

4. Calculate f from Equation 6.47.
5. Compare the values of f from step 4 and step 1. If they do not agree, use the result of step 4 in step 2 and repeat steps 2 through 5 until agreement is reached. Convergence usually requires only two or three trials, at most, unless very unusual conditions are encountered.

Example 6.2 Unknown Flow Rate of a Power Law Slurry

It is desired to determine the flow rate in gpm that would result for a slurry (ρ = 1.6 g/cm³) being pumped through a 300 ft long 2 in. ID pipe, with a pump that develops a discharge pressure of 85 psig. The pressure entering the pump is 10 psig, and the pipe discharges into a tank at atmospheric pressure. The slurry is characterized as a power law fluid with the properties m = 0.6 dyn s^n/cm² and n = 0.8 (Figure E6.2).

Solution:
The system parameters and fluid properties are listed here. We convert the data to cgs units for ease of manipulation:

$$\text{Slurry consistency } (m) = 0.6 \text{ (dyn s}^n\text{/cm}^2) = 0.6 \text{ g/(cm s}^{1.2})$$

$$\text{Slurry flow index } (n) = 0.8$$

$$\text{Slurry density} = 1.6 \text{ g/cm}^3$$

$$\text{Pump pressure} = -\Delta P = (85\,\text{psi})(\text{atm}/14.696\,\text{psi})(1.013 \times 10^6 \text{ dyn/[cm}^2 \text{ atm]})$$

$$= 5.86 \times 10^6 \text{ dyn/cm}^2$$

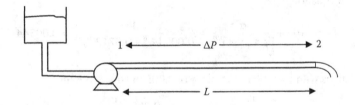

FIGURE E6.2   Unknown flow rate of power law fluid.

$$DF = -\Delta P/\rho = (5.86 \times 10^6 \text{ dyn/cm}^2(\text{g cm/s}^2 \text{ dyn})/(1.6 \text{ g/cm}^3) = 3.66 \times 10^6 \text{ cm}^2/\text{s}^2$$

$$L = 300 \text{ ft } (30.48 \text{ cm/ft}) = 9144 \text{ cm}$$

$$D = 2 \text{ in. } (2.54 \text{ cm/in.}) = 5.08 \text{ cm}$$

$$\text{Assume } f = 0.005.$$

Calculate Q from Equation 6.68:

$$Q = \pi \left(\frac{D^5 DF}{32 fL} \right)^{1/2} = \pi \left[\frac{(5.08 \text{ cm})^5 (3.66 \times 10^6 \text{ cm}^2/\text{s}^2)}{32(0.005)(9144 \text{ cm})} \right]^{1/2} = 9142 \text{ cm}^3/\text{s}$$

Calculate the Reynolds number from Equation 6.71:

$$N_{Re,pl} = \left(\frac{2^{7-3n}}{m\pi^{2-n}} \right) \frac{\rho Q^{2-n}}{D^{4-3n}} \left(\frac{n}{3n+1} \right)^n$$

$$= \left(\frac{2^{7-3(0.8)}}{0.6\pi^{2-(0.8)}} \right) \left(\frac{\text{cm s}^{1.2}}{g} \right) \left(\frac{0.8}{3(0.8)+1} \right)^{0.8} \left(\frac{(1.6 \text{ g/cm}^3)(9,142 \text{ cm}^3/\text{s})^{2-0.8}}{(5.08 \text{ cm})^{4-3(0.8)}} \right) = 21,641$$

Using Equations 6.48 to 6.51:

$$f_L = \frac{16}{N_{Re,pl}} = \frac{16}{21,641} = 0.000739 \tag{6.48}$$

$$f_{Tr} = 1.79 \times 10^{-4} \exp[-5.24n] N_{Re,pl}^{0.414+0.757n}$$

$$= 1.79 \times 10^{-4} \exp[-5.24(0.8)](21,641)^{(0.414+0.757(0.8))} = 0.0712 \tag{6.49}$$

$$f_T = \frac{0.0682 n^{-1/2}}{N_{Re,pl}^{1/(1.87+2.39n)}} = \frac{0.0682(0.8)^{-0.5}}{(21,641)^{1/(1.87+2.39(0.8))}} = 0.00544 \tag{6.50}$$

α is given by

$$\alpha = \frac{1}{1+4^{-\Delta}} = \frac{1}{1+4^{-19366}} = 1 \tag{6.51}$$

where

$$\Delta = N_{Re,pl} - N_{Re,pl} = 21,641 - 2,275 = 19,366$$

and

$$N_{Re,plc} = 2100 + 875(1-n) = 2100 + 875(1-0.8) = 2275$$

Calculate f from Equation 6.47:

$$f = (1-\alpha)f_L + \frac{\alpha}{\left[f_T^{-8} + f_{Tr}^{-8} \right]^{1/8}} = \frac{1}{[0.00544^{-8} + 0.0712^{-8}]^{1/8}} = 0.00544 \tag{6.47}$$

Repeat calculations with $f = 0.00544$ in place of 0.005 in Equation 6.68:

$$Q_2 = Q \left(\frac{f}{f_2} \right)^{1/2} = 9142 \text{ cm}^3/\text{s} \left(\frac{0.005}{0.00544} \right)^{1/2} = 8761 \text{ cm}^3/\text{s}$$

$$(N_{Re,pl})_2 = N_{Re,pl} \left(\frac{Q_2}{Q} \right)^{2-n} = 21{,}641 \left(\frac{8{,}761}{9{,}142} \right)^{1.2} = 20{,}563$$

$$(f_L)_2 = f_L \left(\frac{N_{Re,pl}}{(N_{Re,pl})_2} \right) = 0.000739 \left(\frac{21{,}641}{20{,}563} \right) = 0.000778$$

$$(f_{tr})_2 = f_{tr} \left[\frac{(N_{Re,pl})_2}{N_{Re,pl}} \right]^{(0.414+0.757n)} = 0.07121 \left(\frac{20{,}563}{21{,}641} \right)^{1.096} = 0.0676$$

$$(f_T)_2 = f_T \left[\frac{N_{Re,pl}}{(N_{Re,pl})_2} \right]^{1/(1.87+2.39n)} = 0.00544 \left[\frac{21{,}641}{20{,}563} \right]^{0.2464} = 0.00552$$

$$\Delta_2 = (N_{Re,pl})_2 - N_{Re,plc} = 20{,}563 - 2{,}275 = 18{,}288$$

$$\alpha_2 = \frac{1}{1 + 4^{-\Delta}} = \frac{1}{1 + 4^{-18288}} = 1$$

$$f_2 = (1 - \alpha_2)(f_L)_2 + \frac{\alpha_2}{\left[f_T^{-8} + f_{Tr}^{-8} \right]_2^{1/8}}$$

$$= (1 - 1) + \frac{1}{[0.00552^{-8} + 0.0676^{-8}]^{1/8}} = 0.00552$$

This is close enough to the previous value of 0.0.00544, so that no additional calculations are needed. Thus, the flow rate of $\boxed{8761 \text{ cm}^3/\text{s} = 139 \text{ gpm}}$ is attainable in this system.

---

3. Bingham Plastic

The procedure is very similar to the one that above.

Given: DF, D, L, μ_∞, τ_o, ρ Find: Q

1. Assume $f = 0.005$.
2. Calculate Q from Equation 6.75.
3. Calculate the Reynolds and Hedstrom numbers:

$$N_{Re} = \frac{4Q\rho}{\pi D \mu_\infty}, \quad N_{He} = \frac{D^2 \rho \tau_o}{\mu_\infty^2}. \tag{6.77}$$

4. Calculate f from Equation 6.65.
5. Compare the value of f from step 4 with the assumed value in step 1. If they do not agree, use the value of f from step 4 in step 2 and repeat steps 2 through 5 until they agree.

C.   Unknown Diameter

In this problem, it is desired to determine the size of the pipe (D) that will transport a given fluid (Newtonian or non-Newtonian) at a given flow rate (Q) over a given distance (L) with a given driving force (DF). Because the unknown (D) appears in each of the dimensionless variables, it is appropriate to regroup these variables in a more convenient form for this problem.

1. Newtonian Fluid

The problem statement is as follows:

Given: $DF, Q, L, \varepsilon, \rho, \mu$ *Find*: D

We can eliminate the unknown (D) from two of the three basic groups (N_{Re}, ε/D, and f) as follows:

$$fN_{Re}^5 = \left(\frac{DF\pi^2 D^5}{32LQ^2}\right)\left(\frac{4Q\rho}{\pi D\mu}\right)^5 = \frac{32DF\rho^5 Q^3}{\pi^3 L\mu^5} \tag{6.78}$$

$$N_R = \frac{N_{Re}}{\varepsilon/D} = \frac{4Q\rho}{\pi\mu\varepsilon} \tag{6.79}$$

Thus, the three basic groups for this problem are fN_{Re}^5, N_R, and N_{Re}, with N_{Re} being the dimensionless "unknown" (because it is now the only group containing the unknown D). Because D is unknown, no initial estimate for f can be obtained from the equations because ε/D is also unknown. Thus, the following procedure is recommended for this problem:

1. Calculate fN_{Re}^5 from known quantities using Equation 6.78.
2. Assume $f = 0.005$.
3. Calculate N_{Re} from

$$N_{Re} = \left(\frac{fN_{Re}^5}{0.005}\right)^{1/5}. \tag{6.80}$$

4. Calculate D from N_{Re}:

$$D = \frac{4Q\rho}{\pi\mu N_{Re}}. \tag{6.81}$$

5. Calculate ε/D.
6. Determine f from the Moody diagram or Churchill equation using these values of N_{Re} and ε/D (if $N_{Re} < 2000$, use $f = 16/N_{Re}$).
7. Compare the value of f from step 6 with the assumed value in step 2. If they do not agree, use the result of step 6 for f in step 3 in place of 0.005 and repeat steps 3 through 7 until they agree.

2. Power Law Fluid

The problem statement is as follows:

Given: DF, Q, m, n, ρ, L *Find*: D

The procedure is analogous to that for the Newtonian fluid. In this case, the combined group $fN_{Re,pl}^{5/(4-3n)}$ (which we shall call K, for convenience) is independent of D.

The following procedure can be used to find D:

1. Calculate K from Equation 6.82:

$$fN_{Re,pl}^{5/(4-3n)} = \left(\frac{\pi^2 DFD^5}{32LQ^2}\right)\left[\frac{2^{7-3n}\rho Q^{2-n}}{D^{4-3n}m\pi^{2-n}}\left(\frac{n}{3n+1}\right)^n\right]^{5/(4-3n)} = K \tag{6.82}$$

2. Assume $f = 0.005$.
3. Calculate $N_{Re,pl}$ from

$$N_{Re,pl} = \left(\frac{K}{f}\right)^{(4-3n)/5} \tag{6.83}$$

using $f = 0.005$.

4. Calculate f from Equation 6.47, using the value of $N_{Re,pl}$ from step 3.
5. Compare the result of step 4 with the assumed value in step 2. If they do not agree, use the value of f from step 4 in step 3, and repeat steps 3 through 5 until they agree. The diameter D is obtained from the last (converged) value of $N_{Re,pl}$ from step 3:

$$D = \left[\frac{2^{7-3n} \rho Q^{2-n}}{m\pi^{2-n} N_{Re,pl}} \left(\frac{n}{3n+1} \right)^n \right]^{1/(4-3n)} \quad (6.84)$$

3. Bingham Plastic

The problem variables are as follows:

Given: DF, Q, μ_∞, τ_o, ρ, L *Find*: D

The combined group that is independent of D is equivalent to Equation 6.78, that is,

$$fN_{Re}^5 = \left(\frac{D^5 \pi^2 DF}{32LQ^2} \right) \left(\frac{4Q\rho}{D\pi\mu_\infty} \right)^5 = \left(\frac{32 DF Q^3 \rho^5}{\pi^3 L \mu_\infty^5} \right) \quad (6.85)$$

The following procedure can be used to find D:

1. Calculate fN_{Re}^5 from Equation 6.85.
2. Assume $f = 0.01$.
3. Calculate N_{Re} from

$$N_{Re} = \left(\frac{fN_{Re}^5}{0.01} \right)^{1/5}. \quad (6.86)$$

4. Calculate D from

$$D = \frac{4Q\rho}{\pi\mu_\infty N_{Re}}. \quad (6.87)$$

5. Calculate N_{He} from

$$N_{He} = \frac{D^2 \rho \tau_o}{\mu_\infty^2}. \quad (6.88)$$

6. Calculate f from Equation 6.65 using the values of N_{Re} and N_{He} from steps 3 and 5.
7. Compare the value of f from step 6 with the assumed value in step 2. If they do not agree, insert the result of step 6 for f into step 3 in place of 0.01, and repeat steps 3 through 7 until they agree.

The resulting value of D is determined in step 4.

Example 6.3 Unknown Diameter for a Newtonian Fluid

The flow configuration is the same as for Example 6.2 (see Figure E6.2).
    In this case, the flow rate is given and it is desired to determine the pipe diameter that will deliver 600 gpm of water at 60°F with the given pump.

Solution:

The known quantities are listed here. We will assume that the pump in Example 6.2 operates the same on water as it does with the slurry, so that the pressure developed by the pump is the same.

However, this is NOT a good assumption, as pumps designed to pump slurries are normally significantly different from those designed for clear liquids and have different performance characteristics. Nevertheless, we make this assumption here for the sake of simplicity.

$$\text{Pump pressure} = -\Delta P = (85 \text{ psi})(\text{atm}/14.696 \text{ psi})(1.013 \times 10^6 \text{dyn}/[\text{cm}^2\text{atm}]) = 3.66 \times 10^6 \text{dyn/cm}^2$$

$$DF = -\Delta P/\rho = (3.66 \times 10^6 \text{dyn/cm}^2)(g \text{ cm/s}^2\text{dyn})/(1 \text{ g/cm}^3) = 3.66 \times 10^6 \text{ cm}^2/\text{s}^2$$

$$L = 300 \text{ ft } (30.48 \text{ cm/ft}) = 9\,144 \text{ cm}$$

$$\varepsilon = 0.0018 \text{ in.} = 0.00457 \text{ cm}$$

$$Q = 600 \text{ gpm} = (600 \text{ gpm})(63.09 \text{ cm}^3/(\text{s gpm})) = 37{,}860 \text{ cm}^3/\text{s}$$

$$\mu = 1 \text{ cP} = 0.01 \text{ } P = 0.01 \text{ dyn s/cm}^2 = 0.01 \text{ g/cm s}$$

$$\rho = 1 \text{ g/cm}^3$$

1. Calculate fN_{Re}^5 using the following equation:

$$fN_{Re}^5 = \frac{32DF\rho^5Q^3}{\pi^3L\mu^5}$$

$$= \frac{32(3.66\times10^6 \text{ cm}^2/\text{s}^2)(1\text{g/cm}^3)^5(37{,}860\,\text{cm}^3/\text{s})^3}{\pi^3(9{,}144\,\text{cm})(0.01\,\text{g/cm s})^5} = 2.151\times10^{26} \qquad (6.78)$$

2. Assume $f = 0.005$.
3. Calculate N_{Re} from

$$N_{Re} = \left(\frac{fN_{Re}^5}{0.005}\right)^{1/5} = \left(\frac{2.151\times10^{26}}{0.005}\right)^{1/5} = 5.33\times10^5$$

4. Calculate D from N_{Re}:

$$D = \frac{4Q\rho}{\pi\mu N_{Re}} = \frac{4(37{,}850\,\text{cm}^3/\text{s})(1\text{g/cm}^3)}{\pi(0.01\text{g/cm s})5.33\times10^5} = 9.04\,\text{cm} \qquad (6.81)$$

5. Calculate $\varepsilon/D = 0.00457/9.04 = 5.06 \times 10^{-4}$.
6. Determine f from the Churchill equation using the aforementioned values $f = 0.0044$.

Use this value in step 3, and revise the answer:

$$N_{Re2} = N_{Re1}\left(\frac{f_1}{f_2}\right)^{1/5} = 5.33\times10^5\left(\frac{0.005}{0.0044}\right)^{1/5} = 5.46\times10^5$$

Calculate D_2 from N_{Re2}:

$$D_2 = \frac{4Q\rho}{\pi\mu N_{Re2}} = \frac{4(37{,}850\,\text{cm}^3/\text{s})(1\text{g/cm}^3)}{\pi(0.01\text{g/cm s})5.46\times10^5} = 8.83\,\text{cm} = 3.50\,\text{in.}$$

$$\varepsilon/D = 0.00457/8.83 = 5.18 \times 10^{-4}$$

This gives basically the same value of f from the Churchill equation, so the answer is $D = 8.83$ cm or 3.50 in. It is not reasonable to expect standard commercial pipe to have the exact ID as calculated, so we would choose the pipe with the closest ID that would withstand the maximum pressure (which is 85 psig and small enough that any schedule pipe diameter should suffice).

Actually, consulting the commercially available steel pipe dimensions (see Appendix F), we see that 3½ sch 40s pipe has an ID very close to that required here (3.55 in.). If this were not the case, we would choose the commercial pipe with the next larger standard ID for this system.

D. USE OF TABLES

The relationship between flow rate, pressure drop, and pipe diameter for *water flowing at 60°F in sch 40 horizontal pipe* is tabulated in Appendix G over a range of pipe velocities that cover the most common conditions. For this special case, no iteration or other calculation procedures are required for any of the unknown driving force, unknown flow rate, or unknown diameter problems (although interpolation in the table is usually necessary). Note that the friction loss is tabulated in this table as pressure drop (in psi) per 100 ft of pipe, which is equivalent to $(100 \, \rho e_f / 144 \, L)$ in the Bernoulli equation, where ρ is in ($\mathrm{lb_m/ft^3}$), e_f is in (ft $\mathrm{lb_f /lb_m}$), and L is in (ft).

VII. TUBE FLOW (POISEUILLE) VISCOMETER

In Section II.B of Chapter 3, the tube flow viscometer was described in which the viscosity of any fluid with unknown viscous properties could be determined from measurements of the total pressure gradient $(-\Delta\Phi/L)$ and the volumetric flow rate (Q) in a tube of known dimensions. The viscosity is given by

$$\eta = \frac{\tau_w}{\dot{\gamma}_w} \tag{6.89}$$

where τ_w follows directly from the pressure gradient and Equation 6.4 and the wall shear rate is given by

$$\dot{\gamma}_w = \Gamma\left(\frac{3n'+1}{4n'}\right) \tag{6.90}$$

where
$\Gamma = 4 \, Q/\pi \, R^3 = 8V/D$ and

$$n' = \frac{d \log \tau_w}{d \log \Gamma} = \frac{d \log \Delta\Phi}{d \log Q} \tag{6.91}$$

is the point slope of $-\Delta\Phi$ versus Q at each measured value of Q. Equation 6.90 is completely independent of the specific fluid viscous properties and can be derived from Equation 6.7 as follows. By using Equation 6.4, the independent variable in Equation 6.7 can be changed from r to τ_{rx}, that is,

$$Q = -\pi \int_0^R r^2 \dot{\gamma} \, dr = \frac{\pi R^3}{\tau_w} \int_0^{\tau_w} \tau_{rx}^2 \dot{\gamma} \, d\tau_{rx} \tag{6.92}$$

This can be solved for the shear rate at the tube wall ($\dot{\gamma}_w$) by first differentiating Equation 6.92 with respect to the *parameter* τ_w by the application of Leibniz's rule,* to give

$$\frac{d(\Gamma \tau_w^3)}{d\tau_w} = 4\tau_w^2 \dot{\gamma}_w \tag{6.93}$$

---

* Leibnitz rule:

$$\frac{\partial}{\partial x} \int_{A(x)}^{B(x)} I(x,y) \, dy = \int_{A(x)}^{B(x)} \frac{\partial I}{\partial x} \, dy + I(x,B)\frac{\partial B}{\partial x} - I(x,A)\frac{\partial A}{\partial x}.$$

where $\Gamma = 4Q/\pi R^3$. Solving for $\dot{\gamma}_w$:

$$\dot{\gamma}_w = \frac{1}{4\tau_w^2}\frac{d(\Gamma\tau_w^3)}{d\tau_w} = \frac{\tau_w}{4}\left[\frac{d\Gamma}{d\tau_w} + 3\frac{\Gamma}{\tau_w}\right] = \Gamma\left(\frac{3n'+1}{4n'}\right) \tag{6.94}$$

where $n' = d(\log \tau_w)/d(\log \Gamma)$ is the local slope of the log-log plot of τ_w versus Γ (or $-\Delta\Phi$ vs Q) at each measured value of Q.

VIII.   TURBULENT DRAG REDUCTION

A remarkable effect was observed by Toms during World War II when pumping NAPALM (a "jellied" solution of a polymer in gasoline). He found that the polymer solution could be pumped through pipes in turbulent flow with considerably lower pressure drop (friction loss) than that exhibited by the gasoline at the same flow rate in the same pipe without the polymer. This phenomenon, known as *turbulent drag reduction* (or the *Toms effect*), has been observed for solutions (mostly aqueous) of a variety of very high polymers (e.g., molecular weights on the order of 10^6) and has been the subject of a large amount of research. The effect is very significant because as much as 85% *less* energy is required to pump solutions of some high polymers at concentrations of 100 ppm or less through pipes than is required to pump the solvent alone at the same flow rate through the same pipe. This is illustrated in Figure 6.5, which shows some of Chang's data (Darby and Chang, 1984) for the Fanning friction factor versus Reynolds number (based on the solvent viscosity) for fresh and "degraded" polyacrylamide solutions of concentrations from 100 to 500 ppm in a 2 mm diameter tube. Note that the friction factor at low Reynolds numbers (laminar flow) is much larger than that for the (Newtonian) solvent, whereas it is much lower at high (turbulent) Reynolds numbers. The non-Newtonian viscosity of these solutions is shown in Figure 3.9 in Chapter 3.

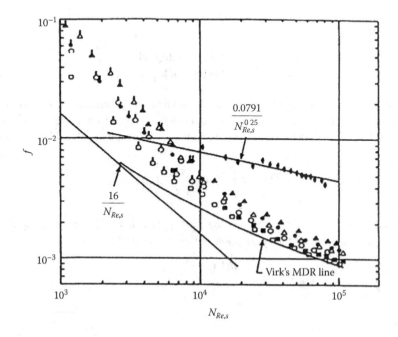

FIGURE 6.5   Drag reduction data for polyacrylamide solutions ($N_{Re,s}$ is the Reynolds number based on solvent properties.) MDR is Virk's maximum drag reduction asymptote (Virk, 1975). (From Darby, R. and Chang, H.D., *AIChE J.*, 30, 274, 1984.)

Although the exact mechanism is debatable, Darby and Chang (1984) and Darby and Pivsa-Art (1991) have presented a model for turbulent drag reduction based on the fact that solutions of very high molecular weight polymers are viscoelastic and the concept that in any unsteady deformation (such as turbulent flow) elastic properties will store energy that would otherwise be dissipated in a purely viscous fluid. Since energy that is dissipated (i.e., the "friction loss") must be made up by adding energy (e.g., by a pump) to sustain the flow, that portion of the energy that is stored by elastic deformations remains in the flow and does not have to be made up by external energy sources. Thus, less energy must be supplied externally to sustain the flow of a viscoelastic fluid, that is, the drag is reduced. This concept is analogous to that of bouncing an elastic ball. If there is no viscosity (i.e., internal friction) to dissipate the energy, the ball will continue to bounce indefinitely with no external energy input needed. However, a viscous ball will not bounce at all, because all of the energy is dissipated by viscous deformation upon contact with the floor and is transformed into "heat." Thus, the greater the fluid elasticity in proportion to the viscosity, the lesser the amount of energy that must be added to replace that which is dissipated by the turbulent motion of the flow.

The model for turbulent drag reduction developed by Darby and Chang (1984) and later modified by Darby and Pivsa-Art (1991) shows that for smooth tubes the friction factor versus Reynolds number relationship for Newtonian fluids (e.g., the Colebrook or Churchill equation) may also be used for drag reducing flows, provided (1) the Reynolds number is defined using the properties (e.g., viscosity) of the Newtonian solvent and (2) the Fanning friction factor is modified as follows:

$$f_p = \frac{f_s}{\sqrt{1 + N_{De}^2}} \tag{6.95}$$

where

f_s is the solvent (Newtonian) Fanning friction factor as predicted for a Newtonian fluid with the viscosity of the solvent using the (Newtonian) Reynolds number

f_p is a "generalized" Fanning friction factor that applies to drag reducing polymer solutions as well as Newtonian fluids

N_{De} is the dimensionless Deborah number, which depends upon the fluid viscoelastic properties and accounts for the storage of energy by the elastic deformations

The replotted data are well represented by the classic Colebrook equation shown as the line on this plot. Figure 6.6 shows the data from Figure 6.5 (and many other data sets as well) replotted in terms of the generalized friction factor.

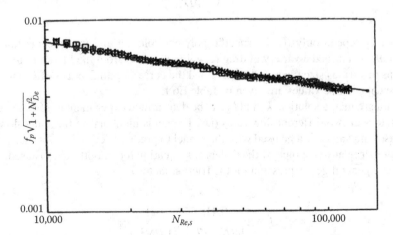

FIGURE 6.6 Drag reduction data replotted in terms of generalized friction factor. (From Darby, R. and Pivsa-Art, S., *Canad. J. Chem. Eng.*, 69, 1395, 1991.)

The complete expression for N_{De} is given by Darby and Pivsa-Art (1991) as a function of the viscoelastic properties of the fluid (i.e., the Carreau parameters η_o, λ, and p). This expression is as follows:

$$N_{De} = \frac{0.0163 N_\varsigma N_{Re,s}^{0.338} (\mu_s/\eta_o)^{0.5}}{\left[1/N_{Re,s}^{0.75} + 0.00476 N_\zeta^2 (\mu_s/\eta_o)^{0.75} \right]^{0.318}} \tag{6.96}$$

where

$$N_\zeta = \left[\left(1 + N_\lambda^2 \right)^p - 1 \right]^{0.5} \tag{6.97}$$

and

$$N_\lambda = \frac{8V\lambda}{D} \tag{6.98}$$

where

$N_{Re,s}$ is the Reynolds number based on the solvent properties
μ_s is the solvent viscosity
D is the pipe diameter
V is the velocity in the pipe
λ is the fluid time constant (from the Carreau model fit of the viscosity curve)

Inasmuch as the rheological properties are very difficult to measure for very dilute solutions (e.g., 100 ppm or less), a simplified expression was developed by Darby and Pivsa-Art (1991) in which these rheological parameters are contained within two "constants," k_1 and k_2:

$$N_{De} = k_2 \left(\frac{8\mu_s N_{Re,s}}{\rho D^2} \right)^{k_1} N_{Re,s}^{0.34} \tag{6.99}$$

where k_1 and k_2 depend only on the specific polymer solution and its concentration. Darby and Pivsa-Art (1991) examined a variety of drag reducing data sets from the literature for various polymer solutions in various size pipes and determined the corresponding values of k_1 and k_2 that fit the model to the data. These values are given in Table E6.1.

For any drag reducing solution, k_1 and k_2 can be determined experimentally from two data points in the laboratory at two different flow rates (i.e., Reynolds numbers) in turbulent flow in any size pipe. The resulting values can be used with the model to predict friction loss for that solution at any Reynolds number in any size pipe. If the Colebrook equation for smooth tubes is used, for example, the appropriate generalized expression for the friction factor is

$$f = \frac{0.41}{\left[\ln(N_{Re,s}/7) \right]^2} \frac{1}{\left(1 + N_{De}^2 \right)^{1/2}} \tag{6.100}$$

TABLE E6.1
Parameters for Equation 6.99 for Various Polymer Solutions

Polymer	Concentration (mg/kg)	Diameter (cm)	k_1 (—)	k_2 (s^{k_1})	References
Guar gum	20	1.27	0.05	0.009	Wang (1972)
(Jaguar A-20-D)	50		0.06	0.014	
	200		0.07	0.022	
	500		0.10	0.029	
	1000		0.16	0.028	
Guar gum	30		0.05	0.008	White (1966)
	60		0.06	0.010	
	240		0.08	0.016	
	480		0.11	0.018	
Polyacrylamide	100	0.176–1.021	0.093	0.0342	Darby and Pivsa-Art (1991)
Separan AP-30	250		0.095	0.0293	
(fresh)	500		0.105	0.0244	
Separan AP-30	100		0.088	0.0431	
(degraded)	250		0.095	0.0360	
	500			0.0280	
AP-273	10	1.090	0.12	0.0420	White and Gordon (1975)
PAM E198	10	0.945	0.21	0.0074	
	280			0.0078	Virk and Baher (1970)
PAA	300	2.0 and 3.0	0.40	0.0050	
	700		0.53	0.0049	Hoffmann and Schummer (1978)
ET-597	125	0.69	0.47	0.00037	Astarita et al. (1969)
	250	1.1 and 2.05	0.39	0.0013	
	500		0.30	0.0061	
Hydroxyethyl	100	2.54	0.10	0.0074	Wang (1972)
cellulose	200		0.16	0.0072	
(OP-100M)	500		0.24	0.0068	
	1000		0.35	0.0063	
(HEC)	2860	4.8, 1.1, and 2.05	0.02	0.0310	Savins (1969)
Polyethylene oxide	10	5.08	0.22	0.017	Goren and Norbury (1967)
WSR 301	20		0.21	0.016	
	50		0.19	0.014	
W205	10	0.945	0.31	0.0022	Virk and Baher (1970)
	105		0.26	0.0080	
Xanthan gum	1000	0.52	0.02	0.046	Bewersdorff and Berman (1988)
(Rhodopol 23)					

Source: Darby, R. and Pivsa-Art, S., *Canad. J. Chem. Eng.*, 69, 1395, 1991.

Example 6.4 Friction Loss in Drag Reducing Solutions

Determine the percentage reduction in the power required to pump water through a 3 in. ID smooth pipe at 300 gpm by adding 100 wppm of "degraded" Separan AP-30.

Solution:
We first calculate the Reynolds number for the solvent (water) under the given flow conditions using a viscosity of 0.01 poise and a density of 1 g/cm³:

$$N_{Re,s} = \frac{4Q\rho}{\pi D\mu} = \frac{4(300\,\text{gpm})[63.1\,\text{cm}^3/\text{s}\,\text{gpm}](1\,\text{g/cm}^3)}{\pi(3\,\text{in.})(2.54\,\text{cm/in.})(0.01\,\text{g/cm}\,\text{s})} = 3.15 \times 10^5$$

Then calculate the Deborah number from Equation 6.99 using $k_1 = 0.088$ and $k_2 = 0.0431$ taken from Table E6.1:

$$N_{De} = k_2\left(\frac{8\mu_s N_{Re,s}}{\rho D^2}\right)^{k_1} N_{Re,s}^{0.34} = 0.0431\left(\frac{8 \times 0.01(\text{g/cm}\,\text{s}) \times 3.15 \times 10^5}{1(\text{g/cm}^3) \times (3\,\text{in.} \times 2.54\,(\text{cm/in.}))^2}\right)^{0.088} (3.15 \times 10^5)^{0.34} = 5.45$$

These values can now be used to calculate the smooth pipe friction factor from Equation 6.100. Excluding the N_{De} term gives the friction factor for the Newtonian solvent (f_s), and including the N_{De} term gives the friction factor for the polymer solution (f_p) under the same flow conditions:

$$f_s = \frac{0.41}{\left[\ln(N_{Re,s}/7)\right]^2} = 0.00357$$

$$f_p = \frac{0.41}{[\ln(N_{Re,s}/7)]^2} \frac{1}{\left(1 + N_{De}^2\right)^{1/2}} = 0.000645$$

The power (HP) required to pump the fluid is given by $-\Delta PQ$. Because $-\Delta P$ is proportional to fQ^2 and Q is the same with and without the polymer, the fractional reduction in power is given by

$$\boxed{DR = \frac{HP_s - HP_p}{HP_s} = \frac{f_s - f_p}{f_s} = (0.00357 - 0.000645)/0.00357 = 0.82}$$

That is, adding the polymer results in an 82% *reduction* in the power required to overcome drag.

SUMMARY

This chapter contains all of the fundamental information and methods for solving virtually any kind of problem involving pipe flows under any and all conditions and for a wide variety of fluids. The major concepts that should be retained from this chapter include the following:

- Understand the principles governing the flow of Newtonian and non-Newtonian in pipes, including the application of the momentum and energy balances (and the similarity thereof).
- Understand the concepts governing turbulent flow, the models that describe turbulent pipe flow, and applications including turbulent drag reduction.
- Be able to solve for either the unknown driving force or the unknown flow rate or the unknown diameter in pipes for Newtonian and non-Newtonian fluids.

PROBLEMS

PIPE FLOWS

1. Show how the Hagen–Poiseuille equation for the steady laminar flow of a Newtonian fluid in a uniform cylindrical tube can be derived starting from the general microscopic equations of motion (e.g., the continuity and momentum equations given in Appendix E).

2. The Hagen–Poiseuille equation (Equation 6.11) describes the laminar flow of a Newtonian fluid in a tube. Since a Newtonian fluid is defined by the relation $\tau = \mu \dot{\gamma}$, rearrange the Hagen–Poiseuille equation to show that the shear rate at the tube wall for a Newtonian fluid is given by $\dot{\gamma}_w = 4Q/\pi R^3 = 8V/D$.

3. Derive the relation between the friction factor and Reynolds number in turbulent flow for smooth pipe (Equation 6.34), starting with the von Karman equation for the velocity distribution in the turbulent boundary layer (Equation 6.26).

4. Evaluate the kinetic energy correction factor α in the Bernoulli equation for turbulent flow assuming the 1/7 power law velocity profile (Equation 6.36) is valid. Repeat this for laminar flow of a Newtonian fluid in a tube, for which the velocity profile is parabolic.

5. A Newtonian fluid with SG = 0.8 is forced through a capillary tube at a rate of 5 cm³/min. The tube has a downward slope of 30° to the horizontal, and the pressure drop is measured between two taps located 40 cm apart on the tube using a mercury manometer, which reads 3 cm. When water is forced through the tube at a rate of 10 cm³/min, the manometer reading is 2 cm.
   (a) What is the viscosity of the unknown Newtonian fluid?
   (b) What is the Reynolds number of the flow for each fluid?
   (c) If two separate pressure transducers, which read the total pressure directly in psig, were used to measure the pressure at each of the pressure taps directly instead of using the manometer, what would be the difference in the transducer readings?

6. A liquid is draining from a cylindrical vessel through a tube in the bottom of the vessel, as illustrated in Figure P6.6 below. If the liquid has a specific gravity of 0.85 and drains out at a rate of 1 cm³/s, what is the viscosity of the liquid? The entrance loss coefficient from the tank to the tube is 0.4, and the system has the following dimensions: $D = 2$ in., $d = 2$ mm, $L = 10$ cm, and $h = 5$ cm

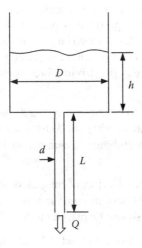

FIGURE P6.6   Fluid draining through tube.

7. You are given a liquid and are asked to find its viscosity. Its density is known to be 0.97 g/cm³. You place the fluid in an open vessel to which a 20 cm long vertical tube with an inside diameter of 2 mm is attached to the bottom (see Figure P6.6). When the depth of the liquid in the container is 6 cm, you find that it drains out through the tube at a rate of 2.5 cm³/s. If the diameter of the open vessel is much larger than that of the tube and friction loss from the vessel to the tube is negligible, what is the fluid viscosity?

8. Repeat problem 7 accounting for the friction loss from the vessel to the tube, assuming a loss coefficient of 0.50 at the contraction.

9. You must measure the viscosity of an oil that has an SG of 0.92. To do this, you put the oil into a large container to the bottom of which a small vertical tube, 25 cm long, has been attached through which the oil can drain by gravity (see Figure P6.6). When the level of the oil in the container is 6 in. above the container bottom, you find that the flow rate through the tube is 50 cm³/min. You run the same experiment with water instead of oil and find that under the same conditions the water drains out at a rate of 156 cm³/min. If the loss coefficient for the energy dissipated in the contraction from the container to the tube is 0.5, what is the viscosity of the oil?

10. You want to transfer No. 3 fuel oil (30°API) from a storage tank to a power plant at a rate of 2000 bbl/day. The diameter of the pipeline is 1½ in. sch 40, with a length of 1200 ft. The discharge of the line is 20 ft higher than the suction end, and both ends are at 1 atm pressure. The inlet temperature of the oil is 60°F, and the transfer pump is 60% efficient. If the specific heat of the oil is 0.5 Btu/(lb$_m$ °F) and the pipeline is perfectly insulated, determine
    (a) The horsepower of the motor required to drive the pump
    (b) The temperature of the oil leaving the pipeline

11. You must specify a pump to deliver 800 bbl/day of a 35° API distillate at 90°F from a distillation column to a storage tank in a refinery. If the level in the tank is 20 ft above that in the column, the total equivalent length of pipe is 900 ft, and both the column and tank are at atmospheric pressure, what horsepower would be needed if you use 1½ in. sch 40 pipe? What power would be needed if you use 1 in. sch 40 pipe?

12. Water is flowing at a rate of 700 gpm through a horizontal 6 in. sch 80 commercial steel pipe at 90°F. If the pressure drops by 2.23 psi over a 100 ft length of pipe, determine the following:
    (a) What is the value of the Reynolds number?
    (b) What is the magnitude of the pipe wall roughness?
    (c) How much driving force (i.e., pressure difference) would be required to move the water at this flow rate through 10 miles of pipe if it were made of commercial steel?
    (d) What size commercial steel pipe would be required to transport the water at the same flow rate over the same distance if the driving force is the static head in a water tower 175 ft above the pipe?

13. A 35° API distillate at 60°F is to be pumped over a distance of 2000 ft through a 4 in. sch 40 horizontal pipeline at a flow rate of 500 gpm. What power must the pump deliver to the fluid if the pipeline is made of (a) drawn tubing, (b) commercial steel, (c) galvanized iron, and (d) PVC plastic?

14. The Moody diagram illustrates the effect of roughness on the friction factor in turbulent flow but indicates no effect of roughness in laminar flow. Explain why this is so. Are there any restrictions or limitations that should be placed on this conclusion? Explain.

15. You have a large supply of very rusty 2 in. sch 40 steel pipe, which you want to use for a pipeline. Because rusty metal is rougher than clean metal, you want to know its effective roughness

before laying the pipeline. To do this, you pump water at a rate of 100 gpm through a 100 ft long section of the pipe and find that the pressure drops by 15 psi over this length. What is the effective pipe roughness in inches?

16. A 32 hp pump (100% efficient) is required to pump water through a 2 in. sch 40 pipeline, 6000 ft long, at a rate of 100 gpm.
    (a) What is the equivalent roughness of the pipe?
    (b) If the pipeline is replaced by new commercial steel 2 in. sch 40 pipe, what power would be required to pump water at a rate of 100 gpm through this pipe? What would be the percentage saving in power compared to the old pipe?

17. You have a piping system in your plant that has gotten old and rusty. The pipe is 2 in. sch 40 steel, 6000 ft long. You find that it takes 35 hp to pump water through the system at a rate of 100 gpm.
    (a) What is the equivalent roughness of the pipe?
    (b) If you replace the pipe with the same size new commercial steel pipe, what percentage savings in the required power would you expect at a flow rate of 100 gpm?

18. Water enters a horizontal tube through a flexible vertical rubber hose that can support no forces. If the tube is 1/8 in. sch 40, 10 ft long, and the water flow rate is 2 gpm, what force (magnitude and direction) must be applied to the tube to keep it stationary? Neglect the weight of the tube and the water in it. The hose ID is the same as that of the tube.

19. A water tower that is 90 ft high provides water to a residential subdivision. The water main from the tower to the subdivision is 6 in. sch 40 steel, 3 miles long. If each house uses a maximum of 50 gal/h (at peak demand) and the pressure in the water main is not to be less than 30 psig at any point, how many homes can be served by the water main?

20. A heavy oil ($\mu = 100$ cP, SG = 0.85) is draining from a large tank through a 1/8 in. sch 40 tube into an open bucket. The level in the tank is 3 ft above the tube inlet, and the pressure in the tank is 10 psig. The tube is 30 ft long, and it is inclined downward at an angle of 45° to the horizontal. What is the flow rate of the oil in gpm? What is the value of the Reynolds number in this problem?

21. SAE 10 lube oil (SG = 0.93) is being pumped upward through a straight 1/4 in. sch 80 pipe that is oriented at 45° angle to the horizontal. The two legs of a manometer using water as the manometer fluid are attached to taps in the pipe wall that are 2 ft apart. If the manometer reads 15 in., what is the oil flow rate in gal/h?

22. Cooling water is fed by gravity from an open storage tank 20 ft above ground, through 100 ft of 1½ in. ID steel pipe, to a heat exchanger at ground level. If the pressure entering the heat exchanger must be 5 psig for it to operate properly, what is the water flow rate through the pipe?

23. A water main is to be laid to supply water to a subdivision located 2 miles from a water tower. The water in the tower is 150 ft above ground, and the subdivision consumes a maximum of 10,000 gpm of water. What size pipe should be used for the water main? Assume Schedule 40 commercial steel pipe. The pressure above the water in the tank is 1 atm and is 30 psig at the subdivision.

24. A water main is to be laid from a water tower to a subdivision that is 2 miles away. The water level in the tower is 150 ft above the ground. The main must supply a maximum of 1000 gpm with a minimum of 5 psig at the discharge end, at a temperature of 65°F. What size commercial steel sch 40 pipe should be used for the water main? If plastic pipe (which is hydraulically smooth) were used instead, would this alter the result? If so, what diameter of plastic pipe should be used?

25. The water level in a water tower is 110 ft above ground level. The tower supplies water to a subdivision, 3 miles away, through an 8 in. sch 40 steel water main. If the minimum water pressure entering the residential water lines at the houses must be 15 psig, what is the capacity of the water main (in gpm)? If there are 100 houses in the subdivision and each consumes water at a peak rate of 20 gpm, how big should the water main be?

26. A hydraulic press is powered by a remote high-pressure pump. The gage pressure at the pump is 20 MPa, and the pressure required to operate the press is 19 MPa (gage) at a flow rate of 0.032 m^3/min. The press and pump are to be connected by 50 m of drawn stainless steel tubing. The fluid properties are those of SAE 10 lube oil at 40°C. What is the minimum tubing diameter that can be used?

27. Water is to be pumped at a rate of 100 gpm from a well that is 100 ft deep, through 2 miles of horizontal 4 in. sch 40 steel pipe, to a water tower that is 150 ft high.
    (a) Neglecting fitting losses, what horsepower will the pump require if it is 60% efficient?
    (b) If the elbow in the pipe at ground level below the tower breaks off, how fast will the water drain out of the tower?
    (c) How fast would it drain out if the elbow at the top of the well gave way instead?
    (d) What size pipe would you have to run from the water tower to the ground in order to drain it at a rate of 10 gpm?

28. A concrete pipe storm sewer, 4 ft in diameter, drops 3 ft in elevation per mile of length. What is the maximum capacity of the sewer (in gpm) when it is flowing full?

29. You want to siphon water from an open tank using a 1/4 in. diameter hose. The discharge end of the hose is 10 ft below the water level in the tank, and the siphon will not operate if the pressure falls below 1 psia anywhere in the hose. If you want to siphon the water at a rate of 1 gpm, what is the maximum height above the water level in the tank that the hose can extend and still operate?

NON-NEWTONIAN PIPE FLOWS

30. Equation 6.43 describes the laminar flow of a power law fluid in a tube. Since a power law fluid is defined by the relation $\tau = m\dot{\gamma}^n$, rearrange Equation 6.43 to show that the shear rate at the tube wall for a power law fluid is given by $\dot{\gamma}_w = (8V/D)(3n+1)/4n$ where $8V/D$ is the wall shear rate for a Newtonian fluid.

31. A large tank contains SAE 10 lube oil at a temperature of 60°F and a pressure of 2 psig. The oil is 2 ft deep in the tank and drains out through a vertical tube in the bottom. The tube is 10 ft long and discharges the oil at atmospheric pressure. Assuming the oil to be Newtonian and neglecting the friction loss from the tank to the tube, how fast will it drain through the tube? If the oil is not Newtonian, but instead can be described as a power law fluid with a flow index of 0.4 and an apparent viscosity of 80 cP at a shear rate of 1 s^{-1}, how would this affect your answer? The tube diameter is 1/2 in.

32. A polymer solution is to be pumped at a rate of 3 gpm through a horizontal 1 in. diameter pipe. The solution behaves as a power law fluid with a flow index of 0.5, an apparent viscosity of 400 cP at a shear rate of 1 s^{-1}, and a density of 60 lb_m/ft^3.
    (a) What is the pressure gradient in psi/ft?
    (b) What is the shear rate at the pipe wall and the apparent viscosity of the fluid at this shear rate?
    (c) If the fluid were Newtonian, with a viscosity equal to the apparent viscosity from (b) above, what would the pressure gradient be?
    (d) Calculate the Reynolds numbers for the polymer solution and for the above Newtonian fluid.

33. A coal slurry that is characterized as a power law fluid has a flow index of 0.4 and an apparent viscosity of 200 cP at a shear rate of 1 s⁻¹. If the coal has a specific gravity of 2.5 and the slurry is 50% coal by weight in water, what pump horsepower will be required to transport 25 million tons of coal per year through a 36 in. ID, 1000 miles long pipeline? Assume that the entrance and exit of the pipeline are at the same pressure and elevation and that the pumps are 60% efficient.

34. A coal slurry is found to behave as a power law fluid, with a flow index of 0.3, a specific gravity of 1.5, and an apparent viscosity of 70 cP at a shear rate of 100 s⁻¹. What volumetric flow rate of this fluid would be required to reach turbulent flow in a 1/2 in. ID smooth pipe, which is 15 ft long? What is the pressure drop in the pipe (in psi) under these conditions?

35. A coal slurry is to be transported by pipeline. It has been determined that the slurry may be described by the power law model, with a flow index of 0.4, an apparent viscosity of 50 cP at a shear rate of 100 s⁻¹, and a density of 90 lb$_m$/ft³. What horsepower would be required to pump the slurry at a rate of 900 gpm through an 8 in. sch 40 pipe that is 50 miles long?

36. A sewage sludge is to be transported a distance of 3 mi through a 12 in. ID pipeline at a rate of 2000 gpm. The sludge is a Bingham plastic with a yield stress of 35 dyn/cm², a limiting viscosity of 80 cP, and a specific gravity of 1.2. What size motor (in horsepower) would be required to drive the pump if it is 50% efficient?

37. A coal suspension is found to behave as a power law fluid, with a flow index of 0.4, a specific gravity of 1.5, and an apparent viscosity of 90 cP at a shear rate of 100 s⁻¹. What would the volumetric flow rate of this suspension be in a 15 ft long, 5/8 in. ID smooth tube, with a driving force of 60 psi across the tube? What is the Reynolds number for the flow under these conditions?

38. A coal-water slurry containing 65% (by weight) coal is pumped from a storage tank at a rate of 15 gpm through a 50 m long 1/2 in. sch 40 pipeline to a boiler where it is burned. The storage tank is at 1 atm pressure and 80°F, and the slurry must be fed to the burner at 20 psig. The specific gravity of coal is 2.5, and it has a heat capacity of 0.5 Btu/(lb$_m$ °F).
    (a) What power must the pump deliver to the slurry if it is assumed to be Newtonian with a viscosity of 200 cP?
    (b) In reality, the slurry is non-Newtonian and can best be described as a Bingham plastic, with a yield stress of 800 dyn/cm² and a limiting viscosity of 200 cP. Accounting for these properties, what would the required pumping power be?
    (c) If the pipeline is well insulated, what will the temperature of the slurry be when it enters the boiler, for both case (a) and case (b)?

39. A sludge is to be transported by pipeline. It has been determined that the sludge may be described by the power law model, with a flow index of 0.6, an apparent viscosity of 50 cP at a shear rate of 1 s⁻¹, and a density of 95 lb$_m$/ft³. What hydraulic horsepower would be required to pump the sludge at a rate of 600 gpm through a 6 in. ID pipe that is 5 miles long?

40. You must design a transfer system to feed a coal slurry to a boiler. However, you don't know the slurry properties, so you measure them in the lab using a cup and bob (Couette) viscometer. The cup has a diameter of 10 cm and a bob diameter of 9.8 cm, and the length of the bob is 8 cm. When the cup is rotated at a rate of 2 rpm, the torque measured on the bob is 2.4 × 10⁴ dyn cm, and at 20 rpm, it is 6.5 × 10⁴ dyn cm.
    (a) If you use the Bingham plastic model to describe the slurry properties, what are the values of the yield stress and the limiting viscosity?
    (b) If the power law model were used instead, what would be the values of the flow index and consistency?

(c) Using the Bingham plastic model for the slurry, with a value of the yield stress of 35 dyn/cm^2, a limiting viscosity of 35 cP, and a density of 1.2 g/cm^3, what horsepower would be required to pump the slurry through a 1000 ft long, 3 in. ID sch 40 pipe at a rate of 100 gpm?

41. A thick slurry with SG = 1.3 is to be pumped through a 1 in. ID pipe that is 200 ft long. You don't know the properties of the slurry, so you test it in the lab by pumping it through a 4 mm ID tube that is 1 m long. At a flow rate of 0.5 cm^3/s, the pressure drop in this tube is 1 psi, and at a flow rate of 5 cm^3/s, it is 1.5 psi. Estimate the pressure drop that would be required to pump the slurry through the 1 in. pipe at a rate of 2 gpm and also at 30 gpm. Clearly explain the procedure you use, and state any assumptions that you make. Comment in detail about the possible accuracy of your predictions. The slurry SG is 1.3.

42. Drilling mud has to be pumped down into an oil well that is 8000 ft deep. The mud is to be pumped at a rate of 50 gpm to the bottom of the well and back to the surface, through a pipe having an effective ID of 4 in. The pressure at the bottom of the well is 4500 psi. What pump head is required to do this? The drilling mud has properties of a Bingham plastic, with a yield stress of 100 dyn/cm^2, a limiting (plastic) viscosity of 35 cP, and a density of 1.2 g/cm^3.

43. A straight vertical tube, 100 cm long and 2 mm ID, is attached to the bottom of a large vessel. The vessel is open to the atmosphere and contains a liquid with a density of 1 g/cm^3 to a depth of 20 cm above the bottom of the vessel.
    (a) If the liquid drains through the tube at a rate of 3 cm^3/s, what is its viscosity?
    (b) What is the largest tube diameter that can be used in this system to measure the viscosity of liquids that are at least as viscous as water, for the same liquid level in the vessel? Assume that the density is the same as water.
    (c) A non-Newtonian fluid, represented by the power law model, is introduced into the vessel with the 2 mm diameter tube attached. If the fluid has a flow index of 0.65, an apparent viscosity of 5 cP at a shear rate of 10 s^{-1}, and a density of 1.2 g/cm^3, how fast will it drain through the tube, if the level is 20 cm above the bottom of the vessel?

44. A non-Newtonian fluid, described by the power law model, is flowing through a thin slit between two parallel planes of width W, separated by a distance H. The slit is inclined upward at an angle θ to the horizontal.
    (a) Derive an equation relating the volumetric flow rate of this fluid to the pressure gradient, slit dimensions, and fluid properties.
    (b) For a Newtonian fluid, this solution can be written in a dimensionless form as

$$f = 24/N_{Re,h}$$

    where the Reynolds number, $N_{Re,h}$, is based on the hydraulic diameter of the channel. Arrange your solution for the power law fluid in dimensionless form and solve for the friction factor, f.
    (c) Set your result from (b) equal to $24/N_{Re,h}$, and determine an equivalent expression for the power law Reynolds number for slit flow.

45. You are drinking a milk shake through a straw that is 8 in. long and 0.3 in. in diameter. The milk shake has the properties of a Bingham plastic, with a yield stress of 300 dyn/cm^2, a limiting viscosity of 150 cP, and a density of 0.8 g/cm^3.
    (a) If the straw is inserted 5 in. below the surface of the milk shake, how hard must you suck to get the shake flowing through the entire straw (e.g., how much vacuum must you pull, in psi)?
    (b) If you pull a vacuum of 1 psi, how fast will the shake flow (in cm^3/s)?

46. Water is to be transferred at a rate of 500 gpm from a cooling lake through a 6 in. diameter sch 40 pipeline to an open tank in a plant that is 30 miles from the lake.
    (a) If the transfer pump is 70% efficient, what horsepower motor is required to drive the pump?
    (b) An injection station is installed at the lake that injects a high polymer into the pipeline to give a solution of 50 ppm concentration with the following properties: a low shear-limiting viscosity of 80 cP, a flow index of 0.5, and a transition point from low shear Newtonian to shear thinning behavior at a shear rate of 10 s^{-1}. What horsepower is now required to drive the same pump to achieve the same flow rate?

47. You measure the viscosity of a sludge in the lab and conclude that it can be described as a power law fluid with a flow index of 0.45, a viscosity of 7 poise at a shear rate of 1 s^{-1}, and a density of 1.2 g/cm^3.
    (a) What horsepower would be required to pump the sludge through a 3 in. sch 40 pipeline, 1000 ft long, at a rate of 100 gpm?
    (b) The viscosity data show that the sludge could also be described by the Bingham plastic model, with a viscosity of 7 poise at a shear rate of 1 s^{-1} and a viscosity of 0.354 poise at a shear rate of 100 s^{-1}. Using this model, what required horsepower would you predict for this pipeline?
    (c) Which answer do you think would be the most reliable and why?

48. An open drum, 3 ft in diameter, contains a mud that is known to be described by the Bingham plastic model, with a yield stress of 120 dyn/cm^2, a limiting viscosity of 85 cP, and a density of 98 lb$_m$/ft^3. A 1 in. ID hose, 10 ft long, is attached to a hole in the bottom of the drum to drain the mud out. How far below the surface of the mud should the end of the hose be lowered in order to drain the mud at a rate of 5 gpm?

49. You would like to determine the pressure drop–flow rate relation for a slurry in a pipeline. To do this, you must determine the rheological properties of the slurry, so you test it in the lab by pumping it through a 1/8 in. ID pipe that is 10 ft long. You find that it takes 5 psi pressure drop in the pipe to produce a flow rate of 100 cm^3/s and that 10 psi results in a flow rate of 300 cm^3/s.
    (a) What can you deduce about the rheological characteristics of the slurry from these data?
    (b) If it is assumed that the slurry can be adequately described by the power law model, what are the values of the fluid properties, as deduced from the data?
    (c) If the Bingham plastic model is used instead of the power law model to describe the slurry, what are its properties?

50. A pipeline is installed to transport a red mud slurry from an open tank in an alumina plant to a disposal pond. The line is 5 in. sch 80 commercial steel, 12,000 ft long, and is designed to transport the slurry at a rate of 300 gpm. The slurry properties can be described by the Bingham plastic model, with a yield stress of 15 dyn/cm^2, a limiting viscosity of 20 cP, and an SG of 1.3. You may neglect any fittings in this pipeline.
    (a) What delivered pump head and hydraulic horsepower would be required to pump this mud?
    (b) What would be the required pump head and horsepower to pump water at the same rate through the same pipeline?
    (c) If 100 ppm of fresh Separan AP-30 polyacrylamide polymer were added to the water in case (b), what would the required pump head and horsepower be?

51. Determine the power required to pump water at a rate of 300 gpm through a 3 in. ID pipeline, 50 mi long, if
    (a) The pipe is new commercial steel
    (b) The pipe wall is hydraulically smooth
    (c) The pipe wall is smooth, and "degraded" Separan AP-30 polyacrylamide is added to the water at a concentration of 100 wppm.

NOTATION

D	Diameter, [L]
DF	"Driving Force," Equation 6.67, $[FL/M = L^2/t^2]$
e_f	Energy dissipated per unit mass of fluid, $[FL/M = L^2/t^2]$
f	Fanning friction factor, [—]
F_x	Force component in x direction, $[F = ML/t^2]$
K_f	Loss coefficient, [—]
L	Length, [L]
m	Power law consistency parameter, $[M/(Lt^{2-n})]$
n	Power law flow index, [—]
N_{Bi}	Bingham number, [—]
N_{De}	Deborah number, [—]
N_{He}	Hedstrom number, Equation 6.62, [—]
N_{Re}	Reynolds number, [—]
$N_{Re,s}$	Solvent Reynolds number, [—]
$N_{Re,pl}$	Power law Reynolds number, [—]
P	Pressure, $[F/L^2 = M/(Lt^2)]$
Q	Volumetric flow rate, $[L^3/t]$
r	Radial direction, [L]
R	Tube radius, [L]
t	Time, [t]
V	Spatial average velocity, [L/t]
v_*	Friction velocity, Equation 6.28, [L/t]
v_x	Local velocity in the x direction, [L/t]
v'_x	Turbulent eddy velocity component in the x direction, [L/t]
v^+	Dimensionless velocity, Equation 6.27, [—]
w	External shaft work (e.g., negative pump work) per unit mass of fluid, $[FL/M = M/Lt^2]$
y^+	Dimensionless distance from wall, Equation 6.27, [—]

GREEK

δ_L	Laminar boundary layer thickness, [L]
δ_T	Turbulent boundary layer thickness, [L]
ε	Roughness, [L]
Φ	Potential $= P + \rho gz$, $[F/L^2 = M/(Lt^2)]$
μ	Viscosity (constant), $[M/(Lt)]$
ρ	Density, $[M/L^3]$
τ_o	Yield stress, $[F/L^2 = M/(Lt^2)]$
τ_{rx}	Shear stress component, force in x direction on r surface, $[F/L^2 = M/(Lt^2)]$
τ'_{rx}	Turbulent (Reynolds) stress component, $[F/L^2 = M/(Lt^2)]$
τ_w	Stress exerted by the fluid on the wall, $[F/L^2 = M/(Lt^2)]$

SUBSCRIPTS

x, y, z, r, θ	Coordinate directions
w	Wall location

REFERENCES

Astarita, G., G. Greco Jr., and L. Nicodemo, A phenomenological interpretation and correlation of drag reduction, *AIChE J.*, 15, 564–567, 1969.

Bewersdorff, H.W. and N.S. Berman, The influence of flow-induced non-Newtonian fluid properties on turbulent drag reduction, *Rheol. Acta*, 27, 130–136, 1988.

Churchill, S.W., Friction factor equation spans all fluid flow regimes, *Chem. Eng.*, 84 (24), 91–92, November 7, 1977.

Darby, R. and H.D. Chang, Generalized correlation for friction loss in drag reducing polymer solutions, *AIChE J.*, 30, 274–280, 1984.

Darby, R. and J. Melson, How to predict the friction factor for flow of Bingham plastics, *Chem. Eng.*, 28, 59–61, 1981.

Darby, R., R. Mun, and D.V. Boger, Predicting friction loss in slurry pipelines, *Chem. Eng.*, 99, 116–119, September 1992.

Darby, R. and S. Pivsa-Art, An improved correlation for turbulent drag reduction in dilute polymer solutions, *Canad. J. Chem. Eng.*, 69, 1395–1400, 1991.

Dodge, D.W. and A.B. Metzner, Turbulent flow of non-Newtonian systems, *AIChE J.*, 5, 189–204, 1959.

Goren, Y. and J.F. Norbury, Turbulent flow of dilute aqueous polymer solutions, *ASME J. Basic Eng.*, 89, 814–822, 1967.

Hoffmann, L. and P. Schummer, Experimental investigation of the turbulent boundary layer in the pipe flow of viscoelastic fluids, *Rheol. Acta*, 17, 98–104, 1978.

Savins, J.G., Contrasts in the solution drag reduction characteristics of polymeric solutions and micellar systems, in *Viscous Drag Reduction*, C.S. Wells (Ed.), Plenum Press, New York, 1969, pp. 183–212.

Virk, P.S., Drag reduction fundamentals, *AIChE J.*, 21, 625–656, 1975.

Virk, P.S. and H. Baher, The effect of polymer concentration on drag reduction, *Chem. Eng. Sci.*, 25, 1183–1189, 1970.

Wang, C.B., Correlation of the friction factor for turbulent pipe flow of dilute polymer solutions, *Ind. Eng. Chem. Fund.*, 11, 546–551, 1972.

White, A., Turbulent drag reduction with polymer additives, *J. Mech. Eng. Sci.*, 8, 452–455, 1966.

White, D. Jr. and R.J. Gordon, The influence of polymer conformation on turbulent drag reduction, *AIChE J.*, 21, 1027–1029, 1975.

7 Internal Flow Applications

"Take the best that exists and make it better. If it does not exist, design it."

—**Henry Royce, 1863–1933, Engineer**

I. NONCIRCULAR CONDUITS

All the relationships presented in Chapter 6 apply directly to circular pipes. However, many of these results caSn also, with appropriate modifications, be applied to conduits with noncircular cross sections. It should be recalled that the derivation of the momentum equation for uniform flow in a tube, for example, Equation 5.44, involved no assumption about the shape of the tube cross section. The result is that the friction loss is a function of a geometric parameter called the "hydraulic diameter":

$$D_h = 4 \frac{A}{W_p} \tag{7.1}$$

where
A is the area of the flow cross section
W_p is the wetted perimeter (i.e., the length of contact between the fluid and the solid boundary in the flow cross section)

For a full circular pipe, $D_h = D$ (the pipe diameter). The hydraulic diameter is the key characteristic geometric parameter for a conduit with any cross-sectional shape.

A. LAMINAR FLOWS

By either integrating the microscopic momentum equations (see Example 5.9) or applying a momentum balance to a "slug" of fluid in the center of the conduit as was done for tube flow, a relationship can be determined between flow rate and driving force for laminar flow in a conduit with a noncircular cross section. This can also be done by the application of the equivalent integral expressions analogous to Equations 6.6 through 6.10. The results for a few examples for Newtonian fluids will be given in the following. These results are the equivalent of the Hagen–Poiseuille equation for a circular tube and are given in both dimensional and dimensionless form.

1. Flow in a Slit

Flow between two flat parallel plates that are closely spaced ($h \ll W$) is shown in Figure 7.1.

The hydraulic diameter for this geometry is $D_h = 4A/W_p = 4hW/2W = 2h$, and the solution for a Newtonian fluid in laminar flow (analogous to Equation 6.11 for tube flow) is

$$Q = -\frac{\Delta\Phi\, W h^3}{12\mu L} \tag{7.2}$$

This can be rearranged into the equivalent dimensionless form

$$f N_{Re,h} = 24 \tag{7.3}$$

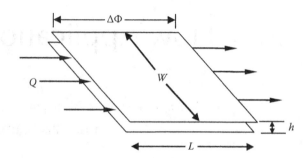

FIGURE 7.1 Flow in a slit.

where

$$N_{Re,h} = \frac{D_h V \rho}{\mu} = \frac{D_h Q \rho}{\mu A} \tag{7.4}$$

Here, $A = Wh$, and the Fanning friction factor is, by definition,

$$f = \frac{e_f}{(V^2/2)(4L/D_h)} = \frac{-\Delta\Phi}{(\rho V^2/2)(4L/D_h)} \tag{7.5}$$

and the Bernoulli equation reduces to $e_f = -\Delta\Phi/\rho$ for this system.

2. Flow in a Film

The flow of a thin film down an inclined plane is illustrated in Figure 7.2. The film thickness is $h \ll W$, and the plate is inclined at an angle θ to the vertical. For this flow, the hydraulic diameter is $D_h = 4hW/W = 4h$ (since only one boundary in the cross section is a wetted surface). The laminar flow solution for a Newtonian fluid is

$$Q = -\frac{\Delta\Phi h^3 W}{3\mu L} = \frac{\rho g h^3 W \cos\theta}{3\mu} \tag{7.6}$$

The dimensionless form of this equation is

$$f N_{Re,h} = 24 \tag{7.7}$$

where the Reynolds number and friction factor are given by Equations 7.4 and 7.5, respectively.

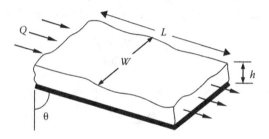

FIGURE 7.2 Flow in a film.

3. Annular Flow

Axial flow in the annulus between two concentric cylinders, as illustrated in Figure 7.3, is frequently encountered in tubular heat exchangers and coating devices. For this geometry, the hydraulic diameter is $D_h = 4(\pi/4)(D_o^2 - D_i^2)/[\pi(D_o + D_i)] = D_o - D_i$, and the Newtonian laminar flow solution is

$$Q = -\frac{\Delta\Phi\pi\left(D_o^2 - D_i^2\right)}{128\mu L}\left(D_o^2 + D_i^2 - \frac{D_o^2 - D_i^2}{\ln\left(D_o/D_i\right)}\right) \qquad (7.8)$$

The dimensionless form of this expression is

$$fN_{Re,h} = 16\alpha \qquad (7.9)$$

where

$$\alpha = \frac{(D_o - D_i)^2}{D_o^2 + D_i^2 - (D_o^2 - D_i^2)/\ln(D_o/D_i)} \qquad (7.10)$$

It can be shown that as $D_i/D_o \to 0$, $\alpha \to 1$ and the flow approaches that for a circular tube. Likewise, as $D_i/D_o \to 1$, $\alpha \to 1.5$ and the flow approaches that for a slit.

It is seen that the value of $fN_{Re,h}$ for laminar flow in a wide variety of geometries varies only by about a factor of 50% or so. This value has been determined for a Newtonian fluid in various geometries, and the results are summarized in Table 7.1. This table gives the expressions for the cross-sectional area and hydraulic diameter for six different conduit geometries and the corresponding values of $fN_{Re,h}$, the dimensionless laminar flow solution. The total range of values for $fN_{Re,h}$ for all of these geometries is seen to be approximately 12–24. Thus, for any completely arbitrary geometry, the dimensionless expression $fN_{Re,h} \approx 18$ would provide an approximate solution for fully developed laminar flow, with an error of about 30% or less.

B. TURBULENT FLOWS

The effect of geometry on the flow field for turbulent flows is much less pronounced than for laminar flows. This is because the majority of the energy dissipation or flow resistance occurs within the boundary layer which, in typical turbulent flows, occupies a relatively narrow region of

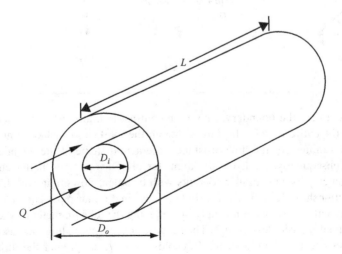

FIGURE 7.3 Flow in an annulus.

TABLE 7.1

Laminar Flow Factors for Noncircular Conduits

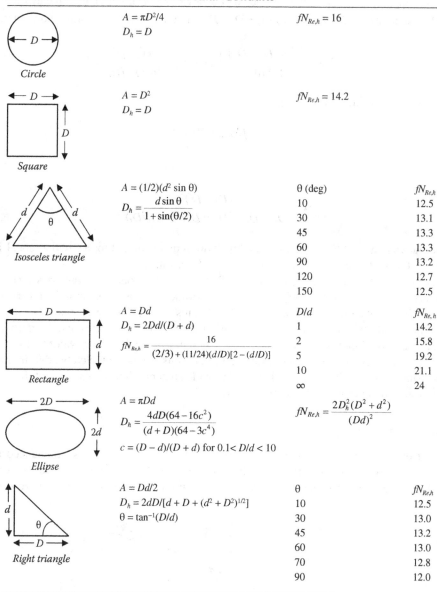

Circle

$A = \pi D^2/4$

$D_h = D$

$fN_{Re,h} = 16$

Square

$A = D^2$

$D_h = D$

$fN_{Re,h} = 14.2$

Isosceles triangle

$A = (1/2)(d^2 \sin\theta)$

$D_h = \dfrac{d\sin\theta}{1+\sin(\theta/2)}$

θ (deg)	$fN_{Re,h}$
10	12.5
30	13.1
45	13.3
60	13.3
90	13.2
120	12.7
150	12.5

Rectangle

$A = Dd$

$D_h = 2Dd/(D+d)$

$fN_{Re,h} = \dfrac{16}{(2/3)+(11/24)(d/D)[2-(d/D)]}$

D/d	$fN_{Re,h}$
1	14.2
2	15.8
5	19.2
10	21.1
∞	24

Ellipse

$A = \pi Dd$

$D_h = \dfrac{4dD(64-16c^2)}{(d+D)(64-3c^4)}$

$c = (D-d)/(D+d)$ for $0.1 < D/d < 10$

$fN_{Re,h} = \dfrac{2D_h^2(D^2+d^2)}{(Dd)^2}$

Right triangle

$A = Dd/2$

$D_h = 2dD/[d+D+(d^2+D^2)^{1/2}]$

$\theta = \tan^{-1}(D/d)$

θ	$fN_{Re,h}$
10	12.5
30	13.0
45	13.2
60	13.0
70	12.8
90	12.0

the total flow field near the boundary. This is in contrast to laminar flows, where the "boundary layer" occupies the entire flow field. Thus, although the total solid surface contacted by the fluid in turbulent flows influences the flow resistance, the actual shape of the boundary surface is not as important. Consequently, the hydraulic diameter provides an even better characterization of the effect of geometry for noncircular conduits with turbulent flows than with laminar flows. The result is that relationships developed for turbulent flows in circular pipes can be applied directly to conduits of noncircular cross section simply by replacing the tube diameter by the hydraulic diameter in the relevant dimensionless groups. The accuracy of this procedure increases with increasing Reynolds number because the higher the Reynolds number, the greater the turbulence intensity, and the thinner the boundary layer, hence the less important the actual shape of the cross section.

It is important to use the hydraulic diameter substitution ($D = D_h$) in the appropriate (original) form of the dimensionless groups (e.g., $N_{Re} = DV\rho/\mu$, $f = e_f/(2LV^2/D)$, ε/D) and not a form that has been adapted for circular tubes (e.g., $N_{Re} = 4Q\rho/\pi D\mu$). That is, the proper modification of the Reynolds number for a noncircular conduit is ($D_h V\rho/\mu$), not ($4Q\rho/\pi D_h\mu$). One clue that the dimensionless group is of the wrong form for a noncircular conduit is the presence of π, which is normally associated only with circular geometries (remember, "*pi are round, cornbread are square*"). Thus, the appropriate dimensionless groups from the tube flow solutions can be modified for noncircular geometries as follows:

$$N_{Re,h} = \frac{D_h V\rho}{\mu} = \frac{4Q\rho}{W_p\mu} \tag{7.11}$$

$$f = \frac{e_f D_h}{2LV^2} = \frac{2e_f}{LQ^2}\left(\frac{A^3}{W_p}\right) \tag{7.12}$$

$$N_R = \frac{N_{Re,h}}{\varepsilon/D_h} = \frac{D_h^2 Q\rho}{\varepsilon A\mu} = \frac{16Q\rho}{\varepsilon\mu}\left(\frac{A}{W_p^2}\right) \tag{7.13}$$

$$fN_{Re,h}^2 = \frac{32e_f\rho^2}{L\mu^2}\left(\frac{A}{W_p}\right)^3 = \frac{e_f\rho^2 D_h^3}{2L\mu^2} \tag{7.14}$$

$$fN_{Re,h}^5 = \frac{2048e_f Q^3\rho^5}{L\mu^5}\left(\frac{A}{W_p^2}\right)^3 \tag{7.15}$$

The circular tube expressions for f and N_{Re} containing π can also be transformed to the equivalent expressions for a noncircular conduit by the substitution

$$\pi \rightarrow \frac{W_p}{D_h} = 4\frac{A}{D_h^2} = \frac{1}{4}\frac{W_p^2}{A} \tag{7.16}$$

II. MOST ECONOMICAL DIAMETER

In the previous chapter, we saw how to determine the driving force (e.g., pumping requirement) required to deliver a specified flow rate through a given pipe size, as well as how to determine the proper pipe size that will deliver a specified flow rate for a given driving force (e.g., pump head). However, when we install a pipeline or piping system, we are normally free to select both the "best" pipe and the "best" pump. The term "best" in this case refers to that combination of pipe and pump that will minimize the total system cost.

The total system cost of a pipeline or piping system includes the fixed capital cost of both the pipe (including valves and fittings) and pumps, as well as the continuous operating costs, that is, the cost of the energy required to drive the pumps, maintenance cost, etc.:

Capital cost of pipe (CCP)
Capital cost of pump stations (CCPS)
Energy cost to power pumps (EC)

Although the energy cost is "continuous" and the capital costs are "one time," it is common to spread out (or amortize) the capital cost over a period of Y years, that is, over the "economic lifetime" of the pipeline. The reciprocal of this ($X = 1/Y$) is the fraction of the total capital cost written off per year. Thus, taking 1 year as the time basis, we can combine the capital cost per year and the energy cost per year to get the total annual cost. We may also factor in a cost for maintenance, which will be a fraction M of the total cost of pipe and pump stations. This is typically about 2% of total installed cost.

Data on the cost of typical pipeline installations of various sizes (including valves and fittings) were reported by Darby and Melson (1982). They showed that these data can be represented by the equation

$$CCP = aD_{ft}^{p}L \tag{7.17}$$

where

D_{ft} is the pipe ID in feet

the parameters a and p depend upon the pipe wall thickness as shown in Table 7.2

Likewise, the capital cost of (installed) pump stations (for 500 hp and over) was shown to be a linear function of the pump power (see Figure 7.4):

$$CCPS = \left(A + B\frac{HP}{\eta_e}\right)(1 + M) \tag{7.18}$$

where

$A = \$172,800$

$B = \$451 \text{ hp}^{-1}$ (in 1980 dollars)

HP/η_e is the horsepower rating of the pump (HP is the "hydraulic power," which is the power delivered directly to the fluid and η_e is the pump efficiency)

M accounts for maintenance costs, which are typically about 2%/year

The energy cost is determined from the power required to drive the pumps, which is the product of the mass flow rate and the pump work per unit mass of fluid, as determined from the Bernoulli equation:

$$-w = \frac{\Delta\Phi}{\rho} + \frac{1}{2}\Delta V^2 + \sum e_f \tag{7.19}$$

TABLE 7.2
Cost of Pipe (in 1980 Dollars)[a]

Parameter	Pipe Grade				
	ANSI 300#	ANSI 400#	ANSI 600#	ANSI 900#	ANSI 1500#
a	23.1	23.9	30.0	38.1	55.3
p	1.16	1.22	1.31	1.35	1.39

Note: The ANSI pipe grades correspond approximately to Sch. 20, 30, 40, 80, and 120 for commercial steel pipe.

[a] Pipe cost ($/ft) = $a(\text{ID}_{ft})^p$.

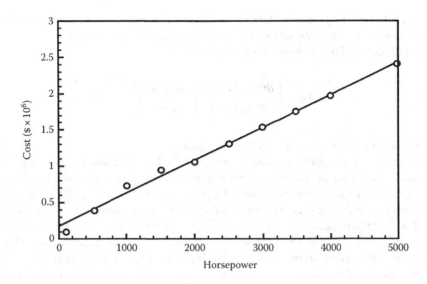

FIGURE 7.4 Cost of pump stations (in 1980 dollars). Pump station cost ($) = CCPS = $A + B$ hp/η_e where $A = 172,800$ and $B = 451$/hp for stations of 500 hp or more.

where

$$\sum e_f = 4f \frac{V^2}{2} \sum \left(\frac{L}{D} \right)_{eq} \tag{7.20}$$

and $\Sigma(L/D)_{eq}$ is assumed to include the equivalent length of any fittings (which are usually a small portion of a long pipeline) as well as a maintenance factor. The required hydraulic pumping power (*HP*) is thus

$$HP = -w\dot{m} = \dot{m} \left[\frac{2fL_{eq}V^2}{D} + \frac{\Delta\Phi}{\rho} \right] = \frac{32fL_{eq}\dot{m}^3}{\pi^2\rho^2D^5} + \dot{m}\frac{\Delta\Phi}{\rho} \tag{7.21}$$

The total pumping energy cost per year is therefore

$$EC = C\frac{HP}{\eta_e} \tag{7.22}$$

where
 C is the unit energy cost (e.g., $/(hp year), ¢/kWh)
 η_e is the pump efficiency

Note that the capital cost increases almost linearly with the pipe diameter, whereas the energy cost decreases in proportion to about the fifth power of the diameter.

The total annual cost of the pipeline is the sum of the capital and energy costs:

$$TC = X(1+M)(CCP + CCPS) + EC \tag{7.23}$$

Substituting Equations 7.17, 7.18, and 7.22 into Equation 7.23 gives

$$TC = X(1+M)(aD^pL + A) + \frac{BX(1+M)+C}{\eta_e} \left[\frac{32fL\dot{m}^3}{\pi^2\rho^2D^5} + \dot{m}\frac{\Delta\Phi}{\rho} \right] \tag{7.24}$$

Now, we wish to find the pipe diameter that minimizes this total cost. To do this, we differentiate Equation 7.24 with respect to D, set the derivative equal to zero, and solve for D (i.e., D_{ec}, the most economical diameter). The resulting expression for D_{ec} is

$$D_{ec} = \left[\frac{B(1+M)+CY}{ap\,\eta_e(1+M)} \left(\frac{160f\dot{m}^3}{\pi^2\rho^2} \right) \right]^{1/(p+5)} \tag{7.25}$$

where $Y = 1/X$ is the "economic lifetime" of the pipeline.

One might question whether the cost information in Table 7.2 and Figure 7.4 could be used today because these data are based on 1980 information and prices have increased greatly since that time. However, as seen from Equation 7.25, the cost parameters (i.e., B, C, and a) appear as a ratio. Because capital costs and energy costs tend to inflate at approximately the same rate (see, e.g., Durand et al., 1999), this ratio is essentially independent of inflation, and conclusions based on 1980 economic data should be valid today. However, this assumes that both capital costs (i.e., B and a) and energy costs (C) are based on the same year. If present-day energy costs are used, an inflation rate can be applied to the capital costs (B and a) to adjust them to present-day values by multiplying B and a by the factor $(1 + i)^{(t-1980)}$ (e.g., Penoncello, 2015), where i is the average inflation rate (in decimal form) from 1980 to the present year t.

A. NEWTONIAN FLUIDS

Equation 7.25 is implicit in the economic diameter, D_{ec}, because the friction factor (f) depends upon D_{ec} through the Reynolds number and the relative roughness of the pipe. It can be solved by iteration in a straightforward manner, however, by applying the procedure used for the "unknown diameter" problem in Chapter 6. That is, first assume a value for f (say, 0.005), calculate D_{ec} from Equation 7.25, and use this diameter to compute the Reynolds number and relative roughness. Then use these values to find f (from the Moody diagram or Churchill equation). If this value differs from the originally assumed value, use it in place of the assumed value and repeat the process until successive values of f agree.

Another approach is to regroup the characteristic dimensionless variables in the problem so that the unknown (D_{ec}) appears in only one group. After rearranging Equation 7.25 for f, we see that the following group will be independent of D_{ec}:

$$fN_{Re}^{p+5} = \left(\frac{4}{\pi} \right)^{p+3} \frac{\rho^2(1+M)ap\eta_e\dot{m}^{p+2}}{10[B(1+M)+CY]\mu^{p+5}} = N_c \tag{7.26}$$

We can call this the "cost group" (N_c) because it contains all of the cost parameters (a, B and C). We can also define a roughness group that does not include the diameter:

$$N_R = \frac{\varepsilon/D_{ec}}{N_{Re}} = \frac{\pi\mu\varepsilon}{4\dot{m}} \tag{7.27}$$

The remaining group is the Reynolds number, which is the dependent group because it alone contains D_{ec}:

$$N_{Re} = \frac{4\dot{m}}{\pi D_{ec}\mu} \tag{7.28}$$

The Moody diagram can be used to construct a plot of N_{Re} versus $N_c = fN_{Re}^{p+5}$ for various values of p and N_R (a double-parametric plot), which permits a direct solution to this problem (see Darby and Melson, 1982). These equations can also be used directly to simplify the iterative solution. Since the value of N_c

is known, assuming a value for f will give N_{Re} directly from Equation 7.26. This, in turn, gives D_{ec} from Equation 7.28 and hence ε/D_{ec}. These values of N_{Re} and ε/D_{ec} are used to find f from the Moody diagram or Churchill equation, and the iteration is continued until successive values of f agree. The most difficult aspect of working with these groups is ensuring a consistent set of units for all the variables (with appropriate use of the conversion factor g_c, if working in engineering units). For this reason, it is easier to work with consistent units in a scientific system (e.g., SI or cgs), which avoids the need for g_c.

Example 7.1 Economic Pipe Diameter

What is the most economical diameter for a pipeline that is required to transport crude oil with a viscosity of 30 cP and an SG of 0.95, at a rate of 1 million bbl/day using ANSI 1500# pipe, if the cost of energy is 5¢/kWh (in 1980 dollars)? Assume that the economical life of the pipeline is 40 years and that the pumps are 50% efficient and the maintenance costs are 2%/year.

Solution:

From Table 7.2, the pipe cost parameters are

$$p = 1.39 \quad a = 55.3\,\frac{\$}{\text{ft}^{2.39}} \times \left(\frac{3.28\,\text{ft}}{\text{m}}\right)^{2.3.9} = 945.5\,\$/\text{m}^{2.39}$$

Using SI units will simplify the problem. After converting, we have

$$\dot{m} = \rho Q = 1748\,\text{kg/s} \quad \mu = 0.03\,\text{Pa}\,\text{s} \quad CY = \$17.52/\text{W}$$

From Figure 7.4, we get the pump station cost factor

$$B = 451\,\$/\text{hp} = 0.604\,\$/\text{W}$$

and the "cost group" is (Equation 7.26)

$$N_c = \left(\frac{4}{\pi}\right)^{p+3} \frac{\rho^2(1+M)ap\eta_e\,\dot{m}^{p+2}}{10[B(1+M)+CY]\,\mu^{p+5}} = 5.07 \times 10^{27} = f N_{Re}^{6.39}$$

Assuming a roughness of 0.0018 in., we can solve for D_{ec} by iteration as follows. First, assume $f = 0.005$ and use the "Cost Group" to get N_{Re} from $N_c = f N_{Re}^{6.39}$. From N_{Re}, we find D_{ec} and thus ε/D_{ec}. Then, using the Churchill equation, we find a value for f, and compare it with the assumed value. This is repeated until convergence is achieved:

Assumed f	N_{Re}	D_{ec} (m)	ε/D_{ec}	f (Churchill)
0.005	4.96×10^4	1.49	3.07×10^{-5}	0.00523
0.00523	4.93×10^4	1.50	3.05×10^{-5}	0.00524

This agreement is close enough. The most economical diameter is $\boxed{1.5\text{ m or }59.2\text{ in.}}$ The "standard pipe size" closest to this value on the high side (or the closest size that can readily be manufactured) would be used.

B. Non-Newtonian Fluids

A procedure analogous to the one followed above for Newtonian fluids can be used for non-Newtonian fluids that follow the power law or Bingham plastic models (Darby and Melson, 1982), as follows:

1. Power Law Fluid

For power law fluids, the basic dimensionless variables are the Reynolds number, the friction factor, and the flow index (n). If the Reynolds number is expressed in terms of the mass flow rate, then

$$N_{Re,pl} = \left(\frac{4}{\pi}\right)^{2-n} \left(\frac{4n}{3n+1}\right)^n \left(\frac{\dot{m}^{2-n} \rho^{n-1}}{D_{ec}^{4-3n} 8^{n-1} m}\right) \tag{7.29}$$

Eliminating D_{ec} from Equations 7.25 and 7.29, the equivalent cost group becomes

$$f^{4-3n} N_{Re,pl}^{5+p} = \frac{(52.4)(10^{3n})(2^{7p-3n(1+p)})}{p^{(2+n)(1+p)} m^{5+p}} \left(\frac{ap\eta_e}{B+CY}\right)^{4-3n} \left(\frac{\rho^{3-p+n(p-1)} \dot{m}^{2(p-1)+n(4-p)}}{[(3n+1)/n]^{n(5+p)}}\right) \tag{7.30}$$

Since all values on the right-hand side of Equation 7.30 are known, assuming a value of f allows a corresponding value of $N_{Re,pl}$ to be determined. This value can then be used to check the assumed value of f using the general expression for the power law friction factor (Equation 6.44 for laminar flow or Equation 6.46 for turbulent flow) and iterating until agreement is attained. (*Note*: The maintenance cost factor has not been included in these equations, but it can easily be accounted for by multiplying the terms a and B by the factor $[1 + M]$).

2. Bingham Plastic

The basic dimensionless variables for the Bingham plastic are the Reynolds number, the Hedstrom number, and the friction factor. Eliminating D_{ec} from the Reynolds number and Equation 7.25 (as mentioned earlier), the cost group is

$$f N_{Re}^{p+5} = \left(\frac{4}{\pi}\right)^{p+3} \frac{\rho^2 ap\eta_e \dot{m}^{p+2}}{10(B+CY)\mu_{\infty¥}^{p+5}} \tag{7.31}$$

D_{ec} can also be eliminated from the Hedstrom number by combining it with the Reynolds number:

$$N_{He} N_{Re}^2 = \left(\frac{4}{\pi}\right)^2 \frac{\tau_0 \rho \dot{m}^2}{\mu_{\infty}^4} \tag{7.32}$$

These equations can readily be solved by iteration, as follows. Assuming a value of f allows N_{Re} to be determined from Equation 7.31. This is then used with Equation 7.32 to find N_{He}. The friction factor is then calculated using these values of N_{Re} and N_{He} and the Bingham plastic pipe friction factor equation (Equation 6.65). The result is compared with the assumed value, and the process is repeated until agreement is attained.

Graphs have been presented by Darby and Melson (1982) that can be used to solve these problems directly without iteration. However, interpolation on double-parametric logarithmic scales is required, so only approximate results can be expected from the precision of reading these plots. As mentioned before, the greatest difficulty in using these equations is that of ensuring consistent units. In many cases, it is most convenient to use cgs units in problems such as these, because fluid properties (density and viscosity) are frequently found in these units and the scientific system (e.g., cgs) does not require the conversion factor g_c. In addition, the energy cost is frequently given in cents per kilowatt-hour, which is readily converted to cgs units (e.g., $/erg).

III. FRICTION LOSS IN VALVES AND FITTINGS

Evaluation of the friction loss in valves and fittings involves the determination of the appropriate loss coefficient (K_f), which in turn defines the energy loss per unit mass of fluid:

$$e_f = \frac{K_f V^2}{2} \tag{7.33}$$

where V is (usually) the velocity in the pipe upstream of the fitting or valve. However, this is not always true and care must be taken to ensure that the value of V that is used is the one that is specified in the defining equation for K_f. The actual evaluation of K_f is done by determining the friction loss e_f from measurements of the pressure drop across the fitting (elbows, tees, valves, etc.). This is not straightforward, however, because the pressure in the pipe is influenced by the presence of the fitting for a considerable distance both upstream and downstream of the fitting. It is not possible, therefore, to obtain accurate values from measurements taken at pressure taps immediately adjacent to the fitting. The most reliable method is to measure the total pressure drop through a long run of pipe both with and without the fitting, at the same flow rate, and determine the fitting loss by difference.

There are several "correlation" expressions for K_f, which are described below (in Sections A through E) in the order of increasing accuracy. The "3-K" method (see Section E) is recommended because it accounts directly for the effect of both Reynolds number and fitting size on the loss coefficient and more accurately reflects the effect of fitting diameter than the 2-K method (Section D). For highly turbulent flow, the Crane method (Section C) agrees well with the 3-K method but is less accurate at low Reynolds numbers and is not recommended for laminar flow. The loss coefficient and $(L/D)_{eq}$ methods are more approximate but give acceptable results at high Reynolds (fully turbulent flow) numbers and when losses in valves and fittings are "minor losses" compared to the pipe friction. They are also appropriate for first estimates in problems that require iterative solutions.

A. LOSS COEFFICIENT

Values of K_f for various types of valves, fittings, etc., are found tabulated in various textbooks and handbooks. The assumption that these values are constant for a given type of valve or fitting is not accurate, however, because in reality the value of K_f varies with both the size (scale) of the fitting and the level of turbulence (Reynolds number). One reason that K_f is not the same for all fittings of the same type (e.g., all 90° elbows) is that all the dimensions of a fitting, such as the diameter and radius of curvature, do not scale by the same factor for large and small fittings. Most tabulated values for constant K_f values are close to the values of K_∞ from the 3-K method.

B. EQUIVALENT L/D METHOD

The basis for the $(L/D)_{eq}$ method is the assumption that there is some length of pipe (L_{eq}) that has the same friction loss as that which occurs in the fitting, at a given (pipe) Reynolds number. Thus, the fittings are conceptually replaced by the equivalent additional length of pipe that has the same friction loss as the fitting:

$$e_f = \frac{4fV^2}{2} \sum \left(\frac{L}{D} \right)_{eq} \tag{7.34}$$

where f is the Fanning friction factor in the pipe at the given pipe Reynolds number and relative roughness. This is a convenient concept because it allows the solution of pipe flow problems with fittings to be carried out in a manner identical to that without fittings if L_{eq} is known. Values of $(L/D)_{eq}$ are tabulated in various textbooks and handbooks for a variety of fittings and valves (and are also listed in Table 7.3 here).

TABLE 7.3

3-K Constants[a] for Loss Coefficients for Valves and Fittings

Fitting		r/D	$(L/D)_{eq}$	K_1	K_i	K_d
Elbows						
90°	Threaded, standard	1	30	800	0.14	4.0
	Threaded, long radius	1.5	16	800	0.071	4.2
	Flanged, welded, bends	1	20	800	0.091	4.0
		2	12	800	0.056	3.9
		4	14	800	0.066	3.9
		6	17	800	0.075	4.2
	Mitered					
	1 weld (90°)		60	1000	0.27	4.0
	2 welds (45°)		30	800	0.136	4.1
	3 welds (30°)		24	800	0.105	4.2
45°	Threaded standard	1	16	500	0.071	4.2
	Long radius	1.5		500	0.052	4.0
	Mitered					
	1 weld (45°)		15	500	0.086	4.0
	2 welds (22.5°)		12	500	0.052	4.0
180°	Threaded, close return bend	1	50	1000	0.23	4.0
	Flanged	1		1000	0.12	4.0
	All	1.5		1000	0.10	4.0
Tees	Through branch (as elbow)					
	Threaded	1	60	500	0.274	4.0
		1.5		800	0.14	4.0
	Flanged	1	20	800	0.28	4.0
	Stub-in branch			1000	0.34	4.0
	Run-through threaded	1	20	200	0.091	4.0
	Flanged	1		150	0.05	4.0
	Stub-in branch			100	0	0
Valves						
Angle valve	Valve					
	45° full line size	$\beta = 1$	55	950	0.25	4.0
	90° full line size	$\beta = 1$	150	1000	0.69	4.0
Globe valve	Standard	$\beta = 1$	340	1500	1.70	3.6
Plug valve	Branch flow		90	500	0.41	4.0
	Straight through		18	300	0.084	3.9
	Three-way (flow through)		30	300	0.14	4.0
Gate valve	Standard	$\beta = 1$	8	300	0.037	3.9
Ball valve	Standard	$\beta = 1$	3	300	0.017	3.5
Diaphragm	Dam type			1000	0.69	4.9
Swing check	$V_{min} = 35[\rho(lb_m/ft^3)]^{-1/2}$ (ft/s)		100	1500	0.46	4.0
Lift check	$V_{min} = 40[\rho(lb_m/ft^3)]^{-1/2}$ (ft/s)		600	2000	2.85	3.8

Note: D_n is the nominal pipe size in inches.

[a] $K_f = \dfrac{K_1}{N_{Re}} + K_i\left(1 + \dfrac{K_d}{D_n^{0.3}}\right).$

The method assumes that (1) sizes of all fittings of a given type can be scaled by the corresponding pipe diameter (D), and (2) the influence of turbulence level (i.e., Reynolds number) on the friction loss in the fitting is identical to that in the pipe (because the pipe f value is used to determine the fitting loss). Neither of these assumptions is accurate (as pointed out earlier), although the approximation provided by this method gives reasonable results at high turbulence levels (fully turbulent flow), especially if fitting losses are minor when compared to the total pipe friction loss.

C. CRANE METHOD

The method given in the Crane Technical Paper 410 (1991) is a modification of the afore-mentioned methods. It is equivalent to the $(L/D)_{eq}$ method except that it recognizes that there is generally a higher degree of turbulence in the fitting than in the pipe at a given (pipe) Reynolds number. This is accounted for by always using the "fully turbulent" value for f (e.g., f_T) in the expression for the friction loss in the fitting, regardless of the actual Reynolds number in the pipe, that is,

$$e_f = \frac{K_f V^2}{2} \quad \text{where } K_f = 4 f_T \left(\frac{L}{D} \right)_{eq} \tag{7.35}$$

The value of f_T can be calculated from the Colebrook equation (Equation 6.40), for example,

$$f_T = \frac{0.0625}{[\log(3.7D/\varepsilon)]^2} \tag{7.36}$$

in which ε is the pipe roughness (0.0018 in. for new commercial steel). This is a two-constant model [f_T and $(L/D)_{eq}$], and values of these constants are tabulated in the Crane paper for a wide variety of fittings, valves, etc. This method gives satisfactory results for high turbulence levels (fully turbulent flow) but is less accurate at low Reynolds numbers and does not scale well with pipe size.

D. 2-K (HOOPER) METHOD

The 2-K method by Hooper (1981, 1988) was based on experimental data from a variety of valves and fittings over a wide range of Reynolds numbers. The effect of both the Reynolds number and scale (fitting size) is reflected in the expression for the loss coefficient:

$$e_f = \frac{K_f V^2}{2}, \quad \text{where } K_f = \frac{K_1}{N_{Re}} + K_\infty \left(1 + \frac{1}{ID_{in.}} \right) \tag{7.37}$$

Here, $ID_{in.}$ is the internal diameter (in inches) of the pipe that contains the fitting. This method is valid over a much wider range of Reynolds numbers than the other methods. However, the effect of pipe size (e.g., $1/ID_{in.}$) in Equation 7.37 does not accurately reflect the scaling with pipe size, as discussed below in Section E.

E. 3-K (DARBY) METHOD

Although the 2-K method applies over a wide range of Reynolds numbers, the scaling term ($1/ID$) does not accurately reflect data over a wide range of sizes for valves and fittings, as reported in a variety of sources (Crane, 1991; CCPS, 1998; Perry and Green, 2007; Darby, 2001; and refer-ences cited therein). Specifically, all the preceding methods tend to underpredict the friction loss for fittings of larger diameters. Darby (2001) has evaluated data from the literature for various valves and fittings and found that they can be represented more accurately by the following "3-K" equation:

$$K_f = \frac{K_1}{N_{Re}} + K_i\left(1 + \frac{K_d}{D_{n,in.}^{0.3}}\right) \tag{7.38}$$

Note that $D_{n,in.}$ is the nominal diameter, in inches. The values of the 3 K's (K_1, K_i, and K_d) are given in Table 7.3 (along with representative values of $(L/D)_{eq}$) for various valves and fittings. These values were determined from combinations of literature values from the references listed earlier and were all found to accurately follow the scaling law given in Equation 7.38. The values of K_1 are mostly those of the Hooper 2-K method, and the values of K_i were mostly determined from the Crane data. However, since there is no single comprehensive data set for many fittings over a wide range of sizes and Reynolds numbers, some estimation was necessary for some values.

Values of K_d are all very close to 4.0, and this value can be used to scale known values of K_f for a given pipe size to apply to other sizes. This method is the most accurate of the methods described for all Reynolds numbers and fitting sizes. Tables 7.4 and 7.5 list values for K_f for expansions and contractions and entrance and exit conditions, respectively (Hooper, 1981). The definition of K_f (i.e., $K_f = 2e_f/V^2$) involves the kinetic energy of the fluid, $V^2/2$. For sections that undergo area changes (e.g., pipe entrance, exit, expansions, or contractions), the entering and leaving velocities will be different. Because the value of the velocity used with the definition of K_f is arbitrary, it is very important to know which velocity is the reference value for a given loss coefficient. Values of K_f are usually based on the larger velocity entering or leaving the fitting (through the smaller cross section), but this should be verified if any doubt exists.

A note is in order regarding the exit loss coefficient, which is listed in Table 7.5 as equal to 1.0. Actually, if the fluid exits the pipe in a free jet into unconfined space, the loss coefficient is zero because the velocity of the fluid exiting the pipe is close to that of the fluid inside the pipe and thus the kinetic energy change is zero. However, when the fluid exits into a confined space so that the fluid leaving the pipe immediately mixes with the same fluid in the receiving vessel, the kinetic energy is dissipated as friction loss in the mixing process so the velocity goes to zero, and thus the loss coefficient is 1.0. In this case, the change in the kinetic energy and the friction loss at the exit cancel out.

IV. NON-NEWTONIAN FLUIDS

There are insufficient data in the literature to enable reliable correlation or prediction of friction loss in valves and fittings for non-Newtonian fluids. As a first approximation, however, it can be assumed that a correlation similar to the 3-K method should apply to non-Newtonian fluids if the (Newtonian) Reynolds number in Equation 7.38 could be replaced by a single corresponding dimensionless group that adequately incorporates the influence of the non-Newtonian properties. For the power law and Bingham plastic fluid models, two rheological parameters are required to describe the viscous properties, which generally results in two corresponding dimensionless groups ($N_{Re,pl}$ and n for the power law and N_{Re} and N_{He} for the Bingham plastic). However, it is possible to define an "effective viscosity" for a non-Newtonian fluid model that has the same significance in the Reynolds number as the viscosity has for a Newtonian fluid and incorporates all of the appropriate parameters for that model, which then can be used to define an equivalent non-Newtonian Reynolds number (see Darby and Forsyth, 1992). For a Newtonian fluid, the Reynolds number can be rearranged as follows:

$$N_{Re} = \frac{DV\rho}{\mu} = \frac{\rho V^2}{\mu V/D} = \frac{\rho V^2}{\tau_w/8} \tag{7.39}$$

TABLE 7.4

Loss Coefficients for Expansions and Contractions

K_f to be used with upstream velocity head, $V_1^2/2$. $\beta = d/D$

Contraction

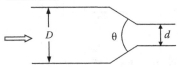

$\theta < 45°$

$N_{Re,1} < 2500$:

$$K_f = 1.6\left[1.2 + \frac{160}{N_{Re,1}}\right]\left[\frac{1}{\beta^4} - 1\right]\sin\frac{\theta}{2}$$

$N_{Re,1} > 2500$:

$$K_f = 1.6\left[0.6 + 1.92f_1\right]\left[\frac{1-\beta^2}{\beta^4}\right]\sin\frac{\theta}{2}$$

$\theta > 45°$

$N_{Re,1} < 2500$:

$$K_f = \left[1.2 + \frac{160}{N_{Re,1}}\right]\left[\frac{1}{\beta^4} - 1\right]\left[\sin\frac{\theta}{2}\right]^{1/2}$$

$N_{Re,1} > 2500$:

$$K_f = \left[0.6 + 1.92f_1\right]\left[\frac{1-\beta^2}{\beta^4}\right]\left[\sin\frac{\theta}{2}\right]^{1/2}$$

Expansion

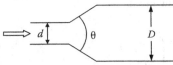

$\theta < 45°$

$N_{Re,1} < 4000$:

$$K_f = 5.2(1-\beta^4)\sin\frac{\theta}{2}$$

$N_{Re,1} > 4000$:

$$K_f = 2.6\left(1+3.2f_1\right)\left(1-\beta^2\right)^2\sin\frac{\theta}{2}$$

$\theta > 45°$

$N_{Re,1} < 4000$:

$$K_f = 2(1-\beta^4)$$

$N_{Re,1} > 4000$:

$$K_f = (1+3.2f_1)(1-\beta^2)^2$$

Source: Hooper, W.B., "Calculate head Loss Caused by Change in Pipe Size", *Chem. Eng.*, 95, pp. 89–92, 1988.

Note: $N_{Re,1}$ is the upstream Reynolds number, and f_1 is the pipe friction factor at this Reynolds number.

TABLE 7.5

Loss Coefficients for Pipe Entrance and Exit

$K_f = K_1/N_{Re} + K_\infty$

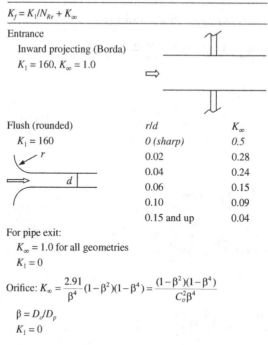

Entrance

   Inward projecting (Borda)

   $K_1 = 160$, $K_\infty = 1.0$

Flush (rounded)	r/d	K_∞
$K_1 = 160$	0 (sharp)	0.5
	0.02	0.28
	0.04	0.24
	0.06	0.15
	0.10	0.09
	0.15 and up	0.04

For pipe exit:

   $K_\infty = 1.0$ for all geometries

   $K_1 = 0$

Orifice: $K_\infty = \dfrac{2.91}{\beta^4}(1-\beta^2)(1-\beta^4) = \dfrac{(1-\beta^2)(1-\beta^4)}{C_o^2\beta^4}$

   $\beta = D_o/D_p$

   $K_1 = 0$

Source: Hooper, W.B., "The 2-K Method Predicts Head Loss in Pipe Fittings" *Chem. Eng.*, 88, pp. 96–100, 1981.

Introducing $\tau_w = m[(8V/D)(3n+1)/4n]^n$ for the power law model results in

$$N_{Re,pl} = \frac{2^{(7-3n)}\rho Q^{(2-n)}}{m\pi^{(2-n)}D^{(4-3n)}}\left(\frac{n}{3n+1}\right)^n \tag{7.40}$$

which is identical to the expression derived in Chapter 6 (see Equation 6.71).

For the Bingham plastic, replacing τ_w for the Newtonian fluid in Equation 7.39 with $\tau_o + \mu_\infty\dot{\gamma}_w$ and using the approximation $\dot{\gamma}_w = 8V/D$, the corresponding expression for the Reynolds number is

$$N_{Re,BP} = \frac{4Q\rho}{\pi D\mu_\infty(1+\pi D^3\tau_o/32Q\mu_\infty)} = \frac{N_{Re}}{1+N_{He}/8N_{Re}} \tag{7.41}$$

The ratio $N_{He}/N_{Re} = D\tau_o/V\mu_\infty$ is also called the Bingham number (N_{Bi}). Darby and Forsyth (1992) showed experimentally that mass transfer in Newtonian and non-Newtonian fluids can be correlated by this method. That is, the same dimensionless correlation can be applied to both Newtonian and non-Newtonian fluids when the Newtonian Reynolds number is replaced by either Equation 7.40 for the power law fluid or Equation 7.41 for the Bingham plastic model. As a first approximation, therefore, we may assume that the same method would apply to friction loss in valves and fittings as described by the 3-K model, Equation 7.38. This approach is in agreement with the scant literature data on fitting losses with power law and Bingham plastic fluids (see, e.g., Chhabra and Richardson, 2008).

V. PIPE FLOW PROBLEMS WITH FITTINGS

The inclusion of significant friction loss in fittings in piping systems requires a somewhat different procedure for the solution of flow problems than that which was used in the absence of fitting losses (see Chapter 6). We will consider the same classes of problems as before, that is, the unknown driving force, the unknown flow rate, and the unknown diameter problems for Newtonian, power law, and Bingham plastic fluids. The governing equation, as before, is the Bernoulli equation written in the form

$$DF = -\left(\frac{\Delta\Phi}{\rho} + w\right) = \sum_{pipe+fittings} e_f + \frac{1}{2}\Delta(\alpha V^2) \qquad (7.42)$$

where

$$\sum_{Pipe+Fittings} e_f = \frac{1}{2} \sum_{Pipe+Fittings} (V^2 K_f) = \frac{8Q^2}{\pi^2} \sum_{Pipe+Fittings} \frac{K_f}{D^4} \qquad (7.43)$$

$$K_{pipe} = \frac{4fL}{D}, \quad K_{fittings} = \frac{K_1}{N_{Re}} + K_i\left(1 + \frac{K_d}{D_n^{0.3}}\right) \qquad (7.44)$$

The summation is over each length of pipe and each fitting of diameter D in the system. The expressions for the loss coefficients for the pipe and fittings are given in Equation 7.44. Substituting Equation 7.43 into 7.42 gives the following form of the Bernoulli equation:

$$DF = \frac{8Q^2}{\pi^2}\left(\sum_{fi} \frac{K_{fi}}{D_i^4} + \frac{\alpha_2}{D_2^4} - \frac{\alpha_1}{D_1^4}\right) \qquad (7.45)$$

Recall that the α's are the KE correction factors at the upstream and downstream positions and that $\alpha = 2$ for laminar flow and α "approx" = 1 for turbulent flow in a circular pipe for a Newtonian fluid. The same should apply (approximately) for non-Newtonian fluids, especially in the turbulent region.

A. UNKNOWN DRIVING FORCE

Here, we wish to calculate the net driving force (pressure head, hydraulic head, or pump work) required to transport a given fluid at a given rate through a given piping system, containing a specified array of valves and fittings.

1. Newtonian Fluid

The knowns and unknowns for this case are as follows:

Given: Q, μ, ρ, D_i, L_i, ε_i, fittings   *Find*: DF

The driving force is given by Equation 7.45, in which the Ks are related to the other variables by the Moody diagram or Churchill equation for each pipe segment, ($K_{pipe,i}$), and by the 3-K method, Equation 7.44, for each valve and fitting (K_{fit}) as a function of the Reynolds number:

$$N_{Re,i} = \frac{4Q\rho}{\pi D_i\mu} \qquad (7.46)$$

The solution procedure is as follows:

1. Calculate $N_{Re,i}$ from Equation 7.46 for each pipe segment, valve, and fitting (i).
2. For each pipe segment of diameter D_i, get f_i from the Churchill equation or Moody diagram using $N_{Re,i}$ and ε_i/D_i, and calculate $K_{pipe,i} = 4(fL/D)_i$.
3. For each valve and fitting, calculate K_{fi} from N_{Rei} and D_i using the 3-K method.
4. Calculate the driving force, DF, from Equation 7.45.

2. Power Law Fluid

The knowns and unknowns for this case are as follows:

Given: Q, D_i, L_i, m, n, ρ, fittings   *Find*: DF

The appropriate expressions that apply are the Bernoulli equation (Equation 7.45), the power law Reynolds number (Equation 7.40), the pipe friction factor as a function of $N_{Re,pl}$ (Equation 6.44), and the 3-K equation for fitting losses (Equation 7.38), with the power law Reynolds number. The solution procedure is as follows:

1. From the given values, calculate $N_{Re,pl}$ (Equation 7.40).
2. Using Equation 6.44, calculate f and the corresponding K_{pipe}, and using the 3-K method, calculate the K_f for each fitting, Equation (7.38).
3. Calculate the driving force, DF, from the Bernoulli equation, Equation 7.45.

3. Bingham Plastic

The procedure is identical to that explained above, except that Equation 7.41 is used for the Reynolds number and Equation 6.65 is used for the pipe friction factor.

B. Unknown Flow Rate

The Bernoulli equation, Equation 7.45, can be rearranged to solve for the flow rate Q as follows:

$$Q = \frac{\pi}{2\sqrt{2}} \left[\frac{DF}{\sum_i \left(K_i/D_i^4\right) + \alpha_2/D_2^4 - \alpha_1/D_1^4} \right]^{1/2} \tag{7.47}$$

The flow rate can be readily calculated if the loss coefficients can be determined. The procedure involves iteration, starting with estimated values for the loss coefficients. These are then inserted into Equation 7.47 to find Q, which is then used to calculate the Reynolds number(s). The result is used to calculate revised values of the K_f's, as follows:

1. Newtonian Fluid

The knowns and unknowns are as follows:

Given: DF, D, L, ε, μ, ρ *Find*: Q

1. A first estimate for the pipe friction factor and the K_f's can be made by assuming that the flow is fully turbulent (and that $\alpha_1 = \alpha_2 = 1$). From the Colebrook equation

$$f_1 = \frac{0.0625}{[\log(3.7D/\varepsilon)]^2} \tag{7.48}$$

and

$$K_{fit} \cong K_i\left(1 + \frac{K_d}{D_{n,in.}^{0.3}}\right) \tag{7.49}$$

2. Use these values to calculate Q from Equation 7.47 and then the Reynolds number from $N_{Re} = 4Q\rho/\pi D\mu$.

3. This Reynolds number is then used to determine a revised pipe friction factor and loss coefficient ($K_{f, pipe} = 4fL/D$) from the Churchill equation or the Moody diagram and the K_{fit} values from the 3-K equation.

4. The solution is the last value of Q calculated from step 2.

2. Power Law Fluid

The knowns and unknowns are as follows:

Given: DF, D, L, m, n, ρ *Find*: Q

The procedure is essentially identical to that explained above for the Newtonian fluid, except that Equation 7.40 is used for the Reynolds number in step 2 and Equation 6.47 is used for the pipe friction factor in step 3.

3. Bingham Plastic

The knowns and unknowns are as follows:

Given: $DF, D, L, \mu_\infty, \tau_o, \rho$ *Find*: Q

The procedure is again analogous to that for the Newtonian fluid, except that the pipe friction factor in step 3 (and thus K_{pipe}) is determined from Equation 6.65 using $N_{Re} = 4Q\rho/\pi D\mu_\infty$ and $N_{He} = D^2\rho\tau_o/\mu_\infty^2$. The values of K_{fit} are determined from the 3-K equation using Equation 7.41 for the corresponding Reynolds number.

C.   UNKNOWN DIAMETER

It is assumed that the system contains only one size (diameter) pipe, which is to be determined. The Bernoulli equation can be arranged to solve for D as follows:

$$D = \left[\frac{8Q^2 \left(\sum_i K_{fi} + \alpha_2 D^4/D_2^4 - \alpha_1 D^4/D_1^4 \right)}{\pi^2 DF} \right]^{1/4} \tag{7.50}$$

This is implicit in D, but the terms involving α's are normally small (or cancel out), so they may be initially neglected. An initial estimate of the K_{fi}'s is made, which then permits calculation of D from Equation 7.50. However, as D is unknown, ε/D is also unknown, so that a "cruder" initial estimate for f and for the K_{fi} values is required. However, because $K_{pipe} = 4fl/D$, an estimate for f still does not give a value of K_{pipe} because D is also unknown. Therefore, an initial estimate for K_{pipe} can be made by neglecting the fittings altogether, as outlined in Chapter 6.

1. Newtonian Fluids

The knowns and unknowns are as follows:

Given: $Q, DF, L, \varepsilon, \mu$, and ρ *Find*: D

If all fittings are neglected, the procedure is the same as that in Section VI.C.1 of Chapter 6. The value of the following group (Equation 6.78) can be calculated from known quantities:

$$fN_{Re}^5 = \frac{32DF\rho^5 Q^3}{\pi^3 L\mu^5} \tag{7.51}$$

The procedure is as follows:

1. For a first estimate, assume $f = 0.005$.
2. Use this value in Equation 7.51 to estimate the Reynolds number:

$$N_{Re} = \left(\frac{fN_{Re}^5}{0.005} \right)^{1/5} = \left(\frac{32DF\rho^5 Q^3}{0.005\pi^3 L\mu^5} \right)^{1/5} \tag{7.52}$$

The first estimate for D can be obtained from

$$D = \frac{4Q\rho}{\pi\mu N_{Re}} \tag{7.53}$$

3. Now calculate ε/D and use the Churchill equation (Equation 6.40), and 3-K equations for f and K_{fit}, respectively, for further iteration.
4. Calculate D from Equation 7.50 using the previous value of D (from step 3) in the α terms.
5. If the values of D from steps 3 and 4 do not agree, calculate N_{Re} using D from step 4, and use these values in step 3.
6. Repeat steps 3 through 5 until the change in D is within acceptable limits.

2. Power Law Fluid

The knowns and unknowns are as follows:

Given: Q, DF, L, m, n, ρ *Find*: D

The procedure is essentially identical to that used for the Newtonian fluid. We get a first estimate for the power law Reynolds number by neglecting fittings and assuming turbulent flow. This is used to estimate the value of f (and hence K_{pipe}) using Equation 6.44 and the values for the K_{fit} from the corresponding 3-K equation. Inserting these into Equation 7.50 gives the first estimate for the diameter, which is then used to revise the Reynolds number. The iteration continues until successive values of the pipe diameter agree to an acceptable tolerance, as follows.

1. Assume $f = 0.005$.
2. Neglecting fittings, the first estimate for $N_{Re,pl}$ is

$$N_{Re,pl} = \left(\frac{fN_{Re,pl}^{5/(4-3n)}}{0.005} \right)^{(4-3n)/5}$$

$$= \left(\frac{\pi^2 DF}{0.16LQ^2} \right)^{(4-3n)/5} \left[\frac{2^{(7-3n)}Q^{(2-n)}}{m\pi^{(2-n)}} \left(\frac{n}{3n+1} \right)^n \right]. \tag{7.54}$$

3. A first estimate for D is obtained from this value of $N_{Re,pl}$, and the definition of the Reynolds number:

$$D = \left[\frac{2^{(7-3n)}\rho Q^{(2-n)}}{m\pi^{(2-n)}N_{Re,pl}} \left(\frac{n}{3n+1} \right)^n \right]^{1/(4-3n)}. \tag{7.55}$$

4. Using the value of $N_{Re,pl}$ and D from step 3, calculate the value of f (and K_{pipe}) from Equation 6.44 and the K_{fit} values from the 3-K method.

5. Insert the K_f values into Equation 7.50 to get a revised value for D.

6. If the value of D from step 5 does not agree with that from step 3, use the value from step 5 to revise $N_{Re,pl}$ and repeat steps 4 through 6 until satisfactory agreement is reached.

3. Bingham Plastic

The knowns and unknowns are as follows:

Given: $Q, DF, L, \tau_o, \mu_\infty, \rho$ *Find*: D

The procedure for the Bingham plastic is similar to that for the Newtonian and power law fluids, using Equation 6.65 for the pipe friction factor and Equation 6.72 for the Reynolds number:

1. Assume $f = 0.02$.
2. Calculate

$$fN_{Re}^5 = \frac{32 DFQ^3\rho^5}{\pi^3 L\mu_\infty^5} \tag{7.56}$$

3. Get a first estimate of N_{Re} from

$$N_{Re} = \left(\frac{fN_{Re}^5}{0.02}\right)^{1/5} \tag{7.57}$$

4. Use this to get a first estimate of D:

$$D = \frac{4Q\rho}{\pi N_{Re}\mu_\infty} \tag{7.58}$$

5. Using these values of D and N_{Re}, calculate $N_{He} = D^2\rho\tau_o/\mu_\infty^2$, the pipe friction factor from Equation 6.65, $K_{pipe} = 4fL/D$, and the K_{fit}s from the 3-K equation using Equation 7.41 for the Bingham plastic Reynolds number.

6. Insert the K_f values into Equation 7.50 to get a revised value of D.

7. Using this value of D, revise the values of N_{Re} and N_{He}, and repeat steps 5 through 7 until successive values are in satisfactory agreement.

VI. SLACK FLOW

A special condition known as "slack flow" can occur when the gravitational driving force exceeds the "full pipe" friction loss. This occurs most frequently when a liquid is pumped up and down over steep hilly terrain, as illustrated in Figure 7.5. Gravity acts against the flow on the upstream side (from point 1 to point 2) and aids the flow on the downstream side (from point 2 to point 3), so the Bernoulli equation must be applied in two stages (1–2 and then 2–3). The pump provides the driving force for moving the fluid up the hill at a flow rate of Q, and gravity is the driving force on the downhill side. The minimum pressure in the system is at point 2 at the top of the hill. The "head" form of the Bernoulli equation from 1 to 2 is

$$H_p = h_{f,1-2} + \frac{\Phi_2 - \Phi_1}{\rho g} \tag{7.59}$$

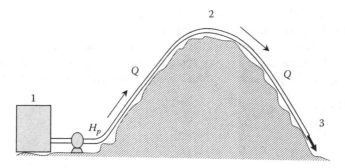

FIGURE 7.5 Conditions for slack flow.

where $H_p = -w/g$ is the required pump head (DF), $h_{f,1-2}$ is the "friction head loss" between points 1 and 2

$$h_{f,1-2} = \frac{4fL_{1-2}}{gD}\left(\frac{V^2}{2}\right) = \frac{32fL_{1-2}Q^2}{g\pi^2 D^5} \tag{7.60}$$

and $\Phi = P + \rho gz$. Now the driving force on the downhill side from 2 to 3 is the pressure from 2 to 3 and gravity from 2 to 3, which must be balanced by the friction loss from 2 to 3:

$$\frac{\Phi_2 - \Phi_3}{\rho g} = h_{f,2-3} = \frac{(P_2 - P_3)}{\rho g} + (z_2 - z_3) \tag{7.61}$$

where
  P_2 is the lowest pressure in the system at point 2, and
  P_3 is the pressure downstream from the hill.
  Both of these pressures will be small relative to the downhill driving force, $(z_2 - z_3)$

In fact, it is not unusual to find that

$$\frac{\Phi_2 - \Phi_3}{\rho g} \approx z_2 - z_3 > h_{f,2-3} \tag{7.62}$$

for a full pipe. This means that the gravity head available is more than enough to overcome the friction loss in a full pipe. Thus, in order to satisfy the Bernoulli equation, $h_{f,2-3}$ must increase to match $(z_2 - z_3)$. The only way that this can occur is if V increases. But since continuity must also be obeyed, that is, $Q = VA = $ constant, so that if V is to increase, then A must decrease, that is, the pipe must flow only partly full. Thus, the pipe flows only partly full on the downside of the hill, while still full on the upside. The vapor space above the fluid on the downside results in a constant pressure in this space. With a constant pressure, the only driving force is gravity. This condition is known as "slack flow." The cross section of the fluid in the partially filled pipe will not be circular (see Figure 7.6), so the methods for flow in a noncircular conduit are applicable, that is, the hydraulic diameter concept applies. Thus, Equation 7.61 becomes

$$z_2 - z_3 = h_{f,2-3} = \frac{2fLQ^2}{gD_h A^2} \tag{7.63}$$

where $D_h = 4A/W_p$. (Recall that formulas for noncircular conduits do not involve velocity directly— only Q/A instead.)

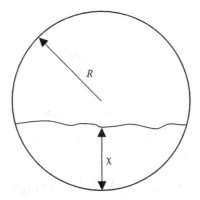

FIGURE 7.6 Pipe flowing less than half full.

The hydraulic diameter can be found as follows. With reference to Figure 7.6, the depth of the fluid in the pipe is χ, which can be either larger or smaller than R. The expressions for the flow cross section and the wetted perimeter are

$$A = R^2 \left\{ \cos^{-1}\left(1 - \frac{\chi}{R}\right) - \left(1 - \frac{\chi}{R}\right)\left[1 - \left(1 - \frac{\chi}{R}\right)^2\right]^{1/2} \right\} \tag{7.64}$$

and

$$W_p = 2R\cos^{-1}\left(1 - \frac{\chi}{R}\right) \tag{7.65}$$

In order to find χ for a given pipe, fluid, and flow rate, an iterative procedure is required:

1. Assume a value of χ/R and calculate A, W_p, and D_h using Equations 7.64 and 7.65.
2. Calculate $N_{Re} = (D_h Q\rho)/A\mu$ and determine f from the Churchill equation or Moody diagram.
3. Calculate the RHS of Equation 7.63. If $(z_2 - z_3) <$ RHS, then increase the value of χ/R and repeat the process. If $(z_2 - z_3) >$ RHS, then decrease the value of χ/R and repeat. The solution is obtained when $(z_2 - z_3) =$ RHS of Equation 7.63.

Slack flow is generally regarded as undesirable, as the increased velocity, as well as the air in the vapor space, tends to increase both erosion and corrosion in the pipe. For this reason, a "choke" or restriction (usually an orifice plate or a valve) is often inserted in the downstream end of the pipe at the bottom of the hill so that the total flow resistance in the pipe on the downside of the hill matches the driving force there and the pipe will flow full everywhere. There will be an increased erosion in the choke, but it is less costly to replace it periodically than the risk of having to replace a large section of pipe.

Example 7.2 Slack Flow

A commercial steel pipeline with a 10 in. ID carries water over a 300 ft high hill. The actual length of the pipe is 500 ft. from the pump station to the top of the hill and 500 ft. on the downhill side. Find (a) the minimum flow rate at which slack flow will not occur in this pipeline and (b) the position of the interface in the pipe when the flow rate is 80% of this value.

Solution:

(a) Slack flow will not occur if

$$z_2 - z_3 \le \frac{2fLQ^2}{gD_hA^2}$$

where $D_h = D$.

Although there are two unknowns in this equation, that is, f and Q, this is a classic unknown flow rate problem that can be solved as previously outlined. Assuming the properties of water are $\rho = 1$ g/cm^3 and $\mu = 0.01$ dyn s/cm^2 and the pipe roughness is $\varepsilon = 0.045$ mm, we can compute

$$fN_{Re}^2 = \frac{g(z_2 - z_3)\rho^2 D^3}{2L_{2-3}\mu^2} = 4.82 \times 10^{10}$$

This is solved iteratively for Q by first assuming $f = 0.005$ and then solving this equation for N_{Re}. This value is then used with the Churchill equation or the Moody diagram to determine f, which is inserted into this equation to determine N_{Re}. This is repeated until successive values are within acceptable limits, and then Q is calculated from

$$\boxed{Q = \frac{\pi D \mu N_{Re}}{4\rho} = 7.44 \times 10^5 \text{ cm}^3\text{/s} = 11,800 \text{ gpm}}$$

(b) If the flow rate is 80% of the value found in (a), slack flow will occur. Determine the position of the fluid interface in the pipe under these conditions

In this case, Equation 7.63 must be satisfied with the fluid flowing through the partially filled pipe (a noncircular conduit). In this case, we cannot calculate either f, A, or $D_h = 4A/W_p$ a priori. Collecting the known quantities together on one side of Equation 7.63, we get

$$\frac{f}{D_hA^2} = \frac{g\Delta z}{2LQ^3} = 2.13 \times 10^{-9} \text{ cm}^{-3}$$

This is used to determine the values of f, A, $D_h = 4A/W_p$, W_p, and R by iteration, using Equations 7.64 and 7.65 and the Churchill equation, as follows:

First, assume a value of χ/R, which permits calculating A and W_p from Equations 7.64 and 7.65 (this gives $D_h = 4A/W_p$).

Then calculate the Reynolds number, $N_{Re} = D_hQ\rho/A\mu$, and ε/D_h, and get f from the Churchill equation. These values are combined to determine the value of f/D_hA^2. This process is repeated until this value does not change, within acceptable limits. The results are as follows:

$$\boxed{\chi/R = 1.37, \quad A = 57.2 \text{ in.}^2, \quad W_p = 19.5 \text{ in.}, \quad N_{Re} = 3.01 \times 10^6, \quad f = 0.0037}$$

It is seen that the fluid interface (χ/R) in the pipe is about 2/3 of the pipe diameter from the bottom of the pipe, that is, it is running about 2/3 full.

VII. PIPE NETWORKS

Piping systems often involve interconnecting segments in various combinations of series and/or parallel arrangements. The principles required to analyze such systems are the same as those we have used for simpler systems, for example, the conservation of mass (continuity) and energy (Bernoulli) equations.

For each pipe junction, or "node" in the network, continuity tells us that the sum of all the flow rates into the node must equal the sum of all the flow rates out of the node. Also, the total driving force (combination of pressure, pump energy, and/or static head) between any two nodes is related to the flow rate and friction loss by the Bernoulli equation, as applied between the two nodes.

If each node in the network is numbered (including the entrance and exit points), then the continuity equation applied at any node i relates the flow rates into and out of this node:

$$\sum_{n=1}^{n} Q_{ni} = \sum_{m=1}^{m} Q_{im} \tag{7.66}$$

where

Q_{ni} represents the flow rate from any upstream node n into node i

Q_{im} is the flow rate out from node i out to any downstream node m

Also, the total driving force in a branch between any two nodes i and j is determined by the Bernoulli equation, (Equation 7.45) as applied to that branch. If the driving force is expressed as the total head loss between nodes (where $h_i = \Phi_i/\rho g$), then

$$h_i - h_j - \frac{w_{ij}}{g} = \frac{8Q_{ij}^2}{g\pi^2 D_{ij}^4} \sum_{i}^{j} K_{fij} \tag{7.67}$$

where

$-w_{ij}/g$ is the pump head (if any) between nodes i and j

D_{ij} is the pipe diameter between nodes i and j

Q_{ij} is the flow rate between nodes i and j

$\sum_{i}^{j} K_{fij}$ represents the sum of the loss coefficients for pipe, valves, fittings, etc., all in the branch between nodes i and j

The latter are determined by the 3-K formula for valves and fittings and the Churchill equation for the pipe segments. These are functions of the flow rates and pipe sizes (Q_{ij} and D_{ij}) in the branch between the nodes i and j. The total number of equations is thus equal to the number of branches plus the number of (internal) nodes, which then equals the number of unknowns that can be determined in the network.

These network equations can be solved for the unknown driving force (across each branch) or the unknown flow rate (in each branch of the network), or an unknown diameter for any one or more of the branches, subject to constraints on the pressure (driving force) and flow rates. Since the solution involves simultaneous coupled nonlinear equations, the process is best accomplished by iteration on a computer, which is easily done using a spreadsheet. Assuming that the overall driving force is known and the various branch flow rates (Q_{ij}) are desired, this is best done by first assuming values for the total head, h_{ij}, at one or more intermediate nodes because these values are bounded by the known upstream and downstream values. Iteration is then accomplished by varying the internal head values.

Example 7.3 Flow in a Manifold

A manifold, or "header," distributes fluid from a common source into various branch lines, as shown in Figure E7.1. The manifold diameter is normally chosen to be significantly larger than that of the branch lines so that the pressure drop in the manifold is negligible compared to that in the branch lines. This insures that the pressure drop in each of the branch lines is nearly the same, which simplifies the calculations. However, these conditions cannot always be satisfied in practice, especially if the total flow rate is large and/or the manifold diameter is not sufficiently larger than that of the branch lines, so these assumption should be verified.

The header in this example is 0.5 in. in diameter and feeds three branch lines, each 0.25 in. in diameter. There are 7 nodes in the problem, as shown in Figure E7.1, with nodes 3, 5, and 7 discharging into the atmosphere at zero elevation. The distance between the branches on the header is 60 ft and the branches are each 200 ft long, with a roughness of 0.0018 in. for each. Water enters

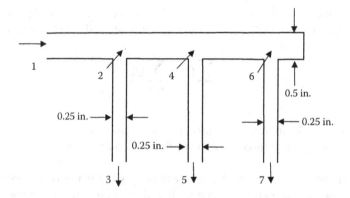

FIGURE E7.1 Flow in a header.

the header at node 1 at 100 psig and exits the branches at points 3, 5, and 7 at atmospheric pressure. We want to know what the flow rate is through the total system and through each of the branches.

This is an "unknown flow rate" problem for which the following form of the Bernoulli equation (Equation 7.67 rearranged) applies:

$$Q_{ij} = \frac{\pi}{2\sqrt{2}} \left[\frac{(P_i - P_j)}{\rho \sum_i^j \left(K_{fij}/D_{ij}^4 \right)} \right]^{1/2} \tag{7.68}$$

The K_{fij} values are those for each pipe segment as well as for the three tees (as elbows), which are determined by the Churchill equation and 3-K equation, respectively.

Continuity tells us that

$$Q_1 = Q_{23} + Q_{45} + Q_{67}$$

$$Q_{23} = Q_1 - Q_{24}$$

$$Q_{45} = Q_{24} - Q_{46} \tag{7.69}$$

$$Q_{67} = Q_{46}$$

It may initially be assumed that the flow is fully turbulent, the pressure drop in the header is negligible, and the losses in the tees corresponded to fully turbulent flow. For fully turbulent pipe flow, the limiting form of the Colebrook can be used, that is,

$$f = \left(\frac{1}{4\log[3.7/(\varepsilon/D)]} \right)^2 \tag{6.40}$$

If the pressure drop in the header is initially neglected, Equation 7.68 can be used to calculate the flow rate in the branches (which will be approximately equal in each branch) and hence the total flow rate from Equation 7.69. These flow rates can then be used to calculate the pressure drop between each node in the header and the branches. These pressure drops are then used to revise the flow rates and the procedure repeated until there is an agreement with the inlet pressure.

The output of a spreadsheet used to solve this problem is shown in Table E7.1. Only the first and the last iterations are shown in this table. In order to initiate the solution, $f = 0.005$ and the value of K_∞ were assumed, as the Reynolds number is unknown. However, all subsequent iterations are based on the actual values of f and K_f calculated from the Churchill equation and the 3-K method. The spreadsheet calculations are carried out by first assuming a value for P_2 (or $h_2 = P_2/\rho g$) and checking the continuity of the flow rates for agreement. Then the value of h_2 is adjusted until these checks are in reasonable agreement.

TABLE E7.1
Spreadsheet Output for Example 7.3

Example: Manifold		Fluid:	Water			

$\rho = 1000$ kg/m^3 = 62.4 lb/ft^3

$\mu = 0.01$ poise = 6.72E–04 lb/ft s

	P_1	100 psig	Assume $f = 0.005$	$\varepsilon = 0.0018$ in.
	P_3	0 psig	Assume $h_2 = 216$ ft	
	P_5	0 psig	For K_{23}, K_{45}, and K_{67}: 1 straight pipe, 2 globe valve, 1 T joint, K_{exit}	
	P_7	0 psig	For K_{12}, K_{24}, and K_{46}: 1 straight pipe only	

Branch	Diameter (in.)		Length (ft)		Loss Coefficient (First Estimation)	
	D_{12}	0.5	L_{12}	60	K_{12}	48.337
	D_{24}	0.5	L_{24}	60	K_{24}	51.3711
	D_{46}	0.5	L_{46}	60	K_{46}	58.3932
	D_{23}	0.25	L_{23}	200	K_{23}	412.733
	D_{45}	0.25	L_{45}	200	K_{45}	413.599
	D_{67}	0.25	L_{67}	200	K_{67}	413.872

Branch	Seg 1–2	Seg 2–3	Seg 2–4	Seg 4–5	Seg 4–6	Seg 6–7	
First iteration							
h_i (ft)	h_1	h_2	h_3	h_4	h_5	h_6	h_7
	230.524	216	0	210.031	0	208.834	0
$Q \times 10^3$ (ft^3/s)	7.763	2.786	4.977	2.748	2.229	2.739	
N_{Re}	22,040.01	15,821.43	14,129.3	15,601.3	6328.649	15,556.76	
ε/D	0.0036	0.0072	0.0036	0.0072	0.0036	0.0072	
A	1.35E+19	3.25E+18	8.38E+18	3.21E+18	2.58E+18	3.21E+18	
B	4,996.524	1,004,907	6,139,431	1,257,436	2.34E+12	1,316,290	
f	0.00812	0.009706	0.008623	0.00972	0.009992	0.009723	
K_{new}[a]	46.7737	397.813	49.6667	398.345	57.5531	398.4599	
$h_{2\,new}$	206.957						
Last iteration							
h_i (ft)	h_1	h_2	h_3	h_4	h_5	h_6	h_7
	230.524	215.93	0	208.948	0	206.817	0
$Q \times 10^3$ (ft^3/s)	6.006	1.977	4.029	1.943	2.087	1.932	
N_{Re}	17,050.7	11,224.41	11,438.51	11,029.88	5923.573	10,969.88	
ε/D	0.0036	0.0072	0.0036	0.0072	0.0036	0.0072	
A	1.04E+19	2.38E+18	6.40E+18	2.34E+18	2.29E+18	2.32E+18	
B	303,510.3	2.44E+08	1.8E+08	3.23E+08	6.74E+12	3.52E+08	
f	0.008392	0.010092	0.008919	0.010114	0.010138	0.010121	
K_{new}	48.337	412.731	51.3711	413.597	58.3932	413.869	
$h_{2\,new}$	215.935						

Note: Values of A and B are calculated from Churchill Equation (6.41).

All values of "h" variables are in units of ft of head.

[a] Use Table 7.3 to calculate updated values of K.

PROBLEMS

1. You must size a pipeline to carry crude oil at a rate of 1 million bbl/day. If the viscosity of the oil is 25 cP and its SG is 0.9, what is the most economical diameter for the pipeline if the pipe costs $3/ft of length and per inch of diameter, the power cost is $0.05/kWh, and the pipeline cost is to be written off over a 3 year period? The oil enters and leaves the pipeline at atmospheric pressure. What would the answer be if the economic lifetime of the pipeline is 30 years?

2. A crude oil pipeline is to be built to carry oil at a rate of 1 million bbl/day (1 bbl = 42 gal). If the pipe cost $12/ft of length per inch of diameter, power to run the pumps costs $0.07/kWh, and the economic lifetime of the pipeline is 30 years, what is the most economical diameter for this pipeline? What total horsepower would be required of the pumps if the line is 800 miles long, assuming 100% efficient pumps? (Oil: viscosity = 35 cP, density = 0.85 g/cm³)?

3. A coal slurry pipeline is to be built to transport 45 million tons/year of slurry over a distance of 1500 miles. The slurry can be approximately described as Newtonian with a viscosity of 35 cP and SG of 1.25. The pipeline is to be built from ANSI 600# commercial steel pipe, the pumps are 50% efficient, energy costs $0.06/kWh, and the economic lifetime of the pipeline is 25 years. What would be the most economical diameter for the pipeline, and what would be the corresponding velocity in the pipe?

4. The Alaskan pipeline was designed to carry crude oil at a rate of 1.2 million bbl/day (1 bbl = 42 gal). If the oil is assumed to be Newtonian, with a viscosity of 25 cP and an SG of 0.85, the cost of energy is $0.10/kWh, and the pipe grade is 600# ANSI, what would be the most economical diameter for the pipeline? Assume that the economic lifetime of the pipeline is 30 years.

5. What is the most economical diameter of a pipeline that is required to transport crude oil (μ = 30 cP, SG = 0.95) at a rate of 1 million bbl/day using ANSI 1500# pipe if the cost of energy is $0.05/kWh (in 1980 dollars), the economic lifetime of the pipeline is 40 years, and the pumps are 50% efficient.

6. Find the most economical diameter of Sch. 40 commercial steel pipe that would be needed to transport a petroleum fraction with a viscosity of 60 cP and SG of 1.3 at a rate of 1500 gpm. The economic life of the pipeline is 30 years, the cost of energy is $0.08/kWh, and the pump efficiency is 60%. The cost of pipe is $20/ft of length, per in. ID. What would be the most economical diameter to use if the pipe is stainless steel at a cost of $85/ft per in. ID, all other things being equal?

7. You must design and specify equipment for transporting 100% acetic acid (density = 1000 kg/m³, μ = 1 mPa s), at a rate of 11.3 m³/h, from a large vessel at ground level into a storage tank that is 6 m above the vessel. The line includes 185 m of pipe and eight flanged elbows. It is necessary to use stainless steel for the system, for which the pipe is hydraulically smooth, and you must determine the most economical diameter to use. You have 38.1 mm (1.5 in.) and 50.2 mm (2 in.) nominal Sch. 40 pipe available. The cost may be determined from the following approximate formulas:

    Pump cost: Cost ($) = 3.1 (m³/s)$^{0.3}$(m of head)$^{0.25}$
    Motor cost: Cost($) = 75(kWh)$^{0.85}$
    Pipe cost: Cost ($)/ft = 2.5(Nom. Dia., in.)$^{3/2}$
    90° elbow: Cost ($) = 5(nom. Dia., in.)$^{3/2}$
    Power cost: = 0.03$/kWh

    (a) Calculate the total pump head required for each size pipe, in ft of head.
    (b) Calculate the motor hp required for each size pipe, assuming 80% efficiency (motors available only in multiples of ¼ hp).
    (c) Calculate the total capital cost for pipe, pump, motor, and fittings for each size pipe.
    (d) Assuming that the useful life of the installation is 5 years, calculate the total operating cost over this period for each size pipe.
    (e) Which size pipe results in the lowest total cost over the 5-year period?

8. A large building has a roof with dimensions 50 ft × 200 ft, which drains into a gutter system. The gutter contains three drawn aluminum downspouts that have a square cross section, 3 in. on a side. The length of the downspouts from the roof to the ground is 20 ft. What is the heaviest rainfall (in./h) that the downspouts can handle before the gutter will overflow?

9. A roof drains into a gutter, which feeds into a downspout with a square cross section (4 in. × 4 in.). The discharge end of the downspout is 12 ft below the entrance and terminates in a 90° mitered (one weld) elbow. The downspout is made of smooth sheet metal.
   (a) What is the capacity of the downspout, in gpm?
   (b) What would the capacity be if there were no elbow at the bottom?

10. An open concrete flume is to be constructed to carry water from a plant unit to a cooling lake by gravity flow. The flume has a square cross section and is 1500 ft long. The elevation at the upstream end is 10 ft higher than the lower discharge end. If the flume is to be designed to carry 10,000 gpm of water when full, what should its size (i.e., width) be? Assume rough cast concrete.

11. An open drainage canal with a rectangular cross section is 3 m wide and 1.5 m deep. If the canal slopes 950 mm in 1 km of length, what is the maximum capacity of the canal in m^3/h?

12. A concrete-lined drainage ditch has a triangular cross section that is an equilateral triangle, 8 ft on each side. The ditch has a slope of 3 ft/mile. What is the flow capacity of the ditch in gpm?

13. An open drainage canal is to be constructed to carry water at a maximum rate of 10^6 gpm. The canal is concrete lined and has a rectangular cross section, with a width that is twice its depth. The elevation of the canal drops 3 ft per mile of length. What should the width and depth of the canal be?

14. A drainage ditch is to be built to carry runoff from a subdivision. The maximum design capacity is to be 1 million gph (gal/h) and it is to be concrete lined. If the ditch has a cross section of an equilateral triangle, open at the top, and if the slope is 2 ft/mile, what should the width at the top be?

15. A drainage canal is to be dug to keep a low-lying area from flooding during heavy rains. The canal would carry the water to a river that is 1 mile away and 6 ft lower in elevation. The canal will be lined with cast concrete and will have a semicircular cross section. If it is sized to drain all of the water falling on a 1 mile^2 area during a rainfall of 4 in./h, what should the diameter of the semicircle be?

16. An open drainage canal with a rectangular cross section and a width of 20 ft is lined with concrete. The canal has a slope of 1 ft/1000 yards. What is the depth of water in the canal when the water is flowing at a rate of 500,000 gpm?

17. An air ventilating system must be designed to deliver air at 20°F and atmospheric pressure at a rate of 150 ft^3/s through 4000 ft of square duct. If the air blower is 60% efficient and is driven by a 30 hp motor, what size duct is required if it is made of sheet metal? Under these conditions, air may be considered as an incompressible fluid.

18. Oil with a viscosity of 25 cP and SG of 0.78 is stored in a large open tank. A vertical tube made of stainless steel with an ID of 1 in. and a length of 6 ft is attached to the bottom of the tank. You want the oil to drain from the tank at a rate of 30 gpm.
   (a) How deep should the oil in the tank be for the oil to drain at this rate?
   (b) If a globe valve is installed in the tube, how deep must the oil in the tank be for it to drain at the same rate, with the globe valve wide open?

19. A vertical tube is attached to the bottom of an open vessel. A liquid with an SG of 1.2 is draining through the tube, which is 10 cm long with an ID of 3 mm. When the depth of the fluid in the tank is 4 cm, the flow rate through the tube is 5 cm^3/s.
   (a) What is the viscosity of the liquid (assuming it is Newtonian)?
   (b) What would your answer be if you neglected the entrance loss from the vessel to the tube?

20. Heat is to be transferred from one process stream to another by means of a double pipe heat exchanger. The hot fluid flows in a 1 in. Sch. 40 tube, which is inside (and concentric with) a 2 in. Sch. 40 tube, with the cold fluid flowing in the annulus between the tubes, in the opposite direction. If both tubes are to carry the fluids at a velocity of 8 ft/s and the total equivalent length of the tubes is 1300 ft, what pump power is required to circulate the colder fluid? The cold fluid properties at an average temperature are $\rho = 55$ lb$_m$/ft^3, $\mu = 8$ cP.

21. A commercial steel pipe ($\varepsilon = 0.0018$ in.) is 1½ in. Sch. 40 diameter, 50 ft long, and includes one globe valve. If the pressure drop across the entire line is 22.1 psi when it is carrying water at a rate of 65 gpm, what is the loss coefficient for the globe valve? The friction factor for the pipe can be calculated from the equation

$$ f = \frac{0.0625}{[\log(3.7D/\varepsilon)]^2} $$

22. Water at 68°F is flowing through a 45° pipe bend at a rate of 2000 gpm. The inlet to the bend is 3 in. ID, and the outlet is 4 in. ID. The pressure at the inlet is 100 psig, and the pressure drop in the bend is equal to half of what it would be in a 3 in. 90° elbow. Calculate the net force (magnitude and direction) that the water exerts on the elbow.

23. What size pump (horsepower) is required to pump an organic product (SG = 0.85 and $\mu = 60$ cP) from tank A to tank B at a rate of 2000 gpm through a 10 in. Sch. 40 pipeline, 500 ft long, containing 20 90° flanged elbows, 1 open globe valve, and 2 open gate valves? The liquid level in tank A is 20 ft below that in tank B, and both are open to the atmosphere.

24. A plant piping system takes a process stream ($\mu = 15$ cP, $\rho = 0.9$ g/cm^3) from one vessel at 20 psig and delivers it to another vessel at 80 psig. The system contains 900 ft of 2 in. Sch. 40 pipe, 24 standard elbows, and 5 globe valves. If the downstream vessel is 10 ft higher than the upstream vessel, what horsepower pump would be required to transport the fluid at a rate of 100 gpm, assuming a pump efficiency of 100%?

25. Crude oil ($\mu = 40$ cP, SG = 0.7) is to be pumped from a storage tank to a refinery through a 10 in. Sch. 20 commercial steel pipeline at a flow rate of 2000 gpm. The pipeline is 50 miles long and contains 35 90° elbows and 10 open gate valves. The pipeline exit is 150 ft higher than the entrance, and the exit pressure is 25 psig. What horsepower is required to drive the pumps in the system if they are 70% efficient?

26. The Alaskan pipeline is 48 in. ID, 800 miles long, and carries crude oil at a rate of 1.2 million bbl/day (1 bbl = 42 gal). Assuming the crude oil to be a Newtonian fluid with a viscosity of 25 cP and an SG of 0.87, what is the pumping horsepower required to operate the pipeline? The oil enters and leaves the pipeline at sea level, and the line contains the equivalent of 150 90° elbows and 100 open gate valves. Assume that the inlet and discharge pressures are both 1 atm.

27. A 6 in. Sch. 40 pipeline carries an intermediate product stream ($\mu = 15$ cP, SG = 0.85) at a velocity of 7.5 ft/s from a storage tank at 1 atm pressure to a plant site. The line contains 1500 ft of straight pipe, 25 90° elbows, and 4 open globe valves. The liquid level in the storage tank is 15 ft above ground level, and the pipeline discharges into a vessel that is 10 ft above ground at a pressure of 10 psig. What is the required flow capacity in gpm and the pressure head to be specified for the pump needed for this job? If the pump is 65% efficient, what horsepower motor is required to drive the pump?

28. An open tank contains 5 ft of water. The tank drains through a piping system containing 10 90° elbows, 10 branched tees, 6 gate valves, and 40 ft of horizontal Sch. 40 pipe. The surface of the water in the tank and the pipe discharge are both at atmospheric pressure. An entrance loss factor of 1.5 accounts for the tank-to-pipe friction loss and kinetic energy change. Calculate the flow rate (in gpm) and Reynolds number for the water draining through the system for nominal pipe diameters of 1/8, ¼, ½, 1, 1.5, 2, 4, 6, 8, 10, and 12 in., including all of the aforementioned fittings, using (a) constant K_f values, (b) $(L/D)_{eq}$ values, and (c) the 3-K method. Constant K_f and $(L/D)_{eq}$ values from the literature are given here for these fittings:

Fitting	Constant K_f	$(L/D)_{eq}$
90° elbow	0.75	30
Branched tee	1.0	60
Gate valve	0.17	8

29. A pump takes water from a reservoir and delivers it to a water tower. The water in the tower is at atmospheric pressure and is 120 ft above the reservoir. The pipeline is composed of 1000 ft of 2 in. Sch. 40 pipe containing 32 gate valves, 2 globe valves, and 14 standard elbows. If the water is to be pumped at a rate of 100 gpm using a pump that is 70% efficient, what horsepower motor is required to drive the pump?

30. You must determine the pump head and power required to transport a petroleum fraction ($\mu = 60$ cP, $\rho = 55$ lb$_m$/ft^3) at a rate of 500 gpm from a storage tank to the feed plate of a distillation column. The pressure in the tank is 2 psig and that in the column is 20 psig. The liquid level in the tank is 15 ft above ground, and the inlet to the column is 60 ft high. If the piping system contains 400 ft of 6 in. Sch. 80 steel pipe, 18 standard elbows, and 4 globe valves, calculate the required pump head (i.e., pressure) and the horsepower required if the pump is 70% efficient.

31. What horsepower pump would be required to transfer water at a flow rate of 100 gpm from tank A to tank B, if the liquid surface in tank A is 8 ft above ground and that in tank B is 45 ft above ground? The piping between tanks consists of 150 ft of 1½ in. Sch. 40 pipe and 450 ft of 2 in. Sch. 40 pipe that includes 16 standard elbows and 4 open globe valves.

32. An additive having a viscosity of 2 cP and a density of 50 lb$_m$/ft^3 is fed from a reservoir into a mixing tank. The pressure in both the reservoir and the tank is 1 atm, and the level in the reservoir is 2 ft above the end of the feed line, in the tank. The feed line consists of 10 ft of ¼ in. Sch. 40 pipe, 4 elbows, 2 plug valves, and 1 globe valve. What will the flow rate of the additive be (in gpm) when all the valves are fully open?

33. The pressure in the water main serving your house is 90 psig. The plumbing between the main and your outside faucet contains 250 ft of galvanized ¾ in. Sch. 40 pipe, 16 elbows, and the faucet, which is an angle valve. When the faucet is wide open, what is the flow rate, in gpm?

34. You are filling your beer mug from a keg. The pressure in the keg is 5 psig, the filling tube from the keg is 3 ft long and ¼ in. ID, and the valve is a diaphragm dam type. The tube is attached to the keg by a (threaded) tee used as an elbow. If the beer leaving the tube is 1 ft above the level of the beer inside the keg and there is a 2 ft long, ¼ in. ID stainless steel tube inside the keg, attached to the valve outside the keg, how long will it take to fill your mug if it holds 500 cm^3? (For beer: $\mu = 8$ cP, $\rho = 64$ lb$_m$/ft^3)

35. You must install a piping system to drain SAE 10 lube oil at 70°F (SG = 0.928) from tank A to tank B by gravity flow. The level in tank A is 10 ft above that in tank B, and the pressure in

tank A is 5 psi greater than that in tank B. The system will contain 200 ft of Sch. 40 pipe, eight standard elbows, two gate valves, and a globe valve. What size pipe should be used if the oil is to be drained at a rate of 100 gpm?

36. A new industrial plant requires a supply of water at a rate of 5.7 m³/min. The pressure in the water main is 800 kPa, and it is 50 m from the plant. The line from the main to the plant will have 65 m of galvanized iron pipe, four standard elbows, and two gate valves. If the water pressure at the plant must be no less than 500 kPa, what diameter pipe must be used?

37. A pump is used to transport water at 72°F from tank A to tank B at a rate of 200 gpm. Both tanks are vented to the atmosphere. Tank A is 6 ft above the ground with a water depth of 4 ft in the tank, and tank B is 40 ft above ground with a water depth of 5 ft in the tank. The water enters the top of tank B, at a point 10 ft above the bottom of the tank. The pipeline between the tanks contains 185 ft of 2 in. Sch. 40 galvanized iron pipe, three standard elbows, and one gate valve.
    (a) If the pump is 70% efficient, what horsepower motor would be required to drive the pump?
    (b) If the pump is driven by a 5 hp motor, what is the maximum flow rate that can be achieved (in gpm)?

38. A pipeline carrying gasoline (SG = 0.72, μ = 0.7 cP) is 5 miles long and is made of 6 in. Sch. 40 commercial steel pipe. The line contains 24 90° elbows, 2 open globe valves, and a pump capable of producing a maximum head of 400 ft is available. The inlet pressure to the line is 10 psig and the exit pressure is 20 psig. The discharge end is 30 ft higher than the inlet end.
    (a) What is the maximum flow rate possible in the line, in gpm?
    (b) What is the horsepower of the motor required to drive the pump if it is 60% efficient?

39. A water tower supplies water to a small community of 800 homes. The level of the water in the tank is 120 ft above ground level, and the water main from the tower to the housing area is 1 mile of Sch. 40 commercial steel pipe. The water system is designed to provide a minimum pressure of 15 psig at peak demand, which is estimated to be 2 gpm per house.
    (a) What nominal size pipe diameter should be used for the water main?
    (b) If this size pipe is installed, what would be the actual flow through the main, in gpm?

40. A 12 in. Sch. 40 pipe, 60 ft long, discharges water at 1 atm pressure from a reservoir. The pipe is horizontal, and the outlet is 12 ft below the surface of the water in the reservoir.
    (a) What is the flow rate, in gpm?
    (b) In order to limit the flow rate to 3500 gpm, an orifice is installed at the end of the line. What should the diameter of the orifice be?
    (c) What size pipe would have to be used to limit the flow rate to 3500 gpm without using an orifice?

41. Crude oil with a viscosity of 12.5 cP and SG = 0.88 is to be pumped through a 12 in. Sch. 30 commercial steel pipe at a rate of 1900 bbl/h. The pipeline is 15 miles long, with a discharge that is 125 ft above the inlet, and contains 10 standard elbows and 4 gate valves.
    (a) What is the total power required to drive the pumps if they are 70% efficient?
    (b) How many pump stations will be required if the pumps develop a discharge pressure of 100 psi each?
    (c) If the pipeline must go over hilly terrain, what is the steepest downslope grade that can be tolerated without creating slack flow in the pipe?

42. A pipeline to carry crude oil at a rate of 1 million bbl/day is constructed with 50 in. ID pipe, and it is 700 miles long. It contains the equivalent of 70 gate valves but no other fittings.
    (a) What is the total power required to drive the pumps if they are 70% efficient?

(b) How many pump stations will be required if the pumps develop a discharge pressure of 100 psig?

(c) If the pipeline must go through hilly terrain, what is the steepest downslope grade that can be tolerated without creating slack flow in the pipe?
(crude oil viscosity = 25 cP, SG = 0.9)

43. You are building a pipeline to transport crude oil (SG = 0.8, viscosity = 30 cP) from a seaport over a mountain to a tank farm. The top of the mountain is 3000 ft above sea level and 1000 ft above the tank farm. The distance from the port to the mountain top is 200 miles and from the mountain top to the tank farm is 75 miles. The oil enters the pumping station at the port at 1 atm pressure and is to be discharged at the tank farm at 20 psig. The pipe diameter is 20 in. Sch. 40 commercial steel, and the oil flow rate is 2000 gpm.

(a) Will slack flow occur in the line? If so, you must install a restriction (orifice) at the downstream end of the line to ensure that the pipe will always be full. What should the pressure drop across the orifice be, in psi?

(b) How much pumping power will be required if the pumps are 70% efficient? What pump head is required, in ft?

44. You want to siphon water from an open tank using a hose. The discharge end of the hose is 10 ft below the level of the water in the tank. The minimum allowable pressure in the hose for proper operation is 1 psia. If you wish the water velocity in the hose to be 10 ft/s, what is the maximum height that the siphon hose can extend above the water level in the tank for proper operation?

45. A liquid is draining from a cylindrical vessel through a tube in the bottom of the vessel, as illustrated in Figure P7.45. The liquid has a specific gravity of 1.2 and a viscosity of 2 cP. The entrance loss from the tank to the tube is 0.4, and the system has the following dimensions:
$D = 2$ in., $d = 3$ mm, $L = 20$ cm, $h = 5$ cm, $\varepsilon = 0.0004$ in.

(a) What is the volumetric flow rate of the liquid, in cm³/s?
What would the answer to (a) be if the entrance loss were neglected?

(b) Repeat part (a) for a value of $h = 75$ cm.

46. Water from a lake is flowing over a concrete spillway at a rate of 100,000 gpm. The spillway is 100 ft wide and is inclined at an angle of 30° to the vertical. If the effective roughness of the concrete is 0.03 in., what is the depth of water in the stream flowing over the spillway?

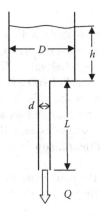

FIGURE P7.45 Fluid draining from tank through tube.

47. A pipeline composed of 1500 ft of 6 in. Sch. 40 pipe containing 25 90° elbows and 4 open gate valves carries oil with a viscosity of 35 cP and SG = 0.85, at a velocity of 7.5 ft/s, from a storage tank to a plant site. The storage tank is at atmospheric pressure and the level in the tank is 15 ft above ground. The pipeline discharge is 10 ft above ground, and the discharge pressure is 10 psig.
    (a) What is the pump capacity (in gpm) and pump head (in ft) required in the pipeline?
    (b) If the pump has an efficiency of 65%, what horsepower motor would be required to drive it?

48. Water is pumped at a rate of 500 gpm through a 10 in. ID pipeline, 50 ft long, with both entrance and exit from the pipe at ground level. The line contains two standard elbows and a swing check valve. The pressure is 1 atm entering and leaving the pipeline. Calculate the pressure drop (in psi) through the pipeline due to friction using (a) the 2-K method, (b) the $(L/D)_{eq}$ method, and (c) the 3-K method.

49. Water at 70°F is flowing in a film down outside a 4 in. ID vertical tube at a rate of 1 gpm. What is the thickness of the film?

50. What diameter pipe would be required to transport a liquid with a viscosity of 1 cP and a density of 1 g/cm³ at a rate of 1500 gpm, the length of the pipe is 213 ft, the wall roughness of the pipe is 0.006 in., and the total driving force is 100 ft lb$_f$/lb$_m$?

51. The ETSI pipeline was designed to carry coal slurry from Wyoming to Texas at a rate of 30×10^6 tons/year. The slurry behaves like a Bingham plastic, with a yield stress of 100 dyn/cm², a limiting viscosity of 40 cP, and a density of 1.4 g/cm³. Using the cost of ANSI 1500# pipe and $0.07/kWh for electricity, determine the most economical diameter for the pipeline if its economic lifetime is 25 year and the pumps are 50% efficient.

52. A mud slurry is being drained from a tank through a 50 ft long plastic hose. The hose has an elliptical cross section with a major axis of 4 in. and a minor axis of 2 in. The open end of the hose is 10 ft below the level in the tank. The mud is a Bingham plastic, with a yield stress of 100 dyn/cm², a limiting viscosity of 50 cP, and a density of 1.4 g/cm³.
    (a) At what rate will the mud drain from the hose, in gpm?
    (b) At what rate would water drain through the hose?

53. A 90° threaded elbow attached to the end of a 3 in. Sch. 40 pipe, and a reducer with an inside diameter of 1 in. is threaded into the elbow. If water is pumped through the pipe and out the reducer into the atmosphere at a rate of 500 gpm, calculate the forces exerted on the pipe at the point where the elbow is attached.

54. A continuous flow reactor vessel contains a liquid reacting mixture with a density of 0.85 g/cm³ and a viscosity of 7 cP at 1 atm pressure. Near the bottom of the vessel is a 1½ in. outlet line containing a safety relief valve. There is 4 ft of pipe with two 90° elbows between the tank and the valve. The relief valve is a spring-loaded lift check valve, which opens when the pressure upstream of the valve reaches 5 psig. Downstream of the valve is 30 ft of horizontal pipe containing four elbows and two gate valves that empties into a vented catch tank. The check valve serves essentially as a level control for the liquid in the reactor because the static head in the reactor is the only source of pressure on the valve. Determine
    (a) The fluid level in the reactor at the point when the valve opens
    (b) When the valve opens, the rate (in gpm) at which the liquid will drain from the reactor into the catch tank

55. A pipeline has been proposed to transport a coal slurry 1200 miles from Wyoming to Texas, at a rate of 50 million tons/year, through a 36 in. diameter pipeline. The coal

slurry has the properties of a Bingham plastic, with a yield stress of 150 dyn/cm², a limiting viscosity of 40 cP, and a SG of 1.5. You must conduct a lab experiment in which the measured pressure gradient can be used to determine the total pressure drop in the pipeline.

(a) Perform a dimensional analysis of the system to determine an appropriate set of dimensionless groups to use (you may neglect the effect of wall roughness for this fluid).

(b) For the lab test fluid, you have available a sample of the above coal slurry and three different muds with the following properties:

	Yield Stress (dyn/cm²)	Limiting Viscosity (cP)	Density (g/cm³)
Mud 1	50	80	1.8
Mud 2	100	20	1.2
Mud 3	250	10	1.4

Which of these would be the best to use in the lab, and why?

(c) What size pipe and what flow rate (in lbm /min) should you use in the lab?

(d) If the measured pressure gradient in the lab is 0.016 psi/ft, what is the total pressure drop in the pipeline?

NOTATION

A	Cross-sectional area, [L²]
A	Pump station cost parameter, Figure 7.4, [$]
a	Pipe cost parameter, [4/L^{p+1}]
B	Pump station cost parameter, Figure 7.4, [$t/FL = $t³/ML²]
C	Energy cost, [$/FL = $t²/ML²]
D	Diameter, [L]
D_h	Hydraulic diameter, [L]
DF	Driving force, Equation 7.42, [FL/M = [L²/t²]
e_f	Energy dissipated per unit mass of fluid, [FL/M = L²/t²]
f	Fanning friction factor, [—]
f_t	Fully turbulent friction factor, Equation 7.36, [—]
g	Acceleration due to gravity [L/t²]
h	Fluid layer thickness, [L], or total head (potential), [L]
H_p	Pump head, [L]
h_f	Friction loss "head," [L]
HP	Power, [FL/t = ML²/t³]
ID_{in}	Pipe inside diameter, in inches, [L]
K_f	Loss coefficient, [—]
K_1, K_∞	2-K loss coefficient parameters, [—]
K_1, K_i, K_d	3-K loss coefficient parameters, [—]
L	Length, [L]
$\dot{m}$	Mass flow rate, [M/t]
M	Maintenance cost factor, [—]
N_c	Cost group, Equation 7.26
N_{He}	Hedstrom number, [—]
$N_{Re,h}$	Reynolds number based on hydraulic diameter, [—]
$N_{Re,pl}$	Power law Reynolds number, [—]

Q	Volumetric flow rate, $[L^3/t]$
R	Pipe radius, $[L]$
V	Spatial average velocity, $[L/t]$
W_p	Wetted perimeter
X	Fraction of capital cost charged per unit time, $[1/t]$
Y	Economic lifetime, $[t]$

GREEK

α	Kinetic energy correction factor, $[—]$
$\Delta()$	$()_2 - ()_1$
ε	Pipe wall roughness, $[L]$
η_e	Efficiency, $[—]$
ρ	Density, $[M/L^3]$
Φ	Potential $(P + \rho g z)$, $[F/L^2] = [M/Lt^2]$
χ	Position of interface in partially full pipe, $[L]$

SUBSCRIPTS

1	Reference point 1 (normally upstream)
2	Reference point 2 (normally downstream)
i	Coordinate direction i
j	Coordinate direction j
n	Nominal pipe size

REFERENCES

CCPS (Center for Chemical Process Safety), *Guidelines for Pressure Relief and Effluent Handling Systems*, AIChE, New York, 1998.

Chhabra, R.P. and J.F. Richardson, *Non-Newtonian Flow and Applied Rheology*, 2nd edn., Butterworth-Heinemann, Oxford, U.K., 2008.

Crane Co., Flow of fluids through valves, fittings and pipe, Technical Paper 410, Crane Co., New York, 1991.

Darby, R., Correlate pressure drops through fittings, *Chem. Eng.*, 108(4), 127–130, 2001.

Darby, R. and J. Forsyth, Salt dissolution in drilling muds - a generalized correlation for mass transfer in non-Newtonian fluids, *Canad. J. Chem. Eng.*, 70, 97–103, 1992.

Darby, R. and J.D. Melson, Direct Determination of Optimum Economic Pipe Diameter for Non-Newtonian Fluids, *J. Pipelines*, 2, 11–21, 1982.

Durand, A.A., J.A. Boy, J.L. Corral, L.O. Barra, J.S. Trueba, and P.V. Brena, Update rules for pipe sizing, *Chem. Eng.*, 106, pp. 153–156, May 1999. Also, see *ibid* 117, pp. 48–50 (January) 2010.

Green, D. and R.H. Perry, Perry's Chemical Engineer's Handbook, 8th ed., Mc'Graw-Hill, New York, 2007.

Hooper, W.B., The two-K method predicts head losses in pipe fittings, *Chem. Eng.*, 88, pp. 96–100, August 24, 1981.

Hooper, W.B., Calculate headloss caused by change in pipe size, *Chem. Eng.*, 95, pp. 89–92, November 7, 1988.

Penoncello, S.G., *Thermal Energy Systems*, CRC Press, Boca Raton, FL, 2015, p. 178.

8 Pumps and Compressors

"There is no right answer, but there is a best answer."

—Anonymous

I. PUMPS

There exists a wide variety of pumps that are designed for various specific applications. However, most of them can be broadly classified into two categories: positive displacement and centrifugal. The primary performance variables for any pump are the pump capacity (e.g., gpm, liters/min, m³/h, etc.) and the pressure or "head" that the pump can develop (e.g., ft or m of fluid). Recall that the head is related to the pressure by $h = \Delta P/\rho g$, which involves the density of the system fluid, so that a specific head value is unique to a specific fluid. The most significant characteristics of each of these two types of pumps are described in the following. More detailed descriptions can be found in specialized books (e.g., Karassik et al., 1976; Karassik and McGuire, 2012).

A. POSITIVE DISPLACEMENT PUMPS

The term *positive displacement pump* is quite descriptive, because such pumps are designed to displace a more-or-less fixed volume of fluid during each cycle of operation. They include piston, diaphragm, screw, gear, progressing cavity, etc. The volumetric flow rate is determined by the displacement per cycle of the moving member (either rotating or reciprocating) times the cycle rate (e.g., rpm). The flow capacity is thus fixed by the design, size, and operating speed of the pump. The pressure (or head) that the pump develops depends upon the flow resistance of the system in which the pump is installed and is limited only by the size of the driving motor and the mechanical strength of the parts. Consequently, the discharge line from the pump should never be closed off without allowing for recycle around the pump or damage to the pump could result.

In general, PD pumps have limited flow capacity but are capable of developing relatively high pressures. These pumps operate at an essentially constant flow rate, with variable head. They are appropriate for high-pressure requirements, very viscous fluids, and applications that require a precisely controlled or metered flow rate.

B. CENTRIFUGAL PUMPS

The term "centrifugal pump" is a very descriptive term, since these pumps operate by the transfer of energy (or angular momentum) from a rotating impeller to the fluid, which is normally inside a casing. A sectional view of a typical centrifugal pump is shown in Figure 8.1. The fluid enters at the axis or "eye" of the impeller (which may be open or closed and usually contains radial curved vanes) and is discharged from the impeller periphery.

The kinetic energy and momentum of the fluid are increased by the angular momentum imparted by the high-speed impeller. This kinetic energy is then converted to pressure energy ("head") in a diverging area (the "volute") between the impeller discharge and the casing before the fluid exits the pump. The head that these pumps can develop depends upon the pump design and the size, shape, and speed of the impeller, and the flow capacity is determined by the fluid resistance of the system in which the pump is installed. Thus, as will be shown, these pumps operate at approximately

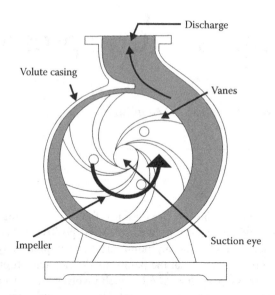

FIGURE 8.1 Typical centrifugal pump.

constant head and variable flow rate, within limits of course, determined by the size and design of the pump and the size of the driving motor.

Centrifugal pumps can be operated in a "closed off" condition (i.e., closed discharge line), because the liquid will recirculate within the pump without causing damage. However, such conditions should be avoided, because energy dissipation within the pump could result in excessive heating of the fluid and/or the pump, or unstable operation with adverse consequences. Centrifugal pumps are most appropriate for "ordinary" (i.e., low to moderate viscosity) liquids under a wide variety of flow conditions and are thus the most common type of pump. The following discussion applies primarily to centrifugal pumps.

II. PUMP CHARACTERISTICS

The Bernoulli equation applied between the suction inlet and the discharge of a pump gives

$$-w = \frac{\Delta P}{\rho} = gH_p \tag{8.1}$$

That is, the *net* energy or work put into the fluid by the pump goes to increasing the fluid pressure, or the equivalent pump head, H_p. However, because pumps are not 100% efficient, some of the energy delivered from the motor to the pump is dissipated or "lost" due to friction in the shear stresses around the high-speed impeller. It is very difficult to separately characterize this friction loss, so it is accounted for in the overall pump efficiency, η_e, which is the ratio of the useful work (or hydraulic work), done by the pump on the fluid $(-w)$ to the work put into the pump by the motor $(-w_m)$:

$$\eta_e = \frac{-w}{-w_m} \tag{8.2}$$

The efficiency of a pump depends upon the pump and impeller design, the size and speed of the impeller, and the conditions under which it is operating, and is determined by tests carried out by the pump manufacturer. This will be discussed in more detail later.

When selecting a pump for a particular application, it is first necessary to specify the flow capacity and head required of the pump. Although many pumps might be able to meet these specifications, the "best" pump is normally the one that has the highest efficiency at the specified operating conditions. The required operating conditions, along with knowledge of the pump efficiency, then allow us to determine the required size (e.g., brake horsepower, *[HP]* or *[BHP]*) of the driving motor for the pump:

$$HP = -w_m \dot{m} = \frac{\Delta P Q}{\eta_e} = \frac{\rho g H_p Q}{\eta_e} \tag{8.3}$$

Now the power delivered from the motor to the pump is also the product of the torque on the shaft driving the pump (Γ) and the angular velocity of the shaft (ω):

$$HP = \Gamma \omega = \frac{\rho g H_p Q}{\eta_e} \tag{8.4}$$

If it is assumed that the fluid leaves the impeller tangentially at the same speed as the impeller (a good approximation), then an angular momentum balance on the fluid in contact with the impeller gives

$$\Gamma = \dot{m} \omega R_i^2 = \rho Q \omega R_i^2 \tag{8.5}$$

where R_i is the radius of the impeller, and the angular momentum of the fluid entering the eye of the impeller has been neglected (a good assumption). By eliminating Γ from Equations 8.4 and 8.5 and solving for the pump head, H_p, we get

$$H_p = \frac{\eta_e \omega^2 R_i^2}{g} \tag{8.6}$$

This shows that the pump head is determined primarily by the size and speed of the impeller and the pump efficiency, independent of the flow rate of the fluid. This is approximately correct for most centrifugal pumps over a wide range of flow rates. However, there is a limit to the flow that a given pump can handle, and as the flow rate approaches this limit the developed head will start to drop off. The maximum efficiency (i.e. the "best efficiency point" or *BEP*) for most pumps occurs near the flow rate where the head starts to drop significantly.

Figure 8.2 shows a typical set of pump characteristic curves as determined by the pump manufacturer. "Size 2 × 3" means that the pump has a 2 in. discharge and a 3 in. suction port. "R&C" and "1⅞ pedestal" are the manufacturer's designations, and 3500 rpm is the speed of the impeller. Performance curves for impellers with diameters from 6¼ to 8¾ in. are shown, and the efficiency is shown as contour lines of constant efficiency. The maximum efficiency for this pump is somewhat above 50%, although some pumps may operate with efficiencies as high as 80% or 90%. Operation at conditions on the right-hand branch of the efficiency contours (i.e., beyond the "maximum normal capacity" line in Figure 8.2) should be avoided, because this could result in unstable operation. The pump with the characteristics in Figure 8.2 is a slurry pump with a semi-open impeller, designed to pump solid suspensions (this pump can pass solid particles as large as 1¼ in. in diameter). Pump characteristic curves for a variety of other pumps are shown in Appendix H.

Such performance curves are normally determined by the manufacturer from operating data using water at 60°F. Note from Equation 8.6 that the head is independent of fluid properties, although from Equation 8.4 the power is proportional to the fluid density (as is the developed pressure). The horsepower curves in Figure 8.2 indicate the motor horsepower required to pump water at 60°F and must be corrected for density when operating with other fluids and/or at other temperatures.

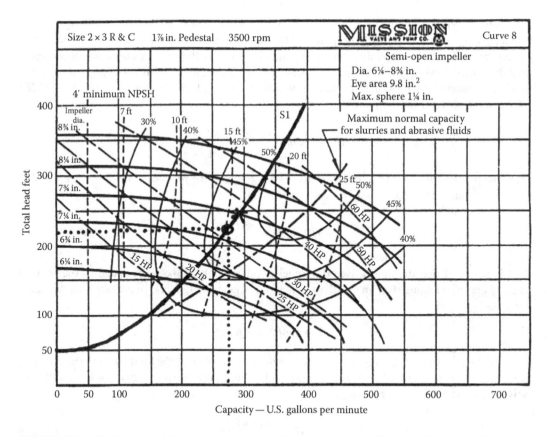

FIGURE 8.2   Typical pump characteristic curves with system operating line superimposed. (From TRW Mission Pump Brochure.)

Actually, it is better to use Equation 8.4 to calculate the motor power from the values of the head, flow rate, fluid density, and efficiency at the operating point. The curves in Figure 8.2 labeled "minimum NPSH" refer to the cavitation characteristics of the pump, which will be discussed later.

III.   PUMPING REQUIREMENTS AND PUMP SELECTION

When selecting a pump for a given application (e.g., a required flow capacity and head), we must specify the appropriate pump type, size and type of impeller, and size (horsepower) and speed (rpm) of the motor that will do the "best" job. "Best" normally means operating in the vicinity of the *best efficiency point* (BEP) on the pump curve (i.e., not lower than about 75% or higher than about 110% of capacity at the BEP). Not only will this condition do the required job at the least cost (i.e., least power requirement), but it also provides the lowest strain on the pump because the pump design is optimum for conditions at the BEP. We will concentrate on these factors and not get involved with the mechanical details of pump design (impeller vane design, casing dimensions, seals, etc.). More details on these topics are given by Karassik et al. (1976) and Karassik and McGuire (2012).

A.   REQUIRED HEAD

A typical piping application starts with a specified flow rate for a given fluid. The piping system is then designed with the necessary valves, fittings, etc. and should be sized for the most economical pipe size, as discussed in Chapter 7. Application of the energy balance (Bernoulli) equation to the entire system, from the upstream end (point 1) to the downstream end (point 2), determines the overall net driving force (DF) in the system required to overcome the frictional resistance:

$$DF = \sum e_f \tag{8.7}$$

where the change in kinetic energy is assumed to be negligible. The total head (DF) is the net sum of the pump head, the total pressure drop, and the elevation drop:

$$\frac{DF}{g} = H_p + \frac{(P_1 - P_2)}{\rho g} + (z_1 - z_2) \tag{8.8}$$

The friction loss (Σe_f) is the sum of all of the losses from point 1 (upstream) to point 2 (downstream):

$$\sum e_f = \sum_i \left(\frac{V^2}{2} K_f \right)_i = \frac{8Q^2}{\pi^2} \sum_i \left(\frac{K_f}{D^4} \right)_i \tag{8.9}$$

where the loss coefficients (K_f's) include all pipes, valves, fittings, contractions, expansions, etc. in the system. Eliminating DF and Σe_f from Equations 8.7 through 8.9 and solving for the pump head, H_p, gives

$$H_p = \frac{P_2 - P_1}{\rho g} + (z_2 - z_1) + \frac{8Q^2}{g\pi^2} \sum_i \left(\frac{K_f}{D^4} \right)_i \tag{8.10}$$

This relates the system pump head requirement to the specified flow rate and the system loss parameters (e.g., the K_f values). Note that H_p is a quadratic function of Q for highly turbulent flow (i.e., constant K_f's). For laminar flow, the K_f values are inversely proportional to the Reynolds number, which results in a linear relationship between H_p and Q. A plot of H_p versus Q from Equation 8.10, illustrated in Figure 8.2 as line S1, is called the *operating line* for the system. Thus, the required pump head and flow capacity are determined by the system requirements as indicated by this line, and we must select the best pump to meet these requirements.

B. COMPOSITE CURVES

Most pump manufacturers provide composite curves, such as those shown in Figure 8.3 that show the operating range of various pumps. Because of the overlap in these curves, more than one pump may be adequate for a given application. For each pump that provides the required flow rate and head, the individual pump characteristics (such as those shown in Figure 8.2 and Appendix H) are then consulted. The intersection of the system curve with the pump characteristic curve for a given impeller determines the pump operating point. The impeller diameter is selected that will produce the required head (or greater). This is repeated for all possible pump, impeller, and speed combinations to determine the combination that results in the highest efficiency (i.e., least power requirement). Note that if the operating point (H_p, Q) does not fall exactly on one of the (impeller) curves, then the actual impeller diameter that produces the higher head at the required flow rate Q is chosen. However, when this pump is installed in the system, the actual operating point will correspond to the intersection of the system curve (Equation 8.10) and the actual pump impeller curve at this point, as indicated by the X in Figure 8.2.

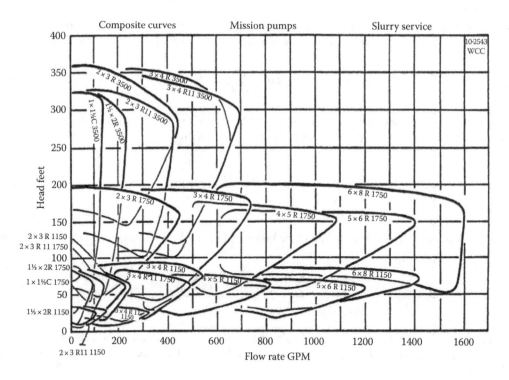

FIGURE 8.3 Typical pump composite curve. (From TRW Mission Pump Brochure [manufacturer's catalog].)

Example 8.1 Pump Selection

Consider a piping system that must deliver water at a rate of 275 gpm from one storage tank to another, both of which are at atmospheric pressure, with the level in the downstream tank being 50 ft higher than in the upstream tank. The piping system contains 65 ft of 2 in. sch 40 pipe, one globe valve, and six elbows. If the pump to be used has the characteristics shown in Figure 8.2, what impeller diameter should be used with this pump, and what motor horsepower would be required?

Solution:

The head requirement for the piping system is given by Equation 8.10. Here, $(z_2 - z_1) = 50$ ft and, since both upstream and downstream pressures are 1 atm, $\Delta P = 0$. The Reynolds number at 275 gpm for water at 60°F is 4.21×10^5, which gives a friction factor of 0.00497 in commercial steel pipe ($\varepsilon/D = 0.0018/2.067 = 0.00087$). The corresponding loss coefficient for the pipe is $K_{pipe} = 4fL/D = 7.51$, and the loss coefficients for the fittings from Table 7.3 are (assuming flanged connections) elbow, $K_1 = 800$, $K_i = 0.091$, $K_d = 4.0$; globe valve, $K_1 = 1500$, $K_i = 1.7$, $K_d = 3.6$. At the pipe Reynolds number, this gives $\Sigma(K_f) = (K_{pipe} + K_{glbv} + 6K_{el}) = 16.4$. The curve labeled S1 in Figure 8.2 is H_p versus Q from Equation 8.10 for this value of the loss coefficients. This neglects the variation of the K_f's over the range of flow rate indicated, which is a good assumption at this Reynolds number. At a flow rate of 275 gpm, the required head from Equation 8.10 is 219 ft.

The point where the flow rate of 275 gpm intersects the system curve in Figure 8.2 (at 219 ft of head) falls between impeller diameters of 7¼ and 7¾ in. as indicated by the point O on the system operating line. Thus, the 7¼ in. diameter would be too small, so we would need the 7¾ in. diameter impeller. However, if the pump with this impeller is installed in the system, the operating point would move to the point indicated by the X in Figure 8.2. This corresponds to a head of almost 250 ft and a flow rate of about 290 gpm (i.e., the excess head provided by the larger impeller results in a higher flow rate than desired, all other things being equal). One way to achieve the desired flow rate of 275 gpm would obviously be to close down on the valve until this value is achieved. This is equivalent to increasing the resistance (i.e., the loss coefficient) for the system, which will shift the system curve upward until it intersects the 7¾ in. impeller curve at the desired flow rate of 275 gpm. The pump will still provide 250 ft of head, but about 30 ft of this head is "lost" (dissipated) due to the additional resistance in the partly closed valve. The pump efficiency at this operating point is about 47%, and the motor power (HP) required to pump water at 60°F at this point is

$$HP = \frac{\rho g H_p Q}{\eta_e} = 37 \text{ hp}$$

A control valve operates in this mode automatically (as described in Chapter 11), but this is not an efficient use of the available energy. A more efficient way of controlling the flow rate, instead of closing the valve, might be to adjust the speed of the impeller by using a variable speed drive. This would save energy because it would not increase the friction loss as does closing down on the valve, but it would require greater capital cost because variable speed drives are more expensive than fixed speed motors.

IV. CAVITATION AND NPSH

A. Vapor Lock and Cavitation

As previously mentioned, a centrifugal pump increases the fluid pressure by first imparting angular momentum (or kinetic energy) to the fluid, which is converted to pressure in the diffuser or volute section. Hence, the fluid velocity in and around the impeller is much higher than that either entering or leaving the pump, and the pressure is the lowest where the velocity is highest. The minimum pressure at which a pump will operate properly must be above the vapor pressure of the fluid; otherwise the fluid will vaporize (or "boil"), a condition known as *cavitation*. Obviously, the higher the temperature, the higher the vapor pressure and the more likely that this condition will occur. When a centrifugal pump contains a gas or vapor, it will still develop the same head, but because the pressure is proportional to the fluid density the developed pressure will be several orders of magnitude lower than the pressure for a liquid at the same head. This condition (when the pump is filled with a gas or vapor) is known as *vapor lock*, and the pump will not function when this occurs.

However, cavitation may result in an even more serious condition than vapor lock. When the pressure at any point within the pump drops below the vapor pressure of the liquid, vapor bubbles will form at that point (this generally occurs on the impeller). These bubbles will then be transported to another region in the fluid where the pressure is greater than the vapor pressure, at which point they will collapse. This formation and collapse of bubbles occurs very rapidly and can create local "shock waves," which can cause erosion and serious damage to the impeller or pump. (It is often obvious when a pump is cavitating, because it may sound as though there are rocks in the pump!)

B. Net Positive Suction Head

To prevent cavitation, it is necessary that the pressure at the pump suction (inlet) be sufficiently high that the minimum pressure anywhere in the pump will always be above the vapor pressure. This required minimum suction pressure (in excess of the vapor pressure) depends upon the pump design, impeller size and speed, and flow rate and is called the *minimum required net positive suction head* (NPSH or NPSHR). Values of the minimum required NPSH for the pump in Figure 8.2 are shown as dashed lines. The NPSH is almost independent of impeller diameter at low flow rates and increases with flow rate as well as with impeller diameter at higher flow rates. A distinction is sometimes made between the minimum NPSH "required" to prevent cavitation (i.e., the NPSHR) and the actual head (e.g., pressure) "available" at the pump suction (NPSHA). A pump will not cavitate if (NPSHA > NPSHR + vapor pressure head).

The NPSH at the operating point for the pump determines where the pump can be installed in a piping system to ensure that cavitation will not occur. The criterion is that the pressure head at the suction (entrance) of the pump (e.g., the NPSHA) must exceed the vapor pressure head by at least the value of the NPSH (or NPSHR) to avoid cavitation. Thus, if the pressure at the pump suction is P_s and the fluid vapor pressure is P_v at the operating temperature, cavitation will be prevented if

$$NPSHA = \frac{P_s}{\rho g} \geq NPSH + \frac{P_v}{\rho g} \tag{8.11}$$

The suction pressure P_s is determined by applying the Bernoulli equation to the suction line upstream of the pump. For example, if the pressure at the entrance to the upstream suction line is P_1, the maximum distance above this point that the pump can be located without cavitating (i.e., the *maximum suction lift*) is determined by applying the Bernoulli equation from P_1 to P_s:

$$h_{max} = \frac{P_1 - P_v}{\rho g} - NPSH + \frac{V_1^2 - V_s^2}{2g} - \frac{\sum (e_f)_s}{g} \tag{8.12}$$

where Equation 8.11 has been used for P_s. V_1 is the velocity entering the suction line, V_s is the velocity at the pump inlet (suction), and $\Sigma(e_f)_s$ is the total friction loss in the suction line from the upstream entrance (point 1) to the pump inlet, including all pipes, fittings, etc. The diameter of the pump suction port is usually bigger than the discharge or exit diameter in order to minimize the kinetic energy head entering the pump, because this kinetic energy decreases the maximum suction lift and enhances the chance of cavitation. Note that if the maximum suction lift (h_{max}) is negative, the pump must be located below the upstream entrance to the suction line to prevent cavitation.

It is best to be conservative when interpreting the NPSH requirements to prevent cavitation. The minimum required NPSH on the pump curves is normally determined using water at 60°F with the discharge line fully open. However, even though a pump will run with a closed discharge line with no bypass, there will be much more recirculation within the pump if this occurs, which increases local turbulence and corresponding dissipative heating, both of which increase the minimum required NPSH. This is especially true with high-efficiency pumps, which have close clearances between the impeller and pump casing.

Example 8.2 Maximum Suction Lift

A centrifugal pump with the characteristics shown in Figure 8.2 is to be used to pump an organic liquid from a reboiler to a storage tank, through a 2 in. sch 40 line, at a rate of 200 gpm. The pressure in the reboiler is 2.5 atm, and the liquid has a vapor pressure of 230 mm Hg, an SG of 0.85, and a viscosity of 0.5 cP at the working temperature. If the suction line upstream of the pump is also 2 in. sch 40 and contains two elbows and one globe valve, and the pump has a 7¾ in. impeller, what is the maximum height above the reboiler that the pump can be located without cavitating?

Solution:

The maximum suction lift is given by Equation 8.12. From Figure 8.2, the NPSH required for the pump at 200 gpm is about 11 ft. The velocity in the reboiler (V_1) can be neglected, and the velocity in the pipe (see Appendix F) is $V_s = 200/10.45 = 19.1$ ft/s.
   The friction loss in the suction line is

$$e_f = \frac{V_s^2}{2} \sum (K_{pipe} + K_{glbv} + 2K_{el})$$

where $K_{pipe} = 4fh/D$ and the fitting losses are given by the 3-K formula and Table 7.3 (elbow: $K_1 = 800$, $K_i = 0.091$, $K_d = 4.0$; globe valve: $K_1 = 1500$, $K_i = 1.7$, $K_d = 3.7$). The value of the Reynolds number for this flow is 5.23×10^5, which, for commercial steel pipe ($\varepsilon/D = 0.0018/2.067 = 0.00087$), gives $f = 0.00493$. Note that the pipe length is h in K_{pipe}, which is the same as the maximum suction length (h_{max}) on the left of Equation 8.12, assuming that the suction line is vertical. The unknown (h) thus appears on both sides of the equation. Solving Equation 8.12 for h gives 17.7 ft.

C. SPECIFIC SPEED

The flow rate, head, and impeller speed at the maximum or "best efficiency point" of the pump characteristic can be used to define a dimensionless group called the *specific speed*:

$$N_s = \frac{N\sqrt{Q}}{H^{3/4}}\left(\frac{\text{rpm}\sqrt{\text{gpm}}}{\text{ft}^{3/4}}\right) \tag{8.13}$$

Although this group is dimensionless, it is common practice to use selected mixed (inconsistent) units when quoting the value of N_s, that is, N in rpm, Q in gpm, and H in feet. The value of the *specific speed represents the ratio of the pump flow rate to the head at the speed corresponding to the maximum efficiency point (BEP)* and depends primarily on the design of the pump and impeller. As previously stated, most centrifugal pumps operate at relatively low heads and high flow rates, i.e., high values of N_s. However, this value depends strongly on the impeller design, which can vary widely from almost pure radial flow to almost pure axial flow (like a fan). Some examples of various types of impeller designs are shown in Figures 8.4 and 8.5. Radial flow impellers have the highest head and lowest flow capacity (low N_s), whereas axial flow impellers have a high flow rate and low head characteristic (high N_s). Thus, the magnitude of the specific speed is a direct indication of the impeller design and performance, as shown in Figure 8.5. Figure 8.5 also indicates the range of flow rates and efficiencies of the various impeller designs, as a function of the specific speed. As seen in Figure 8.5, the maximum efficiency corresponds roughly to a specific speed of about 3000.

D. SUCTION SPECIFIC SPEED

Another "dimensionless" group, analogous to the specific speed that relates directly to the cavitation characteristics of the pump, is the *suction specific speed*, N_{ss}:

$$N_{ss} = \frac{NQ^{1/2}}{(NPSH)^{3/4}} \tag{8.14}$$

The units used in this group are also rpm, gpm, and ft. This identifies the *inlet conditions* that produce similar flow behavior in the inlet for geometrically similar pump inlet passages. Note that the suction specific speed (N_{ss}) relates only to the pump cavitation characteristics as related to the inlet conditions, whereas the specific speed (N_s) relates to the entire pump at the BEP. The suction specific speed can be used, for example, to characterize the conditions under which excessive recirculation may occur at the inlet to the impeller vanes. Recirculation involves flow reversal and reentry resulting from undesirable pressure gradients at the inlet or discharge of the impeller vanes, and its occurrence generally defines the stable operating limits of the pump. For example, Figure 8.6 shows the effect of the suction specific speed on the stable "recirculation-free" operating window, expressed as NPSH versus percent of capacity at BEP, for various values of N_{ss}.

It should be noted that there are conflicting parameters in the proper design of a centrifugal pump. For example, Equation 8.12 shows that the smaller the suction velocity (V_s), the less the tendency to cavitate, that is, the less severe the NPSH requirement. This would dictate that the eye of the impeller should be as large as practical in order to minimize V_s. However, a large impeller eye means a high vane tip speed at the impeller inlet, which is destabilizing with respect to recirculation. Hence, it is advisable to design the impeller with the smallest eye diameter that is practicable.

It should be emphasized that the pump characteristics and performance curves (such as shown in Figure 8.2 and Appendix H) were obtained by the manufacturer using water as the test fluid. Although the pump head (H) values are independent of fluid density, the horsepower curves (HP) are not and must be adjusted for the fluid density. Also, the pump performance will be derated if the fluid viscosity is significantly greater than that of water, often by 10% or more (see, e.g., Wilson et al., 2008; Kalombo et al., 2014).

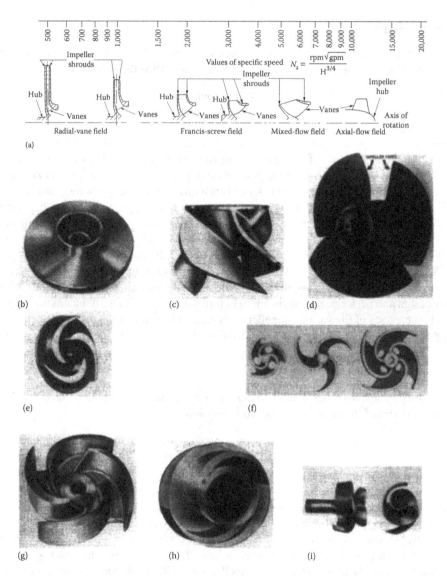

FIGURE 8.4 Impeller designs and specific speed characteristics. (a) Variation in impeller profiles with specific speed and approximate range of specific speed for the various types. (b) Straight-vane, single-suction closed impeller. (c) Open mixed-flow impeller. (d) Axial-flow impeller. (e) Semiopen impeller. (f) Open impellers. Notice that the impellers at left and right are strengthened by a partial shroud. (g) Open impeller with a partial shroud. (h) Phantom view of radialvane nonclogging impeller. (Worthington Pump, Inc.) (i) Paper-pulp impeller. (b–f, h, i: Worthington Pump, Inc.) (From Karassik, I.J. and McGuire, J.T., *Centrifugal Pumps*, 2nd edn., Springer, 2012.)

V. COMPRESSORS

A compressor may be thought of as a high-pressure pump for a compressible fluid. By "high pressure," is meant conditions under which the compressible properties of the fluid (gas) must be considered. This normally occurs when the pressure changes by as much as 30% or more. For "low pressures" (i.e., smaller pressure changes), a fan or blower may be an appropriate "pump" for a gas. Fan operation can be analyzed by using the incompressible flow equations, because the relative pressure difference and hence the relative density change are normally small. As with pumps, compressors may be either PD or centrifugal. The former are suitable for relatively high pressures

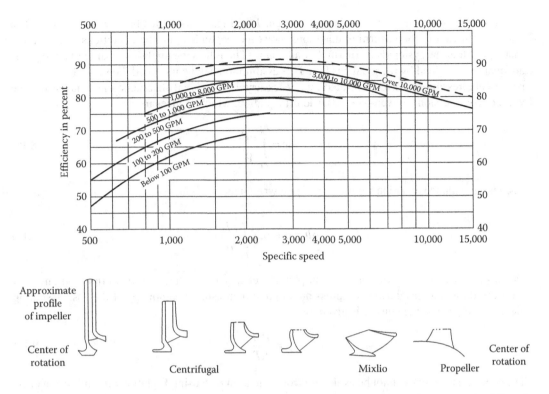

FIGURE 8.5 Correlation between impeller shape, specific speed, and efficiency. (From Karassik, I.J. et al., *Pump Handbook*, McGraw-Hill, New York, 1976.)

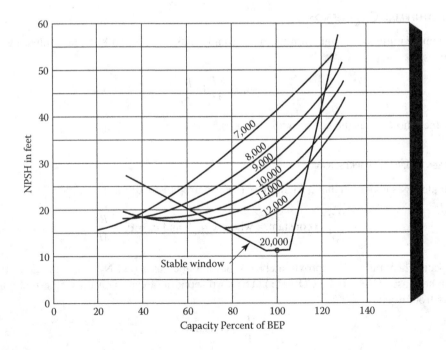

FIGURE 8.6 Effect of suction speed on stable operating window due to recirculation. Numbers on the curves are the values of the suction specific speed, N_{ss}. (From Raymer, R.E., *Chem. Eng. Progr.*, 89(3), 79, March 1993.)

and low flow rates, whereas the latter are designed for higher flow rates but lower pressures. The major distinction in the governing equations, however, depends upon the conditions of operation, that is, whether the system is isothermal or adiabatic. The following analyses assume that the gas is adequately described by the ideal gas law. This assumption can be modified, however, by an appropriate compressibility correction factor, as necessary. For an ideal (frictionless) compression, the work of compression is given by the Bernoulli equation, which reduces to

$$-w = \int_{P_1}^{P_2} \frac{dP}{\rho} \tag{8.15}$$

The energy balance equation for the gas can be written as

$$\Delta h = q + e_f + \int_{P_1}^{P_2} \frac{dP}{\rho} \tag{8.16}$$

which says that the work of compression plus the energy dissipated due to friction and any heat transferred into the gas during compression all go to increasing the enthalpy of the gas. Assuming ideal gas properties, the density is given by

$$\rho = \frac{PM}{RT} \tag{8.17}$$

The compression work cannot be evaluated from Equation 8.15 using Equation 8.17 unless the operating condition or temperature is specified. We will consider two cases: isothermal compression and adiabatic compression.

A. ISOTHERMAL COMPRESSION

If the temperature is constant, eliminating ρ from Equations 8.17 and 8.15 and evaluating the integral gives

$$-w = \frac{RT}{M} \ln \frac{P_2}{P_1} \tag{8.18}$$

where the ratio P_2/P_1 is the *compression ratio* (r).*

B. ISENTROPIC COMPRESSION

For an ideal gas under adiabatic frictionless (i.e., isentropic) conditions:

$$\frac{P}{\rho^k} = \text{constant}, \quad \text{where } k = \frac{c_p}{c_v} \text{ and } c_p = c_v + \frac{R}{M} \tag{8.19}$$

The specific heat ratio k is approximately 1.4 for diatomic gases (O_2, N_2, etc.) and 1.3 for triatomic and higher gases (NH_3, H_2O, CO_2, etc.). The corresponding expression for isothermal conditions follows from Equation 8.17:

$$\frac{P}{\rho} = \text{constant} \tag{8.20}$$

---

* Note that for PD compressors, the compression ratio is defined as the ratio of the volume change during a cycle. However, for centrifugal compressors, the pressure ratio is more appropriate. For an isothermal ideal gas, the two are the same.

Note that the isothermal condition can be considered a special case of the isentropic condition by setting $k = 1$. The "constant" in Equation 8.19 or 8.20 can be evaluated from the known conditions at some point in the system (e.g., P_1 and T_1). Using Equation 8.19 to eliminate the density from Equation 8.15 and evaluating the integral leads to

$$-w = \frac{RT_1 k}{M(k-1)} \left[\left(\frac{P_2}{P_1} \right)^{(k-1)/k} - 1 \right] \tag{8.21}$$

Although it is not obvious by inspection, setting $k = 1$ in Equation 8.21 reduces the equation to Equation 8.18 (this can be confirmed by application of l'Hospital's rule).

If we compare the work required to compress a given gas to a given compression ratio by isothermal and isentropic processes, we see that the isothermal work is always less than the isentropic work. That is, less energy would be required if compressors could be made to operate under isothermal conditions. However, in most cases, a compressor operates under more nearly adiabatic conditions (isentropic, if frictionless) because of the relatively short residence time of the gas in the compressor, which allows very little time for heat generated by compression to be transferred away. However, many compressors are fitted with a cooling jacket to remove the heat of compression and to more closely approach isothermal operation. The temperature rise during an isentropic compression is determined by eliminating ρ from Equations 8.17 and 8.19 to give

$$\frac{T_2}{T_1} = \left(\frac{P_2}{P_1} \right)^{(k-1)/k} = r^{(k-1)/k} \tag{8.22}$$

In reality, most compressor conditions are neither purely isothermal nor purely isentropic but somewhere in between these limits. This can be accounted for in calculating the compression work by using the isentropic equation (Equation 8.21) but replacing the specific heat ratio k by a "polytropic" constant, γ, where $1 < \gamma < k$. The value of γ is a function of the compressor design as well as the properties of the gas.

C. STAGED OPERATION

It is often impossible to reach a desired compression ratio with a single centrifugal compressor. In such cases, multiple compressor "stages" can be arranged in series to increase the overall compression ratio. Furthermore, to increase the overall efficiency, it is common to cool the gas between stages by using "interstage cooling." With interstage cooling to the initial temperature (T_1), it can be shown that as the number of stages increases, the total compression work for isentropic compression approaches that of isothermal compression at T_1.

For multistage operation, there will be an optimum compression ratio for each stage that will minimize the total compression work. This can be easily seen by considering a two-stage compressor with interstage cooling. The gas enters stage 1 at (P_1, T_1), leaves stage 1 at (P_2, T_2), and is then cooled to T_1. It then enters stage 2 at (P_2, T_1) and leaves at P_3. By computing the total isentropic work for both stages (using Equation 8.21), and setting the derivative of this with respect to the interstage pressure (P_2) equal to zero, the value of P_2 that results in the minimum total work can be found. The result is that the optimum interstage pressure that minimizes the total work for a two-stage compression with intercooling to T_1 is

$$P_2 = (P_1 P_3)^{1/2}, \quad \text{or} \quad \frac{P_2}{P_1} = \frac{P_3}{P_2} = r = \left(\frac{P_3}{P_1} \right)^{1/2} \tag{8.23}$$

That is, the total work is minimized if the compression ratio is the same for each stage. This result can easily be generalized to any number (n) of stages (with interstage cooling to the initial temperature), as follows:

$$r = \frac{P_2}{P_1} = \frac{P_3}{P_2} = \cdots = \frac{P_{n+1}}{P_n} = \left(\frac{P_{n+1}}{P_1}\right)^{1/n} \tag{8.24}$$

If there is no interstage cooling or if there is interstage cooling to a temperature other than T_1, it can be shown that the optimum compression ratio for each stage (i) is related to the temperature entering that stage (T_i) by

$$T_i\left(\frac{P_{i+1}}{P_i}\right)^{(k-1)/k} = T_i r_i^{(k-1)/k} = \text{constant} \tag{8.25}$$

D. EFFICIENCY

The foregoing equations apply to ideal (frictionless) compressors. To account for friction losses, the ideal computed work is divided by the compressor efficiency, η_e, to get the total work that must be supplied to the compressor:

$$(-w)_{total} = \frac{(-w)_{ideal}}{\eta_e} \tag{8.26}$$

The energy "lost" due to friction is actually dissipated into thermal energy, which raises the temperature of the gas. This temperature rise is in addition to that due to the isentropic compression, so that the total temperature rise across an adiabatic compressor stage is given by

$$T_2 = T_1 r^{(k-1)/k} + \frac{1-\eta_e}{\eta_e}\left(\frac{-w_{ideal}}{c_v}\right) \tag{8.27}$$

SUMMARY

The main points that are important to retain from this chapter include the following:

- Understand the operation of centrifugal pumps.
- Understand the information presented on the pump characteristic curves and how they interact with the system flow curve.
- Understand *NPSH* and the distinction between *NPSHA* and *NPSHR*, and the conditions for cavitation.
- Understand the use of composite curves.
- Understand the operation of compressors and staged operation.

PROBLEMS

PUMPS

1. The pressure developed by a centrifugal pump for Newtonian liquids that are not highly viscous depends upon the liquid density, the impeller diameter, the rotational speed, and the volumetric flow rate.
    (a) Determine a suitable set of dimensionless groups that should be adequate to relate all of these variables.

(b) You want to know what pressure a pump will develop with a liquid having an SG of 1.4 at a flow rate of 300 gpm using an impeller with a diameter of 12 in. driven by a motor running at 1100 rpm. You have a similar test pump in the lab with a 6 in. impeller driven by an 1800 rpm motor. You want to run a test with the lab pump under conditions that will allow you to determine the pressure developed by the larger pump.

(c) Should you use the same liquid in the lab as in the larger pump, or can you use a different liquid? Why?

(d) If you use the same liquid, at what flow rate will the operation of the lab pump simulate that of the larger pump?

(e) If the lab pump develops a pressure of 150 psi at the proper flow rate, what pressure will the field pump develop at 300 gpm?

(f) What pressure will the field pump develop with water at 300 gpm?

2. The propeller of a speed boat is 1 ft in diameter and is 1 ft below the surface of the water. At what speed (rpm) will cavitation occur at the propeller? Water density = 64 lb_m/ft^3, vapor pressure of water, P_v = 18.65 mmHg.

3. You must specify a pump to be used to transport water at a rate of 5000 gpm through 10 miles of 18 in. sch 40 pipe. The friction loss in valves and fittings is equivalent to 10% of the pipe length, and the pump is 70% efficient. If a 1200 rpm motor is used to drive the pump, determine

(a) The required horsepower and torque rating of the motor

(b) The diameter of the impeller that should be used in the pump

4. You must select a centrifugal pump that will develop a pressure of 40 psi when pumping a liquid with an SG of 0.88 at a rate of 300 gpm. From all the pump characteristic curves in Appendix H, select the best pump for this job. Specify pump head, impeller diameter, motor speed, motor efficiency, and motor horsepower.

5. An oil with a 32.6° API gravity at 60°F is to be transferred from a storage tank to a process unit, that is 10 ft above the tank, at a rate of 200 gpm. The piping system contains 200 ft of 3 in. sch 40 pipe, 25 90° screwed elbows, 6 stub-in tees used as elbows, 2 lift check valves, and 4 standard globe valves. From the pump performance curves in Appendix H, select the best pump for this application. Specify the pump size, motor speed, impeller diameter, operating head, and the efficiency and horsepower of the motor required to drive the pump.

6. You must purchase a centrifugal pump to circulate cooling water that will deliver 5000 gpm at a pressure of 150 psi. If the pump is driven by an 1800 rpm motor, what should the horsepower and torque rating of the motor be, and how large (diameter) should the pump impeller be, assuming an efficiency of 60%?

7. In order to pump a fluid of SG = 0.9 at a rate of 1000 gpm through a piping system, a hydraulic power of 60 hp is required. Determine the required pump head, the torque of the driving motor, and the estimated impeller diameter if an 1800 rpm motor is used.

8. From your prior analysis of pumping requirements for a water circulating system, you have determined that a pump capable of delivering 500 gpm at a pressure of 60 psi is required. If a motor operating at 1800 rpm is chosen to drive the pump, which is 70% efficient, determine

(a) The required horsepower rating of the motor

(b) The required torque rating of the motor

(c) The diameter of the impeller that should be used in the pump

(d) What color the pump should be painted

9. You want to pump water at 70°F from an open well, 200 ft deep, at a rate of 30 gpm through a 1 in. sch 40 pipe, using a centrifugal pump having an NPSH of 8 ft. What is the maximum

distance above the water level in the well that the pump can be located without cavitating? (Vapor pressure of water at 60°F = 18.7 mmHg)

10. Steam condensate at 1 atm and 95°C (P_v = 526 mmHg) is returned to a boiler from the condenser by a centrifugal boiler feed pump. The flow rate is 100 gpm through a 2.5 in. sch 40 pipe. If the equivalent length of the pipe between the condenser and the pump is 50 ft, and the pump has an NPSH of 6 ft, what is the maximum height above the condenser that the pump can be located?

11. Water at 160°F is to be pumped at a rate of 100 gpm through a 2 in. sch 80 steel pipe from one tank to another located 100 ft directly above the first. The pressure in the lower tank is 1 atm. If the pump to be used has a required NPSH of 6 ft of head, what is the maximum distance above the lower tank that the pump may be located?

12. A pump with a 1 in. diameter suction line is used to pump water from an open hot water well at a rate of 15 gpm. The water temperature is 90°C, with a vapor pressure of 526 mmHg and density of 60 lb_m/ft^3. If the pump NPSH is 4 ft, what is the maximum distance above the level of the water in the well that the pump can be located and still operate properly, i.e., without cavitating?

13. Hot water is to be pumped out of an underground geothermally heated aquifer located 500 ft below ground level. The temperature and pressure in the aquifer are 325°F and 150 psig. The water is to be pumped out at a rate of 100 gpm through 2.5 in. pipe using a pump that has a required NPSH of 6 ft. The suction line to the pump contains four 90° elbows and one gate valve. How far below ground level must the pump be located in order to operate properly?

14. You must install a centrifugal pump to transfer a volatile liquid from a remote tank to a point in the plant 500 ft from the tank. To minimize the distance that the power line to the pump must be strung, it is desirable to locate the pump as close to the plant as possible. If the liquid has a vapor pressure of 20 psia, the pressure in the tank is 30 psia, the level in the tank is 30 ft above the pump inlet, and the required pump NPSH is 15 ft, what is the closest that the pump can be located to the plant without the possibility of cavitation? The line is 2 in. sch 40, the flow rate is 100 gpm, and the fluid properties are ρ = 45 lb_m/ft^3 and μ = 5 cP.

15. It is necessary to pump water at 70°F (P_v = 0.35 psia) from a well that is 150 ft deep, at a flow rate of 25 gpm. You do not have a submersible pump, but you do have a centrifugal pump with the required capacity that cannot be submerged. If a 1 in. sch 40 pipe is used, and the NPSH of the pump is 15 ft, how close to the surface of the water must the pump be lowered for it to operate properly?

16. You must select a pump to transfer an organic liquid with a viscosity of 5 cP and SG of 0.87 at a rate of 1000 gpm through a piping system that contains 1000 ft of 8 in. sch 40 pipe, 4 globe valves, 16 gate valves, and 43 std 90° elbows. The discharge end of the piping system is 30 ft above the entrance, and the pressure at both ends is 10 psia.
    (a) What pump head is required?
    (b) What is the hydraulic horsepower to be delivered to the fluid?
    (c) Which combination of pump size, motor speed, and impeller diameter from the pump charts in Appendix H would you choose for this application?
    (d) For the pump selected, what size motor would you specify to drive it?
    (e) If the vapor pressure of the liquid is 5 psia, how far directly above the liquid level in the upstream tank could the pump be located without cavitating?

17. You need a pump that will develop at least 40 psi at a flow rate of 300 gpm of water. What combination of pump size, motor speed, and impeller diameter from the pump characteristics in Appendix H would be the best for this application? State your reasons for the choice you make.

What are the pump efficiency, motor horsepower and torque requirement, and NPSH for the pump you choose at these operating conditions?

18. A centrifugal pump takes water from a well at 120°F (vapor pressure P_v = 87.8 mmHg) and delivers it at a rate of 50 gpm through a piping system to a storage tank. The pressure in the storage tank is 20 psig, and the water level is 40 ft above that in the well. The piping system contains 300 ft of 1.5 in. sch 40 pipe, 10 std 90° elbows, six gate valves, and an orifice meter with a diameter of 1 in.

   (a) What are the specifications required for the pump?

   (b) Would any of the pumps represented by the characteristic curves in Appendix H be satisfactory for this application? If more than one of them would work, which would be the best? What would be the pump head, impeller diameter, efficiency, NPSH, and required horsepower for this pump at the operating point?

   (c) If the pump you select is driven by an 1800 rpm motor, what impeller diameter should be used?

   (d) What should be the minimum torque and horsepower rating of the motor if the pump is 50% efficient?

   (e) If the NPSH rating of the pump is 6 ft at the operating conditions, where should it be located in order to prevent cavitation?

   (f) What is the reading of the orifice meter, in psi?

19. Water at 20°C is pumped at a rate of 300 gpm from an open well in which the water level is 100 ft below ground level into a storage tank that is 80 ft above ground. The piping system contains 700 ft of 3½ in. sch 40 pipe, eight threaded elbows, two globe valves, and two gate valves. The vapor pressure of the water is 17.5 mmHg.

   (a) What pump head and hydraulic horsepower are required?

   (b) Would a pump whose characteristics are similar to those shown in Figure 8.2 be suitable for this job? If so, what impeller diameter, motor speed, and motor horsepower should be used?

   (c) What is the maximum distance above the surface of the water in the well at which the pump can be located and still operate properly?

20. An organic fluid is to be pumped at a rate of 300 gpm, from a distillation column reboiler to a storage tank. The liquid in the reboiler is 3 ft above ground level, the storage tank is 20 ft above ground, and the pump will be at ground level. The piping system contains 14 std elbows, 4 gate valves, and 500 ft of 3 in. sch 40 pipe. The liquid has an SG of 0.85, a viscosity of 8 cP, and a vapor pressure of 600 mmHg. If the pump to be used has characteristics similar to those given in Appendix H, and the pressure in the reboiler is 5 psig, determine

   (a) The motor speed to be used

   (b) The impeller diameter

   (c) The motor horsepower and required torque

   (d) Where the pump must be located to prevent cavitation

21. A liquid with a viscosity of 5 cP, density of 45 lb_m/ft^3, and vapor pressure of 20 psia is transported from a storage tank in which the pressure is 30 psia to an open tank 500 ft downstream, at a rate of 100 gpm. The liquid level in the storage tank is 30 ft above the pump, and the pipeline is 2 in. sch 40 commercial steel. If the transfer pump has a required NPSH of 15 ft, how far downstream from the storage tank can the pump be located without danger of cavitation?

22. You must determine the specifications for a pump to transport water at 60°C from one tank to another at a rate of 200 gpm. The pressure in the upstream tank is 1 atm, and the water level in this tank is 2 ft above the level of the pump. The pressure in the downstream tank is 10 psig, and the water level in this tank is 32 ft above the pump. The pipeline contains 250 ft of 2 in. sch 40 pipe, with 10 standard 90° flanged elbows and six gate valves.

(a) Determine the pump head required for this job.

(b) Assuming your pump has the same characteristics as the one shown in Figure 8.2, what impeller size should be used and what power would be required to drive the pump with this impeller at the specified flow rate?

(c) If the water temperature is raised, the vapor pressure will increase accordingly. Determine the maximum water temperature that can be tolerated before the pump will start to cavitate, assuming that it is installed as close to the upstream tank as possible.

23. A piping system for transporting a liquid ($\mu = 50$ cP, $\rho = 0.85$ g/cm^3 from vessel A to vessel B) consists of 650 ft of 3 in. sch 40 commercial steel pipe containing 4 globe valves and 10 elbows. The pressure is atmospheric in A and 5 psig B, and the liquid level in B is 10 ft higher than that in A. You want to transfer the liquid at a rate of 250 gpm at 80°F using a pump with the characteristics shown in Figure 8.2. Determine

(a) The diameter of the impeller that you would use with this pump

(b) The head developed by the pump and the power (in horsepower) required to pump the liquid

(c) The power of the motor required to drive the pump

(d) The torque that the motor must develop

(e) The NPSH of the pump at the operating conditions

24. You must choose a centrifugal pump to pump a coal slurry. You have determined that the pump must deliver 200 gpm at a pressure of at least 35 psi. Given the pump characteristic curves in Appendix H, tell which pump you would specify (give pump size, speed, and impeller diameter) and why. What is the efficiency of this pump at its operating point, what horsepower motor would be required to drive the pump, and what is the required NPSH of the pump? The specific gravity of the slurry is 1.35.

25. You must specify a pump to take an organic stream from a distillation reboiler to a storage tank. The liquid has a viscosity of 5 cP, an SG of 0.78, and a vapor pressure of 150 mmHg. The pressure in the storage tank is 35 psig, and the inlet to the tank is located 75 ft above the reboiler, which is at a pressure of 25 psig. The pipeline in which the pump is to be located is 2½ in. sch 40, 175 ft long, and there will be two flanged elbows and a globe valve in each of the pump suction and discharge lines. The pump must deliver a flow rate of 200 gpm. If the pump you use has the same characteristics as that illustrated in Figure 8.2, determine:

(a) The proper impeller diameter to use with this pump

(b) The required head that the pump must deliver

(c) The actual head that the pump will develop

(d) The horsepower rating of the motor required to drive the pump

(e) The maximum distance above the reboiler that the pump can be located without cavitating

26. You have to select a pump to transfer benzene from the reboiler of a distillation column to a storage tank at a rate of 250 gpm. The reboiler pressure is 15 psig and the temperature is 60°C. The tank is 5 ft higher than the reboiler and is at a pressure of 25 psig. The total length of piping is 140 ft of 2 in. sch 40 pipe. The discharge line from the pump contains 3 gate valves and 10 elbows, and the suction line has 2 gate valves and 6 elbows. The vapor pressure of benzene at 60°C is 400 mmHg.

(a) Using the pump curves shown in Figure 8.2, determine the impeller diameter to use in the pump, the head that the pump would develop, the power of the motor required to drive the pump, and the NPSH required for the pump.

(b) If the pump is on the same level as the reboiler, how far from the reboiler could it be located without cavitating?

27. A circulating pump takes hot water at 85°C from a storage tank, circulates it through a piping system at a rate of 150 gpm, and discharges it to the atmosphere. The tank is at atmospheric pressure, and the water level in the tank is 20 ft above the pump. The piping consists of 500 ft of 2 in. sch 40 pipe, with one globe valve upstream of the pump and three globe valves and eight threaded elbows downstream of the pump. If the pump has the characteristics shown in Figure 8.2, determine
    (a) The head that the pump must deliver, the best impeller diameter to use with the pump, the pump efficiency and NPSH at the operating point, and the motor horsepower required to drive the pump
    (b) How far the pump can be located from the tank without cavitating
    Properties of water at 85°C: viscosity = 0.334 cP, density = 0.970 g/cm^3, vapor pressure = 433.6 mmHg.

28. A slurry pump must be selected to transport a coal slurry at a rate of 250 gpm from an open storage tank to a rotary drum filter operating at 1 atm. The slurry is 40% solids by volume and has an SG of 1.2. The level in the filter is 10 ft above that in the tank, and the line contains 400 ft of 3 in. sch 40 pipe, two gate valves, and six 90° elbows. A lab test shows that the slurry can be described as a Bingham plastic with μ_∞ = 50 cP and τ_o = 80 dyn/cm^2.
    (a) What pump head is required?
    (b) Using the pump curves in Appendix H, choose the pump that would be the best for this application. Specify the pump size, motor speed, impeller diameter, efficiency, and NPSH. State what criteria you used to make your decision.
    (c) What horsepower motor would you need to drive the pump?
    (d) Assuming the pump you choose has an NPSH of 6 ft at the operating conditions, what is the maximum elevation above the tank that the pump could be located, if the maximum temperature is 80°C? Vapor pressure (P_v) of water is 0.4736 bar at this temperature.

29. A red mud slurry residue from a bauxite processing plant is to be pumped from the plant to a disposal pond at a rate of 1000 gpm, through a 6 in. ID pipeline that is 2500 ft long. The pipeline is horizontal, and the inlet and discharge of the line are both at atmospheric pressure. The mud has properties of a Bingham plastic, with a yield stress of 250 dyn/cm^2, a limiting viscosity of 50 cP, and a density of 1.4 g/cm^3. The vapor pressure of the slurry at the operating temperature is 50 mmHg. You have available several pumps with the characteristics given in Appendix H.
    (a) Which pump, impeller diameter, motor speed, and motor horsepower would you use for this application?
    (b) How close to the disposal pond could the pump be located without cavitating.
    (c) It is likely that none of these pumps would be adequate to pump this slurry. Explain why, and explain what type of pump might be better.

30. A pipeline is installed to transport a red mud slurry from an open tank in an alumina plant to a disposal pond. The line is 5 in. sch 80 commercial steel, 12,000 ft long, and is designed to transport the slurry at a rate of 300 gpm. The slurry properties can be described by the Bingham plastic model, with a yield stress of 15 dyn/cm^2, a limiting viscosity of 20 cP, and an SG of 1.3. You may neglect any fittings in this pipeline.
    (a) What delivered pump head and hydraulic horsepower would be required to pump this slurry?
    (b) What would be the required pump head and horsepower to pump water at the same rate through the same pipeline?
    (c) If 100 ppm of fresh Separan AP-30 polyacrylamide polymer were added to the water in case (b) above, what would be the required pump head and horsepower?
    (d) If a pump with the same characteristics as those illustrated in Figure 8.2 could be used to pump these fluids, what would be the proper impeller size and motor horsepower to use for each of cases (a), (b), and (c) above? Explain your choices.

COMPRESSORS

31. Calculate the work per pound of gas required to compress air from 70°F and 1 atm to 2000 psi with an 80% efficient compressor under the following conditions:
    (a) Single-stage isothermal compression.
    (b) Single-stage adiabatic compression.
    (c) Five-stage adiabatic compression with intercooling to 70°F and optimum interstage pressures.
    (d) Three-stage adiabatic compression with interstage cooling to 120°F and optimum interstage pressures.
    (e) Calculate the outlet temperature of the air for cases (b), (c), and (d) above. For air: $c_p = 0.24$ Btu/(lb$_m$ °F), $k = 1.4$.

32. It is desired to compress ethylene gas [MW = 28, $k = 1.3$, $c_p = 0.357$ Btu/(lb$_m$ °F)] from 1 atm and 80°F to 10,000 psia. Assuming ideal gas behavior, calculate the compression work required per pound of ethylene under the following conditions:
    (a) A single-stage isothermal compressor
    (b) A four-stage adiabatic compressor with interstage cooling to 80°F and optimum interstage pressures
    (c) A four-stage adiabatic compressor with no intercooling, assuming the same interstage pressures as in (b) and 100% efficiency

33. You have a requirement to compress natural gas ($k = 1.3$, MW = 18) from 1 atm and 70°F to 5000 psig. Calculate the work required to do this per pound of gas in a 100% efficient compressor under the following conditions:
    (a) Isothermal single-stage compressor
    (b) Adiabatic three-stage compressor with interstage cooling to 70°F
    (c) Adiabatic two-stage compressor with interstage cooling to 100°F

34. Air is to be compressed from 1 atm and 70°F to 2000 psia. Calculate the work required to do this per pound of air using the following methods:
    (a) A single-stage 80% efficient isothermal compressor.
    (b) A single-stage 80% efficient adiabatic compressor.
    (c) A five-stage 80% efficient adiabatic compressor with interstage cooling to 70°F.
    (d) A three-stage 80% efficient adiabatic compressor with interstage cooling to 120°F. Determine the expression relating the pressure ratio and inlet temperature for each stage for this case by induction from the corresponding expression for optimum operation of the corresponding two-stage case.
    (e) Calculate the final temperature of the gas for cases (b), (c), and (d).

35. It is desired to compress 1000 scfm of air from 1 atm and 70°F to 10 atm. Calculate the total horsepower required if the compressor efficiency is 80% for
    (a) Isothermal compression.
    (b) Adiabatic single-stage compression.
    (c) Adiabatic three-stage compression with interstage cooling to 70°F and optimum interstage pressures.
    (d) Calculate the gas exit temperature for cases (b) and (c).
    Note: $C_p = 7$ Btu/(lbmol°F); Assume ideal gas.

36. You want to compress air from 1 atm, 70°F, to 2000 psig, using a staged compressor with interstage cooling to 70°F. The maximum compression ratio per stage you can use is about 6, and the compressor efficiency is 70%.
    (a) How many stages should you use?
    (b) Determine the corresponding interstage pressures.
    (c) What power would be required to compress the air at a rate of 10^5 scfm?
    (d) Determine the temperature of the air leaving the last stage.
    (e) How much heat (in Btu/h) must be removed by the interstage coolers?

NOTATION

D	Diameter, [L]
DF	Driving force, Equation 8.8, $[L^2/t^2]$
e_f	Energy dissipated per unit mass of fluid, $[FL/M = L^2/t^2]$
g	Acceleration due to gravity, $[L/t^2]$
H_p	Pump head, [L]
HP	Power, $[FL/t = M\ L^2/t^3]$
h_{max}	Maximum suction lift, [L]
k	Isentropic exponent, (= c_p/c_v for ideal gas), [—]
K_f	Loss coefficient, [—]
M	Molecular weight, [M/mol]
N_s	Specific speed, Equation 8.13, [—]
N_{ss}	Suction specific speed, Equation 8.14, [—]
$NPSH$	Net positive suction head, [L]
$\dot{m}$	Mass flow rate, [M/t]
P	Pressure, $[F/L^2 = M/L\ t^2]$
P_v	Vapor pressure, $[F/L^2 = M/L\ t^2]$
Q	Volumetric flow rate, $[L^3/t]$
R	Radius, [L]
r	Compression ratio, [—]
T	Temperature, [T]
w	Work done by fluid system per unit mass of fluid, $[F\ L/M = L^2/t^2]$

GREEK

Γ	Moment or torque, $[F\ L = M\ L^2/t^2]$
$\Delta()$	$()_2 - ()_1$
η_e	Efficiency, [—]
ρ	Density, $[M/L^3]$
ω	Angular velocity, [1/t]

SUBSCRIPTS

1	Reference point 1
2	Reference point 2
i	Impeller, ideal (frictionless)
m	Motor
s	Suction line

REFERENCES

Kalombo, J.J.N., R. Haldenwang, R.P. Chhabra, and V.G. Fester, Centrifugal pump de-rating for non-Newtonian slurries, *ASME J. Fluids Eng.*, 136, 131302-1, 2014.

Karassik, I.J., W.C. Krutzsch, W.H. Fraser, and J.P. Messina, *Pump Handbook*, McGraw-Hill, New York, 1976.

Karassik, I.J. and J.T. McGuire, *Centrifugal Pumps*, 2nd edn., Springer, New York, 2012.

Raymer, R.E., Watch suction specific speed, *Chem. Eng. Progr.*, 89(3), 79–84, March 1993.

Wilson, K.C., G.R. Addie, A. Sellgren, and R. Clift, *Slurry Transport Using Centrifugal Pumps*, 3rd edn., Springer, New York, 2008.

9 Compressible Flows

"Surely it is not knowledge, but learning; not owning but earning; not being there, but getting there; that gives us the greatest pleasure."

—Carl Friedrich Gauss, 1777–1855, Mathematician and Physicist

I. GAS PROPERTIES

The main difference between the flow behavior of incompressible and compressible fluids, and the equations that govern them, is the effect of variable density, for example, the dependence of density upon pressure and temperature. At low velocities (relative to the speed of sound), relative changes in pressure and associated effects on density are often small and the assumption of incompressible flow with a constant (average) density may be reasonable. It is when the gas velocity approaches the speed at which a pressure wave propagates (i.e., the speed of sound) that the effects of compressibility become the most significant. It is this condition of high-speed gas flow ("fast gas") that is of greatest concern to us here.

A. IDEAL GAS

All gases are "nonideal" in that there are conditions under which the density of the gas may not be accurately represented by the ideal gas law:

$$\rho = \frac{PM}{RT} \tag{9.1}$$

However, there are also conditions under which this law provides a very good representation of the density for virtually any gas. In general, the higher the temperature and the lower the pressure relative to the critical temperature and pressure of the gas, the better the ideal gas law represents gas properties. For example, the critical conditions for CO_2 are 304 K, 72.9 atm, whereas for N_2 they are 126 K, 33.5 atm. Thus, at normal atmospheric conditions (300 K, 1 atm), N_2 can be described very accurately by the ideal gas law, whereas CO_2 deviates significantly from this law under such conditions. This is readily discernible from the P–H diagrams for the substance (see, e.g., Appendix D), because ideal gas behavior can be identified with the conditions under which the enthalpy is independent of pressure, that is, the constant temperature lines on the P–H diagram are vertical (see Section III.B of Chapter 5). For the most common gases (e.g., air) at conditions that are not extreme, the ideal gas law provides quite an acceptable representation for most engineering purposes. For gases/vapors and two-phase flows, a thermodynamic database for the fluid properties is generally required for calculations (e.g., NIST–REFPROP Database 23, v. 9.1, 2015).

We will consider the flow behavior of gases under two possible conditions: isothermal and isentropic (or adiabatic). The *isothermal* (constant temperature) condition may be approximated, for example, in a long pipeline in which the residence time of the gas is long enough that there is plenty of time to reach thermal equilibrium with the surroundings. Under these conditions for an ideal gas, Equation 9.1 implies

$$\frac{P}{\rho} = \text{Constant} = \frac{P_1}{\rho_1} = \frac{P_2}{\rho_2}, \text{etc.} \tag{9.2}$$

The *adiabatic* condition occurs, for example, when the residence time of the fluid is short, as for flow through a short pipe, valve, orifice, etc. and/or for well-insulated boundaries. When friction loss is small, the system can also be described as *locally isentropic*. It can readily be shown that an ideal gas under isentropic conditions obeys the relationship:

$$\frac{P}{\rho^k} = \text{Constant} = \frac{P_1}{\rho_1^k} = \frac{P_2}{\rho_2^k}, \text{etc.} \tag{9.3}$$

where $k = c_p/c_v$ is the "isentropic exponent" and, for an ideal gas, $c_p = c_v + R/M$. For diatomic gases, $k \approx 1.4$, whereas for triatomic and higher gases, $k \approx 1.3$. Equation 9.3 is also often used for nonideal gases, with k as a variable. A table of properties of various gases, including the isentropic exponent, is given in Appendix C, which also includes a plot of k as a function of temperature and pressure for non-ideal steam.

B. THE SPEED OF SOUND

Sound is a small amplitude compression pressure wave, and the speed of sound is the velocity at which this wave will travel through a medium. An expression for the speed of sound can be derived as follows. With reference to Figure 9.1, we consider a sound wave moving from left to right with velocity c. If we take the wave as our reference system, this is equivalent to considering a standing wave with the medium moving from right to left with velocity c. Since the conditions are different upstream and downstream of the wave, we represent these differences by ΔV, ΔT, ΔP, and $\Delta \rho$. The conservation of mass principle applied to the flow through the wave reduces to:

$$\dot{m} = \rho A c = (\rho + \Delta \rho) A (c - \Delta V) \tag{9.4}$$

or

$$\Delta V = c \frac{\Delta \rho}{\rho + \Delta \rho} \tag{9.5}$$

Likewise, a momentum balance on the fluid "passing through" the wave is

$$\sum F = \dot{m}(V_2 - V_1) \tag{9.6}$$

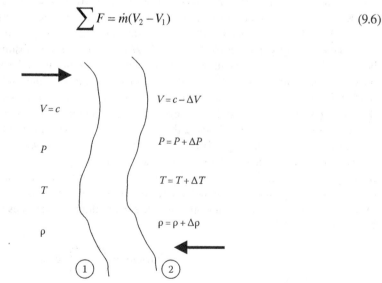

FIGURE 9.1 Sound wave moving at velocity c.

which becomes, in terms of the parameters in Figure 9.1,

$$PA - (P + \Delta P)A = \rho Ac(c - \Delta V - c) \tag{9.7}$$

or

$$\Delta P = \rho c \Delta V \tag{9.8}$$

Eliminating ΔV from Equations 9.5 and 9.8 and solving for c^2 gives

$$c^2 = \frac{\Delta P}{\Delta \rho}\left(1 + \frac{\Delta \rho}{\rho}\right) \tag{9.9}$$

For an infinitesimal wave under isentropic conditions, the term $(1 + \Delta\rho/\rho) \approx 1$, so that Equation 9.9 becomes:

$$c = \left[\left(\frac{\partial P}{\partial \rho}\right)_s\right]^{1/2} = \left[k\left(\frac{\partial P}{\partial \rho}\right)_T\right]^{1/2} \tag{9.10}$$

where the equivalence of the terms in the two radicals follows from Equations 9.2 and 9.3.

For an ideal gas, Equation 9.10 reduces to:

$$c = \left(\frac{kP}{\rho}\right)^{1/2} = \left(\frac{kRT}{M}\right)^{1/2} \tag{9.11}$$

For solids and liquids:

$$\left(\frac{\partial P}{\partial \rho}\right)_s = \frac{K}{\rho} \tag{9.12}$$

where K is the *bulk modulus* (or "compressive stiffness") of the material. It is evident that the speed of sound in a completely incompressible medium would be infinite. From Equation 9.11, we see that the speed of sound in an ideal gas is determined entirely by the nature of the gas (M and k) and the temperature (T).

II. PIPE FLOW

Consider a gas in steady flow in a uniform (constant cross section) pipe. The mass flow rate and mass flux (G) are the same at all locations along the pipe:

$$G = \frac{\dot{m}}{A} = \rho V = \text{Constant} \tag{9.13}$$

Now the pressure drops along the pipe because of energy dissipation (e.g., friction), just as for an incompressible fluid. However, because the density decreases with decreasing pressure and the product of the density and velocity must be constant, the velocity must increase as the gas moves through the pipe. This increase in velocity corresponds to an increase in kinetic energy per unit mass of gas, which also results in a drop in temperature. There is a limit as to how high the velocity can get in a straight pipe, however, which we will discuss shortly.

Because the fluid velocity and properties change from point to point along the pipe, in order to analyze the flow we apply the differential form of the Bernoulli equation to a differential length of pipe (dL):

$$\frac{dP}{\rho} + g\,dz + d\left(\frac{V^2}{2}\right) + \delta e_f + \delta w = 0 \tag{9.14}$$

If there is no shaft work done on the fluid in this system and the elevation (potential energy) change can be neglected, Equation 9.14 can be rewritten using Equation 9.13 as follows:

$$\frac{dP}{\rho} + \frac{G^2}{\rho}d\left(\frac{1}{\rho}\right) = -\delta e_f = -\frac{2fV^2\,dL}{D} = -\frac{2f}{D}\left(\frac{G}{\rho}\right)^2 dL \tag{9.15}$$

where the friction factor f is a function of the Reynolds number:

$$f = \mathrm{fn}\left(N_{Re} = \frac{DG}{\mu} \sim \text{constant}\right) \tag{9.16}$$

Because the gas viscosity is not highly sensitive to pressure, for isothermal flow the Reynolds number and hence the friction factor will be very nearly constant along the pipe. For adiabatic flow, the viscosity may change as the temperature changes, but these changes are usually small. Equation 9.15 is thus valid for any prescribed conditions, and we will apply it to an ideal gas in both isothermal and adiabatic (isentropic) flow.

A.   ISOTHERMAL FLOW

Substituting Equation 9.1 for the density into Equation 9.15, rearranging and integrating from the inlet of the pipe (point 1) to the outlet (point 2), and solving the result for G gives

$$G = \left[\frac{M\left(P_1^2 - P_2^2\right)/(2RT)}{2fL/D + \ln(P_1/P_2)}\right]^{1/2}$$

$$= \sqrt{P_1\rho_1}\left[\frac{1 - P_2^2/P_1^2}{4fL/D - 2\ln(P_2/P_1)}\right]^{1/2} \tag{9.17}$$

If the logarithmic term in the denominator (which comes from the change in kinetic energy of the gas) is neglected, the resulting equation is called the *Weymouth equation*. Furthermore, if the average density of the gas is used in the Weymouth equation, that is,

$$\bar{\rho} = \frac{(P_1 + P_2)M}{2RT} \quad \text{or} \quad \frac{M}{2RT} = \frac{\bar{\rho}}{P_1 + P_2} \tag{9.18}$$

Equation 9.17 reduces identically to the Bernoulli equation for an incompressible fluid in a straight, uniform pipe, which can be written in the form

$$G = \left(\frac{\bar{\rho}(P_1 - P_2)}{2fL/D}\right)^{1/2} = \sqrt{P_1\bar{\rho}}\left(\frac{1 - P_2/P_1}{2fL/D}\right)^{1/2} \tag{9.19}$$

Inspection of Equation 9.17 shows that as P_2 decreases, both the numerator and denominator increase, with opposing effects. By setting the derivative of Equation 9.17 with respect to P_2 (i.e., dG/dP_2) equal to zero, the value of P_2 that maximizes G and the corresponding expression for the maximum G can be found. If the conditions at this state (i.e., the maximum mass flux) are denoted by an asterisk (e.g., P_2^*, G^*), the result is

$$G^* = P_2^* \sqrt{\frac{M}{RT}} = P_2^* \sqrt{\frac{\rho_1}{P_1}} \tag{9.20}$$

or

$$V_2^* = \sqrt{\frac{RT}{M}} = \sqrt{\frac{P_1}{\rho_1}} = c \tag{9.21}$$

That is, as P_2 decreases, the mass velocity in the conduit will increase up to a maximum value of G^*, at which point the velocity at the end of the pipe reaches the speed of sound. Any further reduction in the downstream pressure (P_2) can have no effect on the flow in the pipe, because the speed at which pressure change information can be transmitted is the speed of sound. That is, since pressure changes are transmitted at the speed of sound, they cannot propagate upstream in a gas that is already traveling at the speed of sound. Therefore, the pressure *inside* the downstream end of the pipe will remain at P_2^* regardless of how low the pressure outside the end of the pipe (P_2) may fall. This condition is called *choked flow* and is a very important concept because it establishes the conditions under which maximum gas flow can occur in a conduit. When the flow becomes choked, the mass flow rate in the pipe will be insensitive to the value of the exit pressure (but will still be dependent upon the upstream conditions).

Although Equation 9.17 appears to be explicit for G, it is actually implicit because the friction factor depends upon the Reynolds number, which depends on G. However, the Reynolds number under choked flow conditions is often high enough that fully turbulent flow prevails, in which case the friction factor depends only on the relative pipe roughness ε/D, i.e., (Equation 6.40):

$$\frac{1}{\sqrt{f}} = -4\log\left(\frac{\varepsilon/D}{3.7}\right) \tag{9.22}$$

If the upstream pressure and flow rate are known, the downstream pressure (P_2) can be found by rearranging Equation 9.17, as follows:

$$\frac{P_2}{P_1} = \left\{1 - \frac{G^2}{P_1\rho_1}\left[\frac{4fL}{D} - 2\ln\left(\frac{P_2}{P_1}\right)\right]\right\}^{1/2} \tag{9.23}$$

which is implicit in P_2. A first estimate for P_2 can be obtained by neglecting the logarithmic term on the right (the result corresponding to the Weymouth approximation). This first estimate can then be inserted into the last term in Equation 9.23 to provide a second estimate for P_2, and the process can be repeated as necessary. Alternately, a spreadsheet can be used with the "solve" or "goal seek" function to solve the equation.

Equation 9.23 can be also rewritten in another form in terms of the Mach number, $N_{Ma} = V/c$. From Equations 9.13 and 9.21, $G = \rho V = \rho c N_{Ma} = \sqrt{P\rho}N_{Ma} = $ const and, in combination with Equation 9.2, we have $PN_{Ma} = $ const for isothermal flow. Thus, with Equation 9.23, we have

$$\frac{P_2}{P_1} = \frac{N_{Ma1}}{N_{Ma2}} = \left\{1 - N_{Ma1}^2\left[\frac{4fL}{D} - 2\ln\left(\frac{N_{Ma1}}{N_{Ma2}}\right)\right]\right\}^{1/2} \tag{9.24}$$

or

$$\frac{4fL}{D} = \frac{1}{N_{Ma1}^2} - \frac{1}{N_{Ma2}^2} + 2\ln\left(\frac{N_{Ma1}}{N_{Ma2}}\right) \tag{9.25}$$

B. ADIABATIC FLOW

In the case of adiabatic flow, we can use Equations 9.1 and 9.3 to eliminate density and temperature from Equation 9.15. This can be called the *locally isentropic* approach, because the friction loss is still included in the energy balance. Actual flow conditions are often somewhere between isothermal and adiabatic, in which case the flow behavior can be described by the isentropic equations, with the isentropic constant k replaced by a "polytropic" constant (or "isentropic exponent") γ, where $1 < \gamma < k$, as is done for compressors (the isothermal condition corresponds to $\gamma = 1$, whereas truly isentropic flow corresponds to $\gamma = k$). This same approach can be used for some nonideal gases by using a variable isentropic exponent for k (e.g., for steam, see Figure C.1).

Combining Equations 9.1 and 9.3 leads to the following expressions for density and temperature as a function of pressure:

$$\rho = \rho_1\left(\frac{P}{P_1}\right)^{1/k}, \quad T = T_1\left(\frac{P}{P_1}\right)^{(k-1)/k} \tag{9.26}$$

Using these expressions to eliminate ρ and T from Equation 9.15 and solving for G gives

$$G = \sqrt{P_1\rho_1}\left[\frac{2\left(\dfrac{k}{k+1}\right)\left(1-\left(\dfrac{P_2}{P_1}\right)^{(k+1)/k}\right)}{\dfrac{4fL}{D} - \dfrac{2}{k}\ln\left(\dfrac{P_2}{P_1}\right)}\right]^{1/2} \tag{9.27}$$

If the system contains fittings as well as straight pipe, the term $4fL/D$ ($= K_{f\,pipe}$) can be replaced by ΣK_f, that is, the sum of all loss coefficients in the system.

Just as for isothermal flow, Equation 9.27 can be rewritten in terms of Mach number, as follows. From Equations 9.13 and 9.11, $G = \rho V = \rho c N_{Ma} = \sqrt{kP\rho}\,N_{Ma} = \text{const}$. In combination with Equation 9.3, this gives $P^{(k+1)/(2k)}N_{Ma} = \text{const}$, and

$$\frac{P_2}{P_1} = \left(\frac{N_{Ma1}}{N_{Ma2}}\right)^{2k/(k+1)} \tag{9.28}$$

so that (9.27) can be rewritten in the form:

$$N_{Ma1} = \left[\frac{\left(\dfrac{2}{k+1}\right)\left(1-\left(\dfrac{N_{Ma1}}{N_{Ma2}}\right)^2\right)}{\dfrac{4fL}{D} - \dfrac{4}{k+1}\ln\left(\dfrac{N_{Ma1}}{N_{Ma2}}\right)}\right]^{1/2} \tag{9.29}$$

or

$$\frac{4fL}{D} = \frac{2}{k+1}\left[\frac{1}{N_{Ma1}^2} - \frac{1}{N_{Ma2}^2} + 2\ln\left(\frac{N_{Ma1}}{N_{Ma2}}\right)\right] \tag{9.30}$$

C. Choked Flow

For isentropic flow (just as for isothermal flow), as the downstream pressure drops, the mass velocity increases until it reaches a maximum. When the downstream pressure reaches the point where the velocity becomes sonic at the end of the pipe, the flow is choked. This can be shown by differentiating Equation 9.27 with respect to P_2 (as before) or alternatively as follows, noting that $G = \dot{m}/A = \rho V$:

$$\frac{\partial G}{\partial P} = \frac{\partial(\rho V)}{\partial P} = \rho \frac{\partial V}{\partial P} + V \frac{\partial \rho}{\partial P} = 0 \quad \text{(for max } G\text{)} \tag{9.31}$$

For isentropic conditions, the differential form of the Bernoulli equation is

$$\frac{dP}{\rho} + V\,dV = 0 \quad \text{or} \quad \frac{\partial V}{\partial P} = -\frac{1}{\rho V} \tag{9.32}$$

Substituting this into Equation 9.31 gives

$$-\frac{1}{V} + V \frac{\partial \rho}{\partial P} = 0 \tag{9.33}$$

However, it is noted that

$$c^2 = \left(\frac{\partial P}{\partial \rho}\right)_S \tag{9.34}$$

so that Equation 9.33 can be written as:

$$-\frac{1}{V} + \frac{V}{c^2} = 0 \quad \text{or} \quad V = c \tag{9.35}$$

This shows that when the mass velocity reaches a maximum, the velocity is sonic (i.e., the flow is choked).

1. Isothermal

Under isothermal conditions, choked flow occurs when:

$$V_2 = c = V_2^* = \sqrt{\frac{RT}{M}} = \sqrt{\frac{P_1}{\rho_1}} \tag{9.36}$$

where the asterisk denotes the sonic state. Thus:

$$G^* = \rho_2 V_2^* = \frac{P_2^* M}{RT} \sqrt{\frac{RT}{M}} = \sqrt{P_1 \rho_1}\, \frac{P_2^*}{P_1} \tag{9.37}$$

If G^* is eliminated from Equations 9.17 and 9.37, the result can be solved for ΣK_f to give:

$$\boxed{\sum K_f = \left(\frac{P_1}{P_2^*}\right)^2 - 2\ln\left(\frac{P_1}{P_2^*}\right) - 1} \tag{9.38}$$

where $4fL/D$ in Equation 9.17 has been replaced by ΣK_f. Equation 9.38 shows that the pressure at the inside of the end of the pipe at which the flow becomes sonic (P_2^*) is a unique function of the upstream pressure (P_1) and the sum of the loss coefficients in the system (ΣK_f). Since Equation 9.38 is implicit in P_2^*, it must be solved for P_2^* by iteration for given values of ΣK_f and P_1, or on a spreadsheet using the "solve" or "goal seek" function. Equation 9.38 thus enables the determination of the choke pressure, P_2^*, as a function of ΣK_f and P_1.

2. Adiabatic

For adiabatic (or locally isentropic) conditions, the corresponding expressions are

$$V_2 = c = V_2^* = \left(\frac{kRT_2}{M} \right)^{1/2}, \qquad \frac{T_2}{T_1} = \left(\frac{P_2}{P_1} \right)^{(k-1)/k} \tag{9.39}$$

and

$$G^* = \frac{P_2^* M}{RT_2^*} \left(\frac{kRT_2^*}{M} \right)^{1/2} = \sqrt{P_1\rho_1}\left[k\left(\frac{P_2^*}{P_1} \right)^{(k+1)/k} \right]^{1/2} \tag{9.40}$$

Eliminating G^* from Equations 9.27 and 9.40 and solving for ΣK_f gives

$$\boxed{\sum K_f = \frac{2}{k+1}\left[\left(\frac{P_1}{P_2^*} \right)^{(k+1)/k} - 1 \right] - \frac{2}{k}\ln\left(\frac{P_1}{P_2^*} \right)} \tag{9.41}$$

Just as for isothermal flow, this is an implicit expression for the "choke pressure" (P_2^*) as a function of the upstream pressure (P_1), the loss coefficients (ΣK_f), and the isentropic exponent (k), which is most easily solved by iteration using a spreadsheet. It is very important to realize that once the pressure at the end of the pipe falls to P_2^* and choked flow occurs, all of the conditions within the pipe ($G = G^*$, $P_2 = P_2^*$, etc.) will remain the same regardless of how low the pressure outside the end of the pipe falls. The pressure drop *within the pipe* (which determines the flow rate) is always $P_1 - P_2^*$ when the flow is choked.

D. THE EXPANSION FACTOR

The adiabatic flow equation (Equation 9.27) can be represented in a more convenient form as

$$G = Y\left(\frac{2\rho_1 \Delta P}{\sum K_f} \right)^{1/2} = Y\sqrt{P_1\rho_1}\left[\frac{2(1 - P_2/P_1)}{\sum K_f} \right]^{1/2} \tag{9.42}$$

where
$\rho_1 = P_1 M/RT_1$
$\Delta P = P_1 - P_2$
Y is the *expansion factor*

Note that Equation 9.42 without the Y term is the Bernoulli equation for an incompressible fluid of density ρ_1. Thus, the expansion factor is simply the ratio of the adiabatic mass flux (Equation 9.27) to the corresponding incompressible mass flux ($Y = G_{adiabatic}/G_{incompressible}$) and is a unique function of P_2/P_1, k, and K_f. For convenience, values of Y are shown in Figure 9.2a for $k = 1.3$ and Figure 9.2b

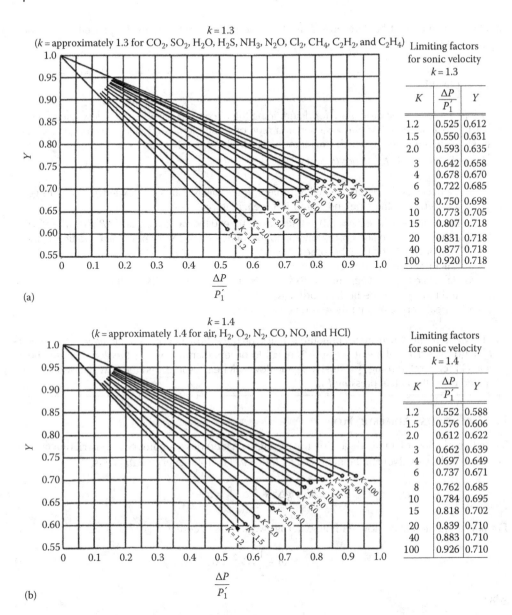

FIGURE 9.2 Expansion factor for adiabatic flow in piping systems. (a) $k = 1.3$ and (b) $k = 1.4$. (From Crane, C., Flow of fluids through valves, fittings and pipe, Technical Manual 410, Crane Co., New York, 1991 [and subsequent issues].)

for $k = 1.4$, as a function of $\Delta P/P_1$ and ΣK_f (which is denoted simply by K on these plots) (Crane, 1991). The conditions corresponding to the lower ends of the lines on the plots (i.e., the "button") represent the sonic (choked flow) state where $P_2 = P_2^*$. These same conditions are given in the table accompanying the plots, which enable the relationships for choked flow to be determined more precisely than is possible from reading the plots. Note that it is not possible to extrapolate beyond the "button" at the end of the lines in Figure 9.2a and b because this represents the choked flow state, in which $P_2 = P_2^*$ (inside the pipe) and is independent of the external exit pressure.

Figure 9.2 provides a convenient way of solving compressible adiabatic flow problems for piping systems. Some iteration is normally required, because the value of K_f depends upon the Reynolds number, which cannot be determined until G is found. An example of the procedure for solving a typical problem is as follows.

Example 9.1

Problem: Outline the procedure for determining the maximum possible mass flow rate of a given gas (molecular weight M and k value), in a pipe of given diameter, length and roughness, and a given upstream pressure which exits into the atmosphere.

Given: $P_1, D, L, \varepsilon, k, M$ *Find*: P_2^* and G^*

1. Estimate ΣK_f by assuming fully turbulent flow. This requires a knowledge of ε/D to get $K_f = 4fL/D$ for the pipe and K_i and K_d for each fitting.
2. From Figure 9.2a (for $k = 1.3$) or b (for $k = 1.4$), read the values of Y and $\Delta P^*/P_1 = (P_1 - P_2^*)/P_1$ at the end of the line corresponding to the value of $K = \Sigma K_f$ (or from the table beside the plot) at which the flow becomes choked.
3. Calculate $G = G^*$ from Equation 9.42.
4. Calculate $N_{Re} = DG/\mu$, and use this to revise the value of $K = \Sigma K_f$ for the pipe ($K_f = 4fL/D$) and fittings (3-K method) accordingly.
5. Repeat steps 2–4 until there is no change in G.

The value of the downstream pressure (P_2) at which the flow becomes sonic ($P_2 = P_2^*$) is given by $P_2^* = P_1(1 - \Delta P^*/P_1)$. If the exit pressure is equal to or less than this value, the flow will be choked and G is calculated using P_2^*. Otherwise, the flow will be subsonic, and the flow rate will be determined using the actual pressure P_2.

E. FRICTIONLESS ADIABATIC FLOW

The adiabatic flow of an ideal gas flowing through a frictionless conduit or a constriction (such as an orifice, nozzle, or valve) can be analyzed as follows. The total energy balance is

$$\Delta h + g\Delta z + \frac{1}{2}\Delta V^2 = q + w \tag{9.43}$$

For horizontal adiabatic flow with no external work, this becomes

$$\Delta h + \frac{1}{2}\Delta V^2 = 0 \tag{9.44}$$

where

$$\Delta h = \Delta(c_p T) = \frac{k}{k-1}\Delta\left(\frac{P}{\rho}\right) \tag{9.45}$$

This follows from the ideal gas relation $c_p - c_v = R/M$, and the definition of k (i.e., $k = c_p/c_v$). Equation 9.44 thus becomes

$$\frac{k}{k-1}\left(\frac{P_2}{\rho_2} - \frac{P_1}{\rho_1}\right) + \frac{V_2^2 - V_1^2}{2} = 0 \tag{9.46}$$

Using the isentropic condition (P/ρ^k = constant) to eliminate ρ_2, this can be written as

$$V_2^2 - V_1^2 = \frac{2k}{k-1}\left(\frac{P_1}{\rho_1}\right)\left[1 - \left(\frac{P_2}{P_1}\right)^{(k-1)/k}\right] \tag{9.47}$$

If V_1 is eliminated using the continuity equation, that is, $(\rho VA)_1 = (\rho VA)_2$, this becomes

$$V_2 = \left\{ \frac{2k}{k-1} \left(\frac{P_1}{\rho_1} \right) \frac{\left[1 - (P_2/P_1)^{(k-1)/k} \right]}{\left[1 - (A_2/A_1)^2 (P_2/P_1)^{2/k} \right]} \right\}^{1/2}$$

(9.48)

Now

$$G = V_2 \rho_2 = V_2 \rho_1 \left(\frac{P_2}{P_1} \right)^{1/k}$$

(9.49)

and assuming that the flow is from a larger conduit through a small constriction, such that $A_1 \gg A_2$ (i.e., $V_1 \ll V_2$), the term in the square brackets in the denominator of Equation 9.48 becomes equal to 1. Substituting the result into Equation 9.49 gives

$$G = \sqrt{P_1 \rho_1} \left\{ \frac{2k}{k-1} \left(\frac{P_2}{P_1} \right)^{2/k} \left[1 - \left(\frac{P_2}{P_1} \right)^{(k-1)/k} \right] \right\}^{1/2}$$

(9.50)

Equation 9.50 represents flow through an "*ideal nozzle*," that is, an isentropic constriction. Setting the derivative of Equation 9.50 to zero (i.e., $\partial G/\partial r = 0$, where $r = P_2/P_1$), it can be shown that the mass flow is a maximum (choked) when

$$\frac{P_2^*}{P_1} = \left(\frac{2}{k+1} \right)^{k/(k-1)}$$

(9.51)

For $k = 1.4$ (e.g., air), this has a value of 0.528. That is, if the downstream pressure is approximately one half or less of the upstream pressure, the flow will be choked. In such a case, the mass velocity can be determined by using Equation 9.40 with P_2^* from Equation 9.51:

$$G^* = P_1 \left(\frac{2}{k+1} \right)^{(k+1)/2(k-1)} \sqrt{\frac{kM}{RT_1}} = \sqrt{kP_1\rho_1} \left(\frac{2}{k+1} \right)^{(k+1)/2(k-1)}$$

(9.52)

For $k = 1.4$, this reduces to

$$G^* = 0.684 P_1 \sqrt{\frac{M}{RT_1}} = 0.684 \sqrt{P_1 \rho_1}$$

(9.53)

The mass flow rate under adiabatic conditions is always somewhat greater than that under isothermal conditions, but the difference is normally <20%. In fact, for long piping systems ($L/D > 1000$), the difference is usually less than 5% (see Figure 9.3).

It should be noted that the adiabatic equations reduce to the isothermal equations by setting $k = 1$. Also, the adiabatic equations can be applied to a nonideal gas by taking the parameter k to be a "polytropic exponent," evaluated as an average value over the range of pressures considered.

The flow of compressible (as well as incompressible) fluids through nozzles and orifices will be considered in the following chapter on flow-measuring devices.

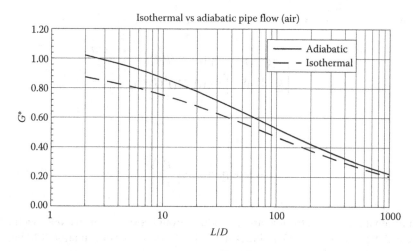

FIGURE 9.3 Dimensionless mass flux versus L/D for adiabatic and isothermal flow in a pipe. (From Holland and Bragg, 1995.)

F. GENERAL FLUID PROPERTIES

The integrated form of the basic differential energy balance (Bernoulli) equation, Equation 9.14 (neglecting the potential energy and work terms), can be written between points 1 and 2 along the pipe and rearranged to solve for G^2:

$$G^2 = \frac{-\int_{P_1}^{P_2} \rho\, dP}{\ln(\rho_1/\rho_2) + \left(\sum_1^2 K_{fi} + 1\right)\Big/2} \tag{9.54}$$

This expression is general and valid for any fluid (liquid, gas, ideal, nonideal, two-phase, etc.). It is only necessary to evaluate the integral of ρ versus P along the pipe to find the mass flux (and hence the mass flow rate). This can easily be done numerically on a spreadsheet, provided a fluid property (thermodynamic) database (such as NIST, 2015) is available to determine point values of ρ versus P over the proper interval, using the numerical approximation for the integral:

$$-\int_{P_1}^{P_2} \rho\, dP \approx \frac{-1}{2} \sum_{P_1}^{P_2} (\rho_i + \rho_{i+1})(P_i - P_{i+1}) \tag{9.55}$$

A suitable pressure interval $[(i + 1) - i]$ is chosen (usually 1 psi), and Equations 9.54 and 9.55 are used to evaluate G versus P_i. The path to be followed over this pressure range should be adiabatic, which is approximated by an isentropic path, with the irreversible loss accounted for by the denominator in Equation 9.54. If the flow is choked, the value of G will exhibit a maximum before P_i reaches P_2, at which point the pressure is at the choke point inside the end of the pipe. If the pressure P_2 falls below that at the pipe exit before the maximum in G is reached, then the flow is not choked. This is actually an iterative process, as the loss coefficients (ΣK_f) depend on G through the Reynolds number. However, an initial estimate for these ΣK_f values can be taken as the fully turbulent values, which in many cases will suffice for all values of G of practical interest.

III. GENERALIZED GAS FLOW EXPRESSIONS: IDEAL GAS

For adiabatic flow of an ideal gas in a constant area duct, the governing equations may be formulated in a more generalized dimensionless form that is useful for the solution of both subsonic and supersonic flows (the condition known as *Fanno flow*). We will present the resulting expressions and illustrate how to apply them here, but we will not show the derivation of all of them. For this, the reader is referred to publications such as that of Shapiro (1953) and Hall (1951) for the ideal gas and to Saad (1992), Cheng et al. (2012), Baltajiev (2012), Oosthuizen and Carscallen (2013), and Korelshteyn (2015) for the case of a general fluid.

A. GOVERNING EQUATIONS

For steady flow of a gas at a constant mass flow rate in a uniform pipe, the pressure, temperature, velocity, density, etc. all vary from point to point along the pipe. The governing equations are the conservation of mass (continuity), conservation of energy, and conservation of momentum, all applied to a differential length of the pipe, as follows:

1. Continuity

$$\frac{\dot{m}}{A} = G = \rho V = \text{Constant} \tag{9.56}$$

or

$$\frac{d\rho}{\rho} + \frac{dV}{V} = 0 \tag{9.57}$$

2. Energy

$$h + \frac{1}{2}V^2 = \text{Constant} = h_o \tag{9.58}$$

or

$$dh + V\,dV = 0 \tag{9.59}$$

Since the fluid properties are defined by the entropy and enthalpy, Equation 9.59 represents a curve on an *h–s* diagram, which is called a *Fanno line*.

3. Momentum

$$\frac{dP}{\rho} + V\,dV = -\frac{4\tau_w}{\rho D_h}\,dL = -\frac{2fV^2}{D_h}\,dL \tag{9.60}$$

By making use of the isentropic condition for an ideal gas (i.e., $P/\rho^k = \text{const.}$), the following relations can be obtained:

$$h = c_p T = \frac{kRT}{(k-1)M} = \frac{P}{\rho}\left(\frac{k}{k-1}\right) \tag{9.61}$$

$$\frac{P}{\rho} = \frac{RT}{M} = \frac{c^2}{k} \tag{9.62}$$

$$N_{Ma} = \frac{V}{c} = \frac{G}{\sqrt{kP\rho}} \tag{9.63}$$

where N_{Ma} is the Mach number.

The behavior of a nonideal gas (or any general homogeneous compressible Newtonian fluid) can be described by several dimensionless (positive) coefficients, or "isentropic exponents", defined as follows:

$$k_{P\rho} = \left(\frac{\partial \ln P}{\partial \ln \rho}\right)_s \tag{9.64}$$

$$k_{TP} = \left(\frac{\partial \ln T}{\partial \ln P}\right)_s \tag{9.65}$$

$$k_{T\rho} = \left(\frac{\partial \ln T}{\partial \ln \rho}\right)_s \tag{9.66}$$

and the dimensionless thermal expansion coefficient

$$\hat{\beta} = -\left(\frac{\partial \ln \rho}{\partial \ln T}\right)_P . \tag{9.67}$$

Each of the above "k's" represents the tangent to the "log–log" plot of the respective property variables, for example, the value of $k_{P\rho}$ is the tangent to the log P versus log ρ plot at constant entropy. This also means that on an isentropic path $(P/\rho^{k_{P\rho}})_s \approx$ constant, $(T/P^{k_{TP}})_s \approx$ constant, $(T/\rho^{k_{T\rho}})_s \approx$ constant, and on isobaric line $(\rho/T^{\hat{\beta}})_P \approx$ constant. These expressions are not exact because the values of $k_{P\rho}$, k_{TP}, $k_{T\rho}$, $\hat{\beta}$ may vary. These coefficients (three of which are independent) describe local thermodynamic behavior of an arbitrary compressible fluid.

For an ideal gas,

$$k_{P\rho} = k, \quad k_{TP} = (k-1)/k, \quad k_{T\rho} = k-1, \quad \hat{\beta} = 1, \tag{9.68}$$

and the classical equations (9.3), (9.39) $(P/\rho^k)_s =$ constant, $(T/P^{(k-1)/k})_s =$ constant, $(T/\rho^{k-1})_s =$ constant, $(\rho T)_P =$ constant apply.

In order to determine the "isentropic exponent" values above for a real gas, a multitude of fluid property databases or thermodynamics libraries may be used. A good inexpensive database containing data for more than a hundred fluids, including some mixtures, is, the NIST–REFPROP Database (NIST, 2010). The value of $k_{P\rho}$ can also be determined from the known (tabulated) value of the speed of sound c, using the equations (under adiabatic conditions) $(P/\rho) = (ZRT/M) = (c^2/k_{P\rho})$ where Z is the compressibility factor, which accounts for deviation from ideal gas behavior. It can be also shown that $k_{P\rho} = (c_p/c_v)\hat{\kappa}$, where the dimensionless isothermal bulk modulus $\hat{\kappa} = (\partial \ln P/\partial \ln \rho)_T$ (for an ideal gas, this is equal to 1). Values of other "isentropic" exponents can also be estimated from the following equations:

$$k_{TP} = \frac{ZR}{Mc_p}\hat{\beta}, \quad k_{T\rho} = \frac{c_p - c_v}{\hat{\beta}c_v} \quad \text{and} \quad \mu_{JT} = \frac{T}{P}k_{TP}\left(1 - \hat{\beta}^{-1}\right) \tag{9.69}$$

where μ_{JT} is the Joule–Thomson coefficient.

For practical calculations, it is important to note that while the values of Z and $\hat{\beta}$ for a real gas can deviate strongly from the ideal gas value of 1 for highly reduced pressure values, especially near the critical point and for supercritical fluids, the "isentropic" exponent values change much more slowly over a wide range of thermodynamic parameters, and their average values can be successfully used for engineering flow analysis of real gases (Istomin, 1997, 1998).

Under adiabatic conditions:

$$N_{Ma} = \frac{V}{c} = \frac{G}{\sqrt{k_{P\rho} P\rho}} \tag{9.70}$$

The above equations can be combined to yield the following dimensionless equations:

$$\frac{dP}{P} = -\frac{k_{P\rho} N_{Ma}^2 \left(1 + k_{T\rho} N_{Ma}^2\right)}{2\left(1 - N_{Ma}^2\right)}\left(4f\frac{dL}{D}\right) \tag{9.71}$$

$$\frac{dV}{V} = -\frac{d\rho}{\rho} = \frac{\left(1 + k_{T\rho}\right) N_{Ma}^2}{2\left(1 - N_{Ma}^2\right)}\left(4f\frac{dL}{D}\right) \tag{9.72}$$

$$\frac{dT}{T} = -\frac{k_{T\rho} N_{Ma}^2}{2}\left(\frac{1 + k_{T\rho} N_{Ma}^2}{1 - N_{Ma}^2} - \frac{1}{\hat{\beta}}\right)\left(4f\frac{dL}{D}\right) \tag{9.73}$$

$$\frac{dN_{Ma}^2}{N_{Ma}^2} = N_{Ma}^2 \frac{A + BN_{Ma}^2}{2\left(1 - N_{Ma}^2\right)} k_{P\rho}\left(4f\frac{dL}{D}\right) \tag{9.74}$$

where

$$A = 1 + 1/k_{P\rho} + k_{TP} + k_{P\rho,Ps} + k_{T\rho}(k_{P\rho,Ps} - k_{P\rho,P\rho})$$

$$B = k_{T\rho}(1 + k_{P\rho,P\rho})$$

$$k_{P\rho,Ps} = \left(\frac{\partial \ln k_{P\rho}}{\partial \ln P}\right)_s, \quad k_{P\rho,P\rho} = \left(\frac{\partial \ln k_{P\rho}}{\partial \ln P}\right)_\rho \tag{9.75}$$

Equation 9.74 can be also written in another, sometimes more convenient, form as

$$\frac{d\left(\hat{N}_{Ma}^2\right)}{\hat{N}_{Ma}^2} = \hat{N}_{Ma}^2 \frac{A_h + B_h \hat{N}_{Ma}^2}{2\left(1 - k_{P\rho}^{-1}\hat{N}_{Ma}^2\right)}\left(4f\frac{dL}{D}\right) \tag{9.76}$$

where

$$\hat{N}_{Ma} = \sqrt{k_{P\rho}} N_{Ma} = \frac{G}{\sqrt{P\rho}}, \quad A_h = 1 + 1/k_{P\rho} + k_{TP}, \quad B_h = k_{TP},$$

For an ideal gas,

$$A = A_h = 2, \quad B = k - 1, \quad B_h = \frac{k - 1}{k} \tag{9.77}$$

Equations 9.71 through 9.76 are general equations valid for adiabatic flow of any compressible Newtonian fluid, including a real gas, liquid, supercritical fluid, and multiphase flow according to the homogeneous equilibrium model (see Chapter 16). Along with fluid thermodynamic models or thermodynamic data, they give a closed set of equations that describe adiabatic flow in pipes.

An "impulse function" (F) is also useful in some problems where the force exerted on solid bounding surfaces is desired:

$$F = PA + \rho AV^2 = PA\left(1 + k_{P\rho}N_{Ma}^2\right) = PA\left(1 + \hat{N}_{Ma}^2\right) \tag{9.78}$$

$$\frac{dF}{F} = \frac{dP}{P} + \frac{\hat{N}_{Ma}^2}{1 + \hat{N}_{Ma}^2}\left[\frac{d\left(\hat{N}_{Ma}^2\right)}{\hat{N}_{Ma}^2}\right] \tag{9.79}$$

From Equations 9.71 and 9.72,

$$\frac{d\ln P}{d\ln\rho} = \frac{k_{P\rho}\left(1 + k_{T\rho}N_{Ma}^2\right)}{1 + k_{T\rho}} \tag{9.80}$$

Equation 9.80 (which also follows directly from Equation 9.69) demonstrates the behavior of the *Fanno line*. For subsonic flow, it lies between isenthalpic flow (which is the same as isothermal flow for the ideal gas) and isentropic flow, approaching isentropic flow with the "polytropic" exponent (see Section III.B) $\gamma = k_{P\rho}$ when $N_{Ma} \to 1$ and approaching isenthalpic flow with the isenthalpic exponent $\gamma = \gamma_h = k_{P\rho}/(1 + k_{T\rho})$ as $N_{Ma} \to 0$. This also means that $A_h = 1 + 1/\gamma_h$. For an ideal gas, $\gamma_h = 1$.

From Equations 9.71 through 9.75, it is easy to derive the following general equations for the change in P, ρ, T as a function of the Mach number:

$$\frac{dP}{P} = -\frac{1 + k_{T\rho}N_{Ma}^2}{A + BN_{Ma}^2}\frac{dN_{Ma}^2}{N_{Ma}^2} = -\frac{1 + k_{TP}\hat{N}_{Ma}^2}{A_h + B_h\hat{N}_{Ma}^2}\frac{d\left(\hat{N}_{Ma}^2\right)}{\hat{N}_{Ma}^2} \tag{9.81}$$

$$\frac{d\rho}{\rho} = -\frac{dV}{V} = -\frac{1 + k_{T\rho}}{A + BN_{Ma}^2}\frac{dN_{Ma}^2}{k_{P\rho}N_{Ma}^2} = -\frac{1 + k_{T\rho}}{k_{P\rho}\left(A_h + B_h\hat{N}_{Ma}^2\right)}\frac{d\hat{N}_{Ma}^2}{\hat{N}_{Ma}^2} \tag{9.82}$$

$$\frac{dT}{T} = -k_{TP}\frac{1 - \hat{\beta}^{-1} + \left(k_{T\rho} + \hat{\beta}^{-1}\right)N_{Ma}^2}{A + BN_{Ma}^2}\frac{dN_{Ma}^2}{N_{Ma}^2}$$

$$= -k_{TP}\frac{1 - \hat{\beta}^{-1} + \left(k_{TP} + k_{P\rho}^{-1}\hat{\beta}^{-1}\right)\hat{N}_{Ma}^2}{A_h + B_h\hat{N}_{Ma}^2}\frac{d\hat{N}_{Ma}^2}{\hat{N}_{Ma}^2} \tag{9.83}$$

For an ideal gas, the above equations take the form

$$\frac{dP}{P} = -\frac{kN_{Ma}^2\left[1 + (k-1)N_{Ma}^2\right]}{2\left(1 - N_{Ma}^2\right)}\left(4f\frac{dL}{D}\right) \tag{9.84}$$

$$\frac{dN_{Ma}^2}{N_{Ma}^2} = \frac{kN_{Ma}^2\left[2 + (k-1)N_{Ma}^2\right]}{2\left(1 - N_{Ma}^2\right)}\left(4f\frac{dL}{D}\right) \tag{9.85}$$

$$\frac{dV}{V} = -\frac{d\rho}{\rho} = \frac{kN_{Ma}^2}{2(1-N_{Ma}^2)}\left(4f\frac{dL}{D}\right) \tag{9.86}$$

$$\frac{dT}{T} = -\frac{k(k-1)N_{Ma}^4}{2(1-N_{Ma}^2)}\left(4f\frac{dL}{D}\right) \tag{9.87}$$

$$\frac{dP_o}{P_o} = -\frac{kN_{Ma}^2}{2}\left(4f\frac{dL}{D}\right) \tag{9.88}$$

$$\frac{dF}{F} = \frac{dP}{P} + \frac{kN_{Ma}^2}{1+kN_{Ma}^2}\left(\frac{dN_{Ma}^2}{N_{Ma}^2}\right) \tag{9.89}$$

and

$$\frac{dP}{P} = -\frac{1+(k-1)N_{Ma}^2}{2+(k-1)N_{Ma}^2}\frac{dN_{Ma}^2}{N_{Ma}^2} \tag{9.90}$$

$$\frac{d\rho}{\rho} = -\frac{dV}{V} = -\frac{1}{2+(k-1)N_{Ma}^2}\frac{dN_{Ma}^2}{N_{Ma}^2} \tag{9.91}$$

$$\frac{dT}{T} = -\frac{(k-1)dN_{Ma}^2}{2+(k-1)N_{Ma}^2} \tag{9.92}$$

$$\frac{dP_o}{P_o} = -\frac{1-N_{Ma}^2}{2+(k-1)N_{Ma}^2}\frac{dN_{Ma}^2}{N_{Ma}^2} \tag{9.93}$$

The subscript o represents the "stagnation" state, that is, the conditions that would prevail if the gas is slowed to a stop and all kinetic energy converted reversibly to internal energy.

For a given gas, these equations show that all conditions in the pipe depend uniquely on the Mach number and dimensionless pipe length. In fact, if $N_{Ma} < 1$ at the pipe entrance, an inspection of these equations shows that as the distance down the pipe (dL) increases, V will increase but P, ρ, and T will decrease. However, if $N_{Ma} > 1$ at the pipe entrance, just the opposite is true, that is, V decreases while P, ρ, and T will increase with distance down the pipe. That is, a flow that is initially subsonic will approach sonic flow (as a limit) as L increases, whereas an initially supersonic flow will also approach sonic flow as L increases. Thus, all flows, regardless of their initial conditions, will tend toward the speed of sound as the gas progresses down a uniform pipe. Therefore, the only way a subsonic flow can be transformed into a supersonic flow is through a converging-diverging nozzle, where the speed of sound is reached at the nozzle throat. We will not be concerned here with supersonic flows, but the interested reader can find this subject treated in many fluid mechanics books such as Hall (1951), Shapiro (1953), Saad (1992) or Oosthuizen and Carscallen (2013).

Real gas flow in all practical known cases follows the same behavior as described earlier. However, there can (theoretically) exist some fluids for which $(A + BN_{Ma}^2)$ in Equation 9.74 changes sign and becomes negative under some thermodynamic conditions. This is controlled by the behavior of the so-called "fundamental derivative":

$$\Gamma = 1 + (\partial \ln c/\partial \ln \rho)_s = (1 + k_{P\rho} + k_{P\rho}k_{P\rho,Ps})/2 \tag{9.94}$$

which is usually positive (Thompson, 1971). It was predicted (using appropriate equations of state) that some fluids (named "Bethe–Zel'dovich–Thompson," or "BZT" fluids) can have $\Gamma < 0$ in some region of thermodynamic parameters and unusual flow behavior in this region. Usage of BZT fluids for energy conversion could provide significant advantages. However, there is still no firm experimental evidence of existence of such fluid behavior.

B. Applications

We will now derive some explicit equations. Suppose the coefficients in the equations above are almost constant so that we can use their average values. In this case, integration of Equation 9.74 gives

$$\frac{4\bar{f}L}{D} = \frac{2}{A\bar{k}_{P\rho}} \left\{ \frac{1}{N_{Ma1}^2} - \frac{1}{N_{Ma2}^2} + (1+\bar{\delta})\ln\left[\frac{N_{Ma1}^2\left(1+\bar{\delta}N_{Ma2}^2\right)}{N_{Ma2}^2\left(1+\bar{\delta}N_{Ma1}^2\right)} \right] \right\} \tag{9.95}$$

and from Equation 9.75

$$\frac{4\bar{f}L}{D} = \frac{2}{\bar{A}_h} \left\{ \frac{1}{\hat{N}_{Ma1}^2} - \frac{1}{\hat{N}_{Ma2}^2} + \left(1/\bar{k}_{P\rho} + \bar{\varepsilon}\right)\ln\left[\frac{\hat{N}_{Ma1}^2\left(1+\bar{\varepsilon}\hat{N}_{Ma2}^2\right)}{\hat{N}_{Ma2}^2\left(1+\bar{\varepsilon}\hat{N}_{Ma1}^2\right)} \right] \right\} \tag{9.96}$$

where $\delta = B/A$ and, $\varepsilon = k_{TP}/A_h$. Usually $\delta \ll 1$ and $\varepsilon \ll 1$.

Also, integration of Equations 9.81 gives

$$\frac{P_2}{P_1} = \left(\frac{N_{Ma1}}{N_{Ma2}}\right)^{2/\bar{A}} \left(\frac{1+\bar{\delta}N_{Ma1}^2}{1+\bar{\delta}N_{Ma2}^2}\right)^{1/(1+\bar{k}_{P\rho,P\rho})-1/\bar{A}} \tag{9.97}$$

or

$$\frac{P_2}{P_1} = \left(\frac{\hat{N}_{Ma1}^2}{\hat{N}_{Ma2}^2}\right)^{1/\bar{A}_h} \left(\frac{1+\bar{\varepsilon}\hat{N}_{Ma1}^2}{1+\bar{\varepsilon}\hat{N}_{Ma2}^2}\right)^{1-1/\bar{A}_h} = \left(\frac{k_{P\rho1}N_{Ma1}^2}{k_{P\rho2}N_{Ma2}^2}\right)^{1/\bar{A}_h} \left(\frac{1+\bar{\varepsilon}k_{P\rho1}N_{Ma1}^2}{1+\bar{\varepsilon}k_{P\rho2}N_{Ma2}^2}\right)^{1-1/\bar{A}_h} \tag{9.98}$$

with similar equations for the other thermodynamic parameters. Equations 9.96 and 9.98 are useful when the change in $k_{P\rho}$ is significant, and should be taken into account.

We will now consider in more detail the ideal gas. In this case:

$$A = A_h = 2, \quad B = k-1, \quad B_h = \frac{k-1}{k}, \quad \delta = \frac{k-1}{2}, \quad \varepsilon = \frac{k-1}{2k}$$

It is convenient to take the sonic state ($N_{Ma} = 1$) as the reference state for application of these equations. Thus, if the upstream Mach number is N_{Ma}, the length of pipe through which this gas must flow to reach the speed of sound ($N_{Ma} = 1$) will be L^*, and Equation 9.90 gives

$$\frac{4\bar{f}L^*}{D} = \frac{1-N_{Ma}^2}{kN_{Ma}^2} + \frac{k+1}{2k}\ln\left[\frac{(k+1)N_{Ma}^2}{2+(k-1)N_{Ma}^2} \right] \tag{9.99}$$

where $\bar{f}$ is the average friction factor over the pipe length L^*. Because the mass flux is constant along the pipe, the Reynolds number (and hence f) will vary only as a result of the variation in the viscosity, which is normally quite small for gases. If $\Delta L = L = L_1^* - L_2^*$ is the pipe length over which the Mach number changes from N_{Ma1} to N_{Ma2}, then

$$\frac{4\bar{f}\Delta L}{D} = \left(\frac{4\bar{f}L^*}{D}\right)_1 - \left(\frac{4\bar{f}L^*}{D}\right)_2 \tag{9.100}$$

Likewise, the following relationships between the problem variables and their values at the sonic (reference) state can be obtained by integrating Equations 9.78 through 9.82 to give

$$\frac{P}{P^*} = \frac{1}{N_{Ma}}\left[\frac{k+1}{2+(k-1)N_{Ma}^2}\right]^{1/2} \tag{9.101}$$

$$\frac{T}{T^*} = \left(\frac{c}{c^*}\right)^2 = \frac{k+1}{2+(k-1)N_{Ma}^2} \tag{9.102}$$

$$\frac{\rho}{\rho^*} = \frac{V^*}{V} = \frac{1}{N_{Ma}}\left[\frac{2+(k-1)N_{Ma}^2}{k+1}\right]^{1/2} \tag{9.103}$$

$$\frac{P_o}{P_o^*} = \frac{1}{N_{Ma}}\left[\frac{2+(k-1)N_{Ma}^2}{k+1}\right]^{(k+1)/2(k-1)} \tag{9.104}$$

With these relationships in mind, the conditions at any two points (e.g., 1 and 2) in the pipe are related by

$$\frac{T_2}{T_1} = \frac{T_2/T^*}{T_1/T^*}, \quad \frac{P_2}{P_1} = \frac{P_2/P^*}{P_1/P^*}, \text{etc.} \tag{9.105}$$

and

$$\frac{4\bar{f}\Delta L}{D} = \frac{4\bar{f}}{D}\left(L_2^* - L_1^*\right) = \left(\frac{4\bar{f}L^*}{D}\right)_2 - \left(\frac{4\bar{f}L^*}{D}\right)_1 \tag{9.106}$$

Also, the mass flux at N_{Ma} and at the sonic state are given by

$$G = N_{Ma}P\sqrt{\frac{kM}{RT}}, \quad G^* = P^*\sqrt{\frac{kM}{RT^*}} \tag{9.107}$$

For a pipe containing fittings, the term $4fL/D$ would be replaced by the sum of all loss coefficients (ΣK_f) for all pipe sections and fittings. These equations apply to adiabatic flow in a constant area duct, for which the sum of the enthalpy and kinetic energy is constant (e.g., Equation 9.58), which defines the *Fanno line*. It is evident that each of the dependent variables at any point in the system is a unique function of the nature of the gas (k) and the Mach number of the flow (N_{Ma}) at that point. Note that although the dimensionless variables are expressed relative to their values at sonic conditions, it is not always necessary to determine the actual sonic conditions to apply these relationships. Because the Mach number is often the unknown quantity, an iterative or trial-and-error procedure for solving the foregoing set of equations is required. However, these relationships may be presented in tabular form (e.g., Appendix I) or in graphical form (e.g., Figure 9.4), which can be used directly for solving various types of problems without iteration, as illustrated in the following.

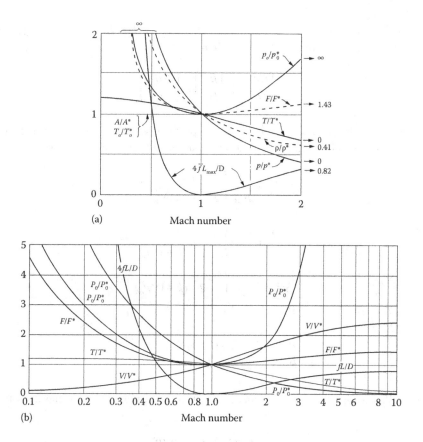

FIGURE 9.4 Fanno line functions for $k = 1.4$. (a) From Hall (1951) and (b) from Shapiro (1953).

For real gases, the "isentropic" exponents' values change with temperature and pressure and differ from the ideal gas values. However, for fluids with $k_{P\rho}$ higher than 1.2, the coefficient $A_h \approx 2$ over a wide region of pressure and temperature, for all temperatures below the critical, except in the very vicinity of the critical point, as well as in the supercritical zone where the reduced pressure $P_r = P/P_c$ does not exceed the reduced temperature $T_r = T/T_c$ ($P_r \leq T_r$). Then from Equations 9.96 and 9.98 it follows that it is possible to use the ideal gas equations (9.85 to 9.101) with an average value of $k_{P\rho}$ instead of k.

In the case of gases with $k_{P\rho}$ close to 1, the coefficient $A \approx 2$ in the same region of reduced temperatures and pressures, and $\delta \ll 1$. Thus, for subsonic flow ($N_{Ma} \leq 1$), we can set $A = 2$, $\delta = 0$ in Equations 9.95 and 9.97, which gives

$$\frac{4\bar{f}L}{D}\bar{k}_{P\rho} = \frac{1}{N_{Ma1}^2} - \frac{1}{N_{Ma2}^2} + 2\ln\left(\frac{N_{Ma1}}{N_{Ma2}}\right) \tag{9.108}$$

$$\frac{P_2}{P_1} = \frac{N_{Ma1}}{N_{Ma2}} \tag{9.109}$$

and

$$\frac{P_2}{P_1} = \left\{1 - \frac{G^2}{P_1\rho_1}\left[\frac{4fL}{D} - \frac{2}{\bar{k}_{P\rho}}\ln\left(\frac{P_2}{P_1}\right)\right]\right\}^{1/2} \tag{9.110}$$

Note that these equations differ from the isothermal equations (9.24) and (9.25) only by the one coefficient k_{P_ρ} and the two coincide for $k_{P_\rho} = 1$. Note also that the isentropic flow Equation (9.27) can be derived from Equations 9.84 and 9.85 by putting $k_{PT} = 0$.

While the isothermal flow model (Equation 9.24) approximates *Fanno* flow for small Mach numbers, the isentropic flow model Equation (9.27) is a good approximation when the flow approaches sonic. The "pseudo-isothermal" flow Equation (9.110) provides a good approximation for *Fanno* subsonic flow for the whole subsonic Mach number range (when $1 \leq k_{P_\rho} \leq 1.2$).

C.   SOLUTION OF HIGH-SPEED IDEAL GAS PROBLEMS

We will illustrate the procedure for solving the three types of pipe flow problems for high-speed ideal gas flows: unknown driving force, unknown flow rate, and unknown diameter.

1.   Unknown Driving Force

The unknown driving force can be either the upstream pressure, P_1, or the downstream pressure, P_2. However, one of these must be known, and the other is determined as follows.

Given: $P_1, T_1, G, D, L, \varepsilon$ *Find*: P_2

1. Calculate $N_{Re} = DG/\mu_1$ and use this to find f_1 (Churchill Eq. or Moody diagram).
2. Calculate $N_{Ma1} = (G/P_1)(RT_1/kM)^{1/2}$. Use this with Equations 9.99, 9.101, and 9.102 or Figure 9.4 or Appendix I to find $(4fL_1^*/D_1)$, P_1/P^*, and T_1/T^*. From these values and the given quantities, calculate L_1^*, P^*, and T^*.
3. Calculate $L_2^* = L_1^* - L$ and use this to calculate $(4f_1L_2^*/D)_2$. Use this with Figure 9.4 or Appendix I or Equations 9.99, 9.101, and 9.102 to determine N_{Ma2}, P_2/P^*, and T_2/T^*. (Note that Equation 9.99 is implicit in N_{Ma2}.) From these values, determine P_2 and T_2.
4. Revise the value of the viscosity by evaluating it at an average temperature, $(T_1 + T_2)/2$, and pressure, $(P_1 + P_2)/2$. Use this to revise N_{Re} and hence f, and repeat steps 3 and 4 until the change is within acceptable limits.

2.   Unknown Flow Rate

The mass velocity (G) is the unknown in this case. This is equivalent to the mass flow rate because the pipe diameter is known. This requires a trial-and-error procedure because neither the Reynolds number nor the Mach numbers can be determined a priori.

Given: $P_1, T_1, L, P_2, D, \varepsilon$ *Find*: G

1. Assume a value for N_{Ma1}. Use Equations 9.101, 9.102, and 9.106 or Figure 9.4 or Appendix I to find P_1/P^*, T_1/T^*, and $(4fL_1^*/D)_1$. From these values and the given quantities, determine P^* and T^*.
2. Calculate $G_1 = N_{Ma1}P_1(kM/RT_1)^{1/2}$ and $N_{Re1} = DG/\mu$. Using the latter, determine f_1 from the Churchill equation or Moody diagram.
3. Calculate $(4fL_2^*/D)_2 = (4fL_1^*/D)_1 - (4fL/D)$. Use this with Equation 9.99 (implicit) and Equations 9.102 and 9.103 or Figure 9.4 or Appendix I to find N_{Ma2}, P_2/P^*, and T_2/T^* at point 2.
4. Calculate $P_2 = (P_2/P^*)P^*$, $T_2 = (T_2/T^*)T^*$, $G_2 = N_{Ma2}P_2(kM/RT)^{1/2}$, and $N_{Re} = DG_2/\mu$. Use the latter to determine a revised value of $f = f_2$.
5. Using $f = (f_1 + f_2)/2$ for the revised friction factor, repeat steps 3 and 4 until the change is within acceptable limits.
6. Compare the given value of P_2 with the calculated value from step 4. If they agree, the answer is the calculated value of G_2 from step 4. If they do not agree, return to step 1 with a new guess for N_{Ma1}, and repeat the procedure until agreement is achieved.

3. Unknown Diameter

The procedure for an unknown diameter involves a trial-and-error process similar to the one for the unknown flow rate.

> *Given*: $P_1, T_1, L, P_2, \dot{m}, \varepsilon$ *Find*: D

1. Assume a value for N_{Ma1} and use Equations 9.101, 9.102, and 9.106 or Figure 9.4 or Appendix I to find P_1/P^*, T_1/T^*, and $(4fL_1^*/D)_1$. Also, calculate $G = N_{Ma1}P_1(kM/RT)^{1/2}$, $D = (4\dot{m}/\pi G)^{1/2}$, and $N_{Re1} = DG/\mu$. Use N_{Re1} and an assumed value of pipe wall roughness of 0.005 in. to find f_1 from the Churchill equation or Moody diagram.
2. Calculate $P_2/P^* = (P_1/P^*)(P_2/P_1)$ and use this with Figure 9.4 or Appendix I or Equations 9.99, 9.101 (implicitly), and 9.102 to find N_{Ma2}, $(4fL_2^*/D)_2$ and T_2/T^*. Calculate $T_2 = (T_2/T^*)(T^*/T_1)T_1$ and use P_2 and T_2 to determine μ_2. Then use μ_2 to determine $N_{Re2} = DG/\mu_2$, which, with a wall roughness of 0.005 in., determines f_2 from the Churchill equation or Moody diagram.
3. Using $f = (f_1 + f_2)/2$, calculate $L = L_1^* - L_2^* = [(4fL_1^*/D)_1 - (4fL_2^*/D)_2](D/4f)$.
4. Compare the value of L calculated in step 3 with the given value, and the value of ε/D with that used in step 1. If they agree, the value of D determined in step 1 and the assumed value of ε are correct. If they do not agree, return to step 1, revise the assumed value of N_{Ma1}, and $\varepsilon = 0.005$ in., and repeat the entire procedure until agreement is achieved.

SUMMARY

The major points covered in this chapter include:

- Understand and be able to apply the basic equations for ideal gas in pipes, for isothermal and adiabatic flow, including the determination of choked flow.
- Know how to use the Expansion Factor approach for problems involving adiabatic flow in pipes.
- Understand and be able to apply the generalized flow equations (under all flow conditions) to an ideal gas in pipes, for any Mach number.
- Understand how to generalize the equations for ideal gases under any conditions, to any compressible fluid.

PROBLEMS

Compressible Flow

1. A 12 in. ID gas pipeline carries methane (molecular weight (M = 16) at a rate of 20,000 scfm. The gas enters the line at a pressure of 500 psia, and a compressor station is located every 100 miles to boost the pressure back up to 500 psia. The pipeline is isothermal at 70°F, and the compressors are adiabatic with an efficiency of 65%. What is the required horsepower for each compressor? Assume ideal gas behavior.

2. Natural gas (CH₄) is transported through a 6 in. ID pipeline at a rate of 10,000 scfm. The compressor stations are 150 miles apart, and the compressor suction pressure is maintained at 10 psig above that at which choked flow would occur in the pipeline. The compressors are each two-stage, operate adiabatically with interstage cooling to 70°F, and have an efficiency of 60%. If the pipeline temperature is 70°F, calculate
   (a) The discharge pressure, interstage pressure, and compression ratio for the compressor stations
   (b) The horsepower required at each compressor station

3. Natural gas (methane) is transported through a 20 in. sch 40 commercial steel pipeline at a rate of 30,000 scfm. The gas enters the line from a compressor at 100 psia and 70°F. Identical

compressor stations are located every 10 miles along the line, and at each station the gas is recompressed to 100 psia and cooled to 70°F.

(a) Determine the suction pressure at each compressor station.

(b) Determine the horsepower required at each station if the compressors are 80% efficient.

(c) How far apart could the compressor stations be located before the flow in the pipeline becomes choked?

4. Natural gas (methane) is transported through an uninsulated 6 in. ID commercial steel pipeline, 1 mile long. The inlet pressure is 100 psia and the outlet pressure is 1 atm. What is the mass flow rate of the gas and the compressor power required to pump it? $T_1 = 70°F$, $\mu_{gas} = 0.02$ cP.

5. It is desired to transfer natural gas (CH_4) at a pressure of 200 psia and a flow rate of 1000 scfs through a 1 mile long uninsulated commercial steel pipeline into a storage tank at 20 psia. Can this be done using either a 6 or 12 in. ID pipe? What diameter pipe would you recommend? $T_1 = 70°F$, $\mu = 0.02$ cP.

6. A natural gas (methane) pipeline is designed to transport the gas at a rate of 5000 scfm. The pipe is 6 in. ID and the maximum pressure that the compressors can develop is 1500 psig. The compressor stations are located in the pipeline at the point at which the pressure drops to 100 psi above that at which choked flow would occur (i.e., the suction pressure for the compressor stations). If the design temperature for the pipeline is 60°F, the compressors are 60% efficient, and the compressor stations each operate with three stages and interstage cooling to 60°F, determine

(a) The proper distance between compressor stations, in miles

(b) The optimum interstage pressure and compression ratio for each compressor stage

(c) The total horsepower required for each compressor station

7. Ethylene gas leaves a compressor at a pressure of 3500 psig and is carried in a 2 in. sch 40 pipeline, 100 ft long, to a unit where the pressure is 500 psig. The line contains two plug valves, one swing check valve, and eight flanged elbows. If the temperature is 100°F, what is the flow rate (in scfm)?

8. A 12 in. ID natural gas (methane) pipeline carries gas at a rate of 20,000 scfm. The compressor stations are 100 miles apart, and the discharge pressure of the compressors is 500 psia. If the temperature of the surroundings is 70°F, what is the required horsepower of each compressor station, assuming 65% efficiency? If the pipeline breaks 10 miles downstream of a compressor station, what will be the flow rate through the broken pipe?

9. The pressure in a reactor fluctuates between 10 and 30 psig. It is necessary to feed air to the reactor at a constant rate of 20 lb_m/h from an air supply at 100 psig, 70°F. To do this, you insert an orifice into the air line that will provide the constant flow rate. What size (diameter) should the orifice be?

10. Oxygen is fed to a reactor at a constant rate of 10 lb_m/s from a storage tank in which the pressure is constant at 100 psig and the temperature is 70°F. The pressure in the reactor fluctuates between 2 and 10 psig, so you want to insert a choke in the line to maintain the flow rate constant. If the choke is a 2 ft length of tubing, what should the ID of the tubing be?

11. Methane is fed to a reactor at a rate of 10 lb_m/min. The methane is available in a pipeline at 20 psia, 70°F, but the pressure in the reactor fluctuates between 2 and 10 psia. To control the flow rate, you want to install an orifice plate that will choke the flow at the desired rate. What should the diameter of the orifice be?

12. Ethylene gas (MW = 28, k = 1.3, μ = 0.1 cP) at 100°F is fed to a reaction vessel from a compressor through 100 ft of 2 in. sch 40 pipe containing 2 plug valves, one swing check valve, and 8 flanged elbows. If the compressor discharge pressure is 3500 psig and the pressure in the vessel is 500 psig, what is the flow rate of the gas, in scfm (1 atm, 60°F)?

13. Nitrogen is fed from a high-pressure cylinder through ¼ ID stainless steel tubing, to an experimental unit. The line ruptures at a point that is 10 ft from the cylinder. If the pressure of the nitrogen in the cylinder is 3000 psig and the temperature is 70°F, what are the mass flow rate of the gas through the line and the pressure in the tubing at the point of the break?

14. A storage tank contains ethylene at 200 psig and 70°F. If a 1 in. ID line that is 6 ft long containing a globe valve on the end is attached to the tank, what would the flow rate of gas be (in scfm) if
    (a) The valve is fully open?
    (b) The line breaks off right at the tank?

15. A 2 in. sch 40 pipeline is connected to a storage tank containing ethylene at 100 psig and 80°C.
    (a) If the pipe breaks at a distance of 50 ft from the tank, determine the rate at which the ethylene will leak out of the pipe (in lb_m/s).
    (b) If the pipe breaks off right at the tank, what would the leak rate be?

16. Saturated steam at 200 psig (388°F, specific volume, ν = 2.13 ft³/lb_m, μ = 0.015 cP) is fed from a header to a direct contact evaporator that operates at 10 psig. If the steam line is 2 in. sch 40 pipe, 50 ft long, and includes four flanged elbows and one globe valve, what is the steam flow rate in lb_m/h?

17. Air is flowing from a tank at a pressure of 200 psia and 70°F through a venturi meter into another tank at a pressure of 50 psia. The venturi meter is mounted in a 6 ID pipe section (that is quite short) and has a throat diameter of 3 in. What is the mass flow rate of the air?

18. A tank containing air at 100 psia and 70°F is punctured with a hole ¼ in. diameter. What is the mass flow rate of the air through the hole?

19. A pressurized tank containing nitrogen at 800 psig is fitted with a globe valve, to which is attached a line with 10 ft of ¼ ID stainless steel tubing and three standard elbows. The temperature of the system is 70°F. If the valve is left wide open, what is the flow rate of nitrogen, in lb_m/s and also in scfm?

20. Gaseous chlorine (M = 71) is transferred from a high-pressure storage tank at 500 psia and 60°F, through an insulated 2 in. sch 40 pipe 200 ft long, into another vessel where the pressure is 200 psia. What is the mass flow rate of the gas and its temperature at the point where it leaves the pipe?

21. A storage tank contains ethylene at a pressure of 200 psig and a temperature of 70°F springs a leak. If the hole through which the gas is leaking is ½ in. diameter, what is the leakage rate of the ethylene, in scfm?

22. A high-pressure cylinder containing N_2 at 200 psig and 70°F is connected by ¼ in. ID stainless steel tubing, 20 ft long, to a reactor in which the pressure is 15 psig. A pressure regulator at the upstream end of the tubing is used to control the pressure in the reactor, and hence the flow rate of the N_2 in the tubing.
    (a) If the regulator controls the pressure entering the tubing at 25 psig, what is the flow rate of N_2 (in scfm)?
    (b) If the regulator fails so that the full pressure from the cylinder is applied at the tubing entrance, what will be the flow rate of the N_2 into the reactor (in scfm)?

23. Oxygen is supplied to an astronaut through an umbilical hose that is 7 m long. The pressure in the oxygen tank is 200 kPa at a temperature of 10°C, and the pressure in the space suit is 20 kPa. If the umbilical hose has an equivalent roughness of 0.01 mm, what should the hose diameter be to supply oxygen at a rate of 0.05 kg/s? If the suit springs a leak and the pressure drops to zero, at what rate will the oxygen escape?

24. Ethylene (MW = 28) is transported from a storage tank, at 250 psig and 70°F, to a compressor station where the suction pressure is 100 psig. The transfer line is 1 in. sch 80, 500 ft long, and contains two ball valves and eight threaded elbows. An orifice meter with a diameter of 0.75 in. is installed near the entrance to the pipeline.
    (a) What is the flow rate of the ethylene through the pipeline, in scfh?
    (b) If the pipeline breaks at a point 200 ft from the storage tank and there are four elbows and one gate valve in the line between the tank and the break, what is the flow rate of the ethylene (in scfh)?
    (c) What is the differential pressure across the orifice for both cases (a) and (b), in inches of water?

25. Air passes from a large reservoir at 70°F through an isentropic converging-diverging nozzle into the atmosphere. The area of the nozzle throat is 1 cm², and that of the exit is 2 cm². What is the reservoir pressure at which the flow in the nozzle just reaches sonic velocity, and what are the mass flow rate and exit Mach number under these conditions?

26. Air is fed from a reservoir through a converging-diverging nozzle into a ½ in. ID drawn steel tube that is 15 ft long. The flow in the tube is adiabatic, and the reservoir temperature and pressure are 70°F and 200 psia.
    (a) What is the maximum flow rate (in lb_m/s) that can be achieved in the tube?
    (b) What is the maximum pressure at the tube exit at which this flow rate will be reached?
    (c) What is the temperature at this point under these conditions?

27. A gas storage cylinder contains nitrogen at 250 psig and 70°F. Attached to the cylinder is a 3 in. long ¼ in. sch 40 stainless steel pipe nipple, and attached to that is a globe valve followed by a diaphragm valve. Attached to the diaphragm valve is a ¼ in. ID copper tubing line. Determine the mass flow rate of nitrogen (in lb_m/s), if
    (a) The copper tubing breaks off at a distance of 30 ft downstream of the diaphragm valve.
    (b) The pipe breaks off right at the cylinder.

28. A compressor supplies natural gas (mainly CH_4) to a pipeline. The compressor suction pressure is 20 psig and the discharge pressure is 1000 psig. The pipe is 5 in. sch 40 and the ambient temperature is 80°F.
    (a) If the pipe breaks at a point 2 miles downstream from the compressor station, determine the rate at which the gas will escape (in scfm).
    (b) If the compressor efficiency is 80%, what power is required to drive it?

29. You have to feed a gaseous reactant to a reactor at a constant rate of 1000 scfm. The gas is stored in a tank, at 80°F and 500 psig, that is located 20 ft from the reactor, in which the pressure fluctuates between 10 and 20 psig. You know that if the flow is choked in the feed line to the reactor, then the flow rate will be independent of the pressure in the reactor, which is what you desire. If the feed line has a roughness of 0.0018 in., what should its diameter be in order to satisfy your requirement? The gas has a MW of 35, an isentropic exponent of 1.35, and a viscosity of 0.01 cP at 80°F.

30. A pressure vessel containing nitrogen at 300°F has a relief valve installed on top of the vessel. The valve is set to open at a set pressure of 125 psig and exhausts its contents at atmospheric

pressure. The valve has a nozzle that has a smoothly rounded entrance and is 1.5 in. in diameter and 4 in. long, which limits the flow through the valve when it is open.

(a) If the flow resistance in the piping between the tank and the valve, and from the valve to the atmospheric discharge, is neglected, determine the mass flow rate through the valve when it opens (in lb_m/s).

(b) In reality, there is a 3 ft length of a 3 in. pipe between the tank and the valve, and a 6 ft length of a 4 in. pipe downstream of the valve discharge. What is the effect of including this piping on the calculated flow rate?

(*Note:* This problem can be solved more accurately using the methods given in Chapter 11)

31. A storage tank contains ethylene at 80°F and has a relief valve that is set to open at a pressure of 250 psig. The valve must be sized to relieve the gas at a rate of 85 lb_m/s when it opens. The valve has a discharge coefficient (the ratio of the actual mass flux to the theoretical mass flux) of 0.975.

(*Note:* This problem can be solved more accurately using the methods given in Chapter 11).

(a) What should be the diameter of the nozzle in the valve? What horsepower would be required to compress the gas from 1 atm to the maximum tank pressure at a rate equal to the flow rate through the valve, for

(b) A single-stage compressor.

(c) A two-stage compressor with intercooling. Assume a 100% efficiency for the compressor.

NOTATION

A	Cross-sectional area, $[L^2]$
A,B,A_h,B_h	Coefficients in general compressible flow equations, [—]
c	Speed of sound, $[L/t]$
D	Diameter, $[L]$
e_f	Energy dissipation per unit mass of fluid, $[FL/M = L^2/t^2]$
f	Fanning friction factor, [—]
F	Force, $[F = ML/t^2]$
G	Mass flux, $[M/tL^2]$
h	Enthalpy per unit mass, $[FL/M = L^2/t^2]$
k	Isentropic exponent ($[= c_p/c_v]$ for ideal gas), [—]
K	Bulk modulus, $[F/L^2 = M/(Lt^2)]$
$k_{P\rho}, k_{TP}, k_{T\rho}$	Isentropic exponents for real gas, [—]
K_f	Loss coefficient, [—]
L	Length, $[L]$
M	Molecular weight, $[M/mol]$
$\dot{m}$	Mass flow rate, $[M/t]$
N_{Ma}	Mach number, [—]
N_{Re}	Reynolds number, [—]
P	Pressure, $[F/L^2 = M/(Lt^2)]$
q	Heat transferred to fluid per unit mass, $[FL/M = L^2/t^2]$
R	Gas constant, $[FL/(t\ mol)]$
T	Temperature, $[T]$
V	Spatial averaged velocity, $[L/t]$
w	Work done by fluid per unit mass, $[FL/M = L^2/t^2]$
Y	Expansion factor, [—]
Z	Compressibility factor, [—]
z	Vertical distance measured upward from reference plane, $[L]$

GREEK

$\hat{\beta}$	Dimensionless thermal expansion coefficient, [—]
Γ	Fundamental derivative, [—]
γ	"Polytropic" exponent, [—]
$\hat{\kappa}$	Dimensionless isothermal bulk modulus, [—]
ρ	Density, $[M/L^3]$
μ	Viscosity, $[Ft/L^2 = M/Lt]$
μ_{JT}	Joule–Thomson coefficient, $[TL^2/F = TLt^2/M]$

SUBSCRIPTS

1	Reference point 1
2	Reference point 2
c	Critical parameters (pressure, temperature)
r	Reduced parameters (pressure, temperature)
s	Constant entropy
T	Constant temperature

SUPERSCRIPTS

*	Sonic state

REFERENCES

Baltajiev, N., An investigation of real gas effects in supercritical CO_2 compressors. PhD Dissertation, MIT, Cambridge, MA, 2012.

Cheng, D., X. Fan, and M. Yang, Quasi-1D compressible flow of hydrocarbon fuel. *Forty-Eighth AIAA/ASME/ SAE/ASEE Joint Propulsion Conference & Exhibit*, Atlanta, GA, 30 July–01 August, 2012.

Crane, C., Flow of fluids through valves, fittings and pipe, Technical Manual 410, Crane Co., New York, 1991 (and subsequent issues).

Hall, N.A., *Thermodynamics of Fluid Flow*, Prentice-Hall, Englewood Cliffs, NJ, 1951.

Holland, F.A. and R. Bragg, *Fluid Flow for Chemical Engineers*, 2nd edn., Edward Arnold, London, U.K., 1995.

Istomin, V.A., Real gas isentropic indexes and their use in thermodynamics of gases and gas dynamics, *Russ. J. Phys. Chem.* [c/c of *Zhurnal Fizicheskoi Khimii*], 72(3), 334–339, 1998.

Istomin, V.A., Real gas isentropic indices: Definition and basic relations, *Russ. J. Phys. Chem.* [c/c of *Zhurnal Fizicheskoi Khimii*], 71(6), 883–888, 1997.

Korelshteyn, L., Choked and near-choked real gas and two-phase flow analysis of discharge piping. *Proceedings of 2015 AIChE Spring Meeting and 11th Global Congress of Process Safety*, Austin, TX, 2015.

NIST (National Institute of Science and Technology), Standard Reference Database 23, Reference Fluid Thermodynamic and Transport Properties—REFPROP, Version 9.0 (and subsequent versions), 2010.

Oosthuizen, P.H. and W.E. Carscallen, *Introduction to Compressible Fluid Flow*, 2nd edn., CRC, Boca Raton, FL, 2013.

Saad, M.A., *Compressible Fluid Flow*, 2nd edn., Prentice Hall, Englewood Cliffs, NJ, 1992.

Shapiro, A.H., *The Dynamics and Thermodynamics of Compressible Fluid Flow*, Vol. I, Wiley, New York, 1953.

Thompson, P.A., A fundamental derivative in gas dynamics, *Phys. Fluids*, 14(9), 1843–1849, 1971.

10 Flow Measurement

"Think three times, measure twice and cut once."

—**Anonymous**

I. SCOPE

In this chapter, we will illustrate and analyze some of the more common methods for measuring flow rate in conduits, including the pitot tube, venturi, nozzle, and orifice meters. This is by no means intended to be a comprehensive or exhaustive treatment, however, as there are a great many other devices in use for measuring flow rate, such as turbine, vane, Coriolis, ultrasonic, and magnetic flow meters, vortex meters, etc., just to name a few (see, e.g., Miller, 1996; Baker, 2009). The examples considered here demonstrate the application of the fundamental conservation principles to the analysis of several of the most common devices. Control valves are typically employed in conjunction with flow metering devices in order to control or regulate the flow rate in a piping system. This chapter is concerned with flow measurement, and the consideration of control valves is presented in Chapter 11.

There is one important caveat regarding meter installation. Most meters assume that the flow profile across the tube is symmetrical. This generally requires that the meter be installed at least 10 pipe diameters downstream and 5 diameters upstream of any flow disturbance elements, such as elbows, valves, bends, contractions/expansions, etc. If this is not possible, it is sometimes possible to install "straightening vanes" in the pipe downstream and/or upstream of the source of the disturbance to minimize this effect.

II. PITOT TUBE

As previously discussed, the volumetric flow rate of a fluid through a conduit may be determined by integrating the local ("point") velocity over the cross section of the conduit:

$$Q = \int_A v \, dA \qquad (10.1)$$

If the conduit cross section is circular, this becomes

$$Q = \int_0^{\pi R^2} v(r) \, d(\pi r^2) = 2\pi \int_0^R v(r) \, r \, dr \qquad (10.2)$$

The pitot tube is a device for measuring $v(r)$, the local velocity at a given position in the conduit, as illustrated in Figure 10.1. The measured values of the local point velocity are then used in Equation 10.2 to determine the volumetric flow rate. The meter includes a differential pressure measuring device (e.g., a manometer, transducer, or DP cell) that measures the pressure difference between the two tubes. One tube is attached to a hollow probe that can be positioned at any radial

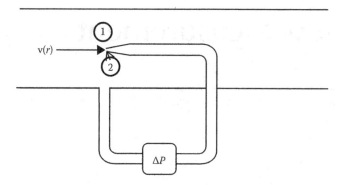

FIGURE 10.1   Pitot tube.

location in the conduit, and the other is attached to the wall of the conduit, in the same axial plane as the end of the probe. The local velocity of the streamline that impinges on the tip of the probe is $v(r)$. The fluid element that impacts the open end of the probe must come to rest at that point, because there is no flow possible through the probe or the DP cell. This is known as the *stagnation point*. The Bernoulli equation can be applied to the fluid streamline that impacts the probe tip:

$$\frac{P_2 - P_1}{\rho} + \frac{1}{2}\left(v_2^2 - v_1^2\right) = 0 \tag{10.3}$$

where
  point 1 is in the free stream just upstream of the probe
  point 2 is just inside the open end of the probe (the stagnation point)

Since the friction loss is negligible in the free stream from 1 to 2, and $v_2 = 0$ because the fluid in the probe is stagnant, Equation 10.3 can be solved for v_1 to give

$$v_1 = \left(\frac{2\left(P_2 - P_1\right)}{\rho}\right)^{1/2} \tag{10.4}$$

The measured pressure difference ΔP is the difference between the "stagnation" pressure in the velocity probe at the point where it connects to the DP cell and the "static" pressure at the corresponding point in the tube connected to the wall. Since there is no flow in the vertical direction, the difference in pressure between any two vertical elevations (at the same horizontal location) is strictly hydrostatic. Thus, the pressure difference measured at the DP cell is the same as that at the elevation of the probe, because the static head between point 1 and the pressure device is the same as that between point 2 and the pressure device, so that $\Delta P = P_2 - P_1$.

We usually want to determine the total flow rate (Q) through the conduit, rather than the velocity at a point. This can be done by using Equation 10.1 or Equation 10.2 if the local velocity is measured at a sufficient number of radial points across the conduit to enable accurate evaluation of the integral. For example, the integral in Equation 10.2 could be evaluated by plotting the measured $v(r)$ values as $v(r)$ versus r^2, or as $rv(r)$ versus r (in accordance with either the first or second form of Equation 10.2), and the area under the curve from $r = 0$ to $r = R$ (the radius of the conduit) can be determined numerically.

The pitot tube is a relatively complex device and requires considerable effort and time to obtain an adequate number of velocity data points, especially close to the wall of the conduit, and to integrate these over the cross section to determine the total flow rate. On the other hand, the probe offers minimal resistance to the flow and hence is very efficient in that it results in negligible friction loss in the conduit. It is also the only practical means for determining the flow rate in very large conduits (such as smokestacks). There are standardized methods for applying this method to determine the total amount of material emitted through a stack, for example.

III. VENTURI AND NOZZLE

There are other devices, however, that can be used to determine the flow rate from a single measurement. These are sometimes referred to as *obstruction meters* because the basic principle involves introducing an "obstruction" (e.g., a constriction) into the flow channel and measuring the pressure drop across the obstruction, which is then related to the flow rate.

Two such devices are the venturi meter and the nozzle, illustrated in Figure 10.2. In both cases, the fluid flows through a reduced area, which results in an increase in the velocity at that point. The corresponding change in pressure between point 1 upstream of the constriction and point 2 at the position of the minimum area (maximum velocity) is measured and is then related to the flow rate through the energy balance (the Bernoulli equation). The velocities are related by the continuity equation

$$\rho_1 V_1 A_1 = \rho_2 V_2 A_2 \tag{10.5}$$

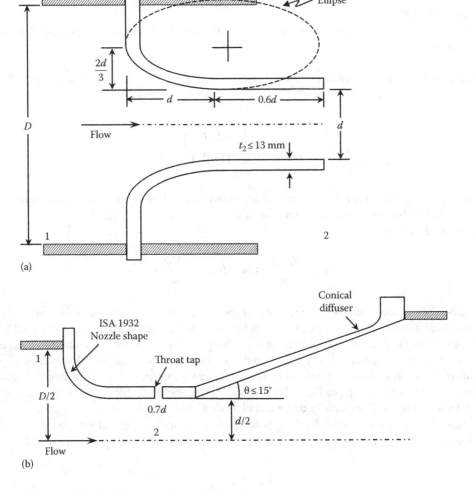

FIGURE 10.2   (a) ASME flow nozzle meter, (b) Venturi meter.

For a constant density fluid $\rho_1 = \rho_2$ which, with Equation 10.5, gives

$$V_1 = V_2 \frac{A_2}{A_1} \tag{10.6}$$

The Bernoulli equation relates the velocity change to the pressure change

$$\frac{P_2 - P_1}{\rho} + \frac{1}{2}\left(V_2^2 - V_1^2\right) + e_f = 0 \tag{10.7}$$

where plug flow has been assumed (so that the kinetic energy correction factor α is assumed to be equal to 1). Using Equation 10.6 to eliminate V_1 and neglecting the friction loss, Equation 10.7 can be solved for V_2:

$$V_2 = \left(\frac{-2\Delta P}{\rho\left(1-\beta^4\right)}\right)^{1/2} \tag{10.8}$$

where
  $\Delta P = P_2 - P_1$
  $\beta = d/D$ (where d is the minimum diameter at the throat of the venturi or nozzle and D is the pipe diameter)

To account for the inaccuracies introduced by assuming plug flow and neglecting friction, Equation 10.8 is written as

$$V_2 = C_d \left(\frac{-2\Delta P}{\rho\left(1-\beta^4\right)}\right)^{1/2} \tag{10.9}$$

where C_d is the *discharge coefficient*, which is determined by calibration as a function of the Reynolds number for a given geometric design of the device. Typical values are shown in Figure 10.3, where

$$N_{Re_D} = \frac{DV_1\rho}{\mu} \quad \text{and} \quad N_{Re_d} = \frac{dV_2\rho}{\mu} = \frac{N_{Re_D}}{\beta}$$

Because the discharge coefficient accounts for the nonidealities in the system such as the friction loss, deviations from plug flow, etc., one would expect it to decrease with increasing Reynolds number, similar to the friction factor in pipe flow, although this is contrary to the trend in Figure 10.3. However, the discharge coefficient also accounts for the deviation from plug flow, which is greater at lower Reynolds numbers. In any event, the coefficient is not greatly different from 1.0, having a value of about 0.985 for (pipe) Reynolds numbers above about 2×10^5, which indicates that these nonidealities are small, especially under turbulent flow conditions.

According to Miller (1996), for $N_{Re_D} > 4000$, the discharge coefficient for the venturi, as well as for the nozzle and orifice, can be correlated as a function of N_{Re_D} and β by the general equation

$$C_d = C_\infty + \frac{b}{N_{Re_D}^n} \tag{10.10}$$

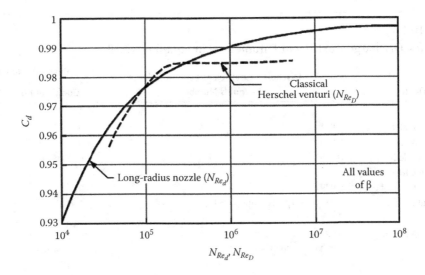

FIGURE 10.3 Venturi and nozzle discharge coefficient versus Reynolds number. (From White, F.M., *Fluid Mechanics*, 3rd edn., McGraw-Hill, New York, 1994.)

where the parameters C_∞, b, and n are given in Table 10.1 as a function of β. The range over which Equation 10.10 applies and its approximate accuracy are given in Table 10.2 (Miller, 1996). Because of the gradual expansion designed into the venturi meter, the pressure recovery is relatively large, so the net friction loss across the entire meter is a relatively small fraction of the measured (maximum) pressure drop, as indicated in Figure 10.4. However, because the flow area changes abruptly downstream of the orifice and nozzle, the expansion is uncontrolled, and considerable eddying occurs downstream. This dissipates more energy, resulting in a significantly higher net friction loss and lower pressure recovery, as also seen in Figure 10.4 and reflected by the relatively lower values of the discharge coefficient for the orifice and nozzle.

The foregoing equations assume that the device is horizontal, that is, that the pressure taps on the pipe are located in the same horizontal plane. If this is not the case, the equations can be easily modified to account for changes in elevation by replacing the pressure P at each point by the total potential $\Phi = P + \rho g z$.

The flow nozzle, illustrated in Figure 10.2, is similar to the venturi meter except that it does not include the diffuser (gradually expanding) section. In fact, one standard design for the venturi meter is basically a flow nozzle with an attached diffuser (see Figure 10.2). The equations that relate the flow rate and measured pressure drop in the nozzle are the same as for the venturi (e.g., Equation 10.9), and the nozzle (discharge) coefficient is also shown in Figure 10.3. It should be noted that the Reynolds number that is used for the venturi coefficient in Figure 10.3 is based on the pipe diameter (D), whereas the Reynolds number used for the nozzle coefficient is based on the nozzle diameter (d) (note that $N_{Re_D} = \beta N_{Re_d}$). There are various "standard" designs for the nozzle, and the reader should consult the literature for details (e.g., Miller, 1996). The discharge coefficient for these nozzles can also be described by Equation 10.10, with the appropriate parameters given in Table 10.1.

IV. ORIFICE METER

The simplest and most common device for measuring flow in a pipe is the orifice meter. This is an "obstruction" meter that consists of a plate with a hole in it that is inserted into the pipe, and the pressure drop across the plate is measured (as illustrated in Figure 10.5). The major

TABLE 10.1
Values for Discharge Coefficient Parameters[a] in Equation 10.10

Primary Device	Discharge Coefficient C_∞ at Infinite Reynolds Number	Reynolds Number Term Coefficient b	Reynolds Number Term Exponent n
Venturi			
Machined inlet	0.995	0	0
Rough cast inlet	0.984	0	0
Rough-welded sheet-iron inlet	0.985	0	0
Universal venturi tube[b]	0.9797	0	0
Lo-Loss tube[c]	$1.05 - 0.471\beta + 0.564\beta^2 + 0.514\beta^3$	0	0
Nozzle			
ASME long radius	0.9975	$-6.53\beta^{0.5}$	0.5
ISA	$0.9900 - 0.2262\beta^{4.1}$	$1{,}708 - 8{,}936\beta + 19{,}779\beta^{4.7}$	1.15
Venturi nozzle (ISA inlet)	$0.9858 - 0.195\beta^{4.5}$	0	0
Orifice			
Corner taps	$0.5959 + 0.0312\beta^{2.1} - 0.184\beta^8$	$91.71\beta^{2.5}$	0.75
Flanged taps (D in in.)			
$D \geq 2.3$	$0.5959 + 0.0312\beta^{2.1} - 0.184\beta^8 + 0.09\dfrac{\beta^4}{D\left(1-\beta^4\right)} - 0.0337\dfrac{\beta^3}{D}$	$91.71\beta^{2.5}$	0.75
$2 \leq D \leq 2.3$[d]	$0.5959 + 0.0312\beta^{2.1} - 0.184\beta^8 + 0.039\dfrac{\beta^4}{1-\beta^4} - 0.0337\dfrac{\beta^3}{D}$	$91.71\beta^{2.5}$	0.75
Flanged taps (D^* in mm)			
$D^* \geq 58.4$	$0.5959 + 0.0312\beta^{2.1} - 0.184\beta^8 + 2.286\dfrac{\beta^4}{D^*(1-\beta^4)} - 0.856\dfrac{\beta^3}{D^*}$	$91.71\beta^{2.5}$	0.75
$50.8 \leq D^* \leq 58.4$	$0.5959 + 0.0312\beta^{2.1} - 0.184\beta^8 + 0.039\dfrac{\beta^4}{1-\beta^4} - 0.856\dfrac{\beta^3}{D^*}$	$91.71\beta^{2.5}$	0.75
D and $D/2$ taps	$0.5959 + 0.0312\beta^{2.1} - 0.184\beta^8 + 0.039\dfrac{\beta^4}{1-\beta^4} - 0.0158\beta^3$	$91.71\beta^{2.5}$	0.75
$2\frac{1}{2}D$ and $8D$ taps[d]	$0.5959 + 0.461\beta^{2.1} - 0.48\beta^8 + 0.039\dfrac{\beta^4}{1-\beta^4}$	$91.71\beta^{2.5}$	0.75

Source: Miller, R.W., *Flow Measurement Engineering Handbook*, 3rd edn., McGraw-Hill, New York, 1996.

[a] Detailed Reynolds number, line size, beta ratio, and other limitations are given in Table 10.2.

[b] From BIF CALC-440/441; the manufacturer should be consulted for exact coefficient information.

[c] Derived from the Badger meter, Inc. Lo-Loss tube coefficient curve; the manufacturer should be consulted for exact coefficient information.

[d] From Stolz (1978).

difference between this device and the venturi and nozzle meters is the fact that the fluid stream leaving the orifice hole contracts to an area considerably smaller than that of the orifice hole. This is called the *vena contracta*, and it occurs because the fluid has considerable radial inward momentum as it converges into the orifice hole, which causes the flow to continue "inward" for a distance downstream of the orifice before it starts to expand to fill the pipe. If the pipe diameter is D, the orifice diameter is d, and the diameter of the vena contracta is d_2, the contraction ratio for the *vena contracta* is defined as $C_c = A_2/A_o = (d_2/d)^2$. For highly turbulent flow, $C_c \approx 0.6$.

TABLE 10.2

Applicable Range and Accuracy of Equation 10.10, with Parameters from Table 10.1

Primary Device	Nominal Pipe Diameter D, in. (mm)	Ratio β	Pipe Reynolds Number, N_{Re_D}, Range	Coefficient Accuracy (%)[a]
Venturi				
Machined inlet	2–10 (50–250)	0.4–0.75	2×10^5 to 10^6	±1
Rough cast inlet	4–32(100–800)	0.3–0.75	2×10^5 to 10^6	±0.7
Rough-welded sheet-iron inlet	8–48 (100–1500)	0.4–0.7	2×10^5 to 10^6	±1.5
Universal venturi tube[b]	≥3 (≥75)	0.2–0.75	$>7.5 \times 10^4$	±0.5
Lo-Loss tube[b]	3–120 (75–3000)	0.35–0.85	1.25×10^5 to 3.5×10^6	±1
Nozzle				
ASME	2–16 (50–400)	0.25–0.75	10^4 to 10^7	±2.0
ISA	2–20 (50–500)	0.3–0.6	10^5 to 10^6	±0.8
		0.6–0.75	2×10^5 to 10^7	$2\beta - 0.4$
Venturi nozzle	3–20 (75–500)	0.3–0.75	2×10^5 to 2×10^6	$\pm 1.2 \pm 1.54\beta^4$
Orifice				
Corner, flange, D and $D/2$	2–36 (50–900)	0.2–0.6	10^4 to 10^7	±0.6
		0.6–0.75	10^4 to 10^7	$\pm\beta$
		0.2–0.75	2×10^3 to 10^4	$\pm 0.6 \pm \beta$
$2\frac{1}{2}D$ and $8D$ (pipe taps)	2–36 (50–900)	0.2–0.5	10^4 to 10^7	±0.8
		0.51–0.7		±1.6
Eccentric				
Flange and vena contracta	4 (100)	0.3–0.75	10^4 to 10^6	±2
	6–14 (150–350)	0.3–0.75	10^4 to 10^6	±1.5
Segmental				
Flange and vena contracta	4–14 (150–350)	0.35–0.75	10^4 to 10^6	±2
Quadrant-edged				
Flanged and corner	1–4 (25–100)	0.25–0.6	250 to 6×10^4	±2 to ± 2.5
Conical entrance				
Corner		0.1–0.3	25 to 2×10^4	±2 to ± 2.5

Source: Miller, R.W., *Flow Measurement Engineering Handbook*, 3rd edn., McGraw-Hill, New York, 1996.

[a] ISO 5167 (1980) and ASME *Fluid Meters* (1971) show slightly different values for some devices.

[b] The manufacturer should be consulted for recommendation.

Typical dimensions for a square-edged orifice are shown in Figure 10.6, and typical pressure tap locations are shown in Figure 10.7.

Note that Figure 10.6 shows the orifice hole to be beveled on the downstream side of the orifice plate. One purpose of this is to approximate the assumption that the orifice hole is in an "infinitely thin plate." The downstream bevel also removes the downstream corner of the orifice hole to ensure that it does not interfere with the fluid stream leaving the orifice. It is not unusual to encounter orifice plates installed backwards, as seems logical by intuition. However, if this occurs, erratic or unreliable pressure readings may result.

The complete Bernoulli equation, as applied between point 1 upstream of the orifice where the diameter is D and point 2 at or near the vena contracta where the fluid stream diameter is d_2, is

$$\int_{P_2}^{P_1} \frac{dP}{\rho} = \frac{1}{2}\left(\alpha_2 V_2^2 - \alpha_1 V_1^2\right) + \frac{K_f}{2} V_1^2 \tag{10.11}$$

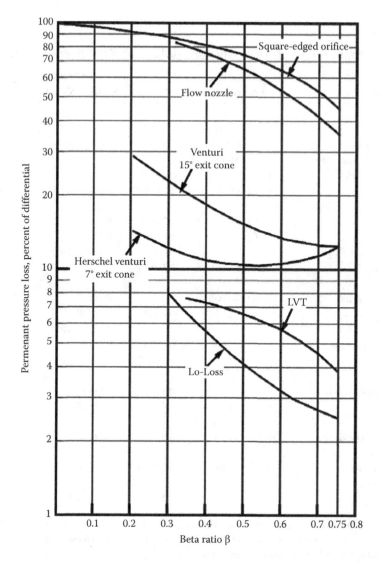

FIGURE 10.4 Unrecovered (friction) loss in various meters as a percentage of measured pressure drop versus β ratio. (From Cheremisinoff, N.P. and Cheremisinoff, P.N., *Instrumentation for Process Flow Engineering*, Technomic Publishing Company, Lancaster, PA, 1987.)

Just as for the other obstruction meters, when the continuity equation is used to eliminate the upstream velocity from Equation 10.11, the resulting expression can be solved for the mass flow rate through the orifice to give

$$\dot{m} = \frac{C_o A_o \rho_2}{\left(1-\beta^4\right)^{1/2}} \left[2\int_{P_2}^{P_1} \frac{dP}{\rho} \right]^{1/2} \tag{10.12}$$

where
  $\beta = d/D$
  C_o is the orifice discharge coefficient

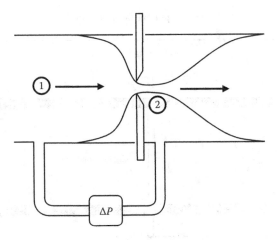

FIGURE 10.5   Orifice meter.

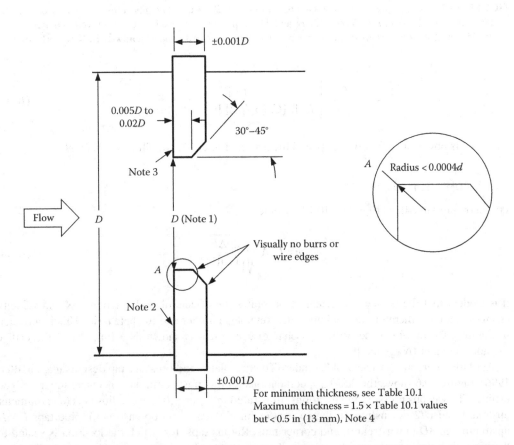

FIGURE 10.6   Concentric square-edged orifice specifications. *Notes*: (1) Mean of four diameters, no diameter >0.05% of mean diameter. (2) Maximum slope less than 1% from perpendicular: relative roughness <$10^{-4}d$ over a circle not less than $1.5d$. (3) Visually does not reflect a beam of light, finish with a fine radial cut from center outward. (4) ANSI/ASME MFC Draft 2 (July 1982), ASME, New York 1982. (From Miller, R.W., *Flow Measurement Engineering Handbook*, 3rd edn., McGraw-Hill, New York, 1996.)

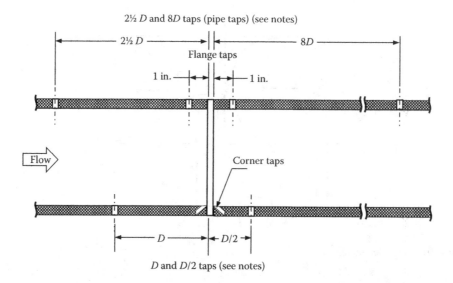

FIGURE 10.7 Orifice pressure tap locations. *Notes*: (1) 2½D and 8D pipe taps are not recommended in ISO 5167 or *ASME Fluid Meters*. (2) D and D/2 taps are now used in place of vena contracta taps. (From Miller, R.W., *Flow Measurement Engineering Handbook*, 3rd edn., McGraw-Hill, New York, 1996.)

$$C_o = \frac{C_c}{\sqrt{\alpha_2}} \left[\frac{1-\beta^4}{1-\beta^4[C_c\,(\rho_2/\rho_1)]^2[(\alpha_1 - K_f)/\alpha_2]} \right]^{1/2} \tag{10.13}$$

where C_o is obviously a function of β, and the loss coefficient K_f (which depends on N_{Re}).

A. INCOMPRESSIBLE FLOW

For incompressible flow, Equation 10.12 becomes

$$\dot{m} = C_o A_o \sqrt{\frac{2\rho\Delta P}{\left(1-\beta^4\right)}} \tag{10.14}$$

It is evident that the orifice coefficient incorporates the effects of both friction loss and velocity changes and must therefore depend upon the Reynolds number and the beta ratio. This is reflected in Figure 10.8, in which the orifice (discharge) coefficient is shown as a function of the orifice Reynolds number (N_{Re_d}) and β.

Actually, there are a variety of "standard" orifice plates and pressure tap designs (e.g., Miller, 1996). Figure 10.6 shows the ASME specifications for the most common concentric square-edged orifice. The various pressure tap locations, illustrated in Figure 10.7, are radius taps (one diameter upstream and $D/2$ downstream); flange taps (1 in. upstream and downstream); pipe taps (2½D upstream and 8D downstream); and corner taps. Radius taps, for which the location is scaled to the pipe diameter, are the most common. Corner taps and flange taps are the most convenient because they can be installed in the flange that holds the orifice plate and so do not require additional taps through the pipe wall. Pipe taps are less commonly used and essentially measure the total unrecovered pressure drop, or friction loss, through the orifice (which is usually quite a bit lower than the maximum pressure drop across the orifice plate). Vena contracta taps are sometimes specified, with the upstream tap one diameter from the plate and the downstream tap at the

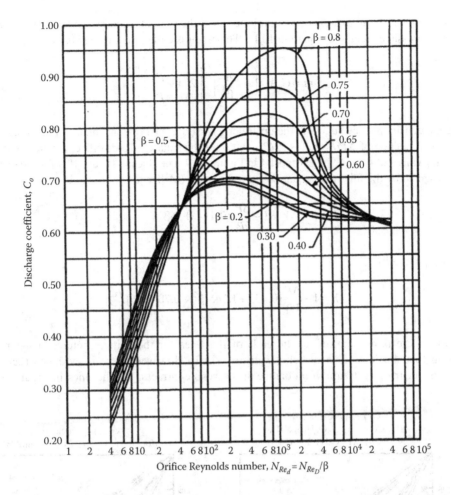

FIGURE 10.8   Orifice discharge coefficient for square-edged orifice and flange, corner, or radius taps. (From Miller, R.W., *Flow Measurement Engineering Handbook*, 3rd edn., McGraw-Hill, New York, 1996.)

vena contracta location, although the latter varies with the Reynolds number and beta ratio and thus is not a fixed position.

Empirical (measured) values of the orifice coefficient are shown in Figure 10.8. This is valid to about 2%–5% (depending upon the Reynolds number) for all pressure tap locations except pipe and vena contracta taps. More accurate values can be calculated from Equation 10.10, with the parameter expressions given in Table 10.1 for the specific orifice and pressure tap arrangement.

B.   COMPRESSIBLE FLOW

Equation 10.14 applies to incompressible fluids such as liquids or gases under small pressure changes. For an ideal gas under adiabatic conditions (i.e., $P/\rho^k = \text{const}$), Equation 10.12 gives

$$\dot{m} = C_o A_o \left(\frac{P_1 \rho_1}{1 - \beta^4} \right)^{1/2} \left\{ \frac{2k}{k-1} \left(\frac{P_2}{P_1} \right)^{2/k} \left[\left(\frac{P_1}{P_2} \right)^{(k-1)/k} - 1 \right] \right\}^{1/2} \tag{10.15}$$

It is more convenient to express this result in terms of the ratio of Equation 10.15 to the corresponding incompressible equation, Equation 10.14. The result is called the *expansion factor Y*:

$$\dot{m} = C_o A_o Y \left(\frac{P_1 \rho_1}{1-\beta^4} \right)^{1/2} \left[2 \left(1 - \frac{P_2}{P_1} \right) \right]^{1/2} \tag{10.16}$$

where the density ρ_1 is evaluated at the upstream pressure (P_1). For convenience, the values of Y are shown as a function of $\Delta P/P_1 = (1 - P_2/P_1)$ and β for the square-edged orifice, nozzles, and venturi meters for values of $k = c_p/c_v$ of 1.3 and 1.4 in Figure 10.9. The lines in Figure 10.9 for the orifice can be represented by the following equation for radius taps (Miller, 1996):

$$Y = 1 - \frac{\Delta P}{kP_1} (0.41 + 0.35\beta^4) \tag{10.17}$$

and for pipe taps by

$$Y = 1 - \frac{\Delta P}{kP_1} [0.333 + 1.145(\beta^2 + 0.7\beta^5 + 12\beta^{13})] \tag{10.18}$$

Note that there is no "button" at the ends of the lines for the orifice meter expansion factor, implying that the flow does not choke in an orifice. However, there is evidence that choked flow does occur downstream of an orifice at the vena contracta, but the lines in Figure 10.9 and

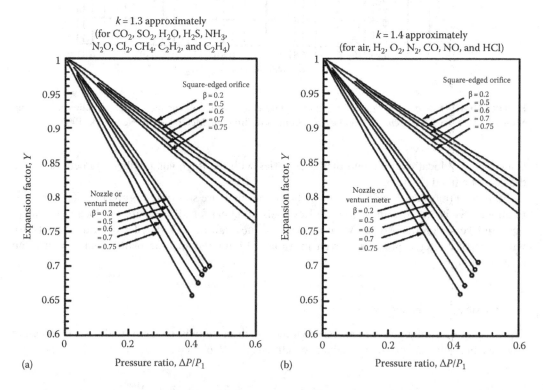

FIGURE 10.9 Expansion factor for orifice, nozzle, and venturi meter (a) $k = 1.3$, (b) $k = 1.4$. (From Crane Company, *Flow of Fluids Through Valves, Fittings, and Pipe*, Technical Manual 410, Crane Company, King of Prussia, PA, 1978.)

Equations 10.17 and 10.18 do not apply beyond the range shown in Figure 10.9. A study by Kirk (2005) and Cunningham (1951) has indicated that the expansion factor for flange and radius taps follows Equation 10.17 for $1 > P_2/P_1 > 0.63$ (or $0 < \Delta P/P_1 < 0.37$) and the following equation for $0.63 > P_2/P_1 > 0$ (or $0.37 < \Delta P/P_1 < 1.0$):

$$Y = Y_{0.63} - (0.49 + 0.45\beta^4)(0.63 - P_2/P_1)/k \tag{10.19}$$

C. FRICTION LOSS COEFFICIENT

The total friction loss in an orifice meter, after all pressure recovery has occurred, can be expressed in terms of a loss coefficient, K_f, as follows. With reference to Figure 10.10, the total friction loss is $(P_1 - P_3)$. By taking the system to be the fluid in the region from a point just upstream of the orifice plate (P_1) to a downstream position beyond the vena contracta where the stream has expanded to fill the pipe (P_3), the momentum balance becomes

$$\sum_{on\ system} F = \dot{m}(V_3 - V_1) = 0 = P_1 A_o + P_2 (A_1 - A_o) - P_3 A_1 \tag{10.20}$$

The orifice equation (Equation 10.14) can be solved for the pressure drop $(P_1 - P_2)$ to give

$$P_1 - P_2 = \frac{\rho V_o^2}{2}\left(\frac{1-\beta^4}{C_o^2}\right) \tag{10.21}$$

Eliminating P_2 from Equations 10.20 and 10.21 and solving for $(P_1 - P_3)$ provides a definition for K_f based on the pipe velocity (V_1):

$$P_1 - P_3 = \frac{\rho V_o^2(1-\beta^4)(1-\beta^2)}{2C_o^2} = \rho e_f = \frac{\rho V_1^2 K_f}{2} \tag{10.22}$$

Thus, the loss coefficient is

$$K_f \approx \frac{(1-\beta^4)(1-\beta^2)}{C_o^2\beta^4} \tag{10.23}$$

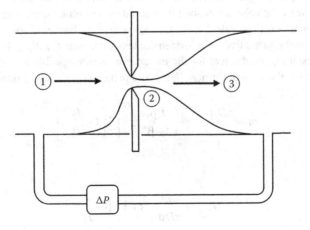

FIGURE 10.10 Friction loss in an orifice meter.

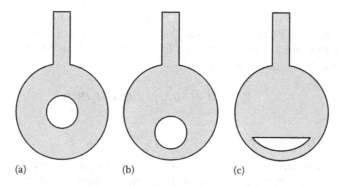

FIGURE 10.11   Other orifice geometries (a) Concentric, (b) eccentric, (c) segmental.

If the loss coefficient is based upon the velocity through the orifice (V_o) instead of the pipe velocity, the β^4 term in the denominator of Equation 10.23 doesn't appear:

$$K_f \approx \frac{(1-\beta^4)(1-\beta^2)}{C_o^2} \tag{10.24}$$

Equation 10.22 represents the net total (unrecovered) pressure drop due to friction in the orifice. This is expressed as a percentage of the maximum (orifice) pressure drop in Figure 10.4.

D.   OTHER GEOMETRIES

While the concentric orifice (hole in the center) is the most common design, eccentric and segmental orifice designs (Figure 10.11) are also used for the vapors and gases laden with small amount of contaminants (dust, condensed vapors, or other impurities heavier than the carrier such as dilute slurries, etc.). The designs shown in Figure 10.11 minimize the tendency for the deposition of impurities in front of the orifice plate. All such modifications necessitate the introduction of additional corrections to the formulae derived in this chapter. The discussion in this chapter is, however, restricted to concentric orifice plates.

V.   ORIFICE PROBLEMS

Three classes of problems involving orifices (or other obstruction meters) that the engineer might encounter are similar to the types of problems encountered in pipe flows. These are the "unknown pressure drop," "unknown flow rate," and "unknown orifice diameter" problems. Each involves relationships between the same five basic dimensionless variables: C_d, N_{Re_D}, β, $\Delta P/P_1$, and Y, where C_d represents the discharge coefficient for the meter. For incompressible liquids, this list reduces to four variables, because $Y = 1$ by definition. The basic orifice equation that relates these variables is

$$\dot{m} = \frac{\pi D^2 \beta^2 Y C_d}{4} \left(\frac{P_1 \rho_1}{1-\beta^4} \right)^{1/2} \left[2\left(1 - \frac{P_2}{P_1}\right) \right]^{1/2} \tag{10.25}$$

with

$$N_{Re_D} = \frac{4\dot{m}}{\pi D \mu} \quad \text{and} \quad \beta = \frac{d}{D} \tag{10.26}$$

and $Y = \text{fn}(\beta, \Delta P/P_1)$ (as given by Equation 10.17 or 10.18 or Figure 10.9), and $C_d = \text{fn}(\beta, N_{Re_D})$ (as given by Equation 10.10 or Figure 10.8). The procedure for solving each of these problems is as follows.

A. UNKNOWN PRESSURE DROP

In the case of an unknown pressure drop, we want to determine the pressure drop to be expected when a given fluid flows at a given rate through a given orifice meter:

Given: $\dot{m}, \mu, \rho_1, D, d (\beta = d/D), P_1$ *Find:* ΔP

The procedure is as follows:

1. Calculate N_{Re_D} and $\beta = d/D$ from Equation 10.26.
2. Get $C_d = C_o$ from Figure 10.8 or Equation 10.10.
3. Assume $Y = 1$, and solve Equation 10.25 for $(\Delta P)_1$:

$$(\Delta P)_1 = \left(\frac{4\dot{m}}{\pi D^2 \beta^2 C_o} \right)^2 \left(\frac{1 - \beta^4}{2\rho_1} \right) \tag{10.27}$$

4. Using $(\Delta P)_1/P_1$ and β, get Y from Equation 10.17 or 10.18 or Figure 10.9.
5. Calculate $\Delta P = (\Delta P)_1/Y^2$.
6. Use the value of ΔP from step 5 in step 4, and repeat steps 4–6 until there is no change.

B. UNKNOWN FLOW RATE

In the case of an unknown flow rate, the pressure drop across a given orifice is measured for a fluid with known properties, and the flow rate is to be determined.

Given: $\Delta P, P_1, D, d (\beta = d/D), \mu, \rho_1$ *Find:* $\dot{m}$

1. Using $\Delta P/P_1$ and β, get Y from Equation 10.17 or 10.18 or Figure 10.9.
2. Assume $C_o = 0.61$.
3. Calculate $\dot{m}$ from Equation 10.25.
4. Calculate N_{Re_D} from Equation 10.26.
5. Using N_{Re_D} and β, get C_o from Figure 10.8 or Equation 10.10.
6. If $C_o \neq 0.61$, use the value from step 5 in step 3, and repeat steps 3–6 until there is no change.

C. UNKNOWN DIAMETER

For design purposes, the proper size orifice (d or β) must be determined for a specified (maximum) flow rate of a given fluid in a given pipe with a ΔP device having a given (maximum) range.

Given: $\Delta P, P_1, \mu, \rho, D, \dot{m}$ *Find:* d (i.e., β)

1. Solve Equation 10.25 for β,

$$\beta = \left(\frac{X}{1+X} \right)^{1/4} \quad \text{where } X = \frac{8}{\rho_1 \Delta P} \left(\frac{\dot{m}}{\pi D^2 Y C_o} \right)^2 \tag{10.28}$$

2. Assume that $Y = 1$ and $C_o = 0.61$.

3. Calculate $N_{Re_d} = N_{Re_D}/\beta$, and get C_o from Figure 10.8 or Equation 10.10 and Y from Figure 10.9 or Equation 10.17 or 10.18.

4. Use the results of step 3 in step 1, and repeat steps 1–4 until there is no change.

The required orifice diameter is $d = \beta D$.

VI. NONINVASIVE TECHNIQUES

The discussion so far has been restricted to the so-called invasive methods (such as obstruction meters, venturi meter, etc.) of measuring local velocity and flow rate. The major disadvantage of this class of measuring method is that the flow is disturbed due to the insertion of the measuring sensor, like the pitot tube, orifice, or venturi device in the pipe. There are situations such as hostile temperature and pressure conditions, corrosive or abrasive nature of fluid, or issues with the contamination of the products such as in food, pharma, and health-care product sectors where it is not possible to use such devices. Further difficulties arise if the fluid is multiphase, such as slurries or suspensions of solids, gases, or immiscible liquids, or exhibits non-Newtonian flow characteristics (Fyrippi et al., 2004). Thus, over the years, a range of noninvasive techniques has been developed which can measure the flow rate of liquids or slurries or suspensions in pipes without disturbing the flow. Some of these are briefly described here, including their underlying principles.

A. Vortex-Shedding Flow Meter

When a fluid stream impinges on a bluff body (e.g., a cylinder) placed in its path, the boundary layers develop over the object and a wake (vortices) is formed in the rear of the bluff body. At a critical Reynolds number, these vortices begin to shed alternatively giving rise to the so-called von Karman vortex street (Figure 10.12). These vortices are shed at a well-defined frequency, which is proportional to the fluid velocity and which can be detected by oscillations of the bluff body. Vortex flow meters are quite versatile and can be used with clean gases and liquids, as described by Ginesi (1991) and Livelli (2013). However, this device cannot be used to measure low flow rates, as the vortex-shedding pattern is irregular under these conditions. The recommended minimum value of the pipe Reynolds number is of the order of ~30,000.

B. Magnetic Flow Meter

This device measures volumetric flow rate directly and its operation hinges on well-known electromagnetic principles. Electrical current is induced in a conductor (the fluid) when it moves through a magnetic field. The voltage induced is directly proportional to the width of the conductor, the intensity

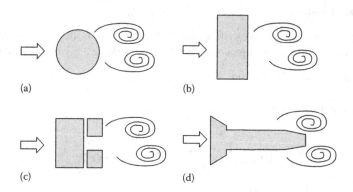

FIGURE 10.12   Schematics of vortex shedding sensors for a range of bluff body cross-sectional shapes.

of the magnetic fluid, and the velocity of the conductor. The process stream passes through the magnetic field induced by electromagnetic coils or permanent magnets built around a short length of pipe (called the metering section). Due to the magnetic field, the metering section must be made of nonmagnetic material or it must be lined with an electrically insulating material isolating it from the flowing liquid, which must be conducting. Commercial electromagnetic flow meters are now available for all sizes of pipes and have been found particularly suitable for metering flow rates in pharmaceutical and biological process engineering applications. Modern instruments can be used with conductivity levels as small as 0.5 µS/mm. These have also been shown to work reasonably well with non-Newtonian solutions and slurries (Heywood and Mehta, 1999; Fyrippi et al., 2004), with small loss in precision near the laminar-turbulent transition. This device is sensitive to the nature of the velocity profile in the pipe.

C. ULTRASONIC FLOW METER

Within this general class of flow meters, there are three types of devices based on different underlying principles, namely, time of flight method, frequency difference, and the Doppler effect. In the time of flight method, a high-frequency pressure wave is transmitted at an acute angle to the walls of the pipe which impinges on a receiver on the other side of the pipe (Figure 10.13). The elapsed time between transmission and reception depends on the velocity and velocity profile across the pipe, the speed of sound in the fluid, and the angle at which the wave is transmitted. In practice, a pair of transducers is used to exploit the fact that the velocity of an ultrasound pulse traveling against the flow direction is reduced, a and is increased when the pulse travels in the same direction. These meters are eminently suited for the flow measurements of water, clean liquids, liquefied gases, and natural gas.

The so-called frequency-difference ultrasonic flow meters employ two independent measuring paths, each having a transmitter–receiver pair (Figure 10.14). The arrival of a transmitted pulse at

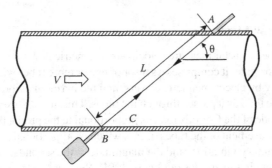

FIGURE 10.13 Ultrasonic flow meter. In the transit time meter two transducers (*A* and *B*) each act alternatively as the receiver and transmitter.

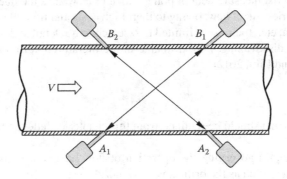

FIGURE 10.14 Frequency-difference ultrasonic flow meter.

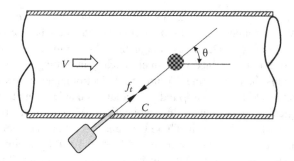

FIGURE 10.15 Doppler ultrasonic flow meter.

a receiver triggers the transmission of a further pulse thereby leading to the establishment of two distinct frequencies (in the upstream and downstream directions). The difference between these two frequencies is directly proportional to the velocity of the medium. The flow measurement is thus independent of the velocity of sound in the medium.

The Doppler ultrasonic flow meter may be used when small particles (solids, gas bubbles, droplets, or even strong eddies) are dispersed in the fluid. These are assumed to move essentially at the same velocity as the fluid medium to be metered. A continuous ultrasonic wave is transmitted at an acute angle to the wall of the pipe. The shift in frequency between the transmitted wave, which is scattered when incident on the particles, and that part of the wave which is reflected back to the receiver is measured (shown schematically in Figure 10.15). The shift in frequency between the transmitted and scattered waves is directly proportional to the flow velocity. However, the flow velocity must be much smaller than the velocity of sound in the fluid medium. This technique has gained wide spread acceptance for metering the flow of dilute slurries, dispersions, emulsions, etc.

D. THE CORIOLIS FLOW METER

The flow meters described thus far measure velocity or volumetric flow rate, whereas the Coriolis flow meter measures mass flow rate. It comprises a sensor of one or more tubes which are made to vibrate at their resonant frequency by electromagnetic drivers, and the harmonic vibrations of the tubes impart an angular motion to the fluid as it passes through the tubes. This, in turn, exerts a Coriolis force on the tube walls. The magnitude of the Coriolis force is proportional to the product of density and velocity of the fluid. A secondary movement of the tubes occurs which is proportional to the mass flow rate which superimposes on the primary vibration, and the magnitude of the secondary oscillations is measured. More details can be found in the articles of Ginesi (1991), Reizner (2004), and Livelli (2013).

In addition to the bulk flow meters, highly sophisticated techniques like laser Doppler anemometer (LDA), magnetic resonance imaging (MRI), and particle image velocimetry (PIV) are also available which allow the measurement of point velocities as well as the fluctuating components in complex flow geometries. As of now, owing to their high costs and limited range of applicability in terms of fluid medium, etc., their use is limited to research only. Detailed descriptions together with their advantages and disadvantages are available in the literature (Adrian and Westerweel, 2010; Zhang, 2010; Smits and Lim, 2012).

SUMMARY

The significant points that should be retained from this chapter include the following:

- The basic principles governing the operation of flow meters such as the pitot tube and obstruction meters such as the orifice, venturi, nozzle, etc.
- Application of the orifice meter for incompressible and compressible flows

- Determination of the unknown pressure drop, unknown flow rate, or unknown diameter for the orifice meter or other obstruction meter
- Awareness of the large variety of other non-invasive flow meters such as vortex shedding, magnetic, Coriolis, ultrasonic, etc. meters and where to find information on them

PROBLEMS

FLOW MEASUREMENT

1. An orifice meter with a hole of 1 in. diameter is inserted into a 1½ in. sch 40 line carrying SAE 10 lube oil at 70°F ($SG = 0.93$). A manometer using water as the manometer fluid is used to measure the orifice pressure drop reads 8 in. What is the flow rate of the oil in gallons per minute (gpm)?

2. An orifice with a 3 in. diameter hole is mounted in a 4 in. diameter pipeline carrying water. A manometer containing an immiscible fluid with a specific gravity (SG) of 1.2 connected across the orifice reads 0.25 in. What is the flow rate in the pipe in gpm?

3. An orifice with a 1 in. diameter hole is installed in a 2 in. sch 40 pipeline carrying SAE 10 lube oil at 100°F. The pipe section where the orifice is installed is vertical, with the flow being upward. Pipe taps that are 10.5 pipe diameters apart are used, which are connected to a manometer containing mercury to measure the pressure drop. If the manometer reading is 3 in., what is the flow rate of the oil in gpm?

4. The flow rate in a 1.5 in. ID line can vary from 100 to 1000 bbl/day, and you must install an orifice meter to measure it. If you use a differential pressure (DP) cell with a range of 10 in. H_2O to measure the pressure drop across the orifice, what size orifice should you use? After this orifice is installed, you find that the DP cell reads 0.5 in. H_2O. What is the flow rate in barrels per day (bbl/day)? The fluid in the pipe is an oil with an $SG = 0.89$ and $\mu = 1$ cP.

5. A 4 in. sch 80 pipe carries water from a storage tank on top of a hill to a plant at the bottom of the hill. The pipe is inclined at an angle of 20° to the horizontal. An orifice meter with a diameter of 1 in. is inserted in the line, and a mercury manometer across the meter reads 2 in. What is the flow rate in gpm?

6. You must size an orifice meter to measure the flow rate of gasoline ($SG = 0.72$) in a 10 in. ID pipeline at 60°F. The maximum flow rate expected is 1000 gpm, and the maximum pressure differential across the orifice is to be 10 in. of water. What size orifice should you use?

7. A 2 in. sch 40 pipe carries SAE 10 lube oil at 100°F ($SG = 0.928$). The flow rate may be as high as 55 gpm, and you must select an orifice meter to measure the flow rate.
   (a) What size orifice should be used if the pressure difference is measured using a DP cell having a full-scale range of 100 in. H_2O?
   (b) Using this size orifice, what is the flow rate of oil in gpm when the DP cell reads 50 in. H_2O?

8. A 2 in. sch 40 pipe carries a 35° API distillate at 50°F ($SG = 0.85$). The flow rate is measured by an orifice meter, which has a diameter of 1.5 in. The pressure drop across the orifice plate is measured by a water manometer connected to flange taps.
   (a) If the manometer reading is 1 in., what is the flow rate of the oil in gpm?
   (b) What would the diameter of the throat of a venturi meter be that would give the same manometer reading at this flow rate?
   (c) Determine the unrecovered pressure loss for both the orifice and the venturi in psi under these conditions.

9. An orifice having a diameter of 1 in. is used to measure the flow rate of SAE 10 lube oil (SG = 0.928) in a 2 in. sch 40 pipe at 70°F. The pressure drop across the orifice is measured by a mercury (SG = 13.6) manometer, which reads 2 cm.
    (a) Calculate the volumetric flow rate of the oil in liters/second.
    (b) What is the temperature rise of the oil as it flows through the orifice in °F? ($C_v = 0.5$ Btu/(lb$_m$ °F)
    (c) How much power (in horsepower) is required to pump the oil through the orifice? (*Note:* this is the same as the rate of energy dissipated in the flow.)

10. An orifice meter is used to measure the flow rate of CCl_4 in a 2 in. sch 40 pipe. The orifice diameter is 1.25 in., and a mercury (SG = 13.6) manometer attached to the pipe taps across the orifice reads 1/2 in. Calculate the volumetric flow rate of CCl_4 in ft³/s (SG of $CCl_4 = 1.6$). What is the permanent energy loss in the flow due to the presence of the orifice in ft lb$_f$/lb$_m$? Express this also as a total overall "unrecovered" pressure loss in psi.

11. An orifice meter is installed in a 6 in. ID pipeline that is inclined upward at an angle of 10° from the horizontal. Benzene is flowing in the pipeline at the flow rate of 10 gpm. The orifice diameter is 3.5 in., and the orifice pressure taps are 9 in. apart.
    (a) What is the pressure drop between the pressure taps in psi?
    (b) What would be the reading of a water manometer connected to the pressure taps?

12. You are to specify an orifice meter for measuring the flow rate of a 35° API distillate (SG = 0.85) flowing in a 2 in. sch 160 pipe at 70°F. The maximum flow rate expected is 2000 gph, and the available instrumentation for a differential pressure measurement has a limit of 2 psi. What size hole should the orifice plate have?

13. You must select an orifice meter for measuring the flow rate of an organic liquid (SG = 0.8, μ = 15 cP) in a 4 in. sch 40 pipe. The maximum flow rate anticipated is 200 gpm, and the orifice pressure difference is to be measured with a mercury (SG = 13.6) manometer having a maximum reading range of 10 in. What size should the orifice be?

14. An oil with an SG of 0.9 and viscosity of 30 cP is transported in a 12 in. sch. 20 pipeline at a maximum flow rate of 1000 gpm. What size orifice should be used to measure the oil flow rate if a DP cell with a full-scale range of 10 in. H_2O is used to measure the pressure drop across the orifice? What size venturi would you use in place of the orifice in the pipeline, everything else being the same?

15. You want to use a venturi meter to measure the flow rate of water, up to 1000 gpm, through an 8 in. sch 40 pipeline. To measure the pressure drop in the venturi, you have a DP cell with a maximum range of 15 in. H_2O pressure difference. What size venturi (i.e., throat diameter) should you specify?

16. Gasoline is pumped through a 2 in. sch 40 pipeline upward into an elevated storage tank at 60°F. An orifice meter is mounted in a vertical section of the line, which uses a DP cell with a maximum range of 10 in. H_2O to measure the pressure drop across the orifice at radius taps. If the maximum flow rate expected in the line is 10 gpm, what size orifice should you use? If a water manometer with a maximum reading of 10 in. is used instead of the DP cell, what would the required orifice diameter be?

17. You have been asked by your boss to select a flow meter to measure the flow rate of gasoline (SG = 0.85) at 70°F in a 3 in. sch 40 pipeline. The maximum expected flow rate is 200 gpm, and you have a DP cell (which measures differential pressure) with a range of 0–10 in. H_2O available.
    (a) If you use a venturi meter, what should the diameter of the throat be?
    (b) If you use an orifice meter, what diameter orifice should you use?

(c) For a venturi meter with a throat diameter of 2.5 in., what would the DP cell read (in inches of water) for a flow rate of 150 gpm

(d) For an orifice meter with a diameter of 2.5 in., what would the DP cell read (in inches of water) for a flow rate of 150 gpm.

(e) How much power (in hp) is consumed by friction loss in each of the meters under the conditions of (c) and (d)?

18. A 2 in. sch 40 pipe is carrying water at a flow rate of 8 gpm. The flow rate is measured by means of an orifice with a 1.6 in. diameter hole. The pressure drop across the orifice is measured using a manometer containing an oil with an SG of 1.3.
    (a) What is the manometer reading in inches?
    (b) What is the power (in hp) consumed as a consequence of the friction loss due to the orifice plate in the fluid?

19. The flow rate of CO_2 in a 6 in. ID pipeline is measured by an orifice meter with a diameter of 5 in. The pressure upstream of the orifice is 10 psig, and the pressure drop across the orifice is 30 in. H_2O. If the temperature is 80°F, what is the mass flow rate of CO_2?

20. An orifice meter is installed in a vertical section of a piping system, in which SAE 10 lube oil is flowing upward (at 100°F). The pipe is 2 in. sch 40, and the orifice diameter is 1 in. The pressure drop across the orifice is measured by a manometer containing mercury as the manometer fluid. The pressure taps are pipe taps (2½ ID upstream and 8 ID downstream), and the manometer reading is 3 in. What is the flow rate of the oil in the pipe in gpm?

21. You must install an orifice meter in a pipeline to measure the flow rate of 35.6° API crude oil at 80°F. The pipeline diameter is 18 in. sch 40, and the maximum expected flow rate is 300 gpm. If the pressure drop across the orifice is limited to 30 in. H_2O or less, what size orifice should be installed? What is the maximum permanent pressure loss that would be expected through this orifice in psi?

22. You are to specify an orifice meter for measuring the flow rate of a 35° API distillate (SG = 0.85) flowing in a 2 in. sch 160 pipe at 70°F. The maximum flow rate expected is 2000 gph (gal/h), and the available instrumentation for the differential pressure measurement has a limit of 2 psi. What size orifice should be installed?

23. A 6 in. sch 40 pipeline is designed to carry SAE 30 lube oil at 80°F (SG = 0.87) at a maximum velocity of 10 ft/s. You must install an orifice meter in the line to measure the oil flow rate. If the maximum pressure drop to be permitted across the orifice is 40 in. H_2O, what size orifice should be used? If a venturi meter is used instead of an orifice, everything else being the same, how large should the venturi throat be?

24. An orifice meter with a diameter of 3 in. is mounted in a 4 in. sch 40 pipeline carrying an oil with a viscosity of 30 cP and an SG of 0.85. A mercury manometer attached to the orifice meter reads 1 in. If the pumping stations along the pipeline operate with a suction (inlet) pressure of 10 psig and a discharge (outlet) pressure of 160 psig, how far apart should the pump stations be, if the pipeline is horizontal?

25. A 35° API oil at 50°F is transported in a 2 in. sch 40 pipeline. The oil flow rate is measured by an orifice meter that is 1.5 in. in diameter, using a water manometer.
    (a) If the manometer reading is 1 in., what is the oil flow rate (in gpm)?
    (b) If a venturi meter is used instead of the orifice meter, what should the diameter of the venturi throat be to give the same reading as the orifice meter at the same flow rate?
    (c) Determine the unrecovered pressure loss for both the orifice and venturi meters.

26. A 6 in. sch 40 pipeline carries a petroleum fraction (viscosity 15 cP, SG 0.85) at a velocity of 7.5 ft/s, from a storage tank at 1 atm pressure to a plant site. The line contains 1500 ft of straight pipe, 25 90° flanged elbows, and four open globe valves. The oil level in the storage tank is 15 ft above ground, and the pipeline discharge is at a point 10 ft above ground, at a pressure of 10 psig.
    (a) What is the required flow capacity (in gpm) and the head (pressure) to be specified for the pump needed to move the oil?
    (b) If the pump is 85% efficient, what horsepower motor is required to drive it?
    (c) If a 4 in. diameter orifice is inserted in the line to measure the flow rate, what would the pressure drop reading across it be at the specified flow rate?

27. Water drains by gravity out of the bottom of a large tank, through a horizontal 1 cm ID tube, 5 m long that has a venturi meter mounted in the middle of the tube. The level in the tank is 4 ft above the tube, and a single open vertical tube is attached to the throat of the venturi. What is the smallest diameter of the venturi throat for which no air will be sucked through the tube attached to the throat? What is the flow rate of the water under this condition?

28. Natural gas (CH_4) is flowing in a 6 in. sch 40 pipeline at 50 psig and 80°F. A 3 in. diameter orifice is installed in the line, which reads a pressure drop of 20 in. H_2O. What is the gas flow rate, in lb_m/h and scfm?

29. A solvent (SG = 0.9, μ = 0.8 cP) is transferred from a storage tank to a process unit through a 3 in. sch 40 pipeline that is 2000 ft long. The line contains 12 elbows, four globe valves, an orifice meter with a diameter of 2.85 in., and a pump having the characteristics shown in Figure 8.2 with a 7¼ in. impeller. The pressures in the storage tank and the process unit are both 1 atm, and the process unit is 60 ft higher than the storage tank. What is the pressure reading across the orifice meter in in. H_2O?

NOTATION

A	Cross-sectional area, $[L^2]$
C_d	Discharge coefficient for a flow meter, [—]
C_o	Orifice coefficient, [—]
D	Pipe diameter, [L]
d	Orifice, nozzle, or venturi throat diameter, [L]
F	Force, $[F = ML/t^2]$
k	Isentropic coefficient, c_p/c_v for ideal gas, [—]
K_f	Loss coefficient, [—]
$\dot{m}$	Mass flow rate, [M/t]
N_{Re}	Reynolds number, [—]
P	Pressure, $[F/L^2 = M/Lt^2]$
Q	Volumetric flow rate, $[L^3/t]$
R	Radius, [L]
r	Radial position, [L]
SG	Specific gravity, [—]
V	Spatial average velocity, [L/t]
v	Local velocity, [L/t]
Y	Expansion factor, [—]

GREEK

α	Kinetic energy correction factor, [—]
β	d/D, [—]
$\Delta(\ )$	$(\ )_2 - (\ )_1$

ρ Density, $[M/L^3]$
v Kinematic viscosity, $[L^2/t]$ also Specific volume, $[ft^3/lb_m]$

SUBSCRIPTS

1   Reference point 1
2   Reference point 2
d   Of orifice, nozzle, or venturi throat
D   Of pipe
o   Orifice
scfh   Standard cubic feet per hour
v   Venturi, viscosity correction, vapor pressure

REFERENCES

Adrian, R.J. and J. Westerweel, *Particle Image Velocimetry*, Cambridge University Press, New York, 2010.

Baker, R.C., *Flow Measurement Handbook*, Cambridge University Press, Cambridge, U.K., 2009.

Cheremisinoff, N.P. and P.N. Cheremisinoff, *Instrumentation for Process Flow Engineering*, Technomic Publishing Company, Lancaster, PA, 1987.

Crane Company, *Flow of Fluids Through Valves, Fittings, and Pipes*, Technical Manual 410, Crane Company, King of Prussia, PA, 1978.

Cunningham, R., Orifice meters with supercritical compressible flow, *Trans. ASME*, 73, 625–633, July 1951.

Fyrippi, I.O. and M.P. Escudier, Flowmetering of non-Newtonian liquids, *Flow Meas. Instrum.*, 15, 131–138, 2004.

Ginesi, D., A raft of flowmeters on tap, *Chem. Eng.*, 98, 147–155, May 1991.

Heywood, N.I. and K.B. Mehta, Performance evaluation of eight electromagnetic flow meters with fine sand slurries. *Proceedings of the Hydrotransport 14*, Maastricht, the Netherlands, 1999, pp. 413–431.

Kirk, D., Tech memo—Y factor.doc, Dennis Kirk Engineering, Mount Lawley, WA, 2005.

Livelli, G., Selecting flow meters to minimize energy costs, *Chem. Eng. Prog.*, 109, 34–39, May 2013.

Miller, R.W., *Flow Measurement Engineering Handbook*, 3rd edn., McGraw-Hill, New York, 1996.

Reizner, J.R., Exposing Coriolis mass flowmeters' dirty little secret, *Chem. Eng. Prog.*, 100, 24–30, March 2004.

Smits, A.J. and T.T. Lim, *Flow Visualization: Techniques and Examples*, 2nd edn., Imperial College Press, London, U.K., 2012.

White, F.M., *Fluid Mechanics*, 3rd edn., McGraw-Hill, New York, 1994.

Zhang, Z., *LDA Application Methods: Laser Doppler Anemometry for Fluid Dynamics*, Springer, New York, 2010.

11 Safety Relief and Control Valves

"If I have ever made any valuable discoveries, it has been owing more to patient attention, than to any other talent."

—Isaac Newton, 1643–1727, Scientist

I. SAFETY RELIEF VALVES

A. BACKGROUND

It is imperative that all pressure vessels be equipped with a relief device to prevent the pressure from exceeding the maximum allowable working pressure (MAWP) of the vessel, with the consequent rupture of the vessel and loss of containment of the contents. A typical installation is illustrated in Figure 11.1. Typical "worst case" scenarios that could result in excessive pressure buildup include runaway reactions, blockage of discharge lines, external fires, etc. A safety relief valve (SRV) is designed to open at a preset "set pressure" and close when the pressure drops below that pressure by a set amount ("blowdown") in order to minimize the loss of containment. A rupture disk is sometimes used to relieve the vessel pressure, but when it ruptures virtually all of the vessel contents is lost. More detailed discussion of the selection, design, and operation of safety relief and control valves can be found in several books, e.g., CCPS, AIChE (1998), Cheremisinoff and Cheremisinoff (1987), Liptak (2006), etc.

The design of a relief system begins with an energy balance on the system (i.e., the fluid in the vessel) to determine the rate of energy input from the source (e.g., a runaway reaction, blocked line, external fire, etc.). Then the relief valve must be sized so that the mass flow rate through the valve releases this energy at a rate which is equal to or greater than it is input to the vessel from the source, at the desired pressure. The valve relief pressure is normally set at 10% above the MAWP (Maximum Allowed Working Pressure) of the vessel (depending on the energy source), and the valve must be sized for the required mass flow at this pressure. Thus, it is assumed that this determination of the required mass flow rate has been done prior to calculating the required valve size.

B. VALVE SIZING

A typical spring-loaded pop action SRV is shown in Figure 11.2a. The spring determines the pressure at which the valve opens and the disk lifts. The relieving fluid flows through the nozzle and out of the valve through discharge piping. The nozzle provides the smallest flow area and thus controls the flow rate.

The determination of the proper size of the relief valve (nozzle or orifice) required to properly discharge the contents of a pressure vessel at a mass flow rate sufficient to prevent excessive pressure buildup in the vessel requires the following three types of information:

- A model for the flow configuration in the valve (the flow model).
- A model for the properties of the fluid in the valve (a fluid property model).
- Data for the flow capacity of the actual valve with the type of fluid of interest (valve flow data, i.e. the valve discharge coefficient).

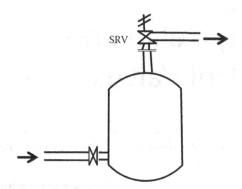

FIGURE 11.1 Typical safety relief valve installation.

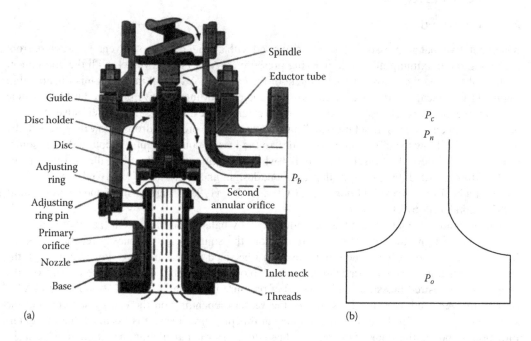

FIGURE 11.2 (a) A typical spring-loaded safety valve and (b) ideal nozzle geometry. (a: Courtesy of Consolidated Valve Co.)

1. Flow Model

The reference geometrical configuration for the flow model for relief valve flow analysis is the isentropic nozzle, an idealized representation of which is shown in Figure 11.2b.

The basic equation for the mass flux through an isentropic nozzle follows directly from the frictionless differential Bernoulli equation:

$$u\,du = -v\,dP = -\frac{dP}{\rho} \tag{11.1}$$

Integration of this equation for the velocity u from the stagnation state in the vessel ($P = P_o$, $u = 0$) to the nozzle exit (P_n), and expressing the result in terms of the mass flux through the nozzle, $G_o = \rho u = \dot{m}/A$, gives

$$G_o = \rho_n \left(-2 \int_{P_o}^{P_n} \frac{dP}{\rho} \right)^{1/2} \tag{11.2}$$

Once the mass flux is known, the nozzle or "orifice" area is determined from the known mass flow rate, that is, $A = \dot{m}/G$. If this area falls between two standard-size available orifices, the larger of the two is selected. The frictional pressure drop in the entrance line from the vessel to the valve is limited by the API recommendations to 3% of the valve set pressure or less, so if this is satisfied this pressure drop is commonly neglected in practice.

For adiabatic flow, Equation 11.1 can also be written as

$$u\,du = -dH \tag{11.3}$$

Integrating this expression across the nozzle gives

$$G_o = \rho_n \sqrt{2\left(H_n - H_o\right)} \tag{11.4}$$

where H is the specific enthalpy (per unit mass) of the fluid. Equation 11.4 is more general than Equation 11.2 in that Equation 11.4 is not restricted to isentropic processes as is the case for Equation 11.2. However, the mass flux calculated by Equation 11.4 is very sensitive to slight variations in the values of the enthalpy (because of the large conversion factor from thermal to mechanical energy units (e.g., 778 ft lb$_f$/Btu). This requires highly accurate enthalpy values for determining reliable mass flux values. Such accuracy is not widely available, so that Equation 11.2 is preferred in general.

2. Fluid Property Model

The fluid property model is the relation between the fluid density (single- or two-phase) and pressure, that is, $\rho(P)$, which is necessary to evaluate the integral in Equation 11.2. This can be the actual measured or calculated density or values estimated from an appropriate equation of state (EOS). As with any thermodynamic property, the density is a function of two independent parameters such as $\rho(P, S)$ or $\rho(P, H)$ or $\rho(P, T)$. For an ideal nozzle, the fluid follows an isentropic path so the density should be expressed simply as $\rho(P)_S$, or density as a function of pressure at constant entropy. In some simple cases, such as the flow of an incompressible fluid or the flow of an ideal gas, the density can be determined accurately. In more complex systems, such as two-phase flashing flow, the density versus pressure may be determined from data or an EOS model and tabulated in a thermodynamic property database (e.g. steam tables, NIST (2015), etc.).

3. Flow Data

The valve flow data are represented by a discharge coefficient, defined as

$$K_D = \frac{G_n}{G_o} \tag{11.5}$$

where
  G_n is the measured mass flux for a test fluid through a representative valve of a particular type, for a specific fluid, under specific conditions
  G_o is the theoretically predicted mass flux from the isentropic nozzle and fluid property models

Thus, the value of K_D accounts for any discrepancies between the isentropic nozzle and fluid property models, and the flow in the actual valve. It is therefore a correction factor that multiplies the theoretical isentropic model mass flux to give the actual mass flux. If the fluid property model is accurate, K_D should be a function only of the valve geometry and flow conditions (e.g., turbulence level, choking, etc., similar to the dependence of the pipe friction factor or fitting loss coefficients on geometry and turbulence level) and independent of fluid properties. It sometimes happens that a reference fluid is used to calculate the mass flux, in which case the discharge coefficient must also reflect the difference between the properties of the reference fluid and the actual fluid. For a given valve, the values of K_D for gas flows (K_{DG}) and for liquid flows (K_{DL}) are significantly different, which will be explained later.

There are exact solutions for the mass flux for liquids and ideal gases, as shown below. There are also many "models" for nonideal and two-phase fluids, all of which assume that the fluid is *homogeneous*, that is, each of the two phases is highly turbulent and sufficiently well mixed that they can be described as a "pseudo-single-phase" fluid, with a suitable average density. The following gives the most significant of these results (according to the judgment of the authors).

It is emphasized that G_o in the following equations is the theoretical calculated value for the isentropic nozzle and must be corrected by applying the empirical K_D (e.g., K_{DL} for liquids or K_{DG} for gases). The significance of K_D is discussed later.

C. FLUID MODELS

1. Incompressible Fluids

Most liquids under ordinary conditions can be considered "incompressible" or isochoric, with a constant density. For these fluids, evaluation of the integral in Equation 11.2 is direct and gives

$$G_o = \sqrt{2\rho\left(P_o - P_n\right)} \tag{11.6}$$

As the pressure at the nozzle (P_n) is not normally known, it is assumed to be the same as that at the exit of the valve, or the "backpressure," under the assumption that the pressure drop in the valve body is negligible (more on this later). The above equation assumes fully turbulent flow (high Reynolds number), where the effect of the fluid viscosity is negligible.

If the fluid is sufficiently viscous that the Reynolds number through the nozzle is about 10,000 or less, a viscosity "correction factor," K_v, is required. A curve for this correction factor is presented in API 520-1, which was actually adapted from a study by Stiles (1964) on control valves. More recently, Darby and Molavi (1997) presented a CFD (Computational Fluid Dynamics) solution for a viscous fluid in a typical valve geometry to determine the viscosity correction factor. The results are shown in Figure 11.3, along with the API curve.

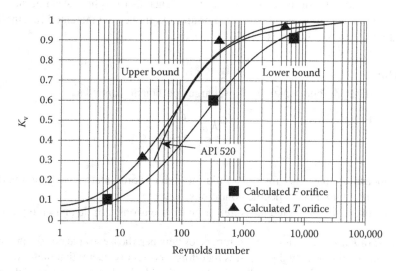

FIGURE 11.3 Viscosity correction factor. (From Darby and Molavi, 1997.)

The equation that represents the curves shown on the plot is

$$K_v = 0.975 \sqrt{\frac{\beta^{0.1}}{\left[950(1-\beta)^{1.4} / N_{Re}\right] + 0.9}} \tag{11.7}$$

where $\beta = D_n/D_p$ is the ratio of the nozzle diameter to the inlet pipe diameter, and the Reynolds number is based on the nozzle throat diameter.

The calculations were performed for two orifice sizes, F and T, with β values of 0.422 and 0.757, respectively.

2. Ideal Gases

Many gases can be described by the ideal gas law, that is, $\rho = PM/RT$, over a fairly wide range of conditions, which under isentropic conditions is $P/\rho^k = $ constant, where k is the "isentropic exponent" (for an ideal gas, $k = \gamma$ is the specific heat ratio, c_p/c_v). This model is quite accurate for most gases sufficiently far removed from the thermodynamic critical point (such as air or nitrogen under reasonable conditions). The value of k for air under standard conditions is about 1.4, and it varies only slightly with temperature and pressure. These relations can be used to evaluate the integral in Equation 11.2 to give

$$G_o = \sqrt{P_o \rho_o} \left[\left(\frac{2k}{k-1}\right) \left(\frac{P_n}{P_o}\right)^{2/k} \left\{ 1 - \left(\frac{P_n}{P_o}\right)^{(k-1)/k} \right\} \right]^{1/2} \tag{11.8}$$

It can be seen from this equation that as the nozzle pressure, P_n, decreases, the mass flux increases until it reaches a maximum value at

$$G_c = \sqrt{k P_o \rho_o} \left(\frac{2}{k+1}\right)^{(k+1)/2(k-1)} \tag{11.9}$$

At this point, the flow is choked and is independent of the nozzle downstream pressure (P_n). If no maximum is reached before P_n drops to the valve exit pressure, then the flow is not choked. The criterion for choked flow is

$$P_n \le P_c = P_o \left[\frac{2}{k+1}\right]^{k/(k-1)} \tag{11.10}$$

For real gases (pure or mixtures), k is a function of pressure as well as temperature and may be evaluated from an appropriate EOS, or from a thermodynamic database as described below for two-phase flows. For a nonideal gas, it is often adequate to use Equations 11.9 and 11.10, with an average value of k over the range of T, P of interest.

Example 11.1

A relief valve must be selected to discharge a vessel at a rate of 75 lb_m/s when fully open from a vessel for which the maximum allowed pressure is 150 psia. The fluid is a vapor with a molecular weight of 45, which can be considered an ideal gas, with a specific heat ratio of 1.3, a temperature of 70°F, and it is discharged into the atmosphere. Determine the diameter of the nozzle in the valve that should be selected (*Note*: if the calculated nozzle area does not match any of the standard sizes available, the next larger size is selected. An additional 10% is also usually added to the calculated nozzle area as a "safety factor.")

Solution:

For an ideal gas, the theoretical expression Equation 11.8 applies:

$$G_o = \sqrt{P_o \rho_o} \left[\left(\frac{2k}{k-1} \right) \left(\frac{P_n}{P_o} \right)^{2/k} \left\{ 1 - \left(\frac{P_n}{P_o} \right)^{(k-1)/k} \right\} \right]^{1/2}$$

The expression for the exit pressure at which the flow is choked is given by Equation 11.10:

$$P_c = P_o \left[\frac{2}{k+1} \right]^{k/(k-1)} = 150 \text{ psia} \left[\frac{2}{1.3+1} \right]^{1.3/(1.3-1)} = 81.86 \text{ psia}$$

Thus, because $P_c > P_b$, the flow is choked and $P_n = 81.86$ psia. Thus, Equation 11.9 applies:

$$G_c = \sqrt{kP_o \rho_o} \left(\frac{2}{k+1} \right)^{(k+1)/2(k-1)}$$

$$= \sqrt{1.3(150 \text{ lb}_f/\text{in.}^2)1.19 \left(\text{lb}_m/\text{ft}^3 \right)} \left(\frac{2}{1.3+1} \right)^{(1.3+1)/2(1.3-1)} \left[12 \text{ in./ft} \sqrt{32.17 \text{ ft lb}_m/\text{lb}_f \text{ s}^2} \right] = 606 \text{ lb}_m/\text{s ft}^2$$

Assume $K_G = 0.975$, and applying the 10% safety factor:

$$A = \dot{m}/K_G G_c(0.9) = (75 \text{ lb}_m/\text{s})/\left[(0.975)(606 \text{ lb}_m/\text{s ft}^2)(0.9) \right] = 0.141 \text{ ft}^2 = 20.3 \text{ in.}^2$$

$$d_o = \sqrt{4A/\pi} = \sqrt{4(20.3 \text{ in.}^2)/\pi} = 5.09 \text{ in.}$$

3. The Homogeneous Direct Integration Method for Any Single- or Two-Phase Flow

There are a large number of "models" for two-phase flows, both "frozen" and "flashing," for which the two phases are assumed to be in thermodynamic equilibrium. "Frozen" flows involve nonvolatile liquids and noncondensable gases (e.g., air and water) with constant quality (i.e., mass fraction of gas, x), whereas flashing flows involve a volatile liquid and a change of quality as the liquid flashes. Most of these models require a knowledge of the physical/thermodynamic properties sufficient to evaluate the function $\rho = fn(P)$ for at least one or more conditions (e.g., P, T) within the nozzle. If this information is available, it is easier and much more general and accurate to evaluate this function at a series of points from P_o to P_n (or to the choke pressure, P_c) and to use a finite difference procedure to evaluate the integral in Equation 11.2, such as:

$$G_o = \rho_n \left[-4 \sum_{P_o}^{P_n} \left(\frac{P_{i+1} - P_i}{\rho_{i+1} + \rho_i} \right) \right]^{1/2} \tag{11.11}$$

Here, the density as a function of pressure is evaluated over an isentropic path from P_o to P_n using an appropriate physical property database. This method is referred to as the homogeneous direct integration (HDI) method (Darby et al., 2002; Darby, 2004).

The mass flux will increase as P_n is decreased until a maximum in G_o (i.e., G_c) is reached, at which point the flow is choked and $P_n = P_c$. If P_n is reached before the mass flux reaches a maximum, then the flow is not choked. The procedure is to first select an appropriate pressure increment (1% of the difference between P_o and P_b is usually more than adequate). Then, utilizing the thermodynamic property data for the fluid (or mixture), determine the density of the fluid (single- or two-phase), and hence the mass flux, G_o, at successive increments of P_n over a constant entropy path, starting with $P_n = P_o$. As P_n is decreased, G_o will increase until it reaches a maximum, at which point the flow is choked at pressure P_c, which is the lowest pressure that can exist inside the nozzle exit. If no maximum is reached before $P_n = P_b$ (the back pressure on the valve), then the flow is not choked. As long as a suitable property database is available (and there are many), this procedure works well for any fluid (single- or two-phase, frozen or flashing, single- or multicomponent, etc.) under equilibrium conditions.

The density of the two-phase mixture is related to the density of each phase and the quality, or mass fraction of the vapor/gas phase, (x), by

$$\rho = \alpha \rho_G + (1 - \alpha) \rho_L \tag{11.12}$$

where

$$\alpha = \frac{x}{x + S(1-x)\rho_G/\rho_L} \tag{11.13}$$

is the volume fraction of vapor/gas in the flowing mixture and S is the "slip factor." The slip factor is the ratio of the velocity of the vapor/gas to that of the liquid at any point. For homogeneous flow, $S = 1$. The quality, x, is related to the specific volumes (v) of the two phases by

$$x = \frac{v - v_L}{v_G - v_L} = \frac{v - v_L}{v_{GL}} = \frac{v - v_s}{v_{GLs}} \tag{11.14}$$

where the subscripts represent
   L = liquid
   G = gas/vapor
   GL = gas − liquid
   s = saturation

(Note that for an initially subcooled liquid, $v < v_s$ so that x is negative, which should be interpreted as zero.)

This procedure utilizes a database (such as steam tables, or one of many computerized property databases (such as REFPROP, NIST (2015), etc.). For these fluids, this method gives accurate results over all possible phase regions from pure liquid, subcooled flashing, saturated flashing, two-phase flashing, or pure vapor, either near or far from the thermodynamic critical point, with no modification of the method or equation. For multicomponent mixtures, an isentropic flash calculation is appropriate, as the basic flow model is the isentropic nozzle. However, an adiabatic flash calculation at each pressure increment can also be used to determine the liquid and vapor compositions, phase ratio or quality (x), and the two-phase densities.

The error introduced by using Equation 11.11 to approximate Equation 11.2 is normally negligible for pressure increments as large as 10% of the total pressure difference (i.e., 10 or more pressure increments). The accuracy of the method is limited solely by the accuracy of the property data used to determine the two-phase $\rho(P)$ function. This procedure is known as the HDI method.

4.   Nonequilibrium (Flashing) Flows

For flashing liquids, thermodynamic equilibrium is not achieved unless the residence time (e.g., the flow path) is long enough, for example, a nozzle length greater than about 10 cm (Henry and Fauske, 1971). This is because flashing is a rate process involving nucleation and heat and mass transfer from the liquid to the vapor, requiring a finite amount of time ("boiling delay") to reach equilibrium (e.g., a few milliseconds). The residence time in a typical nozzle less than about 10 cm long near sonic velocity is insufficient to achieve equilibrium. In this case, the actual mass fraction of the vapor (or quality, x) will be less than it would be under equilibrium conditions. Assuming that nonequilibrium conditions exist for nozzles less than 10 cm long (see Henry and Fauske, 1971), nonequilibrium can be accounted for by replacing the quality x in Equations 11.12 through 11.14 by the nonequilibrium quality:

$$x = x_o + \left(x_e - x_o\right)L/10 \qquad (11.15)$$

where
   x_o is the initial (entering) quality
   x_e is the local quality assuming thermodynamic equilibrium
   L is the nozzle length, in cm ($L \leq 10$ cm)

If the entering fluid is a subcooled liquid, the initial quality is zero until the pressure in the nozzle drops to the saturation pressure. An "effective quality" can be calculated from Equation 11.14. This gives a negative value for x at pressures above the saturation pressure, in which case the quality should be set to zero. Further drop in pressure results in a corresponding increase in quality. However, for significant subcooling, the flow will likely be all liquid and may flash only at or near the nozzle exit, at which point it will be choked if ($P_n = P_c > P_b$). If the actual nozzle is stepped or tapered, the length of the minimum diameter straight section at the end should be used in Equation 11.15, as this is the geometrically controlling area for pressure drop and flow (Darby, 2004, 2005, 2010). The validity of this method has been verified by comparison with data on steam-water flashing flows and air-water frozen flows (Lenzing et al., 1997, 1998; Darby, 2004) in actual relief valves. If the entering fluid is a two-phase mixture with quality (x_o) greater than about 0.05, the mixture will be in equilibrium in the nozzle.

D. THE DISCHARGE COEFFICIENT

The discharge coefficient K_D (i.e., K_{DG} or K_{DL}) accounts for any deviation of the actual flow in the SRV from that of the theoretical ideal nozzle prediction. If the fluid property model is exact, then the value of the dimensionless discharge coefficient should be a function only of the valve geometry and flow regime (i.e., laminar or turbulent, choked or non-choked, etc.), and independent of fluid properties. The values of K_D are measured under highly controlled test conditions for each type of valve, using air (or nitrogen) for K_{DG}, or water for K_{DL}. The test conditions for K_D result in fully turbulent flow, so that deviations from this condition (e.g., nozzle Reynolds numbers below about 1000) must be accounted for by including the viscosity correction factor K_v (see Figure 11.3).

The measured values of K_{DG} for gases are generally on the order of 0.975, whereas those for K_{DL} for liquids are much smaller, on the order of 0.6–0.7. This difference can be readily understood when it is recognized that the test conditions for gases are invariably conducted under choked flow conditions, whereas liquid flows do not choke (unless the pressure drops below the vapor pressure, in which case flashing will occur with the likely result of choking). For gases, choking occurs at the nozzle discharge, so that nothing downstream of this point (i.e. in the valve body) can affect the flow. Inspection of Figure 11.1 shows clearly that the nozzle geometry in many valves closely approximates the ideal nozzle geometry, whereas for non-choked flow the entire internal geometry of the valve influences the flow. This flow path deviates much more from the ideal nozzle geometry flow path and accounts for the difference between K_{DG} and K_{DL}. The result is that for choked flow (whether in gas–vapor or two-phase service) the K_{DG} value should be used, whereas for non-choked flow (whether in gas–vapor or two-phase service) the value of K_{DL} should be used. This has been confirmed by comparison with data in the literature (Lenzing et al., 1997, 1998) by Darby (2004) for both frozen (air-water) and flashing (steam-water) in non-equilibrium flow in several different valves.

Example 11.2

An SRV must be sized to relieve hot water from a water heater at a rate of 20 lb_m/s. If the water is at 220°F and 45 psia, what nozzle area in the valve would be required? For the valve, $K_{DL} = 0.65$ and $K_{DG} = 0.98$.

Solution:

Assuming the valve discharges into the atmosphere at 14.7 psia, the water (at 220°) will reach its boiling point before leaving the nozzle and will flash in the nozzle. Therefore, we will have a case of all liquid entering the nozzle and a 2-phase liquid/vapor mixture leaving the nozzle. For the solution, we will utilize the HDI method (Equation 11.11), which requires data for the density of the fluid (both water and the water/water vapor mixture) at constant entropy over the range from $P_o = 45$ psia to $P_n = 14.7$ psia. Such data are available in the steam tables, which can be found in any number of computerized databases. We shall use the NIST Property Database 23, which is easy to use.

The method employs the application of Equation 11.11:

$$G_o = \rho_n \left[-4 \sum_{P_o}^{P_n} \left(\frac{P_{i+1} - P_i}{\rho_{i+1} + \rho_i} \right) \right]^{1/2}$$

to compute the mass flux, assuming local equilibrium (i.e., a nozzle length over 10 cm long). The procedure is as follows (this is most easily done on a spreadsheet):

1. Use the property database to determine the entropy at the nozzle entrance ($P_o = 45$ psia, $T_o = 220°F$, entropy, $s = 0.32433$ Btu/lb_m).
2. Select an appropriate pressure interval (e.g., 1 psia), and use the property database to determine the fluid density (single- or two-phase mixture) at each pressure step, keeping the entropy constant at each step.
3. Decrease P_n in steps and use the database to determine the density at each step.
4. Use the formula from Equation 11.11 to calculate the value of G_o at each pressure step, starting with $P_n = P_o$.
5. As P_n is decreased, G_o will increase. If the flow is choked, G_o will reach a maximum at $P_n > P_b = P_c$, and $G_o = G_c$. This is the maximum mass flux that can be obtained, regardless of the downstream exit pressure, P_b. If the exit pressure (P_b) is reached before the maximum in G_o is reached, then the flow is not choked and will be sensitive to changes in P_b.

Following this procedure on the spreadsheet, the following results are obtained:

1. A maximum in G_o is reached at $P_n = 31$ psia, where $G_o = G_c = 1497$ lb_m/s ft² (shown in bold in Table 11.1). Because the flow is choked as it leaves the nozzle, the gas/vapor value of $K_{DG} = 0.98$ should be used to calculate the required nozzle area/diameter:

$$A_n = \frac{\dot{m}}{G_c K_{DG}} = \frac{(20 \text{ lb}_m/s)(144 \text{ in.}^2/\text{ft}^2)}{(1497 \text{ lbm}_m/\text{sft}^2)(0.98)} = 1.96 \text{ in.}^2, \text{ or } d_n = 1.60 \text{ in.}$$

In practice, the standard nozzle size that is closest to the resulting calculated value, on the high side, is selected (Table E11.2).

TABLE E11.2

Summary of numerical integration of Equation 11.11

P_o (psia)	P_n (psia)	Density (lb_m/ft^3)	x	i	Sum( )	G_o (lb_m/sft^2)
45	45	59.632		0.008385	0.0084	0
45	44	59.631		0.008385	0.0168	776
45	43	59.631		0.008385	0.0252	929
45	42	59.631		0.008385	0.0335	1048
45	41	59.631		0.008385	0.0419	1143
45	40	59.631		0.008385	0.0503	1222
45	39	59.63		0.008385	0.0587	1287
45	38	59.63		0.008385	0.0671	1340
45	37	59.63		0.008385	0.0755	1384
45	36	59.63		0.008385	0.0838	1420
45	35	59.63		0.008385	0.0922	1448
45	34	59.63		0.008385	0.1006	1469
45	33	59.629		0.008385	0.1090	1484
45	32	59.629		0.008385	0.1174	1493
45	**31**	**59.629**		**0.008385**	**0.1258**	**1497**
45	30	59.629		0.008385	0.1342	1496
45	29	59.629		0.008385	0.1425	1491
45	28	59.628		0.008385	0.1509	1481
45	27	59.628		0.008385	0.1593	1468
45	26	59.628		0.008385	0.1677	1450
45	25	59.628		0.008385	0.1761	1429
45	24	59.628		0.008385	0.1845	1404
45	23	59.628		0.008385	0.1929	1376
45	22	59.627		0.008385	0.2012	1344
45	21	59.627		0.008385	0.2096	1309
45	20	59.627		0.008385	0.2180	1272
45	19	59.627		0.008385	0.2264	1231
45	18	59.627		0.010887	0.2373	1194
45	17	32.228	0.00061	0.024293	0.2616	1184
45	16	8.9359	0.003847	0.072581	0.3342	1260
45	15	4.8418	0.007224	0.033178	0.3673	1238
45	14.7	4.2004	0.008266	3.499667	3.8670	3937

II.   CONTROL VALVES

Control valves are used to maintain a desired flow rate, level, temperature or pressure in a process. Control is achieved by automatically adjusting the valve (opened or closed) continuously to achieve a desired flow rate, or other process variable that depends on flow. The valve is controlled by a computer that senses the output signal from a flow meter and adjusts the control valve by pneumatic or electrical signals in response to deviations of the measured flow rate (or other process variable) from a desired set point. A representative flow control system is illustrated in Figure 11.4.

The control valve acts as a variable resistance in the flow line, because closing down on the valve is equivalent to increasing the flow resistance (i.e., the K_f) in the line and vice versa.

The nature of the relationship between the valve stem or plug position (which is the manipulated variable) and the flow rate through the valve, or other process variable, (which is the

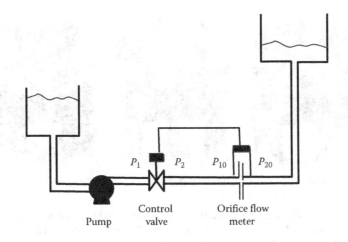

FIGURE 11.4 Schematic of a flow control piping system.

desired variable) is a nonlinear function of the pressure-flow characteristics of the piping system, the driver (i.e., pump), and the valve trim characteristic, which is determined by the design of the valve plug. This will be illustrated shortly.

A. VALVE CHARACTERISTICS

There are a large number of types of control valves, and the reader should consult the manufacturers' literature for details (e.g. Emerson, 2005). A typical globe valve type is illustrated in Figure 11.5.

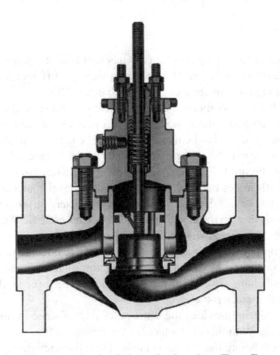

FIGURE 11.5 Valve body with cage-style trim, balanced valve plug. (From Emerson Process Management, *Fisher Controls, Control Valve Handbook*, 4th edn., Emerson, Process Management, Marshalltown, IA, 2005.)

(a)   W0958/IL   (b)   W0959/IL   (c)   W0957/IL

FIGURE 11.6   Characteristic cages for globe-style valve bodies: (a) Quick opening, (b) linear, and (c) equal percentage. (From Emerson Process Management, *Fisher Controls, Control Valve Handbook*, 4th edn., Emerson, Process Management, Marshalltown, IA, 2005.)

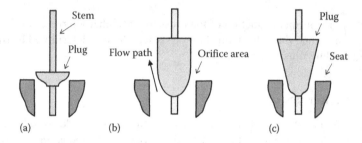

FIGURE 11.7   Valve plug shapes for various trim characteristics: (a) Quick opening, (b) linear, and (c) equal percentage.

For globe valves, different valve plugs (or "cages" that surround the plug) are available for a given valve with a cage-style trim (see Figure 11.6), each providing a different flow response characteristics when the valve setting (i.e., the stem position) is changed. Other valves achieve the desired trim by the shape of the valve plug. A specific valve characteristic must be chosen that, when matched to the response of the pump and the flow system, will give the desired valve stem—system flow response (which is usually desired to be linear). This will be demonstrated later.

Figure 11.6 shows the various "trim cages" that can be inserted in the valve shown in Figure 11.5 to achieve the desired trim. Figure 11.7 shows various valve plug shapes that result in different trim characteristics. Figure 11.8 illustrates the flow versus valve stem travel characteristic for various typical valve trim functions (Emerson, 2005). The *quick opening* characteristic provides the maximum change in flow rate at low opening or stem travel with a fairly linear characteristic. As the valve approaches the wide-open position, the change in flow with stem travel approaches zero. This is best suited for on-off control but is also appropriate for some applications where a linear response is desired.

The *linear flow* characteristic has a constant "valve gain," that is, the incremental change in flow rate with the percentage change in the valve plug position is the same at all flow rates. This is the most commonly desired characteristic.

The *equal percentage* trim provides the same percentage change in flow for equal increments of the valve plug position. With this characteristic, the change in flow is always proportional to the value of the flow rate just before the change is made. This characteristic is used in pressure control applications and where a small pressure drop across the valve is required relative to that in the rest of the system.

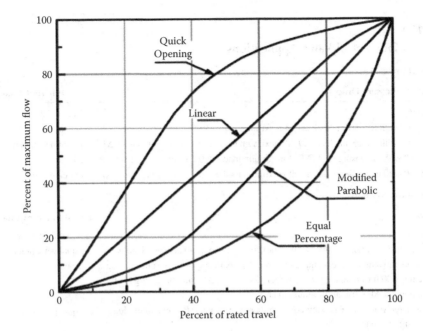

FIGURE 11.8   Various valve trim characteristics.

The *modified parabolic* characteristic is intermediate to the linear and equal percentage characteristics and can be substituted for equal percentage valve plugs in many applications, with some loss in performance.

Some general guidelines for the application of the proper valve characteristics are shown in Table 11.1. These are rules of thumb, and the proper valve response can be determined only by a complete analysis of the system in which the valve is to be used (see also Baumann (1991) for simplified guidelines). We will illustrate later how the valve trim characteristic interacts with the pump and system characteristics to affect the flow rate in the system and how to use this information to select the most appropriate valve trim.

B.   OVERVIEW OF CONTROL VALVE SIZING

The basic principle that governs the flow through a control valve is the same as that through an orifice, that is, the engineering Bernoulli equation as applied to an orifice. However, there are several important characteristics of control valves that are fundamentally different from an orifice that must be accounted for, namely:

- There is no one specific flow area in a control valve that can serve as a reference area, such as that in an orifice or relief valve. Consequently, the "flow area" and the coefficient of discharge are combined into one parameter, which is not dimensionless.
- The reference pressure drop through the valve is the total net pressure drop, instead of the pressure drop to the vena contracta as in the orifice. This total net pressure drop $(P_1 - P_2)$ is also referred to as the "service pressure drop." This requires that the pressure recovery in the valve outlet body be accounted for, as it is the pressure drop to the *vena contracta* that determines the flow rate, and not the service pressure drop.
- For compressible flow, the choke pressure has a unique definition, which is not necessarily the same as the actual choke pressure.
- The definition of the expansion coefficient (Y) is a variation of that used for the orifice.

TABLE 11.1
Guidelines for Control Valve Applications

Liquid Level Systems

Control Valve Pressure Drop	Best Inherent Characteristic
Constant ΔP	Linear
Decreasing ΔP with increasing load, ΔP at maximum load >20% of minimum load ΔP	Linear
Decreasing ΔP with increasing load, ΔP at maximum load <20% of minimum load ΔP	Equal percentage
Increasing ΔP with increasing load, ΔP at maximum load <200% of minimum load ΔP	Linear
Increasing ΔP with increasing load, ΔP at maximum load >200% of minimum load ΔP	Quick opening

Pressure Control Systems

Application	Best Inherent Characteristic
Liquid process	Equal percentage
Gas process, small volume, less than 10 ft of pipe between control valve and load valve	Equal percentage
Gas process, large volume (process has a receiver, distribution system, or transmission line exceeding 100 ft of nominal pipe volume), decreasing ΔP with increasing load, ΔP at maximum load >20% of minimum load ΔP	Linear
Gas process, large volume, decreasing ΔP with increasing load, ΔP at maximum load <20% of minimum load ΔP	Equal percentage

Flow Control Processes

Flow Measurement Signal to Controller	Location of Control Valve in Relation to Measuring Element	Best Inherent Characteristics	
		Wide Range of Flow Set Point	Small Range of Flow but Large ΔP Change at Valve with Increasing Load
Proportional to flow	In series	Linear	Equal percentage
	In bypass[a]	Linear	Equal percentage
Proportional to flow squared	In series	Linear	Equal percentage
	In bypass[a]	Equal percentage	Equal percentage

Source: Emerson Process Management, *Fisher Controls, Control Valve Handbook*, 4th edn., Emerson, Process Management, Marshalltown, IA, 2005.

[a] When control valve closes, flow rate increases in the measuring element.

The general approach to sizing a control valve is to consider it as an orifice and then apply various "correction factors" to account for the deviation of the valve flow from flow through an actual orifice. This involves lumping parameters for the valve that are not uniquely defined and applying empirical correction factors to account for them. Clearly, from inspection of Figure 11.5, these correction factors must be very important because of the significant deviation of the valve geometry from that of the orifice.

Just as for the relief valve, the starting point for all geometries, incompressible, compressible, and two-phase flows, is the general steady-state Bernoulli equation (Equation 11.2) for the mass flux through the valve:

$$G_o = \rho_2 \left(-2\int_{P_1}^{P_2} \frac{dP}{\rho} \right)^{1/2} = \frac{\dot{m}}{A_o} \tag{11.2}$$

This can be rearranged to solve for the mass flow rate, with inclusion of the "discharge coefficient" (C_d) to account for deviations from ideal orifice flow:

$$\dot{m} = \frac{C_d A_o \rho_2}{\left(1-\beta^4\right)^{1/2}} \left[2 \int_{P_2}^{P_1} \frac{dP}{\rho}\right]^{1/2} \quad \text{or}$$

$$\dot{m}\ (\text{lb}_m/\text{h}) = A_o\ (\text{in.}^2) K_d \rho_2\ (\text{lb}_m/\text{ft}^3)(2407) \left[\int_{P_2}^{P_1} \frac{dP\ (\text{psia})}{\rho\ (\text{lb}_m/\text{ft}^3)}\right]^{1/2}$$

(11.16)

Here, $\beta = d_o/D$, where d_o is the orifice diameter and D is the pipe diameter. For the valve, d_o, and hence A_o, is not uniquely defined so that the discharge coefficient and the beta geometrical term are combined into a single "flow coefficient," K_d or C_o:

$$K_d = C_o = C_d/\sqrt{1-\beta^4}$$

(11.17)

The "effective area" in Equation 11.16 is $A_o K_d = F(C_v/38)$ where C_v is the valve coefficient and $F = F_L$ for liquid flow and $F = F_G$ for gas flow. F is a dimensionless correction factor that is applied to account for the pressure recovery from the vena contracta (P_{vc}) to P_2 (hence to the service pressure drop $(P_1 - P_2)$, as discussed below).

Notation: Because of the variety of notation used by different references, there is no single universal standard. The two most definitive references for control valve sizing and selection are those of Emerson (2005) and ISA (2012) as well as IEC (2011). Because of its longer standing in the field and common usage in earlier publications, the Emerson notation is still frequently used. However, the IEC/ISA notation is gradually becoming the accepted standard. The most common terms and their definitions are given in the section "Notation" at the end of this chapter.

C. The Equation Constant

Because of the complex combinations of units, as well as the different systems of units employed, most of the equations require various combinations of conversion factors to give the desired units. To facilitate the conversion of units, many of the equations from IEC/ISA and Emerson (2005) include an "equation constant," N, which incorporates the various conversion factors needed in that equation for the desired net units. This applies to "American (English) engineering" units, as well as metric units. A listing of these equation constants is given in Table 11.2. This table is not exhaustive, as there are other combinations of units that may be encountered, such as those in Equation 11.16.

1. Incompressible Fluids

For incompressible fluids, Equation 11.16 can be written as

$$\dot{m}\ (\text{lb}_m/\text{h}) = C_v \sqrt{\rho(P_1 - P_2)}$$
$$= A_o K_d\ (\text{in.}^2)(2407)\sqrt{\rho\ (\text{lb}_m/\text{ft}^3)(P_1 - P_2)\ (\text{psi})}$$

(11.18)

TABLE 11.2

Equation Constants[a]

		N	$\dot{m}$	Q	P^b	ρ	T	D, d
N_1		0.0865	—	m³/h	kPa	—	—	—
		0.865	—	m³/h	bar	—	—	—
		1.00		m³/h	psia			
N_2		0.00214	—	—	—	—	—	mm
		890	—	—	—	—	—	in.
N_5		0.00241	—	—	—	—	—	mm
		1000	—	—	—	—	—	in.
N_6		2.73	kg/h	—	kPa	kg/m³	—	—
		27.3	kg/h	—	bar	kg/m³	—	—
		63.3	lb$_m$/h		psia	lb$_m$/ft³		
$N_7{}^c$	Normal conditions $T_N = °C$	3.94	—	m³/h	kPa	—	°K	—
		394	—	m³/h	bar	—	°K	—
	Standard conditions $T_S = 15.5°C$	4.17	—	m³/h	kPa	—	°K	—
		417	—	m³/h	bar	—	°K	—
	Standard conditions $T_S = 60°F$	1360	—	scfh	psia	—	°R	—
N_8		0.948	kg/h	—	kPa	—	°K	—
		94.8	kg/h	—	bar	—	°K	—
		19.3	lb$_m$/h		psia		°K	—
$N_9{}^c$	Normal conditions $T_N = °C$	21.2	—	m³/h	kPa	—	°K	—
		2120	—	m³/h	bar	—	°K	—
	Standard conditions $T_S = 15.5°C$	22.4	—	m³/h	kPa	—	°K	—
		2240	—	m³/h	bar	—	°K	—
	Standard conditions $T_S = 60°F$	7320	—	scfh	psia	—	°R	—

Source: Emerson Process Management, *Fisher Controls, Control Valve Handbook*, 4th edn., Emerson, Process Management, Marshalltown, IA, 2005.

[a] Many of the equations used in the sizing procedures contain a numerical constant, N, along with a numerical subscript. These numerical constants provide a means for using different units in the equations. Values of the numerical and the appropriate units are given in this table. For example, if the flow rate is given in the U.S. gpm and the pressure in psia, N_1 has a value of 1.00. If the flow rate is m³/hr and the pressure is in kPa, the N_1 constant is 0.0865.

[b] All pressures are absolute.

[c] The base pressure is 101.3 kPa or 1.013 bar or 14.7 psia.

where C_v is a combination of the coefficients and parameters in Equation 11.16 and is called the *valve sizing coefficient*. This is explained in the following section. The term $A_o K_d$ is assumed to be an "effective flow area" for the valve, which when compared with the equivalent IEC/ISA expression, shows that $A_o K_d = F_L C_v / 38$. There are a variety of correction factors or coefficients that are applied to the flow equation for incompressible fluids to account for deviation from the ideal orifice model, and these are shown in Table 11.2 and described in the following.

D. Valve Coefficients

1. The Valve Sizing Coefficient

The coefficient C_v in Equation 11.18 is the *valve sizing coefficient* and accounts for deviations from ideal incompressible orifice flow in the valve. Note that C_v is *not dimensionless*, as it incorporates all of the numerical coefficients in Equation 11.16, including the effective flow area $A_o K_d$, which is an unknown geometrical parameter for the valve. This coefficient is also identified by other symbols in various references (e.g., C, K_v, A_v, etc.). It is determined empirically by the manufacturer, and representative values are given in the manufacturers' handbook (e.g., Emerson, 2005) or the *ISA Handbook* (1971), as shown in Table 11.3. If Equation 11.18 is written for the volumetric flow rate (Q) instead of the mass flow rate, it can be expressed as

$$Q\ (gpm) = C_v \sqrt{\frac{(P_1 - P_2)\ (psi)}{SG_L}} = C_v \sqrt{\rho_w g h_v} = 0.658 C_v \sqrt{h_v}\ (ft) \qquad (11.19)$$

where

SG_L is the specific gravity of the flowing fluid relative to a standard (i.e., water at $60°F = 62.3\ \mathrm{lb_m/ft^3}$)
h_v is the "head loss" across the valve, in ft (of fluid)

The value of the reference fluid density is included in the definition of C_v.

The units normally used in the United States are the typical "American (English) Engineering" units, as follows:

$\dot{m}$ is the mass flow rate (liquid or gas) $(\mathrm{lb_m/h})$.
Q is the volumetric flow rate (gpm for liquids or scfh for gas or steam).

TABLE 11.3
Example Flow Coefficient Values for a Control Valve with Various Trim Characteristics

Valve Size (in.)	Valve Plug Style	Flow Characteristic	Port Dia. (in.)	Rated Travel (in.)	C_v	F_L	X_T	F_D
½	Post guided	Equal percentage	0.38	0.50	2.41	0.90	0.54	0.61
¾	Post guided	Equal percentage	0.56	0.50	5.92	0.84	0.61	0.61
1	Micro-form	Equal percentage	3/8	3/4	3.07	0.89	0.66	0.72
			1/2	3/4	4.91	0.93	0.80	0.67
		Linear	3/4	3/4	8.84	0.97	0.92	0.62
	Cage guided		1 5/16	3/4	20.6	0.84	0.64	0.34
		Equal percentage	1 5/16	3/4	17.2	0.88	0.67	0.38
1 ½	Micro-form	Equal percentage	3/8	3/4	3.20	0.84	0.65	0.72
			1/2	3/4	5.18	0.91	0.71	0.67
			3/4	3/4	10.2	0.92	0.80	0.62
	Cage guided	Linear	1 7/8	3/4	39.2	0.82	0.66	0.34
		Equal percentage	1 7/8	3/4	35.8	0.84	0.68	0.38
2	Cage guided	Linear	2 5/16	1 1/8	72.9	0.77	0.64	0.33
		Equal percentage	2 5/16	1 1/8	59.7	0.85	0.69	0.31
3	Cage guided	Linear	3 7/16	1 1/2	148	0.82	0.62	0.30
		Equal percentage			136	0.82	0.68	0.32
4	Cage guided	Linear	4 3/8	2	236	0.82	0.69	0.28
		Equal percentage			224	0.82	0.72	0.28
6	Cage guided	Linear	7	2	433	0.84	0.74	0.28
		Equal percentage			394	0.85	0.78	0.26
8	Cage guided	Linear	8	3	846	0.87	0.81	0.31
		Equal percentage			818	0.86	0.81	0.26

Source: ISA, (2012).

SG_L, SG_G is the specific gravity [relative to water for liquids (62.3 lb_m/ft^3) or air at 60° and 1 atm for gases (0.0764 lb_m/ft^3)].

ρ_1 is the density at upstream conditions (lb_m/ft^3).

P_1 is the upstream pressure (psia).

ΔP is the total (net recovered "service") pressure drop across valve (psi).

There are a number of combinations of other "engineering" or "metric" units that are also used with these equations (e.g., *gpm, psi, scfh, lb_m/h, lb_m/ft^3, m^3/h, kPa, bar*, etc.), and for each combination of units there is a specific value of the combined conversion factors that satisfy the equations. For example, the most common units associated with C_v in "American (English) engineering units" are $gpm/(psi)^{1/2}$. If the fluid density were to be included in Equation 11.19 instead of the specific gravity (SG_L), the dimensions of C_v would be $[L^2]$, reflecting the inclusion of the unknown factor A_o in C_v (i.e., $C_v = 38A_oK_d/F_L$). The "usual American engineering units" will be indicated in most of the working equations, along with the net numerical values of the conversion factors required to satisfy the equation.

2. F_P: The Piping Geometry Factor

It often occurs that the nominal size of the control valve that is best for a given application is smaller than the pipe in which it is installed (i.e., there is a pipe reducer upstream of the valve and a pipe expander downstream), and/or there may be other fittings such as elbows, tees, etc., attached to the valve. In this case, the *piping geometry factor, F_P*, must be applied. It is a dimensionless factor and is often determined empirically by the manufacturer and reported in the tables such as Table 11.3. Otherwise, it is determined from the following equation (Emerson, 2005):

$$F_P = \left[1 + \sum K_f \left(\frac{C_v}{d^2}\right)\right]^{-1/2} \tag{11.20}$$

where

d is the nominal valve size

C_v is the valve sizing coefficient at 100% stem travel

The term $\sum K_f$ is the sum of all friction loss factors for any upstream and downstream fittings and also includes the upstream and downstream kinetic energy loss factors:

$$K_{ke1} = 1 - \left(\frac{d}{D_1}\right)^4, \quad K_{ke2} = -\left[1 - \left(\frac{d}{D_2}\right)^4\right] \tag{11.21}$$

where D is the pipe *ID*. Obviously, if the upstream and downstream piping are the same size, these terms cancel. For other combinations of fittings, see Emerson (2005) or Chapter 7 in this book.

3. F_L^2 (K_m): The Liquid Pressure Recovery Factor

The total (measured) pressure drop across the valve, or service pressure drop ($P_1 - P_2$) in Equation 11.19 and Figure 11.9 omit) is the overall (recovered) pressure drop. This is not the maximum pressure drop across the valve (or orifice), however, which is ($P_1 - P_{vc}$) where P_{vc} is the pressure at the *vena contracta*.

In order to account for the difference between the (measured) pressure downstream of the valve (P_2) and the lower pressure just downstream at the *vena contracta*, (P_{vc}), as well as the

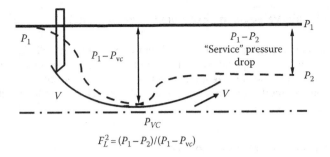

FIGURE 11.9 Vena contracta correction factor.

TABLE 11.4

Representative Full-Open $K_m = F_L^2$ Values for Various Valves

Body Type	$K_m = F_L^2$
Globe: single port, flow opens	0.70–0.80
Globe: double port	0.70–0.80
Angle: flow closes	
Venturi outlet liner	0.20–0.25
Standard seat ring	0.50–0.60
Angle: flow opens	
Maximum orifice	0.70
Minimum orifice	0.90
Ball valve	
v-notch	0.40
Butterfly valve	
60° open	0.55
90° open	0.30

pressure recovery downstream, a dimensionless correction factor F_L called the *liquid recovery factor* is defined as

$$F_L^2 = K_m = \frac{(P_1 - P_2)}{(P_1 - P_{vc})} \tag{11.22}$$

Values of the liquid recovery factor are determined by the valve manufacturer, as shown in Table 11.3, and are assumed to be constant regardless of the pressure drop. Note that different symbols are used in various references for the same correction factor (e.g., $F_L^2 = K_m$). The notation F_L is from the ANSI/ISA 75.01.01/IEC-60534-2-1 (2012) standard and K_m is that of Emerson/Fisher Controls (1990). This dimensionless factor is determined by the manufacturer for each control valve, and is assumed to be constant regardless of the pressure drop. Typical values of F_L are shown in Tables 11.3 and 11.4.

4. F_d: The Valve Style Modifier

F_d is the dimensionless ratio of the hydraulic diameter of a single flow passage to the diameter of a circular orifice, the area of which is equivalent to the sum of areas of all identical flow passages at a given travel. It is normally given by the manufacturer as a function of stem travel.

Example 11.3 Liquid Valve Sizing Example (Emerson, 2005)

Consider an installation that, upon start-up, will not be operating at full capacity. The lines are sized for the ultimate system capacity, but it is desired to install a control valve sized for currently anticipated requirements. The line size is 8 in., and a Class 300 globe valve with an equal percentage cage has been specified. Standard concentric reducers will be used to install the valve in the line. Determine the appropriate valve size.

Solution:

The following operating conditions are specified:

- *Valve design*: Class 300 globe valve with equal percentage cage and an assumed size of 3 in.
- *Fluid*: Liquid propane
- *Service conditions*:
  - $Q = 800$ gpm
  - $P_1 = 300$ psig $= 314.7$ psia
  - $P_2 = 275$ psig $= 289.7$ psia
  - $\Delta P = 25$ psi
  - $T_1 = 70°F$
  - $SG_f = 0.5$
  - $P_v = 124.3$ psia
  - $P_c = 616.3$ psia

From Table 11.2, $N_1 = 1$.

Determine the piping geometry factor, F_P. This accounts for the fitting losses for the 3 in. valve installed in an 8 in. line:

$$F_P = \left[1 + \frac{\Sigma K}{N_2} \left(\frac{C_v}{d^2} \right)^2 \right]^{1/2}$$

where $N_2 = 890$ from Table 11.2, and $d = 3$ in.

From the flow coefficient table for a Class 300 3 in. globe valve with equal percentage cage, $C_v = 121$. For a valve installed between two identical reducers:

$$\Sigma K = K_1 + K_2 = 1.5 \left(1 - \frac{d^2}{D^2} \right)^2$$

$$= 1.5 \left(1 - \frac{3^2}{8^2} \right)^2 = 1.11$$

Therefore,

$$F_P = \left[1 + \frac{1.11}{890} \left(\frac{121}{3^2} \right)^2 \right]^{1/2} = 0.90$$

Determine ΔP_{max} (the allowable sizing pressure drop). Based on the small required pressure drop, the flow will not be choked ($\Delta P_{max} > \Delta P$).

Solve the liquid flow equation for C_v:

$$\boxed{C_v = \frac{Q}{N_1 F_P \sqrt{(P_1 - P_2)/SG_f}} = \frac{800}{(1.0)(0.90)\sqrt{25/0.5}} = 125.7}$$

The valve size based on this value of C_v and the tables for the flow coefficient show that it is larger than the assumed value of $C_v = 121$. Thus, the calculation is repeated assuming the next larger valve size of 4 in., having a C_v of 203. This yields the following results: $\Sigma K = 0.84$, $F_P = 0.93$, and $C_v = 116.2$. Thus, the proper valve size would be between 3 and 4 in., so the 4 in. valve opened to about 75% of total travel should be adequate.

E. CAVITATING AND FLASHING LIQUIDS

1. Introduction

The minimum pressure in the valve (P_{vc}) generally occurs at the vena contracta, just downstream of the flow orifice. The pressure then rises downstream to P_2 (as shown in Figure 11.9), with the amount of pressure recovery depending upon the valve design. If the downstream choke pressure (P_c) is less than the fluid vapor pressure (P_v), the liquid will partially vaporize forming bubbles. This is sometimes described as "liquid choking," although liquids do not "choke" as such. Anytime the local pressure falls below the vapor pressure of the liquid it will "flash" or vaporize, resulting in vapor or two-phase flow, which may choke. At the "choke point," a further drop in the downstream pressure will not increase the flow rate any further. An increased pressure drop ($P_1 - P_c$) may be due to an increase in temperature or velocity (i.e., cavitation).

If the pressure recovers to a value greater than P_v, these bubbles may collapse suddenly, setting up local shock waves, which can result in considerable damage to the valve. This condition is referred to as *cavitation* (similar to that seen in centrifugal pumps), as opposed to *flashing* that occurs if the recovered pressure ($P_2 = P_c$) remains below P_v so that the vapor does not condense. After the first vapor cavities form, the flow rate will no longer be proportional to the square root of the pressure difference across the valve because of the decreasing density of the mixture. If sufficient vapor forms, the flow can become choked, at which point the flow rate will be independent of the downstream pressure as long as P_1 remains constant.

2. $F_F = r_c$: The Liquid Critical Pressure Ratio

The critical pressure ratio $P_2/P_v = P_c/P_v = r_c$ is that at which the liquid will vaporize resulting in possibly damaging cavitation or flashing, resulting in choked flow. The corresponding maximum flow rate through the valve at that point is

$$Q_{max} = F_L C_v \sqrt{\frac{P_1 - r_c P_v}{SG_L}} \tag{11.23}$$

where
  P_v is the vapor pressure of the fluid
  r_c represents the choke pressure
  SG_L is its specific gravity (relative to water)

Note that the reference density of water must then be included in the "conversion factors" required for the units used in Equation 11.23.

For pure components, the choke pressure may be estimated as $r_c P_v$ and values for $F_F = r_c$ can be determined from the following correlation:

$$F_F = r_c = 0.96 - 0.28 \sqrt{\frac{P_v}{P_{tc}}} \tag{11.24}$$

where P_{tc} is the thermodynamic (absolute) critical pressure for the fluid. Figure 11.10 shows the correlation of r_c for water, and Figure 11.11 shows the correlation for other fluids. Table 11.5 gives the

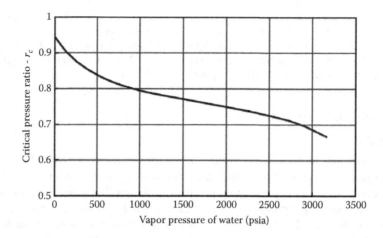

FIGURE 11.10 Critical pressure ratio for water. The abscissa is the water vapor pressure at the valve inlet. The ordinate is the corresponding critical pressure ratio, $r_c = F_F$. (From Emerson Process Management, *Fisher Controls, Control Valve Handbook*, 4th edn., Emerson, Process Management, Marshalltown, IA, 2005.)

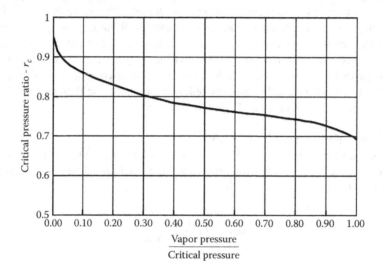

FIGURE 11.11 Critical pressure for cavitating and flashing liquids other than water. The abscissa is the ratio of the liquid vapor pressure at the valve inlet divided by the thermodynamic critical pressure of the liquid. The ordinate is the corresponding critical pressure ratio, $r_c = F_F$ (From Emerson Process Management, *Fisher Controls, Control Valve Handbook*, 4th edn., Emerson, Process Management, Marshalltown, IA, 2005.)

thermodynamic critical temperature for various fluids. However, the correlation in Equation 11.24 was derived for pure components and it may overstate the choke pressure for multicompound fluids, thus resulting in an undersized control valve. The determination of the choke pressure for multicomponent systems (or any system) may be determined more easily using the HDI method (see Section 4).

To determine if cavitation will occur, the cavitation pressure is determined from the choke pressure (P_c):

$$\Delta P_{cav} = F_L^2 \left(P_1 - P_c \right) \qquad (11.25)$$

TABLE 11.5
Thermodynamic Critical Pressure of Various Fluids (psia)

Ammonia	1636
Argon	705.6
Butane	550.4
Carbon dioxide	1071.6
Carbon monoxide	507.5
Chlorine	1118.7
Dowtherm A	465
Ethane	708
Ethylene	735
Fluorine	808.5
Helium	33.2
Hydrogen	188.2
Hydrogen chloride	1198
Isobutane	529.2
Isobutylene	580
Methane	673.3
Nitrogen	492.2
Nitrous oxide	1047.6
Oxygen	736.5
Phosgene	823.2
Propane	617.4
Propylene	670.3
Refrigerant 11	635
Refrigerant 12	596.9
Refrigerant 22	716
Water	3206.2

The proper pressure drop to use in the liquid sizing equation depends on cavitation, as follows:

- If the overall "service" pressure drop $(P_1 - P_2)$ is less than ΔP_{cav}, or if the pressure at the vena contracta (P_{vc}) is greater than the choke pressure (P_c), then the fluid will not cavitate or choke
- If the overall service pressure drop $(P_1 - P_2)$ is greater than ΔP_{cav}, or if the pressure at the vena contracta (P_{vc}) is less than the choke pressure (P_c), then the fluid will choke and the cavitation pressure drop is to be used in the valve equation to determine Q:

$$Q = F_L C_v \sqrt{\frac{\Delta P_c}{SG}} \qquad (11.26)$$

- Otherwise, the total pressure drop $(P_1 - P_2)$ is used.

The notation used here is that of the Emerson/Fisher literature (e.g., Emerson, 2005). The ANSI/ISAS75.01 standard for control valves uses the same equations, except that it uses the notation $F_L = K_m^{1/2}$ and $F_F = r_c$ in place of the factors K_m and r_c, respectively.

3. x_T: The Pressure Differential Ratio Factor for a Control Valve without Attached Fittings at Choked Flow

This parameter provides the same information as $F_F = r_c$. Specifically, x_T is the dimensionless pressure drop ratio $[x_T = (P_1 - P_c)/P_1]$ required to produce critical or maximum (flashing) flow through the valve when $F_k = 1$ (where $F_k = k/1.4$ is the specific heat ratio factor). Values of x_T are provided by the manufacturer, as shown in Table 11.4.

F. VISCOUS FLUIDS

1. F_v: The Viscosity Correction Factor

A correction for fluid viscosity must be applied to the flow coefficient (C_v) for liquids other than water. This viscosity correction factor (F_v) is obtained from Figure 11.12 by the following procedure, depending upon whether the objective is to find the valve size for a given Q and ΔP, to find Q for a given valve and ΔP, or to find ΔP for a given valve and Q.

1. To find valve size
   For the given Q and ΔP, calculate the required C_v as follows:

$$C_v = \frac{Q\ (gpm)}{\sqrt{\Delta P\ (psi)/SG_L}}$$ (11.27)

   Then determine the Reynolds number from the equation:

$$N_{Rev} = 17,250 \frac{Q}{v_{cs}\sqrt{C_v}}$$ (11.28)

   where
   Q is in *gpm*
   ΔP is in psi
   v_{cs} is the fluid kinematic viscosity ($v_{cs} = \mu/\rho$, in centistokes)

   The viscosity correction factor, F_v, is then read from the middle line on Figure 11.12 and used to correct the value of C_v as follows:

$$C_{vc} = C_v F_v$$ (11.29)

2. To predict flow rate
   For a given valve (i.e., a given C_v) and given ΔP, the maximum flow rate (Q_{max}) is determined from

$$Q_{max} = C_v \sqrt{\frac{\Delta P}{SG}}$$ (11.30)

   The Reynolds number is then calculated from Equation 11.28, and the viscosity correction factor, F_v, is read from the bottom curve in Figure 11.12. The corrected flow rate is then

$$Q_c = \frac{Q_{max}}{F_v}$$ (11.31)

3. To predict pressure drop
   For a given valve (C_v) and given flow rate (Q), calculate the Reynolds number from Equation 11.28 and read the viscosity correction factor, F_v, from the top line of Figure 11.12. The predicted pressure drop across the valve is then

$$\Delta P = SG \left(\frac{QF_v}{C_v}\right)^2$$ (11.32)

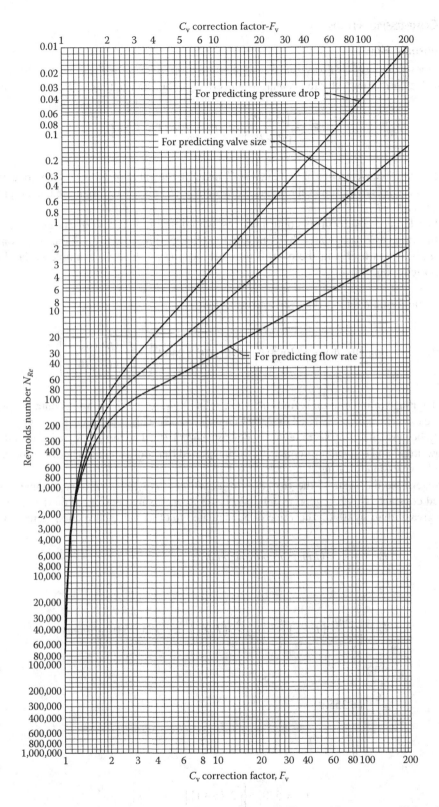

FIGURE 11.12 Viscosity correction factor for C_v. (From Emerson Process Management, *Fisher Controls, Control Valve Handbook*, 4th edn., Emerson, Process Management, Marshalltown, IA, 2005.)

G. COMPRESSIBLE FLUIDS

1. Subsonic Flow

Just as for incompressible flows, the overall pressure drop ($\Delta P = P_1 - P_2$) is used in the sizing equation, with a correction factor to account for the effect of the vena contracta (P_{vc}), defined as

$$F_G^2 = \frac{(P_1 - P_2)}{(P_1 - P_{vc})} \quad \text{or} : P_{vc} = P_1 - (P_1 - P_2)/F_G^2 \tag{11.33}$$

- For high-recovery control valves (low C_v and X_T), the value of F_G is less than one and the choking pressure drop at the vena contracta ($P_1 - P_{vc}$) is greater than the overall pressure drop ($P_1 - P_2$).
- For large-pressure drop control valves with minimum recovery (large C_v and X_T), the value of F_G is greater than one and the choking pressure drop at the vena contracta ($P_1 - P_{vc}$) is less than the overall pressure drop ($P_1 - P_2$). An example is the double-ported globe valve.

The general formula for gas flow valve sizing is thus

$$\dot{m} \ (\text{lb}_m/\text{h}) = A_o \ (\text{in.}^2) K_d \rho_2 \ (\text{lb}_m/\text{ft}^3)(2407)\left(-\int_{P_{vc}}^{P_{vc}} \frac{dP \ (\text{psi})}{\rho \ (\text{lb}_m/\text{ft}^3)}\right)^{1/2} \tag{11.34}$$

where the range of integration is from the valve inlet pressure, P_1, to the vena contracta pressure at the valve exit, P_{vc}, which is related to P_2 by Equation (11.33). This is less than the downstream pressure, P_2, because of the pressure recovery downstream of P_{vc}. The term $A_o K_d$ represents the "effective flow area," and is found by comparing the integrated Bernoulli equation with the corresponding IEC/ISA expression, yielding $A_o K_d = F_G C_v / 38$.

To determine the value of the *gas pressure-adjustment coefficient* F_G, the parameter ($A_o K_d$) is assumed constant. This is a good assumption if the Reynolds number based on the upstream flow area is greater than 100,000. The value of F_G is related to other valve correction factors by

$$F_G = \frac{C_1}{28.9} \quad \text{(\textit{Emerson parameters})}$$

or

$$F_G = \left(\frac{490}{C_\gamma}\right)\left(F_\gamma X_T\right)^{0.5} = \left(\frac{1.38}{C_2}\right)\left(F_\gamma X_T\right)^{0.5}, \quad (\textit{ISA-75.01 parameters}) \tag{11.35}$$

since

$$\left(\frac{C_\gamma}{C_2}\right) = 355.9$$

where $\gamma = C_p/C_v = k$, and

$$C_\gamma = 520\left[\gamma\left\{\left[2/(\gamma+1)\right]^{(\gamma+1)/(\gamma-1)}\right\}\right]^{0.5} \tag{11.36}$$

F_G is not strongly dependent upon the value of γ, as a variation in γ of 50% results in a variation in F_G of approximately 4%. Here, C_γ and F_γ are a consolidation of terms that are functions only of γ:

$$F_\gamma = \gamma/1.4, \tag{11.37}$$

and

$$C_2 = 2.0665 \left[\left\{ \frac{\gamma}{(\gamma+1)} \right\} \left\{ \left[\frac{2}{(\gamma+1)} \right]^{2/(\gamma-1)} \right\} \right]^{0.5} \tag{11.38}$$

The factor X_T is based on air near atmospheric pressure as the flowing fluid with a specific heat ratio of 1.40. If the specific heat ratio for the flowing fluid is not 1.40, the factor $F_\gamma = \gamma/1.4$ is used to adjust X_T, as in Equation 11.35.

Equation 11.34 evolved from the assumption of perfect gas behavior and extension of the orifice plate model, based on air and steam testing on control valves. Analysis of that model over a range of $1.08 < \gamma < 1.65$ has led to the adoption of the current linear model embodied in Equation 11.34. The difference between the original orifice model, other theoretical models, and Equation 11.34 is small within this range. However, the differences become significant outside of the indicated range. For maximum accuracy, the flow calculations based on this model should be restricted to a specific heat ratio within this range and to ideal gas behavior.

For relatively low pressure drops (i.e., less than about 30%), the effect of compressibility is negligible and the general incompressible flow equation (Equation 11.18) may be applied. Introducing the conversion factors to give the flow rate in standard cubic feet per hour (scfh) and the density of air at standard conditions (1 atm, 520°R), Equation 11.18 becomes

$$Q_{scfh} = 1362 C_v P_1 \left(\frac{\Delta P}{P_1 SG_G T_1} \right)^{1/2} \tag{11.39}$$

The effect of variable density can be accounted for by an expansion factor Y (Hutchinson, 1971), as has been done for flow in pipes and orifice meters (see Chapter 10), in which case Equation 11.39 can be written as

$$Q_{scfh} = 1362 C_v P_1 Y \left(\frac{X}{SG_G T_1} \right)^{1/2} \tag{11.40}$$

where

$$X = \frac{\Delta P}{P_1} = \frac{P_1 - P_2}{P_1} \tag{11.41}$$

Here, SG is defined as $MW/28.9655$, and $Q_{scfh} = W(\text{lb}_m/\text{h})$ [379.49 scf/lb$_{mol}$]/$M(\text{lb}_m/\text{lb}_{mol})$ at 14.696 psia and 60°F.

The expansion factor Y depends on the pressure drop X, the dimensions (clearance) in the valve, the gas specific heat ratio $k = \gamma$, and the Reynolds number (the effect of which is often negligible). It has been found from measurements (Hutchison, 1971) that the expansion factor for a given valve can be represented, to within about $\pm 2\%$, by the expression

$$Y = 1 - \frac{X}{3 F_\gamma X_T} \tag{11.42}$$

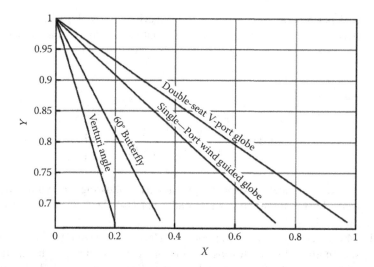

FIGURE 11.13 Expansion factor (Y)—as a function of pressure drop ratio (X) for four different types of control valves. (From Hutchinson, J.W., *ISA Handbook of Control Valves*,1st edn., Instrument Society of America, Durham, NC, 1971.)

where X_T is specific to the valve at the choke point, as illustrated in Figure 11.13. Deviations from the ideal gas law can be incorporated by multiplying T_1 in Equation 11.39 or 11.40 by the compressibility factor, Z, for the gas.

2. Choked Flow

When the gas velocity reaches the speed of sound, choked flow occurs and the mass flow rate reaches a maximum. It can be shown from Equation 11.42 that this is equivalent to a maximum in $YX^{1/2}$, which occurs at $Y = 0.667$ and corresponds to the terminus of the lines in Figure 11.13. That is, X_T is the pressure ratio across the valve at which choking occurs, which is determined empirically. At the choke point, any further increase in X (e.g., ΔP) due to lowering P_2 can have no effect on the flow rate.

The flow coefficient C_v is determined by calibration with water, and it is not entirely satisfactory for predicting the flow rate of compressible fluids under choked flow conditions. This has to do with the fact that different valves exhibit different pressure recovery characteristics with gases and hence will choke at different pressure ratios, which does not apply to liquids. For this reason, another flow coefficient, C_g, is often used for gases and is determined by calibration with air under critical flow conditions (Emerson, 2005). The corresponding flow equation for gas flow is

$$Q_{critical} = C_g P_1 \left(\frac{520}{SG_G T} \right)^{1/2} \tag{11.43}$$

H. GENERAL (HDI) METHOD FOR ALL FLUIDS AND ALL CONDITIONS

There are many other factors or coefficients that may be applied to the basic equation to account for gas flow, cavitation, flashing, etc. However, regardless of whether the fluid is a liquid, gas or vapor, or single- or two-phase flashing flow, the flow rate can be calculated directly by a simple numerical integration (HDI) of the basic integral form of the Bernoulli equation, provided a thermodynamic fluid property database is available for the fluid properties. All that is needed is the density of the fluid or the mixture as a function of pressure over the pressure range of interest. As mentioned for safety valves, there are many such databases available. A handy and easy-to-use inexpensive database containing hundreds of fluids (and some two-phase systems) is the NIST REFPROP, (2015) database.

The method is based on the integration of the differential form of the Bernoulli equation. For liquids, the equation is

$$\dot{m} \ (lb_m/h) = A_o \ (in.^2) K_d \rho_{vc} (lb_m/ft^3)(2407) \left[\int_{P_{vc}}^{P_1} \frac{dP(lb_f/in.^2)}{\rho(lb_m/ft^3)} \right]^{1/2}$$

$$= \left(\frac{F_L C_v}{38} \right) (in.^2) \ \rho_{vc} (lb_m/ft^3)(2407) \left[\int_{P_{vc}}^{P_1} \frac{dP(lb_f/in.^2)}{\rho(lb_m/ft^3)} \right]^{1/2} \tag{11.44}$$

where $A_o \ (in.^2) K_d = (F_L C_v/38) \ (in.^2)$. The equivalent ISA and Fisher notation for the gas flow coefficients is shown in Table 11.6.

Note that the range of integration is from P_1 to P_{vc}, the pressure at the vena contracta, which is assumed to exist at the valve exit. This is lower than the pressure P_2, which is the "recovered" pressure downstream of the valve. The value of the vena contracta pressure for liquids can be determined from the parameter $F_L^2 = K_m$ as defined by Equation 11.22:

$$F_L^2 = K_m = \left(P_1 - P_2 \right) / \left(P_1 - P_{vc} \right) \tag{11.22}$$

which can be used to convert $(P_1 - P_{vc})$ to the service pressure drop, $(P_1 - P_2)$.

Equation 11.44 can be written in numerical form to apply from the upstream pressure (P_1) to the minimum exit pressure at the *vena contracta*:

$$\dot{m} \left[lb_m/h \right] = \left(A_o K_d \right) (in.^2) \rho_{vc} \left[lb_m/ft^3 \right] (2407) \left[\int_{P_{vc}}^{P_1} \frac{dP}{\rho} \right]^{1/2} \left[lb_f \ ft^3/lb_m \ in.^2 \right]^{1/2}$$

$$\cong \left(\frac{C_v F_L}{38} \right) (in.^2) F_L \rho_{vc} \left[lb_m/ft^3 \right] (2407) \left[2 \sum_{P_{vc}}^{P_1} \frac{P_i - P_{i+1}}{\rho_i + \rho_{i+1}} \right]^{1/2} \left[lb_f \ ft^3/lb_m \ in.^2 \right]^{1/2} \tag{11.45}$$

where the summation is along an isentropic path from P_1 to P_{vc}. In practice, the summation starts with a vena contracta pressure just below P_1 and proceeds until the value of P_{vc} is reached. If the flow is choked, the mass flow rate will reach a maximum before the pressure reaches P_{vc}, in which case no further reduction of P_{vc} will result in an increase in flow rate.

TABLE 11.6

Equivalent Parameters for $A_o K_d$ for Gases

ISA parameters

$\{A_o K_d\} = 12.873 \ [C_v/C_\gamma] \ [(X_T \ F_\gamma)^{1/2}]$

$\{A_o K_d\} = [(C_v/38)] \ [489.174 \ (X_T \ F_\gamma)^{1/2}/C_\gamma]$

Fisher parameters

$\{A_o K_d\} = [C_g/1100]$

$\{A_o K_d\} = [C_v * C_1/1100]$

$\{A_o K_d\} = [C_v/38] \ [C_1/28.9]$

For compressible fluids, an analogous form of Equation 11.44 is

$$\dot{m} \text{ (lb}_m/\text{h}) = A_o \text{ (in.}^2) K_d \rho_{vc} (2407) \left(\int_{P_{vc}}^{P_1} \frac{dP \text{ (psi)}}{\rho \text{ (lb}_m/\text{ft}^3)} \right)^{1/2} \tag{11.46}$$

where the vena contracta pressure can be determined by using the parameter F_G^2, given by Equation 11.33:

$$F_G^2 = (P_1 - P_2)/(P_1 - P_{vc}) \tag{11.33}$$

which is analogous to Equation 11.22 for liquids. The effective flow area is:

$$A_o K_d \left[in.^2 \right] = F_G C_v / 38 \quad \text{and} \quad F_G = C_1 / 28.9 \tag{11.47}$$

The pressure drop $(P_1 - P_{vc})$ can be converted to the service pressure drop $(P_1 - P_2)$ by Equation 11.33. This method can be applied directly to all cases, including incompressible, compressible, single- and two-phase flows, flashing and cavitating flows, etc., as long as the requisite fluid properties (i.e., $\rho(P)$) are available. For two-phase flow, if the flow is choked the value of $A_o K_d$ for gas flow should be used, but if the flow is not choked the liquid value should be used.

It should be noted that the *vena contracta* for gases is significantly different than for liquids, due to the rapid expansion of the gas as it leaves the valve. Also, if the flow is choked as it leaves the valve (as is the usual case), no further pressure drop or any change in conditions downstream of that point will affect the flow rate, so it is important to initially determine if the flow is choked. Although the methods outlined previously are recommended in the literature, it is generally easier and more accurate to use the HDI method for these more complex situations when possible.

In the case of two-phase flow, the question arises as to whether to use F_L or F_G to relate P_{vc} to P_2. For many valves, there is little difference between these two, but where there is, logic would indicate a preference for using F_G.

Example 11.4 Valve Sizing for Compressible Fluids Using the Fisher/Emerson Method (Emerson 2005)

Steam is to be supplied to a process that is to operate at 250 psig. The steam is supplied from a header at 500 psig and 500°F. A 6 in. line from the header to the process is planned. If the required control valve calculated is less than 6 in., it will be installed using concentric reducers. Determine the appropriate Design ED valve with a linear cage.

Specify the Necessary Design Variables

a. *Desired valve*: Class 300 Design ED valve with a linear cage. Assume a 4 in. valve.
b. *Process fluid*: Superheated steam.
c. *Conditions*: $\dot{m} = 125,000$ lb$_m$/h, $P_1 = 500$ psig $= 514.7$ psia, $P_2 = 250$ psig $= 264.7$ psia, $\Delta P = 250$ psi, $x = \Delta P/P_1 = 250/514.7 = 0.49$, $T_1 = 500°F$, $\gamma_1(\rho_1) = 1.0434$ lb$_m$/ft^3 (from steam tables), $k = 1.28$ (from steam tables).
d. Determine equation constants from equation constants (*N*) table—because the flow rate and density are in mass units (*lb$_m$/h* and *lb$_m$/ft^3*), and the pressure is in psi, the appropriate equation constant is $N_6 = 63.3$.
e. Determine the piping geometry factor

$$F_p = \left[1 + \frac{\Sigma K_f}{N_2} \left(\frac{C_v}{d^2} \right)^2 \right]^{-1/2}$$

where

$N_2 = 890$ (from equation constants table)

$d = 4$ in.

$C_v = 236$ (from the manufacturer's flow coefficient table for a 4 in. Design ED valve at 100% travel).

and

$$\Sigma K_f = K_1 + K_2 = 1.5\left(1 - \frac{d^2}{D^2}\right)^2 = 1.5\left(1 - \frac{4^2}{6^2}\right)^2 = 0.463$$

Also

$$F_p = \left[1 + \frac{\Sigma K_f}{N_2}\left(\frac{C_v}{d^2}\right)^2\right]^{-1/2} = \left[1 + \frac{0.463}{890}\left(\frac{(236)}{4^2}\right)^2\right]^{-1/2} = 0.95$$

f. Determine the expansion factor, Y:

$$Y = 1 - \frac{x}{3F_k x_{TP}}$$

where $F_k = k/1.40 = 1.28/1.40 = 0.91$, $x = 0.49$ (from given pressures).

Because the 4 in. valve is to be installed in a 6 in. line, the x_{TP} is given by

$$x_{TP} = \frac{x_T}{F_p^2}\left[1 + \frac{x_T K_i}{N_5}\left(\frac{C_v}{d^2}\right)^2\right]^{-1}$$

where

$N_5 = 1000$, from equation constants table

$d = 4$ in., $F_p = 0.95$ (step e.)

$x_T = 0.688$ (from the manufacturer's flow coefficient table)

$C_v = 236$ (from step e.) and

$$K_i = K_1 + K_{B1} = 0.5\left(1 - \frac{d^2}{D^2}\right)^2 + \left[1 - \left(\frac{d}{D}\right)^4\right]$$

$$= 0.5\left(1 - \frac{4^2}{6^2}\right)^2 + \left[1 - \left(\frac{4}{6}\right)^4\right] = 0.96$$

where $D = 6$ in.

So

$$x_{TP} = \frac{x_T}{F_p^2}\left[1 + \frac{x_T K_i}{N_5}\left(\frac{C_v}{d^2}\right)^2\right]^{-1} = \frac{0.69}{0.95^2}\left[1 + \frac{(0.69)(0.96)}{1000}\left(\frac{236}{4^2}\right)^2\right]^{-1} = 0.67$$

$$\text{Therefore } Y = 1 - \frac{x}{3F_k x_{TP}} = 1 - \frac{0.49}{(3)(0.91)(0.67)} = 0.73$$

g. Calculate C_v:

$$C_v = \frac{\dot{m}}{N_6 F_p Y \sqrt{x P_1 \gamma_1}} = \frac{125,000}{(63.3)(0.95)(0.73)\sqrt{(0.49)(514.7)(1.0434)}} = 176$$

h. Select the valve size from the manufacturer's flow coefficient table for Design ED valves with a linear cage and the calculated value of C_v. The assumed 4 in. valve has a C_v of *236* at *100%* travel and the next smaller size (3 in.) has a C_v of *148*, so the assumed size is correct. If the calculated value of C_v had been small enough to be handled by the 3 in. valve, or larger than the rated C_v for the assumed size, it would be necessary to rework the problem with a new assumed size.

Example 11.5 Valve Sizing for Compressible Fluids using the HDI Method

A control valve is to be sized to control the flow rate of air through a pipeline. The pressure upstream of the valve is 100 psia at a temperature of 70°F, and the discharge pressure is 14.7 psia. The valve to be used is a 3 in. high recovery Venturi ball valve, with the following parameters:

$$C_v = 202, \; C_g = 2848, \; C_1 = 14.1, \; X_T = (C_g/C_v)^2/1584.6 = 0.1254, \; F_G = C_1/28.9 = 0.4871$$

$$A_oK_d = C_g/1100 = 2.5891, \; F_\gamma = 1.0$$

Determine the maximum flow rate of air that this valve can handle, in lb_m/h, using the HDI method.

Solution: The basic equation that governs this system is Equation 11.46:

$$\dot{m}[lb_m/h] = A_o\left[in.^2\right]K_d\rho_{vc}\left[\frac{lb_m}{ft^3}\right](2407)\left[2\sum_{P_{vc}}^{P_1}\frac{P_i - P_{i+1}}{\rho_i + \rho_{i+1}}\right]^{1/2}\left[\frac{ft^3 lb_f}{lb_m in.^2}\right]^{1/2}$$

$$= (C_v/38)(C_1/28.9)\left[in.^2\right]\rho_{vc}\left[\frac{lb_m}{ft^3}\right](2407)\left[\sum_{P_{vc}}^{P_1}2\frac{P_i - P_{i+1}}{\rho_i + \rho_{i+1}}\right]^{1/2}\left[\frac{ft^3 lb_f}{lb_m in.^2}\right]^{1/2}$$

where the "hypothetical" mass flux (G) through the valve is given by $G = \dot{m}/A_oK_d$.

The first step is to determine the density of air at 70°F and *100 psia*, and over the pressure range of interest at constant entropy, using a fluid property database such as NIST REFPROP, (2015 or later). A suitable pressure interval (e.g. 2 psi) is used, starting at $P_1 = 100$ *psi*. This will give the density of air from $P_1 = 100$ *psia* to *14.7 psia* in increments of *2 psi*. These data can now be used to calculate the summation term in the above equation, for each pressure increment from P_1 to $(P_{vc})_i$. Thus the mass flux and the flow rate can be calculated from the above equation for each value of $(P_{vc})_i$. These values are shown in bold face in Table E11.5.

It is seen that the flow is choked at about $P_{vc} = 54$ psia, with a mass flux of 343 lb_m/s ft.^{2}, or a total mass flow rate of:

$$\dot{m} = G(A_oK_d) = (342.7\,lb_m/s\text{-}ft^2)(2.59\,in.^2)(3600\,s/hr)/(144\,in.^2/ft^2) = 2.2\times10^4\,lb_m/hr\text{-}in.^2$$

The pressure at the vena contracta can be converted to the equivalent discharge "service" pressure using the factor F_G from Equation 11.33:

$$F_G^2 = (P_1 - P_2)/(P_1 - P_{vc}) = 0.4871^2 = 0.2373$$

or :

$$P_2 = P_1 - F_G^2(P_1 - P_{vc}) = P_1 - 0.2373(P_1 - P_{vc})$$

Given that $P_1 = 100$ psia, a discharge vena contracta pressure of $P_{vc} = 14.7$ psia corresponds to a discharge service pressure of $P_2 = 79.8$ psia, and a choke pressure of $P_{vcchoke} = 54$ psia corresponds to $P_{2c} = 89.1$ psia.

TABLE E11.5

Spreadsheet Output for Example 11.5

P_{vc} psia	T °F	ρ_{vc} lb$_m$/ft^3	i	sum( )	G lb$_m$/s-ft^2
100	70	0.51067	−1.97	0.00	0.00
98	66.931	0.5034	−2.00	−3.97	136.59
96	63.821	0.49609	−2.03	−6.00	165.47
94	60.665	0.48873	−2.06	−8.07	188.94
92	57.459	0.48133	−2.09	−10.16	208.84
90	54.204	0.47388	−2.13	−12.29	226.11
88	50.897	0.46638	−2.16	−14.45	241.32
86	47.535	0.45883	−2.20	−16.65	254.83
84	44.118	0.45123	−2.24	−18.88	266.90
82	40.64	0.44357	−2.27	−21.16	277.72
80	37.104	0.43586	−2.31	−23.47	287.44
78	33.503	0.42809	−2.36	−25.83	296.15
76	29.835	0.42027	−2.40	−28.23	303.96
74	26.098	0.41238	−2.45	−30.68	310.92
72	22.288	0.40443	−2.50	−33.18	317.10
70	18.402	0.39642	−2.55	−35.72	322.53
68	14.435	0.38833	−2.60	−38.33	327.26
66	10.385	0.38018	−2.66	−40.99	331.32
64	6.2462	0.37196	−2.72	−43.70	334.73
62	2.0141	0.36366	−2.78	−46.49	337.52
60	−2.3165	0.35528	−2.85	−49.34	339.69
58	−6.7512	0.34682	−2.92	−52.25	341.27
56	−11.296	0.33827	−2.99	−55.25	342.26
54	**−15.959**	**0.32963**	**−3.07**	**−58.32**	**342.68**
52	−20.746	0.3209	−3.16	−61.48	342.52
50	−25.666	0.31207	−3.25	−64.73	341.79
48	−30.728	0.30314	−3.35	−68.08	340.49
46	−35.943	0.2941	−3.45	−71.54	338.61

I. VALVE–SYSTEM INTERACTION

In normal operation, a linear relation between the manipulated variable (valve stem position) and the desired variable (flow rate) is desired. However, the valve is normally a component of a flow system that includes a pump or other driver, pipe, and fittings characterized by loss coefficients, as illustrated in Figure 11.4. In such a system, the flow rate is a nonlinear function of the component loss coefficients. Thus, the control valve must have a nonlinear response (i.e., trim) to compensate for the nonlinear system characteristics if a linear response is to be achieved. Selection of the proper size and trim of the valve to be used for a given application requires matching the valve, piping system, and pump characteristics, all of which interact (Darby, 1997). The operating point for a piping system depends upon the pressure-flow behavior of both the system and the pump, as described in Chapter 8 and illustrated in Figure 8.2 (see also Example 8.1). The control valve acts like a variable resistance in the piping system. That is, as the valve is closed, the valve loss coefficient K_f increases and the valve sizing coefficient

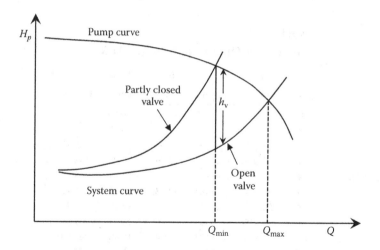

FIGURE 11.14 Effect of control valve on system operating point.

C_v decreases. The operating point for the system is where the pump head (H_p) characteristic intersects the system head requirement (H_s):

$$H_s = \frac{\Delta P}{\rho g} + \Delta z + \frac{Q^2}{g}\left[\frac{8}{\pi}\sum_i\left(\frac{K_f}{D^4}\right)_i + \frac{1}{\rho_w C_v^2}\right] = H_p \qquad (11.48)$$

where the last term in the brackets is the head loss through the control valve, h_v, from Equation 11.19, and C_v depends upon the valve stem travel, X (e.g., Figure 11.8):

$$C_v = C_{v,max}\, f(X) \qquad (11.49)$$

A typical situation is illustrated in Figure 11.14, which shows the pump curve and a system curve with no control valve, and the same system curve with a valve that is partially closed. Closing down on the valve (i.e., reducing X) decreases the valve C_v and increases the head loss (h_v) through the valve. The result is to shift the system curve upward by an amount h_v at a given flow rate (note that h_v also depends on the flow rate). The range of possible flow rates for a given valve (also known as the "turndown" ratio) lies between the intersection on the pump curve of the system curve with a "fully open" valve (Q_{max}, corresponding to $C_{v,max}$) and the intersection of the system curve with the (partly) closed valve. Of course, the minimum flow rate is zero when the valve is fully closed. The desired operating point should be as close as possible to Q_{max}, because this corresponds to an open valve with minimum flow resistance. The flow is then controlled by closing down on the valve (i.e., reducing X and C_v and thus raising h_v). The minimum operating flow rate (Q_{min}) is established by the turndown ratio (i.e., the operating range) required for proper control. These limits set the size of the valve (e.g., the required $C_{v,max}$), and the head flow rate characteristic of the system (including pump and valve) over the desired flow range. This determines the proper trim for the valve, as follows. As the valve is closed (i.e., reducing X), the system curve is shifted up by an amount h_v:

$$h_v = \frac{Q^2}{\rho_w g C_{v,max}^2 f^2(X)} \qquad (11.50)$$

where $f(X)$ represents the valve trim characteristic function. Equation 11.50 follows directly from Equations 11.19 and 11.49. Thus, as X (the relative valve stem travel) is reduced, $f(X)$ and C_v are also reduced. This increases h_v, with the result that the system curve now intersects the pump curve further to the left, at a lower value of Q. Substituting Equation 11.50 into Equation 11.48 gives

$$H_s = \frac{\Delta P}{\rho g} + \Delta Z + Q^2 \left[\frac{8}{g\pi^2} \sum_i \left(\frac{K_f}{D^4} \right)_i + \frac{1}{\left[0.658 C_{v,max} f(X) \right]^2} \right] = H_p \tag{11.51}$$

which relates the system required head (H_s) to the valve stem position (X).

J. MATCHING VALVE TRIM TO THE SYSTEM

The valve trim function is chosen to provide the desired relationship between the valve stem travel (X) and flow rate (Q). This is usually a linear relation and/or a desired sensitivity of the flow rate to a change in valve stem position. For example, if the operating point were far to the left on the diagram where the pump curve is fairly flat, then h_v would be nearly independent of flow rate. In this case, Q would be proportional to $C_v = C_{v,max} f(X)$, and a linear valve characteristic [$f(X)$ vs. X] would be desired. However, the operating point usually occurs where both curves are nonlinear, so that h_v depends strongly on Q, which in turn is a nonlinear function of the valve stem position, X. In this case, the most appropriate valve trim can be determined by evaluating Q as a function of X for various trim characteristics (e.g., Figure 11.8) and choosing the trim that provides the most linear response over the desired operating range. For a given valve (e.g., $C_{v,max}$), and a given trim response (e.g., $f(X)$), this can be done by calculating the system curve (e.g., Equation 11.51) for various valve settings (X) and determining the corresponding values of Q from the intersection of these curves with the pump curve (e.g., the operating point). The trim that gives the most linear (or most sensitive) relation between X and Q is then chosen. This process can be aided by fitting the trim function (e.g., Figure 11.8) by an empirical equation such as the following (Darby, 1997):

$$Linear\,trim{:}\quad f(X) = X \tag{11.52}$$

$$Parabolic\,trim{:}\quad f(X) = X^2 \quad or \quad X^n \tag{11.53}$$

$$Equal\,percentage\,trim{:}\quad f(X) = \frac{exp(aX^n) - 1}{exp(a) - 1} \tag{11.54}$$

$$Quick\,opening\,trim{:}\quad f(X) = 1 - \left[a(1-X) - (a-1)(1-X^n) \right] \tag{11.55}$$

where a and n are parameters that can be adjusted to give the best fit to the trim curves.

Likewise, the pump characteristic can usually be described by a quadratic equation of the form

$$H_p = H_o - cQ - bQ^2 \tag{11.56}$$

The operating point is where the pump head (Equation 11.56) intersects the system head requirement (Equation 11.51). Thus, if Equation 11.51 for the system head is set equal to Equation 11.56 for the pump head, the result can be solved for $1/f(X)^2$ to give

$$\left(\frac{1}{f(X)} \right)^2 = 0.433 C_{v,max}^2 \left[\frac{H_o - cQ - bQ^2 - \Delta z - \Delta P / \rho g}{Q^2} - \frac{0.00259 \sum K_f}{D_{in}^4} \right] \tag{11.57}$$

where conversion factors have been included for units of Q in *gpm*, H_o, Δz and $\Delta P/\rho g$ in *ft*, D_{in} in *in.*; and $C_{v,max}$ in ($gpm/psi^{1/2}$). The valve position X corresponding to a given flow rate Q is determined by equating the value of $f(X)$ obtained from Equation 11.57 to that corresponding to a specific valve trim, for example, Equations 11.52 through 11.55. This procedure is illustrated by the following example.

Example 11.6 Control Valve Trim Selection

It is desired to find the trim for a control valve that gives the most linear relation between the stem position (X) and flow rate (Q) when used to control the flow rate in the fluid transfer system shown in Figure 11.14. The fluid is water at 60°F, through 100 ft of 3 in. sch 40 pipe containing 12 standard threaded elbows in addition to the control valve. The fluid is pumped from tank 1 upstream at atmospheric pressure to tank 2 downstream, which is also at atmospheric pressure and at an elevation $z_2 = 20$ ft higher than tank 1. The pump is a 2 × 3 centrifugal pump with an 8 3/4 in. diameter impeller, for which the head curve can be represented by Equation 11.56 with $H_o = 360$ ft, $c = 0.0006$ ft of head/gpm, and $b = 0.0005$ ft of head/(gpm)^2 (with Q in gpm). A 3 in. control valve with the following possible trim plug characteristics will be considered.

Equal percentage (EP)	$C_v = 51 \dfrac{exp(0.5X^{2.5}) - 1}{exp(0.5) - 1}$
Modified parabolic (MP)	$C_v = 60X^{1.6}$
Linear (L)	$C_v = 64X$
Quick opening (QO)	$C_v = 70\left\{1 - \left[0.1(1 - X) - (0.1 - 1)(1 - X^{2.5})\right]\right\}$

The range of flow rates possible with the control valve can be estimated by inserting the linear valve trim (i.e., $C_{v,max} f(X) = 64X$) into Equation 11.51 and calculating the system curves for the valve open, half open, and one-fourth open ($X = 1, 0.5, 0.25$). The intersection of these system curves with the pump curve shows that the operating range with this valve is approximately 150–450 gpm, as shown in Figure E11.1.

The Q versus X relation for various valve trim functions can be determined as follows. First, a flow rate (Q) is assumed, which is used to calculate the Reynolds number and hence the pipe friction factor and the loss coefficients for the pipe and fittings. Then a valve trim characteristic function is assumed and, using the pump head function parameters, the right-hand side of

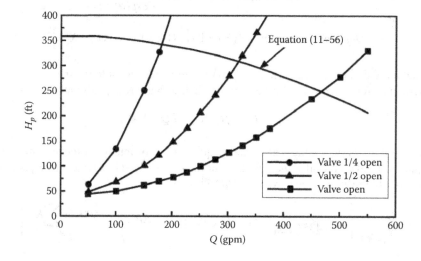

FIGURE E11.1   Flow range with linear trim.

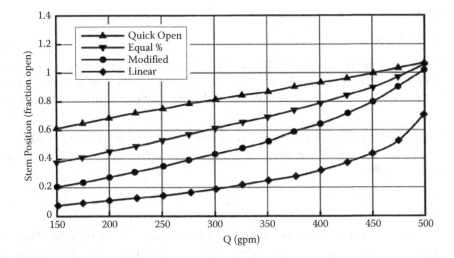

FIGURE E11.2   Flow rate versus stem position for various trim functions.

Equation 11.57 is evaluated. This gives the value for $f(X)$ for that trim, which corresponds to the assumed flow rate. This is then equated to the appropriate trim function for the valve given above (e.g., *EP, MP, L, QO*), and the resulting equation is solved for X (this may require an iteration procedure or the use of a nonlinear equation solver). The procedure is repeated over a range of assumed Q values for each of the given trim functions, giving the Q versus X response for each trim as shown in Figure E11.2. Aside from the *quick open* response, it is evident that the *equal percentage* trim results in the most linear response and also results in the greatest range or "turndown" (which is inversely proportional to the slope of the line).

In addition to the ISA/ANSI standards and the manufacturers' handbooks (e.g., Emerson/Fisher), there are numerous other references that offer guidelines and practical tips for selecting a suitable control valve for a specific application (e.g., Driskill et al. 1969, 1970, Christy 1996, Singleton 2013, etc.).

SUMMARY

The following are the most significant points covered in this chapter:
Safety relief valves

- The isentropic nozzle as the theoretical basis for sizing an SRV, and how it differs from the actual valve geometry
- The three elements needed to size an SRV for a given mass relief flow rate
- The equations and data needed to size an SRV for liquids, ideal gases, two-phase (or any) fluids, as well as flashing equilibrium and nonequilibrium flows using the HDI and HNDI methods.
- Selection of the proper discharge coefficient

Control valves

- The function of a control valve and the various trim characteristics
- The orifice equation as the theoretical basis for sizing control valves
- The *valve sizing coefficient*, and the various correction factors for attached piping, the flow coefficient, flashing and cavitating liquids, viscous liquids.
- Equations for subsonic and choked gas flow.
- General numerical (HDI) method for two-phase or any flow regime
- How to match the control valve trim to the piping system characteristics

PROBLEMS

SAFETY RELIEF VALVES

1. Rework Example 11.1 using the HDI method and tables for the properties of air, and compare the results with the example. Which answer do you think is the best, and why? (You should have access to a computer database file for the thermodynamic properties of various gases and vapors.)

2. An SRV must be sized to relieve hot water from a water heater at a rate of 20 lb_m/s. If the water is at 220°F and 45 psia, what nozzle area would be required? For the valve, $K_{DL} = 0.65$ and $K_{Dg} = 0.98$.

CONTROL VALVES

3. Air at 100 psia and 60°F is being controlled through a venturi control valve. The backpressure at the valve discharge is $P_2 = 30$ psia. The flow rate through the valve at its maximum opening is to be found, using the equations from ISA-75.01.01 and numerical integration. The parameters provided by the valve company for a wide-open valve are $C_1 = 14.1$ and $X_T = 0.1254$ (the low value of X_T indicates that it is a high-recovery valve). Additionally, $F_\gamma = 1.0$ (value for air); $C_\gamma = 355.9$ (value for air); molecular weight for air = 29; ideal gas compressibility $Z = 1.0$; and specific heat ratio, $\gamma = 1.4$ calculated at the inlet pressure and temperature.

4. You want to control the flow rate of a liquid in a transfer line at 350 gpm. The pump in the line has the characteristics shown in Figure 8.2, with a 5 1/4 in. impeller. The line contains 150 ft of 3 in. sch 40 pipe, 10 flanged elbows, four gate valves, and a 3 × 3 control valve. The pressure and elevation at the entrance and exit of the line are the same. The valve has an equal percentage trim. What should the value of C_v be for the valve to achieve the desired flow rate? The fluid has a viscosity of 5 cP and an SG of 0.85.

5. A liquid with a viscosity of 25 cP and an SG of 0.87 is pumped from an open tank to another tank in which the pressure is 15 psig. The line is 2 in. sch 40 diameter, 200 ft long, and contains eight flanged elbows, two gate valves, a control valve, and an orifice meter.
    What should be the diameter of the orifice in the line for a flow rate of 100 gpm, if the pressure across the orifice is not to exceed 80 in. H_2O?

6. Water at 60°F is to be transferred at a rate of 250 gpm from the bottom of a storage tank to the bottom of a process vessel. The water level in the storage tank is 5 ft above ground level, and the pressure in the tank is 10 psig. In the process vessel, the level is 15 ft above ground and the pressure is 20 psig. The transfer line is 150 ft of 3 in. sch 40 pipe, containing eight flanged elbows, three 80% reduced trim gate valves, and a 3 × 3 control valve. The pump in the line has the same characteristics as those shown in Figure 8.2 with an 8 in. impeller, and the control valve has a linear characteristic. If the stem on the control valve is set to provide the desired flow rate under the specified conditions, what should be the required value of the valve coefficient to achieve this?

7. A piping system takes water at 60°F from a tank at atmospheric pressure to a plant vessel at 25 psig that is 30 ft higher than the upstream tank. The transfer line contains 300 ft of 3 in. sch 40 pipe, 10 90° elbows, an orifice meter, a 2 × 3 pump with a 7 3/4 in. impeller (with the characteristic as given on Figure 8.2), and a 3 × 2 equal percentage control valve. A constant flow rate of 200 gpm is required in the system.
    (a) What size orifice should be installed if the DP (differential pressure) cell used to measure the pressure drop across the orifice has a maximum range of 25 in. H_2O?
    (b) What is the C_v of the valve that gives the required flow rate?

NOTATION

A	Flow cross-sectional area, $[L^2]$
A_o	Theoretical flow area for control valve, $[—]$
$A_o K_d$	Effective flow area for fully open control valve, $[L^2]$
c_p	Specific heat at constant pressure, $[FL/M]$
c_v	Specific heat at constant volume, $[FL/M]$
C_1	C_g/C_v, $[M/L^7]^{1/2}$
C_2	Correction factor for isentropic exponent, Equation 11.40 $[—]$
C_d	Control valve discharge coefficient $\left(= C_o\sqrt{1-\beta^2}\right)$, $[—]$
C_g	Flow correction factor for gases $[—]$
C_o	Flow coefficient for control valve $(= K_d)$, $[—]$
C_v	Control valve sizing coefficient, $[L^7/M]^{1/2}$
C_v	Control valve viscosity correction coefficient, Figure 11–12 $[—]$
d	Control valve nominal valve size, $[L]$
EP	Equal percentage control valve trim, $[—]$
F_d	Control valve style modifier, $[—]$
F_F	Control valve liquid critical pressure ratio $(= r_c)$, $[—]$
F_G	Vena contracta correction for gas flow, Equations 11.35, $[—]$
F_L	Control valve liquid pressure recovery factor $(= K_m)^{1/2}$, $[—]$
F_p	Control valve piping geometry factor, $[—]$
F_v	Control valve viscosity correction factor, $[—]$
G_c	Maximum (choked) mass flux, $[M/L^2 t]$
G_n	Measured mass flux through actual nozzle, $[M/L^2 t]$
G_o	Theoretical mass flux through isentropic nozzle, $[M/tL^2]$
h_v	Head loss in ft of fluid, $[L]$
H	Specific enthalpy (per unit mass), $[L^2/t^2]$
H_P	Pump head, $[L]$
H_s	System head (Figure 11.14), $[L]$
k	Isentropic exponent $= \gamma$ (ideal gas), $[—]$
K_d	Flow coefficient for control valve $(= C_o)$, $[—]$
K_D	SRV discharge coefficient, $[—]$
K_f	Friction loss coefficient (pipe, fittings, etc.), $[—]$
K_{DG}	SRV gas discharge coefficient, $[—]$
K_{DL}	SRV liquid discharge coefficient, $[—]$
K_m	Control valve liquid recovery coefficient $(= F_L^2)$, $[—]$
K_v	Viscosity correction factor, Equation 11.7, $[—]$
L	Linear control valve trim, $[—]$
L	Nozzle length, $[L]$
$\dot{m}$	Mass flow rate, $[M/t]$
MP	Modified equal percentage control valve trim, $[—]$
M_w	Molecular weight, $[M/mol]$
N_{Rev}	Control valve Reynolds number, $[—]$
N	"Equation constant" for unit conversions, $[—]$
P	Pressure, $[F/L^2]$
P_b	Back pressure on valve, $[F/L^2]$
ΔP_{cav}	Pressure at which cavitation appears, $[F/L^2]$
P_c	Critical (choked) pressure, $[F/L^2]$
P_{tc}	Thermodynamic critical pressure (absolute), $[F/L^2]$
P_v	Fluid vapor pressure, $[F/L^2]$

Q	Volumetric flow rate, gpm, [L^3/t]
QO	Quick opening control valve trim, [—]
R	Ideal gas constant, [ML^2/mol t^2 °]
r_c	Critical pressure ratio at which choking will occur (= $P_{2c}/P_v = F_F$), [—]
S	Slip, ratio of gas velocity to liquid velocity in a mixture, [—]
SRV	Safety relief valve
T	Absolute temperature, [°]
u	Velocity, [L/t]
x	Quality, or mass fraction of gas/vapor in mixture, [—]
X	Fraction of fully open stem for control valve, [—]
X	Dimensionless pressure crop across control valve, Equation 11.43, [—]
X_T	Maximum value of X for a specific valve, [—]
x_T	Control valve pressure differential ratio factor without fittings at choked flow, [—]
Y	Control valve expansion factor, [—]

GREEK

α	Volume fraction of gas/vapor in mixture, [—]
β	$D_n/D_p = d_o/D$, Ratio of nozzle or orifice diameter to inlet pipe diameter, [—]
γ	$C_p/C_v = k$, [—]
ν	Specific volume, [L^3/M]
ρ	Density, [M/L^3]

SUBSCRIPTS

b	Valve exit
c	Critical or choked value
e	Equilibrium
G	Gas/vapor
L	Liquid
n	Nozzle exit
o	Nozzle entrance
s	Saturation
vc	Vena contracta

REFERENCES

Baumann, H.D., *Control Valve Primer*, Instrument Society of America, Durham, NC, 1991.

Cheremisinoff, N.P. and P.N. Cheremisinoff, *Instrumentation for Process Flow Engineering*, Technomic Publishing Company, Lancaster, PA, 1987.

Center for Chemical Process Safety (CCPS), *Guidelines for Pressure Relief and Effluent Handling Systems*, American Institute of Chemical Engineers, New York (1998) (in revision).

Christy, J.R.E., On selecting appropriate control valves for pipework systems, *Chemical Engineering Education*, Winter 1996, pp. 54–57.

Darby, R., Control valves: Match the trim to the selection, *Chem. Eng.*, 104, 147–152, June 1997.

Darby, R and K Molavi, "Viscosity Correction Factor for Safety Relief Valves", *Process Safety Progress*, 16(2), 80–83, **16,** no.2 (1997)

Darby, R., F.E. Self, and V.H. Edwards, Properly size pressure-relief valves for two-phase flow, *Chem. Eng.*, 109(6), 68–74, June 2002.

Darby, R., On two-phase frozen and flashing flows in safety relief valves, *J. Loss Prev. Process Ind.*, 17, 255–259, 2004.

Darby, R., Size safety-relief valves for any conditions, *Chem. Eng.*, 112(9), 42–50, September 2005.

Darby, R., *A Review of the HDI/HNDI Valve Sizing Method and Appropriate Discharge Coefficient*, DIERS Users Group, Reno, NV, October 2010.

Driskell, L.R., New approach to control valve sizing, *Hydrocarb. Process.*, 48(7), 131–134, July 1969.

Driskell, L.R., Sizing valves for gas flow, *ISA Trans.*, 9(4), 325–331, 1970. (*Note*: Both Driskell papers define AK = Cv FL/38 where K is the "flow coefficient including velocity of approach".)

Emerson Process Management, *Fisher Controls, Control Valve Handbook*, 4th edn., Emerson Process Management, Marshalltown, IA, 2005.

Fisher Controls, Catalog 10, Ch. 2, Marshalltown, IA, 1987.

Fisher Controls, *Control Valve Source Book*, Fisher Controls, International, Marshalltown, IA, 1990.

Henry, R.E. and H.K. Fauske, The two-phase critical flow of one-component mixtures in nozzles, orifices and short tubes, *Trans. Am. Soc. Mech. Eng.*, *J. Heat Transfer*, 93, 179–187, May 1971.

Hutchison, J.W., *ISA Handbook of Control Valves*, 1st edn., Instrument Society of America, Durham, NC, 1971.

International Electrotechnical Commission (IEC), IEC-60534-2-1. *Industrial Process Control Valves— Part 2-1: Flow Capacity—Sizing Equations for Fluid Flow under Installed Conditions*, 2011.

Instrument Society of America (ISA), ISA-75.01.01 (60534-2-1 MOD). *Industrial Process-Control Valves— Part 2-1: Flow Capacity—Sizing Equations for Fluid Flow under Installed Conditions*, 2012.

Lenzing, F., J. Friedel, J. Cremers, and M. Alhusein, Prediction of the maximum full lift safety valve two-phase flow capacity, *J. Loss Prev. Proc. Ind.*, 11, 307–321, 1998.

Lenzing, T., J. Schecker, L. Friedel, and J. Cremers, Safety relief valve critical mass flux as a function of fluid properties and valve geometry, Presented at *ISO TC 185 WG1 Meeting*, Paper N103, Rome, Italy, October 29–31 1997.

Lipták, B.G., C.G. Langford, F.M. Cain, and H.D. Baumann, Sizing, Chapter 16, in *Process Control and Optimization, vol. 2 of Instrument Engineers' Handbook*, 4th edn., B.G. Liptaak, (Ed.), CRC Press, Boca Raton, FL, 2006, p. 1234

NIST (National Institute of Science and Technology), Standard Reference Database 23, Reference Fluid Thermodynamic and Transport Properties—REFPROP, Version 8.0 (and subsequent versions), 2015.

Singleton, E., *A Specifier's Guide to Control Valves*, KCI Publishing, 2013.

Stiles, G.F., Liquid viscosity effects on control valve sizing, *14th Annual Symposium on Instrumentation for the Process Industries*, Texas A&M University, College Station, TX 1964.

12 External Flows

"You have to learn the rules of the game. And then you have to play better than anyone else."

—Albert Einstein, 1879–1955, Physicist

I. THE DRAG COEFFICIENT

When a fluid flows past a solid body, or the body moves through the fluid (e.g., Figure 12.1), the force (F_D) exerted on the body by the fluid is proportional to the relative rate of momentum (mass flow rate × velocity = $\rho V^2 A$) transported by the fluid. This can be expressed in terms of a dimensionless *drag coefficient* (C_D), which is defined by the following equation:

$$\frac{F_D}{A} = \frac{C_D}{2}\rho V^2 \tag{12.1}$$

Here

ρ is the density of the fluid

V is the relative velocity between the fluid and the solid body

A is the cross-sectional area of the body normal to the velocity vector V. For example, for a sphere of diameter d it is $\pi d^2/4$

Note that the definition of the drag coefficient from Equation 12.1 is analogous to that of the friction factor for flow in a conduit, that is,

$$\tau_w = \frac{f}{2}\rho V^2 \tag{12.2}$$

where τ_w is the force exerted by the moving fluid on the wall of the pipe per unit area. In the case of τ_w, however, the area is the total contact area between the fluid and the conduit wall, as opposed to the cross-sectional area normal to the flow direction (or projected area) in the case of C_D. One reason for this is that the fluid interaction with the tube wall is uniform over the entire surface for fully developed one-dimensional flow, whereas for a body immersed in a moving fluid, the nature and degree of interaction varies with the position around the body, which is two- or three-dimensional flow.

A. STOKES FLOW

If the relative velocity is sufficiently low, the fluid streamlines can follow the contour of the body almost completely all the way around without incurring any loss of momentum (this is called *creeping flow*). For this case, the microscopic momentum balance equations in spherical coordinates for the two-dimensional flow [$v_r(r, \theta)$, $v_\theta(r, \theta)$] of a Newtonian fluid were solved by Stokes for the distribution of the pressure, velocity, and local stress components. These equations can then be integrated over the surface of the sphere to determine the total drag F_D acting on the sphere in the direction of flow, 2/3 of which results from viscous drag and 1/3 from the nonuniform pressure distribution (form drag). That is:

$$F_D = \underset{\text{(viscous drag)}}{2\pi\mu dV} + \underset{\text{(form drag)}}{\pi\mu dV} \tag{12.3}$$

333

FIGURE 12.1   Drag on a sphere.

This result can be expressed in dimensionless form using Equation 12.1, as a theoretical expression for the drag coefficient:

$$C_D = \frac{24}{N_{Re}} \tag{12.4}$$

where

$$N_{Re} = \frac{dV\rho}{\mu} \tag{12.5}$$

This is known as *Stokes flow*, and Equation 12.4 has been found to be accurate for flow over a sphere in an unconfined expanse of fluid for $N_{Re} < 0.1$ (or to within about 5% for $N_{Re} < 1$). Note the similarity between Equation 12.4 and the dimensionless *Hagen–Poiseuille* equation for laminar tube flow, that is, $f = 16/N_{Re}$. However, note that the limiting value of the Reynolds number for the validity of these equations varies from 2100 (for a pipe) to ~0.1 for a sphere. This is due to the difference in the nature of the transition to turbulent flow, as explained below.

B. FORM DRAG

As the fluid flows over the forward part of the sphere, the velocity increases because the available flow area decreases, and the pressure must decrease in order to satisfy the conservation of energy. Conversely, as the fluid flows around the back side of the body, the velocity decreases due to the expanding flow area, and the pressure increases. This is not unlike the flow in a diffuser or a converging-diverging duct. The flow behind the sphere thus moves from a low-pressure point near the equator to a higher pressure, that is, into a region of "adverse pressure gradient" behind the sphere. This is inherently unstable so that, as the relative velocity (and N_{Re}) increases, it becomes more difficult for the streamlines to follow the contour of the body and they eventually break away from the surface. This condition is called *separation*, although it is the smooth streamline that is separating from the surface, not the fluid itself. When separation occurs, eddies or vortices form behind the body, as illustrated in Figure 12.1, which form a "wake" behind the sphere.

As the velocity (and N_{Re}) increases, the point of streamline separation from the surface moves further upstream and the wake gets larger, both in the lateral and axial directions. The wake region contains circulating eddies of a three-dimensional turbulent nature, so it is a region of relatively high velocity and hence low pressure. Thus, the pressure in the wake is lower than that on the front of the sphere, and the product of this pressure difference and the projected area of the wake results in a net force acting on the sphere in the direction of the flow, that is, in the same direction as the drag force. This additional force resulting from the low pressure in the wake is called *form drag* (i.e., that component of the drag due to the pressure distribution, in excess of the viscous drag). The total drag is thus a combination of Stokes drag and wake drag, and consequently the drag coefficient is greater than that given by Equation 12.4 for $N_{Re} > 0.1$. This is illustrated in Figure 12.2, which shows C_D versus N_{Re} for spheres (as well as for cylinders and disks oriented normal to the flow direction). For $1000 < N_{Re} < 1 \times 10^5$, the sphere drag coefficient is approximately $C_D = 0.45$. In this region,

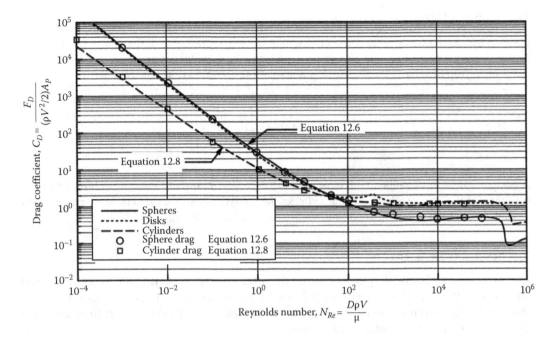

FIGURE 12.2 Drag coefficients for spheres, disks, and cylinders. (From Perry, J.H., Ed., *Chemical Engineers' Handbook*, 6th edn., McGraw-Hill, New York, 1984.)

the wake, and consequently the wake drag, is a maximum, and the streamlines actually separate slightly ahead of the equator of the sphere. The drag at this point is completely dominated by the wake, which is actually larger in diameter than the sphere itself (see Figure 12.4).

C. ALL REYNOLDS NUMBERS

For $N_{Re} > 0.1$ (or >1, within ~5%), a variety of expressions for C_D versus N_{Re} (mostly empirical) have been proposed in the literature (Clift et al., 1978; Chhabra, 2006). However, a simple and very useful equation, which represents the entire range of C_D versus N_{Re} reasonably well (within experimental error) up to about $N_{Re} = 2 \times 10^5$, has been given by Dallavalle (1948):

$$C_D = \left(0.632 + \frac{4.8}{\sqrt{N_{Re}}} \right)^2 \tag{12.6}$$

(Actually, according to Coulson et al. [1991], this equation was first presented by Wadell in 1934.) A comparison of Equation 12.6 with the measured values is shown in Figure 12.2. A somewhat more accurate equation, although more complex, has been proposed by Khan and Richardson (1987):

$$C_D = \left(\frac{2.25}{N_{Re}^{0.31}} + 0.358 N_{Re}^{0.06} \right)^{3.45} \tag{12.7}$$

Although Equation 12.7 is more accurate than Equation 12.6 at intermediate values of N_{Re}, Equation 12.6 provides a sufficiently accurate prediction for most applications. Also it is simpler to manipulate, so we will prefer it as the analytical expression describing the drag coefficient for a sphere.

D. Cylinder Drag

For flow past a circular cylinder normal to the cylinder axis, with $L/d \gg 1$, the flow is similar to that for the sphere. An equation that adequately represents the cylinder drag coefficient over the entire range of N_{Re} (up to about 2×10^5), which is analogous to the Dallavalle equation, is

$$C_D = \left(1.05 + \frac{1.9}{\sqrt{N_{Re}}}\right)^2 \tag{12.8}$$

A comparison of this equation with measured values is also shown in Figure 12.2. It may be noted here that for the case of a long circular cylinder, no analytical expression akin to Equation 12.3 is possible in the limit of low Reynolds numbers (see, e.g., Lamb, 1945).

E. Boundary Layer Effects

As seen in Figure 12.2, the drag coefficient for the sphere exhibits a sudden drop from 0.45 to about 0.15 (almost 70%) at a Reynolds number of $2–5 \times 10^5$. For the cylinder, the drop is from about 1.1 to about 0.35. This drop is due to the transition of the flow inside the boundary layer from laminar to turbulent flow and can be explained as follows.

As the fluid encounters the solid boundary and proceeds along the surface, a boundary layer forms as illustrated in Figure 12.3. The boundary layer is the region of the fluid near a solid boundary in which viscous forces dominate and the velocity varies with the distance, from zero at the wall to a maximum value (V) at the edge of the boundary layer. Outside the boundary layer, the fluid velocity is that of the free stream that is unaffected by the presence of the wall. Adjacent to the wall in the boundary layer, the velocity is small and and the flow is laminar and stable. However, the boundary layer thickness (δ) grows along the plate (in the x direction), in proportion to $N_{Rex}^{1/2}$ (where $N_{Rex} = xV\rho/\mu$). As the boundary layer grows, inertial forces increase and the boundary layer becomes less stable until it reaches a point (at $N_{Rex} \approx 2 \times 10^5$) where it becomes unstable, that is, turbulent. Within the turbulent boundary layer, the flow streamlines are no longer parallel to the boundary but break up into three-dimensional eddy structures.

With regard to the flow over an immersed body (e.g., a sphere), the boundary layer grows from the impact (stagnation) point along the front of the body and remains laminar until $N_{Rex} \approx 2 \times 10^5$, where x is the distance traveled along the surface of the sphere, at which point it becomes turbulent. If the boundary layer is laminar at the point where streamline separation occurs, the separation point can be ahead of the equator of the sphere, resulting in a wake diameter

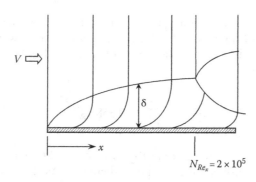

FIGURE 12.3 Boundary layer over a flat plate.

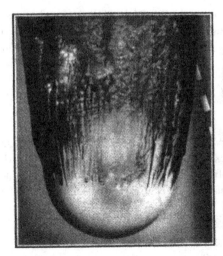 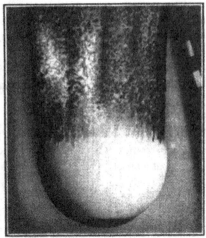

FIGURE 12.4 Two bowling balls falling in still water at 25 ft/s. The ball on the left is smooth and the one on the right has a patch of sand on the nose. (From Coulson, J.M. et al., *Chemical Engineering*, Vol. 2, 4th edn., Pergamon Press, 1991.)

which is larger than that of the sphere. However, if the boundary layer becomes turbulent before separation occurs, the three-dimensional eddy structure in the turbulent boundary layer carries momentum components inward toward the surface, which delays the separation of the streamline and tends to stabilize the wake. This delayed separation results in a smaller wake and a corresponding reduction in the form drag, which is the cause of the sudden drop in C_D at $N_{Re} \approx 2 \times 10^5$.

This shift in the size of the wake can be rather dramatic, as illustrated in Figure 12.4, which shows two pictures of a bowling ball falling in water with the wake clearly visible. The ball on the left shows a large wake since the boundary layer is laminar and the separation point is ahead of the equator. The ball on the right has a rougher surface that promotes turbulence, and the boundary layer has become turbulent before separation occurs resulting in a much smaller wake due to the delayed separation. The primary effect of surface roughness on the flow around immersed objects is to promote transition to a turbulent boundary layer and to delay the separation of the streamlines, and thus to slightly lower the value of N_{Re} at which the sudden drop (or "kink") in the $C_D - N_{Re}$ curve occurs. This apparent paradox—wherein the promotion of turbulence actually results in lower drag—has been exploited in various ways, such as the dimples on golf balls and the "boundary layer spoilers" on airplane wings and automobiles.

II. FALLING PARTICLES

Many engineering operations involve the separation of solid particles from fluids, in which the motion of the particles is the result of a gravitational force (or other potential such as centrifugal force in a centrifuge or electric field in electrostatic precipitators). To illustrate this, consider a spherical solid particle with diameter d and density ρ_s, surrounded by a fluid of density ρ and viscosity μ, which is released and begins to fall (in the $x = -z$ direction) under the influence of gravity. A momentum balance on the particle is simply $\Sigma F_x = ma_x$, where the forces include gravity acting on the solid particle (F_g), the buoyant force due to the fluid displaced by the solid particle (F_b), and the drag exerted by the fluid (F_D). The inertial term involves the product of the acceleration ($a_x = dV_x/dt$) and the effective mass (m) of the particle. The mass that is accelerated includes that of the solid (m_s) as well as the "virtual mass" (m_f) of the fluid, which is displaced by the body as it accelerates. It can be shown that the latter is equal to one-half of the total mass of the displaced fluid (Clift et al., 1978), that is, $m_f = \frac{1}{2} m_s (\rho/\rho_s)$. Thus, the x component of the momentum balance becomes

$$\frac{g(\rho_s - \rho)\pi d^3}{6} - \frac{C_D \rho \pi d^2 V^2}{8} = \frac{\pi d^3 (\rho_s + \rho/2)}{6} \frac{dV}{dt} \tag{12.9}$$

At $t = 0$, $V = 0$ and the drag force is zero. As the particle accelerates, the drag force increases, which decreases the acceleration. This process continues until the acceleration drops to zero, at which time the particle falls at a constant velocity resulting from the balance of forces due to drag and gravity (buoyancy). This steady-state *terminal velocity* of the body is given by the solution to Equation 12.9 with the acceleration equal to zero:

$$V_t = \left(\frac{4g\Delta\rho d}{3\rho C_D} \right)^{1/2} \tag{12.10}$$

where $\Delta\rho = (\rho_s - \rho)$. It is evident that the velocity cannot be determined until the drag coefficient is known, which in turn depends on the velocity. If the Stokes flow regime prevails, then $C_D = 24/N_{Re}$, and Equation 12.10 becomes

$$V_t = \frac{g\Delta\rho d^2}{18\mu} \tag{12.11}$$

However, the criterion for Stokes flow (i.e., $N_{Re} < 1$) cannot be tested until V_t is known, and if it is not valid, then Equation 12.11 will not apply. This will be addressed shortly. It is, however, useful to note here that under these conditions, the terminal falling velocity of a spherical particle varies as $V_t \propto d^2$. Indeed, this forms the basis of classifying mixtures of particles of different sizes in settling ponds, centrifuges, thickeners, etc.

There are several types of problems that we may encounter with falling particles, depending upon what is known and what is to be found. All of these problems involve the two primary dimensionless variables C_D and N_{Re} or combinations thereof. The former is determined, for gravity-driven motion, by Equation 12.10, that is,

$$C_D = \frac{4g\Delta\rho d}{3\rho V_t^2} \tag{12.12}$$

and C_D can be related to N_{Re} by the Dallavalle equation (Equation 12.6) over the entire practical range of N_{Re}. The following procedures for the various types of problems apply to Newtonian fluids under all flow conditions.

A. Unknown Velocity

In this case, the unknown velocity (V_t) appears in the definitions of both the drag coefficient C_D (Equation 12.12) and the Reynolds number N_{Re} (Equation 12.5). Hence, a suitable dimensionless group that does not contain the unknown V_t can be formulated as follows:

$$C_D N_{Re}^2 = \frac{4d^3 \rho g \Delta\rho}{3\mu^2} = \frac{4}{3} N_{Ar} \tag{12.13}$$

where N_{Ar} is known as the *Archimedes* number (also sometimes called the *Galileo* number). The most appropriate set of dimensionless variables to use for this problem is therefore N_{Ar} and N_{Re}. An equation for N_{Ar} can be obtained by multiplying both sides of the Dallavalle equation (Equation 12.6) by N_{Re}, and the result can then be rearranged for N_{Re} to give

$$N_{Re} = \left[\left(14.42 + 1.827\sqrt{N_{Ar}} \right)^{1/2} - 3.798 \right]^2 \tag{12.14}$$

The procedure for determining the unknown velocity is, therefore, as follows:

Given: d, ρ, ρ_s, μ *Find*: V_t

1. Calculate the Archimedes number from Equation 12.13:

$$N_{Ar} = \frac{d^3 \rho g \Delta \rho}{\mu^2} \tag{12.15}$$

2. Insert this value into Equation 12.14 and calculate N_{Re}.
3. Determine V_t from N_{Re}, that is, $V_t = N_{Re}\mu/d\rho$.

If $N_{Ar} < 15$, then the system is within the Stokes law range and the terminal velocity is given by Equation 12.11.

B.   UNKNOWN DIAMETER

It often happens that we know or can measure the particle velocity and wish to know the size of the falling particle. In this case, we may form a dimensionless group that does not contain d as follows:

$$\frac{C_D}{N_{Re}} = \frac{4\mu\Delta\rho g}{3\rho^2 V_t^3} \tag{12.16}$$

This group can be related to the Reynolds number by dividing Equation 12.6 by N_{Re} and then solving the resulting equation for $1/(N_{Re})^{1/2}$ to give

$$\frac{1}{\sqrt{N_{Re}}} = \left(0.00433 + 0.208\sqrt{\frac{C_D}{N_{Re}}} \right)^{1/2} - 0.0658 \tag{12.17}$$

The two appropriate dimensionless variables are now C_D/N_{Re} and N_{Re}. The procedure is as follows:

Given: V_t, ρ_s, ρ, μ *Find*: d

1. Calculate C_D/N_{Re} from Equation 12.16.
2. Insert the result into Equation 12.17 and calculate $1/(N_{Re})^{1/2}$, and hence N_{Re}.
3. Calculate $d = \mu N_{Re}/V_t\rho$.

If $C_D/N_{Re} > 30$, the flow is within the Stokes law range, and the diameter can be calculated directly from Equation 12.11:

$$d = \left(\frac{18\mu V_t}{g\Delta\rho} \right)^{1/2} \tag{12.18}$$

This is often used to estimate the size of small particles by measuring their terminal velocity in a container much larger than the particles, with a fluid of known properties.

C. UNKNOWN VISCOSITY

The viscosity of a Newtonian fluid can be determined by measuring the terminal velocity of a sphere of known diameter and density if the fluid density is known. If the Reynolds number is low enough for Stokes flow to apply ($N_{Re} < 0.1$), then the viscosity can be determined directly by rearrangement of Equation 12.11 as follows:

$$\mu = \frac{d^2 g \Delta\rho}{18 V_t} \tag{12.19}$$

This equation forms the basis of commercially available falling ball viscometers (Gupta, 2014). The Stokes flow criterion is rather stringent e.g., a 1 mm diameter sphere would have to fall at a rate of 1 mm/s or slower in a fluid with a viscosity of 10 cP and SG = 1 to be in the Stokes range, which means that the density of the solid would have to be within 2% of the density of the fluid!

However, with only a slight loss in accuracy, the Dallavalle equation can be used to extend the useful range of this measurement to a much higher Reynolds number, as follows. From the known quantities, C_D can be calculated from Equation 12.12. The Dallavalle equation (Equation 11.6) can be rearranged to give N_{Re}:

$$N_{Re} = \left(\frac{4.8}{C_D^{1/2} - 0.632} \right)^2 \tag{12.20}$$

The viscosity is then determined from the known value of N_{Re}:

$$\mu = \frac{d V_t \rho}{N_{Re}} \tag{12.21}$$

Note that when $N_{Re} > 1000$, $C_D \approx 0.45$ (constant), which, from Equation 12.20, gives $\mu = 0$! Although this may seem strange, it is consistent because in this range the drag is dominated by form (wake) drag and viscous forces are negligible. It should be evident that one cannot determine the viscosity from measurements made under conditions that are insensitive to viscosity, which means that the utility of Equation 12.20 is limited in practice to approximately $N_{Re} < 50–100$.

III. CORRECTION FACTORS

A. WALL EFFECTS

All expressions presented so far have assumed that the particles are surrounded by an infinite sea of fluid, that is, that the boundaries of the fluid container are far enough from the particle that their influence is negligible. For a falling particle, this might seem to be a reasonable assumption if $d/D < 0.01$ (say), where D is the container diameter. However, the presence of the wall is felt by the particle over a much greater distance than one might expect. This is because as the particle falls it must displace an equal volume of fluid, which must then flow upward around the particle to fill the space just vacated by the particle. Thus, the relative velocity between the particle and the adjacent fluid is much larger in a confined space than it would be in an infinite fluid, that is, the effective "free stream" (relative) velocity is no longer zero, as it would be for an infinite stagnant fluid. A variety of analyses of this problem have been performed, as reviewed by Clift et al. (1978) and Chhabra (2002, 2006). These can be represented by a wall correction factor (K_w) that is a multiplier for the "infinite fluid" terminal velocity and corrects for the wall effect. (This is also equivalent to correcting the Stokes law drag force by a factor of K_w.) The following equation due to Francis (1933) is found to be valid for $d/D < 0.97$ and $N_{Re} < 1$:

$$K_{w_o} = \left(\frac{1 - d/D}{1 - 0.475 d/D} \right)^4 \tag{12.22}$$

For larger Reynolds numbers, the following expression is found to be satisfactory for $d/D < 0.8$ and $N_{Re} > 1000$:

$$K_{w_\infty} = 1 - \left(\frac{d}{D}\right)^{1.5} \qquad (12.23)$$

Although these wall correction factors appear to be independent of the Reynolds number for small (Stokes) and large values of N_{Re} (> 1000), the value of K_w is a function of both N_{Re} and (d/D) for intermediate Reynolds numbers (Chhabra, 2002, 2006). Further corrections to the measured falling velocities are required for off-center settling and/or in short tubes due to bottom effects (Wham et al., 2002; Chhabra, 2006).

B. Effect of Particle Shape

It is readily conceded that nonspherical particles like cylinders, disks, thin wires, cones, prisms, oblates and prolates, or of irregular shape, etc. are encountered much more frequently than the idealized case of spherical particles. Naturally, the drag and terminal settling velocity of such particles in fluids are strongly influenced by their shape and orientation. For instance, the settling velocity of a given cylinder (known length and diameter) can vary by a large factor depending upon its orientation during its sedimentation. Therefore, the need to estimate the terminal settling velocity (or drag force) often arises in process design calculations for slurry pipe lines, fluidized beds, hydrocyclones, settling chambers, etc. Most of the results available in the literature (Chhabra et al., 1999; Chhabra, 2006) fall into two distinct categories: in the first approach, the particle shape and orientation are maintained constant and drag coefficient is correlated with the Reynolds number by numerical solution of the Navier–Stokes equations or experiments. Naturally, this approach is accurate but has limited appeal due to its applicability to a fixed shape and orientation. In the second approach, the results for variously shaped particles are consolidated via a single correlation. The following discussion is based mainly on the latter approach.

At least two attributes of a nonspherical particle are needed to account for the settling behavior, namely, size and shape. A sphere is unique in so far that one only needs to specify its radius or diameter as it exhibits perfect symmetry. This is not so for a nonspherical particle, for example, diameter and length of a cylinder, minor and major axes of spheroids, etc. are needed to describe their "size." One of the simplest definitions of the effective diameter of a nonspherical particle is the so-called Stokes diameter, which is the diameter of the sphere that has the same terminal velocity as the actual nonspherical particle. This can be determined from a direct measurement of the settling rate of the particles and provides the best value of equivalent diameter for use in applications involving fluid drag on particles. Another widely used measure of the size of a nonspherical particle is the diameter of the sphere with a volume equal to that of the particle, d_s. For a nonspherical particle of volume V_p, the equivalent diameter, d_s, is

$$d_s = \left(\frac{6V_p}{\pi}\right)^{1/3} \qquad (12.24)$$

Next, it is conceivable that a cylinder and a spheroid can have the same d_s value, yet their terminal falling velocity (and drag force experienced) differs from each other significantly due to intrinsic difference in their shapes. The simplest measure of the shape (or deviation from a sphere) is the so-called sphericity of a particle, ψ. The sphericity is defined as

$$\psi = \frac{\text{surface area of an equal volume sphere}}{\text{surface area of the actual particle}} \qquad (12.25)$$

Equation 12.25 can also be expressed in terms of the volume (V_p) and surface area (A_p) of the particle as

$$\psi = \frac{\pi d_s^2}{A_p} = \frac{\pi(6V_p/\pi)^{2/3}}{A_p} = (6^2\pi)^{1/3}\frac{V_p^{2/3}}{A_p} = \frac{4.84V_p^{2/3}}{A_p} \tag{12.26}$$

Obviously for a sphere, $\psi = 1$ and, since for a given volume, the sphere has the minimum surface area/volume, $\psi < 1$ for all other shapes. It is also possible to define an effective particle diameter by $d = \psi d_s = 6/a_s$, where $a_s = $ *particle surface area/particle volume*. If the particles are spherical with diameter d, then $a_s = 6/d$. This definition is commonly used in the context of packed and fluidized beds, as will be seen in Chapters 15 and 16. These two definitions of ψ are equivalent.

Representative values of ψ for some familiar shapes are given in Table 12.1. The smaller the value of ψ, the greater is the departure from a spherical shape.

Using the two parameters d_s and ψ, Haider and Levenspiel (1989) collated much of the literature data for drag behavior of nonspherical particles, as shown in Figure 12.5. While the general trend for a fixed value of ψ is similar to that seen for a sphere in Figure 12.2 (except for the sudden drop in the value of the drag coefficient at $N_{Re} \sim 2 \times 10^5$), the drag experienced by a nonspherical particle is always higher than that of a sphere at the same Reynolds number. Further inspection of Figure 12.5 shows that the onset of fully turbulent regime (constant value of drag coefficient) occurs at lower Reynolds numbers for smaller values of ψ. For thin disks oriented normal to the flow, this limiting behavior is approached at about $N_{Re} \sim 10$ as opposed to that for a sphere, that is, $N_{Re} \sim 10^3$.

Another interesting observation is that the drag is seen to be influenced by shape much more at high Reynolds numbers than that at low Reynolds numbers. Generally, the size of the particle (surface area) matters more at small Reynolds numbers, whereas the shape of the particle (ψ) is more relevant at high Reynolds numbers. Based on the literature data, Haider and Levenspiel (1989) put forth the following correlation for calculating the drag of a nonspherical particle:

$$C_D = \frac{24}{N_{Re}}\left[1 + 8.172e^{-4.066\psi}N_{Re}^{0.0964+0.557\psi}\right] + \frac{73.7e^{-5.075\psi}N_{Re}}{N_{Re} + 5.38e^{6.21\psi}} \tag{12.27}$$

Note that in both Figure 12.5 and Equation 12.27, $d = d_s$ is used in the definitions of the Reynolds number and drag coefficient. Equation 12.27 purports to reproduce the literature data with a root mean square (RMS) error of 5.8%.

TABLE 12.1

Values of Sphericity

Shape	ψ
Sphere	1
Cube	$(\pi/6)^{1/3}$
Circular cylinder[a] (length L, diameter d)	$\left\{\dfrac{(3/2)(L/d)}{(L/d)+(1/2)}\right\}^{1/2}$

[a] For large values of (L/d), the factor ($1/2$) in the denominator arising from the two ends can be neglected.

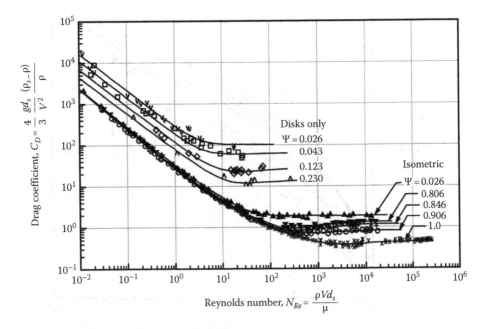

FIGURE 12.5 Drag coefficients for nonspherical particles. (Redrawn from A. Haider and O. Levenspiel, *Powder Technol.*, 58, 63, 1989.)

In order to facilitate the estimation of the unknown velocity or the unknown diameter (Sections II.A and II.B), Haider and Levenspiel (1989) introduced a dimensionless falling velocity $V*$ and a dimensionless diameter $d*$ defined as follows:

$$V* = \left(\frac{4}{3}\frac{N_{Re}}{C_D}\right)^{1/3} = \left(\frac{\rho^2 V_t^3}{\mu(\Delta\rho)g}\right)^{1/3} \tag{12.28}$$

which is similar to Equation 12.16, and

$$d* = \left(\frac{3}{4}C_D N_{Re}^2\right)^{1/3} = (N_{Ar})^{1/3} \tag{12.29}$$

Using the same experimental data, Haider and Levenspiel (1989) rearranged these results in terms of $d*$ and $V*$ as follows:

$$V* = \left[\frac{18}{(d*)^2} + \frac{2.335 - 1.74\psi}{\sqrt{d*}}\right]^{-1} \tag{12.30}$$

Figure 12.6 shows a graphical representation of Equation 12.30. It is clearly seen that for a given particle volume (i.e., constant value of d_s), nonspherical particles settle slower than the same size sphere under identical conditions, that is, a nonspherical particle experiences more drag than a sphere.

It should be noted here that although both d_s and ψ are measures of the particle size and shape, respectively, this approach does not account for the orientation of the particle. Thus, this approach will not distinguish whether a cylinder falls with its long axis aligned with or transverse to the direction of gravity. This issue has been discussed by Ganser (1993) and Rajitha et al. (2006), among others. In essence, this necessitates introducing another ratio (d_n/d_s), where d_n is the diameter of

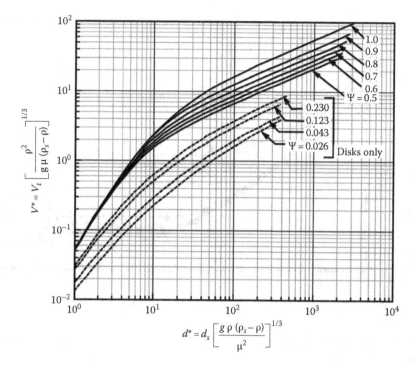

FIGURE 12.6 Graphical representation of Equation 12.30. (Based on A. Haider and O. Levelspiel, Drag coefficient and terminal velocity of spherical and non-spherical particles, Powder Technology, 58, 63, 1989.)

the circle with the same area as the projected area of the particle during its fall. Obviously, $d_n/d_s = 1$ for a sphere. However, even this measure will not distinguish the cases of a cone falling with its apex upward or downward. Some correlations are available that tend to be far more complex than Equation 12.27, which account for the orientation of the particle during settling. Another weakness of the discussion presented here is that it is applicable only for regular shaped particles for which the surface area can be calculated or measured. Some ideas for treating irregular shaped particles are also available in the literature (Zakhem et al., 1992; Tran-Cong et al., 2004).

Example 12.1

Calculate the terminal falling velocity of a 1 mm diameter sphere made of plastic ($\rho_s = 1100$ kg/m³), glass ($\rho_s = 2500$ kg/m³), and steel ($\rho_s = 7870$ kg/m³) in air, water, and oil ($\rho = 875$ kg/m³, $\mu = 27$ mPa s) at 25°C.

Solution:

First, we look up density and viscosity values for air and water in Appendix A:

At 25°C	Air	Water	Oil
μ (mPa s)	0.01849	0.8904	27
ρ (kg/m³)	1.184	1000	875

For the 1 mm diameter plastic sphere ($\rho_s = 1100$ kg/m³) falling in air, from Equation 12.13 $N_{Ar} = d^3 \rho g \Delta\rho/\mu^2$

$$N_{Ar} = \frac{(1\times10^{-3}\,\text{m})^3(1.184\,\text{kg/m}^3)(9.81\,\text{m/s}^2)(1100-1.184)\text{kg/m}^3}{(0.01849\times10^{-3}\,\text{kg/m s})^2} = 37{,}331$$

From Equation 12.14, $N_{Re} = \left[\left(14.42 + 1.827\sqrt{N_{Ar}}\right)^{1/2} - 3.798\right]^2 = 236.2$

Now, $N_{Re} = \rho V d / \mu$

$$\therefore V = \frac{N_{Re}\mu}{\rho d} = \frac{(236.2)(0.01849 \times 10^{-3}\ \text{kg/m s})}{(1.184\ \text{kg/m}^3)(0.001\ \text{m})} = 3.333\ \text{m/s}$$

Similarly, calculations for the other spheres yield the following results:

		Plastic	Glass	Steel
Air	N_{Ar}	3.73×10^4	8.66×10^4	2.67×10^5
	N_{Re}	236	388	738
	V (m/s)	3.69	6.06	11.5
Water	N_{Ar}	1.24×10^3	1.92×10^4	9.74×10^4
	N_{Re}	25.7	158	415
	V (m/s)	0.0229	0.1404	0.3698
Oil	N_{Ar}	2.65	19.7	82.4
	N_{Re}	0.139	0.901	3.13
	V (m/s)	0.00428	0.02779	0.09666

Example 12.2

Calculate the terminal falling velocity of the following particles of different shapes but equal volume made of plastic (ρ_s = 1100 kg/m³) in air and water at 25°C.

(a) 5 mm diameter sphere
(b) Cube
(c) Cylinder with L/d = 1, 2, 5
(d) Spheroids of aspect ratios 0.5 and 2

Physical properties of air and water from Appendix A:

At 25°C	Air	Water
μ (mPa s)	0.01849	0.8904
ρ (kg/m³)	1.184	1000

Solution:

For the particle of plastic (ρ_s = 1100 kg/m³) falling in air from Equation 12.29,

$$d^* = \left(\frac{3}{4} C_D N_{Re}^2\right)^{1/3} = (N_{Ar})^{1/3}$$

For equal volume particles, the value of d_s will be the same for each shape, though the value of ψ will vary with shape. Thus, $d = d_s = 5$ mm.

$$\therefore d^* = \left(\frac{d^3 \rho g (\rho_s - \rho)}{\mu^2}\right)^{1/3} = \left(\frac{(0.005\ \text{m})^3 (1.184\ \text{kg/m}^3)(9.81\ \text{m/s}^2)(1100 - 1.184)\text{kg/m}^3}{(0.01849 \times 10^{-3}\ \text{kg/ms})^2}\right)^{1/3} = 167.12$$

(a) For the 5 mm diameter sphere, from Equation 12.30,

$$V^* = \left[\frac{18}{(d^*)^2} + \frac{2.335 - 1.74\psi}{\sqrt{d^*}} \right]^{-1}$$

For the sphere, $\psi = 1$

$$\therefore V^* = 21.43$$

The terminal falling velocity of the particle, from Equation 12.28, is

$$V^* = \left[\frac{\rho^2 V_t^3}{\mu(\rho_s - \rho)g} \right]^{1/3}$$

$$\therefore V_t = V^* \left[\frac{\mu(\rho_s - \rho)g}{\rho^2} \right]^{1/3} = (21.43) \times \left[\frac{(0.01849 \times 10^{-3}\ \text{kg/m s})(1100 - 1.184)(\text{kg/m}^3)(9.81\ \text{m/s}^2)}{(1.184\ \text{kg/m}^3)^2} \right]^{1/3}$$

$$\boxed{\therefore V_t = 11.18\ \text{m/s}}$$

The corresponding value of the Reynolds number is

$$N_{Re} = \frac{\rho V d_s}{\mu} = \frac{(1.184\ \text{kg/m}^3)(11.18\ \text{m/s})(0.005\ \text{m})}{(0.01849 \times 10^{-3}\ \text{kg/m s})} = 3580$$

which is below the critical Reynolds number.

(b) Cube

For a cube of equal volume to a sphere,

Volume of cube = Volume of sphere

$$l^3 = \left(\frac{\pi}{6} \right) d_s^3$$

$$l = \left(\frac{\pi}{6} \right)^{1/3} (0.005\ \text{m}) = 0.004\ \text{m}$$

For a cube,

$$\psi = \frac{\pi d_s^2}{6l^2} = 0.82$$

$$V^* = \left[\frac{18}{(d^*)^2} + \frac{2.335 - 1.74\psi}{\sqrt{d^*}} \right]^{-1} = \left[\frac{18}{(167.12)^2} + \frac{2.335 - 1.74(0.82)}{\sqrt{167.12}} \right]^{-1} = 14.05$$

$$\therefore V_t = V^* \left[\frac{\mu(\rho_s - \rho)g}{\rho^2} \right]^{1/3} = (14.05) \times \left[\frac{(0.01849 \times 10^{-3}\ \text{kg/m s})(1100 - 1.184)(\text{kg/m}^3)(9.81\ \text{m/s}^2)}{(1.184\ \text{kg/m}^3)^2} \right]^{1/3}$$

$$\boxed{\therefore V_t = 7.34\ \text{m/s}}$$

(c) For cylinders with $L/d = 1, 2, 5$

$$L/d = 1 \therefore L = d$$

For a cylinder of volume equal to the sphere,
Volume of cylinder = Volume of sphere

$$\frac{\pi}{4}d^2L = \frac{\pi}{6}d_s^3$$

$$d = \left(\frac{2}{3}\right)^{1/3} d_s = \left(\frac{2}{3}\right)^{1/3} (0.005\text{ m}) = 0.0044\text{ m}$$

$$\therefore L = 0.0044\text{ m}$$

For the cylinder,

$$\psi = \frac{\pi d_s^2}{\pi dL + 2((\pi/4)d^2)} = 0.87$$

$$V^* = \left[\frac{18}{(d^*)^2} + \frac{2.335 - 1.74\psi}{\sqrt{d^*}}\right]^{-1} = \left[\frac{18}{(167.12)^2} + \frac{2.335 - 1.74(0.87)}{\sqrt{167.12}}\right]^{-1} = 15.69$$

$$\therefore V_t = V^* \left[\frac{\mu(\rho_s - \rho)g}{\rho^2}\right]^{1/3} = (15.69) \times \left[\frac{(0.01849 \times 10^{-3}\text{ kg/m s})(1100 - 1.184)(\text{kg/m}^3)(9.81\text{ m/s}^2)}{(1.184\text{ kg/m}^3)^2}\right]^{1/3}$$

$$\boxed{\therefore V_t = 8.19\text{ m/s}}$$

Similarly, for a cylinder with $L/d = 2$,

$$\psi = 0.85, V^* = 14.58, V_t = 7.58\text{ m/s}$$

$$L/d = 5: \psi = 0.74, V^* = 12.23, \boxed{V_t = 6.38\text{ m/s}}$$

(d) For a spheroid with an aspect ratio of 0.5,

$$\text{Aspect ratio} = b/a = 0.5, \therefore b = 0.5a$$

For a spheroid of equal volume to a sphere,
Volume of spheroid = Volume of sphere

$$\frac{4}{3}\pi ab^2 = \frac{\pi}{6}d_s^3$$

$$a = \left(\frac{1}{2}\right)^{1/3} d_s = \left(\frac{1}{2}\right)^{1/3} (0.005\text{ m}) = 0.004\text{ m}$$

$$\therefore b = 0.002\text{ m}$$

For the spheroid,

$$\psi = \frac{\pi d_s^2}{2\pi b^2 + (2\pi ab/e)\sin^{-1}e} = 0.92$$

where

$$e = \left[1 - \left(\frac{b}{a}\right)^2\right]^{1/2} = 0.87$$

$$V^* = \left[\frac{18}{(d^*)^2} + \frac{2.335 - 1.74\psi}{\sqrt{d^*}}\right]^{-1} = \left[\frac{18}{(167.12)^2} + \frac{2.335 - 1.74(0.92)}{\sqrt{167.12}}\right]^{-1} = 17.28$$

$$\therefore V_t = V^* \left[\frac{\mu(\rho_s - \rho)g}{\rho^2}\right]^{1/3} = (17.28) \times \left[\frac{(0.01849 \times 10^{-3}\,\text{kg/m s})(1100 - 1.184)\,(\text{kg/m}^3)(9.81\,\text{m/s}^2)}{(1.184\,\text{kg/m}^3)^2}\right]^{1/3}$$

$$\boxed{\therefore V_t = 9.02\,\text{m/s}}$$

For the spheroid with an aspect ratio of 2,

$$\text{Aspect ratio} = a/b = 2, \therefore a = 2b$$

For the spheroid of equal volume to a sphere,
  Volume of spheroid = Volume of sphere

$$\frac{4}{3}\pi a^2 b = \frac{\pi}{6}d_s^3$$

$$b = \left(\frac{1}{32}\right)^{1/3} d_s = \left(\frac{1}{32}\right)^{1/3} (0.005\,\text{m}) = 0.0016\,\text{m}$$

$$\therefore a = 0.0032\,\text{m}$$

For the spheroid,

$$\psi = \frac{\pi d_s^2}{2\pi b^2 + (\pi a^2/e)\ln((1+e)/(1-e))} = 0.71$$

where

$$e = \left[1 - \left(\frac{b}{a}\right)^2\right]^{1/2} = 0.87$$

$$V^* = \left[\frac{18}{(d^*)^2} + \frac{2.335 - 1.74\psi}{\sqrt{d^*}}\right]^{-1} = \left[\frac{18}{(167.12)^2} + \frac{2.335 - 1.74(0.71)}{\sqrt{167.12}}\right]^{-1} = 11.68$$

$$\therefore V_t = V^* \left[\frac{\mu(\rho_s - \rho)g}{\rho^2}\right]^{1/3} = (11.68) \times \left[\frac{(0.01849 \times 10^{-3}\,\text{kg/m s})(1100 - 1.184)\,(\text{kg/m}^3)(9.81\,\text{m/s}^2)}{(1.184\,\text{kg/m}^3)^2}\right]^{1/3}$$

$$\boxed{\therefore V_t = 6.10\,\text{m/s}}$$

Summary of results

			V_t (m/s)	
		ψ	Air	Water
Sphere		1	11.18	0.109
Cube		0.82	7.32	0.073
Cylinder	$L/d = 1$	0.87	8.19	0.081
	$L/d = 2$	0.84	7.58	0.076
	$L/d = 5$	0.74	6.38	0.064
Spheroid	$b/a = 0.5$	0.92	9.02	0.089
	$a/b = 2$	0.71	6.10	0.061

As the value of ψ increases, that is, the deviation from a spherical shape decreases, the terminal falling velocity of nonspherical particles is seen to approach that of the sphere, as expected.

C. DROPS AND BUBBLES

Because of the surface tension forces, very small drops and bubbles are nearly rigid spheres and behave in a manner similar to rigid particles. However, larger fluid drops or bubbles may experience a considerably different settling behavior because the shear stress on the drop surface can be transmitted to the fluid inside the drop, which, in turn, results in circulation of the internal fluid. This internal circulation dissipates energy, which is extracted from the energy of the bubble motion and is equivalent to an additional drag force. For example, in Stokes flow of spherical drops or bubbles (e.g., $N_{Re} < 1$), it has been shown by Hadamard and Rybczynski (see, e.g., Clift et al., 1978; Grace, 1983) that the drag coefficient can be corrected for this effect as follows:

$$C_d = \frac{24}{N_{Re}}\left(\frac{\kappa + 2/3}{\kappa + 1}\right) \tag{12.31}$$

where $\kappa = \mu_i/\mu_o$, μ_i is the viscosity of dispersed ("inside") fluid and μ_o is the viscosity of the continuous ("outside") fluid.

For larger Reynolds numbers ($1 < N_{Re} < 500$), Rivkind and Ryskind (Grace, 1983) have proposed the following equation for the drag coefficient for spherical drops and bubbles:

$$C_D = \frac{1}{\kappa + 1}\left[\kappa\left(\frac{24}{N_{Re}} + \frac{4}{N_{Re}^{1/3}}\right) + \frac{14.9}{N_{Re}^{0.78}}\right] \tag{12.32}$$

As the size of the drop or bubble increases, however, it will become distorted due to the unbalanced forces around it. The viscous shear stresses tend to elongate the shape, whereas the pressure distribution tends to flatten it out in the direction normal to the flow. Thus, the shape tends to progress from spherical, to ellipsoidal, to a "spherical cap" form as the size increases. Above a certain size, the deformation is so great that the drag force is approximately proportional to the volume, and the terminal velocity becomes nearly independent of size. More detailed discussion can be found in the books of Clift et al. (1978) and Michaelides (2006).

IV. NON-NEWTONIAN FLUIDS

The motion of solid particles, drops, or bubbles through non-Newtonian fluid media is encountered frequently and has been the subject of considerable research (e.g., Chhabra, 2006). We will present some relations here that are applicable to purely viscous non-Newtonian fluids, although there is also much interest and activity in viscoelastic fluids as well. Despite the relative large amount of work that has been done in this area, there is still no general agreement as to the "right," or even the "best," description of the drag on a sphere in non-Newtonian fluids. This is due not only to the complexity of the momentum equations that must be solved for the various models but also the difficulty in obtaining good, reliable, representative data for fluids with well-characterized unambiguous rheological properties.

A. POWER LAW FLUIDS

The usual approach for non-Newtonian fluids is to start with the known results for Newtonian fluids and modify them to account for the non-Newtonian properties. For example, the definition of the Reynolds number for a power law fluid can be obtained by replacing the viscosity in the Newtonian definition by an appropriate shear rate–dependent viscosity function. If the characteristic shear rate for flow over a sphere is taken to be (V/d), for example, the power law viscosity function becomes

$$\mu \to \eta(\dot{\gamma}) \cong m\left(\frac{V}{d}\right)^{n-1} \tag{12.33}$$

and the corresponding expression for the Reynolds number is

$$N_{Re\,pl} = \frac{\rho V d}{m(V/d)^{n-1}} = \frac{\rho V^{2-n} d^n}{m} \tag{12.34}$$

The corresponding (creeping flow) drag coefficient can be characterized by a correction factor (X) applied to the Stokes law drag coefficient:

$$C_D = \frac{24}{N_{Re,pl}} X \tag{12.35}$$

A variety of theoretical expressions, as well as experimental values, for the correction factor X as a function of the power law flow index (n) have been summarized by Chhabra (2006). In a series of papers, Chhabra (1995, 2006) and Tripathi et al. (1994, 1995) presented the results of numerical calculations for the drag on spheroidal particles in a power law fluid in terms of $C_D = fn(N_{Re}, n)$. Darby (1996) analyzed these results and has shown that this function can be expressed in a form equivalent to the Dallavalle equation, which applies over the entire range of n and N_{Re}. This equation is

$$C_D = \left(C_1 + \frac{4.8}{\sqrt{N_{Re,pl}/X}}\right)^2 \tag{12.36}$$

where both X and C_1 are functions of the flow index, n. These functions were determined by Darby (1996) by empirically fitting the following equations to the values given by Chhabra (1995):

$$\frac{1}{C_1} = \left[\left(\frac{1.82}{n}\right)^8 + 34\right]^{1/8} \tag{12.37}$$

$$X = \frac{1.33 + 0.37n}{1 + 0.7n^{3.7}} \tag{12.38}$$

The agreement between these values of C_1 and X and the values given by Chhabra is shown in Figures 12.7 and 12.8 up to $N_{Re,pl} \leq 100$. Equations 12.36 through 12.38 are equivalent to the Dallavalle equation for a sphere in a power law fluid. A comparison of the values of C_D predicted by

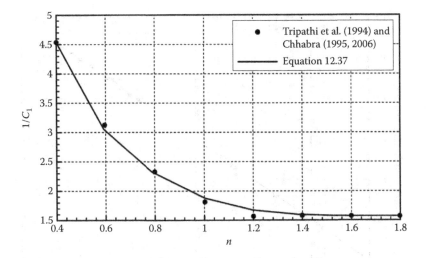

FIGURE 12.7 Plot of $1/C_1$ versus n for power law fluid.

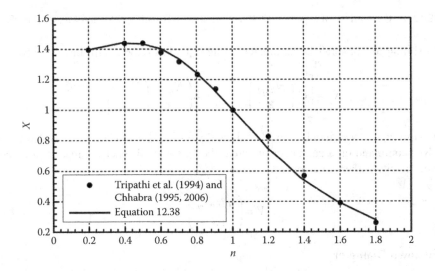

FIGURE 12.8 Plot of X versus n for power law fluid.

Equation 12.36 with the values given by Chhabra is shown in Figure 12.9. The deviation is the greatest for highly dilatant fluids, in the Reynolds number range of about 5–50, although the agreement is quite reasonable above and below this range, and for pseudoplastic fluids over the entire range of Reynolds number. We will illustrate the application of these equations by outlining the procedure for solving the "unknown velocity" and the "unknown diameter" problems.

1. Unknown Velocity

The expressions for C_D and $N_{Re,pl}$ can be combined to give a group that is independent of V, as follows:

$$C_D^{2-n}\left(\frac{N_{Repl}}{X}\right)^2 = \left(\frac{\rho}{Xm}\right)^2 \left(\frac{4g\Delta\rho}{3\rho}\right)^{2-n} d^{n+2} = N_d \tag{12.39}$$

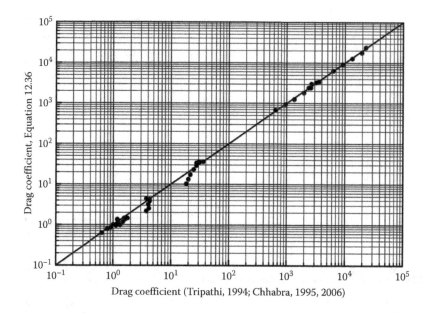

FIGURE 12.9 Comparison of Equation 12.36 with results of Tripathi (1994) and Chhabra (1995, 2006).

which is similar to Chhabra's D⁺ parameter. Using Equation 12.36 to eliminate C_D gives

$$N_d = \left[C_1 \left(\frac{N_{Re,pl}}{X} \right)^{1/(2-n)} + 4.8 \left(\frac{N_{Re,pl}}{X} \right)^{n/2(2-n)} \right]^{2(2-n)} \tag{12.40}$$

Although this equation cannot be solved analytically for $N_{Re,pl}$, it can be solved by iteration (or using the "solve" command on a calculator or spreadsheet), since all other parameters are known. The unknown velocity is then given by

$$V = \left(\frac{m N_{Re,pl}}{\rho d^n} \right)^{1/(2-n)} \tag{12.41}$$

2. Unknown Diameter

The diameter can be eliminated from the expressions for C_D and $N_{Re,pl}$ as follows:

$$\frac{C_D^n X}{N_{Re,pl}} = \left(\frac{4g\Delta\rho}{3\rho} \right)^n \left(\frac{Xm}{\rho} \right) V^{-(n+2)} = N_V \tag{12.42}$$

Again using Equation 12.36 to eliminate C_D gives

$$N_V = \left[C_1 \left(\frac{X}{N_{Re,pl}} \right)^{1/2n} + 4.8 \left(\frac{X}{N_{Re,pl}} \right)^{(1+n)/2n} \right]^{2n} \tag{12.43}$$

As before, everything in this equation is known except for $N_{Re,pl}$, which can be determined by iteration (or the "solve" spreadsheet or calculator command). When this is found, the unknown diameter is given by

$$d = \left(\frac{m N_{Re,pl}}{\rho V^{2-n}} \right)^{1/n} \tag{12.44}$$

Example 12.3: Unknown Velocity and Unknown Diameter of a Sphere Settling in a Power Law Fluid

Table E12.1 summarizes the procedure and shows the results of a spreadsheet calculation for an application of this method to the same three examples given by Chhabra (1995). Examples 1 and 2 are "unknown velocity" problems, and Example 3 is an "unknown diameter" problem (Table E12.2). The line labeled "Equation" refers to Equation 12.40 for the unknown velocity cases and Equation 12.43 for the unknown diameter case. The "Stokes" value is from Equation 12.10, which only applies for $N_{Re,pl} < 1$ (e.g., Example 1). It is seen that the solutions for Examples 1 and 2 are virtually identical to Chhabra's, and that for Example 3 is within 5% of Chhabra's. The values labeled "Using Data" were obtained by iteration using the data from Figure 4 of Tripathi et al. (1994). These values are only approximate, since they were obtained by interpolating from the

TABLE E12.1
Procedure for Determining Unknown Velocity or Unknown Diameter for Particles Settling in a Power Law Fluid

Problem	Unknown Velocity	Unknown Diameter
Given:	Particle diameter (d) and fluid properties (m, n, and ρ)	Particle settling velocity (V) and fluid properties (m, n, and ρ)
Step 1	Using value of n, calculate C_1 and X from Equations 12.37 and 12.38	Using value of n, calculate C_1 and X from Equations 12.37 and 12.38
Step 2	Calculate N_d from Equation 12.39	Calculate N_v from Equation 12.42
Step 3	Solve Equation 12.40 for $N_{Re,pl}$ by iteration (or using "solve" function)	Solve Equation 12.43 for $N_{Re,pl}$ by iteration (or using "solve" function)
Step 4	Get V from Equation 12.41	Get d from Equation 12.44

TABLE E12.2
Comparison of Calculated Settling Properties Using Equation 12.35 with Literature Values, Chhabra (1995)

	Example 1	Example 2	Example 3
Given data:	$d = 0.002$ m	$d = 0.002$ m	$V = 0.2$ m/s
	$m = 1.3$ Pa s^n	$m = 0.015$ Pa s^n	$m = 0.08$ Pa s^n
	$n = 0.6$	$n = 0.8$	$n = 0.5$
	$\rho = 1002$ kg/m^3	$\rho = 1050$ kg/m^3	$\rho = 1005$ kg/m^3
	$\rho_s = 7780$ kg/m^3	$\rho_s = 2500$ kg/m^3	$\rho_s = 8714$ kg/m^3
Calculated values:	$X = 1.403$	$X = 1.244$	$X = 1.438$
	$C_1 = 0.329$	$C_1 = 0.437$	$C_1 = 0.275$
	$N_d = 15.4$	$N_d = 2830$	$N_d = 0.064$
	$N_{Re,pl} = 0.082$	$N_{Re,pl} = 55.7$	$N_{Re,pl} = 29.8$
Darby (1996)	$V = 0.0208$ m/s	$V = 0.165$ m/s	$d = 7.05 \times 10^{-4}$ m
Chhabra (1995)	$V = 0.020$ m/s	$V = 0.167$ m/s	$d = 6.67 \times 10^{-4}$ m
Stokes' law	$V = 0.0206$ m/s	$V = 0.514$ m/s	$d = 5.31 \times 10^{-4}$ m
Tripathi et al. (1994)		$V = 0.167$ m/s	$d = 7.18 \times 10^{-4}$ m
		$N_{Re,pl} = 57$	$N_{Re,pl} = 30$
		$C_D = 1.3$	$C_D = 1.8$

(very compressed) log scale of the plot. The method shown here has several advantages over that reported by Chhabra (1995), namely:

(1) All expressions are given in equation form, and it is not necessary to read any plots to solve the problems (i.e., the empirical data are represented analytically by curve-fit equations).
(2) The method is more general, in that it is a direct extension of the technique of solving similar problems for Newtonian fluids and applies over all values of the Reynolds number.
(3) Only one calculation procedure is required, regardless of the value of the Reynolds number for the specific problem.
(4) The calculation procedure is simple and straightforward and can be done quickly using a spreadsheet.

B. WALL EFFECTS

The wall effect for particles settling in power law non-Newtonian fluids appears to be significantly smaller than for Newtonian fluids. For power law fluids, the wall correction factor in creeping flow, as well as for very high Reynolds numbers, appears to be independent of the Reynolds number, similar to the behavior seen in Newtonian fluids. For creeping flow, the wall correction factor given by Chhabra (2002, 2006) is

$$K_{w_o} = 1 - 1.6\left(\frac{d}{D}\right) \tag{12.45}$$

whereas for high Reynolds numbers, he gives

$$K_{w_\infty} = 1 - 3\left(\frac{d}{D}\right)^{3.5} \tag{12.46}$$

For intermediate Reynolds numbers, the wall factor depends upon the Reynolds number as well as d/D. Over a range of $10^{-2} < N_{Re,pl} < 10^3$, $0 < d/D < 0.5$, and $0.53 < n < 0.95$, the following equation describes the Reynolds number dependence of the wall factor quite well:

$$\frac{(1/K_w) - (1/K_{w_\infty})}{(1/K_{w_o}) - (1/K_{w_\infty})} = \left(1 + 1.3N_{Re,pl}^2\right)^{-0.33} \tag{12.47}$$

It may be concluded that the predictions of Equation 12.47 are in good agreement with the numerical predictions of Missirlis et al. (2001) in the creeping flow regime, whereas the correspondence between the predictions and experiments deteriorates with the increasing Reynolds number and/or decreasing value of the power law index (Song et al., 2009). Also, note that Equations 12.45 through 12.47 implicitly include the dependence on the power law index, whereas the numerical results (Song et al., 2009) suggest a stronger influence of the power law index (n) than these equations.

C. CARREAU FLUIDS

As discussed in Chapter 3, the Carreau viscosity model is one of the most general and useful and reduces to many of the common two-parameter models (power law, Ellis, Sisko, Bingham, etc.) as special cases. This model can be written as

$$\eta = \frac{\eta_\infty + (\eta_o - \eta_\infty)}{[1 + (\lambda\dot{\gamma})^2]^{(n-1)/2}} \tag{12.48}$$

where $n = (1 - 2p)$ is the flow index for the power law region (p is the shear thinning parameter in the form of this model given in Equation 3.27). Since the shear conditions surrounding particles virtually never reaches the levels corresponding to the high shear viscosity (η_∞), this parameter can be neglected and the parameters reduced to three: η_o, λ, and n. Chhabra and Uhlherr (1980) and Bush and Phan-Thien (1984) have determined the Stokes flow correction factor, X, for this model, which is a function of the dimensionless parameters n and $N_\lambda = \lambda V/d$. The following equation represents their results for the C_D correction factor over a wide range of data to $\pm 10\%$, for ($0.4 < n < 1$) and ($0 < N_\lambda < 400$):

$$X = \frac{1}{[1 + (0.275 N_\lambda)^2]^{(1-n)/2}} \tag{12.49}$$

where the Stokes equation uses η_o for the viscosity in the definition of the Reynolds number.

D. BINGHAM PLASTICS

A particle will not fall through a fluid with a yield stress unless the weight of the particle is sufficient to overcome the yield stress. Because the stress is not uniform around the particle and the distribution is very difficult to determine, it is not possible to determine the critical "yield" criterion exactly. However, it should be possible to characterize this state by a dimensionless "gravity yield" parameter, at least for a sphere falling under its own weight:

$$Y_G = \frac{\tau_o}{gd\Delta\rho} \tag{12.50}$$

By equating the vertical component of the yield stress acting over the surface of the sphere to the weight of the particle, a critical value of $Y_G = 0.17$ results (Chhabra, 2006; Chhabra and Richardson, 2008). Experimentally, however, the results appear to fall into groups: one, for which $Y_G \approx 0.2$ and the other for which $Y_G \approx 0.04$–0.08. There seems to be no consensus as to the correct value, and the difference may well be due to the fact that the yield stress is not an unambiguous parameter, in as much as values determined from "static" measurements can differ significantly from the values determined from "dynamic" measurements, that is, whether the yield point is approached from above or below.

With regard to the drag on a sphere that is moving in a Bingham plastic medium, the drag coefficient (C_D) must be a function of the Reynolds number as well as either the Hedstrom number or the Bingham number ($N_{Bi} = N_{He}/N_{Re} = \tau_o d/\mu_\infty V$). One approach is to reconsider the Reynolds number from the perspective of the ratio of inertial to viscous momentum flux. For a Newtonian fluid in a tube, this is equivalent to

$$N_{Re} = \frac{DV\rho}{\mu} = \frac{8\rho V^2}{\mu(8V/D)} = \frac{8\rho V^2}{\tau_w} \tag{12.51}$$

which follows from the Hagen–Poiseuille equation, since $\tau_w = \mu(8V/D)$ is the drag per unit area of pipe wall, and the shear rate at the wall of the pipe is $\dot\gamma_w = 8V/D$. By analogy, the drag force per unit area on a sphere is $F/A = C_D\rho V^2/2$, which, for Stokes flow (i.e., $C_D = 24/N_{Re}$), becomes $F/A = 12\mu V/d$. If F/A for the sphere is considered to be analogous to the "wall stress" (τ) on the sphere, the corresponding "effective wall shear rate" is $12V/d$. Thus, the sphere Reynolds number could be written as

$$N_{Re} = \frac{dV\rho}{\mu} = \frac{12\rho V^2}{\mu(12V/d)} = \frac{12\rho V^2}{\tau} \tag{12.52}$$

For a Bingham plastic, the corresponding expression would be

$$\frac{12\rho V^2}{\tau} = \frac{12\rho V^2}{\mu_\infty(12V/d) + \tau_o} = \frac{N_{Re}}{1 + N_{Bi}/12} \tag{12.53}$$

Equation 12.53 is thus the effective Reynolds number for correlating the drag coefficient.

Another approach is to consider the effective shear rate over the sphere to be V/d, as was done in Equation 12.33 for the power law fluid. If this approach is applied to a sphere in a Bingham plastic, the result is

$$N_{ReBP} = \frac{N_{Re}}{1 + N_{Bi}} \tag{12.54}$$

This is similar to the analysis obtained by Ansley and Smith (1967) using the slip line theory from soil mechanics, which results in a dimensionless group called the *plasticity number*:

$$N_{Pl} = \frac{N_{Re}}{1 + 2\pi N_{Bi}/24} \tag{12.55}$$

A finite element analysis (Blackery and Mitsoulis, 1997) resulted in an equivalent Stokes law correction factor $X (= C_D N_{Re}/24)$ that is a function of N_{Bi} for $N_{Bi} < 1000$, as follows:

$$X = 1 + aN_{Bi}^b \tag{12.56}$$

where $a = 2.93$ and $b = 0.83$ for a sphere in an unbounded fluid, and $2.93 > a > 1.63$ and $0.83 < b < 0.95$ for $0 < d/D < 0.5$. Also, based upon available data, Chhabra and Uhlherr (1988) found that the "Stokes flow" relation ($C_D = 24X/N_{Re}$) applies up to $N_{Re} \leq 100N_{Bi}^{0.4}$ for Bingham plastics. Equation 12.56 is equivalent to a Bingham plastic Reynolds number (N_{ReBP}) of

$$N_{ReBP} = \frac{N_{Re}}{1 + 2.93N_{Bi}^{0.83}} \tag{12.57}$$

Unfortunately, there are insufficient experimental data reported in the literature to verify or confirm any of these expressions. Thus, for the lack of any other information, Equation 12.57 is recommended, because it is based on the most detailed analysis.

This can be extended beyond the Stokes flow region by incorporating Equation 12.57 into the equivalent Dallavalle equation:

$$C_D = \left(0.632 + \frac{4.8}{\sqrt{N_{ReBP}}}\right)^2 \tag{12.58}$$

which can be used to solve the "unknown velocity" and "unknown diameter" problems as previously discussed. However, in this case, rearrangement of the dimensionless variables C_D and N_{ReBP} into an alternate set of dimensionless groups in which the unknown appears in only one group is not possible due to the form of N_{ReBP}. Thus, the procedure would be to equate Equations 12.58 and 12.12 and solve the resulting equation directly by iteration for the unknown V or d (as the case requires).

More detailed discussions concerning the combined effects of shear thinning and yield stress on drag and wake phenomena and on wall effects for a sphere can be found in the literature (Nirmalkar et al., 2013; Das et al., 2015). The corresponding results for a cylinder are also available in the literature (Nirmalkar and Chhabra, 2014).

SUMMARY

The key concepts that should be retained from this chapter include

- Similarities and differences between fluid-particle drag and fluid-wall drag in a pipe
- Cause and effect of "form drag"
- Using a falling particle in a fluid to determine (a) particle velocity, (b) particle diameter, and (c) fluid viscosity
- Influence of the container walls on the velocity of a falling particle
- Effect of geometry of nonspherical particles on the terminal falling velocity
- The flow behavior of drops and bubbles
- The falling velocity of particles in non-Newtonian fluids

PROBLEMS

1. By careful streamlining, it is possible to reduce the drag coefficient of an automobile from 0.4 to 0.25. How much power would this save at 40 mph and 60 mph, assuming the effective projected area of the car is 25 ft²?

2. If your pickup truck has a drag coefficient equivalent to a 5 ft diameter disk, and the same projected frontal area, how much horsepower is required to overcome wind drag at 40 mph? What horsepower is required at 70 mph?

3. You take a tumble while water skiing. The handle attached to the tow rope falls beneath the water and remains perpendicular to the direction of the boat's heading. If the handle is one inch in diameter and 1 ft long, and the boat is moving at 20 mph, how much horsepower is required to pull the handle through the water?

4. Your new car is reported to have a drag coefficient of 0.3. If the cross-sectional area of the car is 20 ft², how much horsepower is used to overcome wind resistance at 40 mph, 55 mph, 70 mph, and 100 mph? ($T = 70°F$)

5. The supports for a tall chimney must be designed to withstand a 120 mph wind. If the chimney is 10 ft in diameter and 40 ft high, what is the wind force on the chimney at this speed? $T = 50°F$.

6. A speedboat is propelled by a water jet motor that takes water in at the bow through a 10 cm diameter duct and discharges it through a 50 mm diameter nozzle at a rate of 80 kg/s. Neglecting friction in the motor and internal ducts and assuming that the drag coefficient for the boat hull is the same as for a 1 m diameter sphere, determine
   (a) The static thrust developed by the motor when it is stationary
   (b) The maximum velocity attainable by the boat
   (c) The power (kW) required to drive the motor
   (Assume seawater density to be 1030 kg/m³, viscosity to be 1.2 cP.)

7. After blowing up a balloon, you release it without tying off the opening, and it flies out of your hand. If the diameter of the balloon is 6 in., the pressure inside it is 1 psig, and the opening is 1/2 in. in diameter, what is the balloon velocity? You may neglect friction in the escaping air and the weight of the balloon and assume that an instantaneous steady state (i.e., a "pseudo" steady state) applies.

8. A mixture of titanium (SG = 4.5) and silica (SG = 2.65) particles, with diameters ranging from 50 to 300 μm, is dropped into a tank in which water is flowing upward. What is the velocity of the water if all of the silica particles are carried out with the water?

9. A small sample of ground coal is introduced at the top of a column of water 30 cm high, and the time required for the particles to settle out is measured. If it takes 26 s for the first particle to reach the bottom and 18 h for all particles to settle, what is the range of particle sizes in the sample? ($T = 60°F$, $SG_{coal} = 1.4$)

10. You want to determine the viscosity of an oil which has an SG of 0.9. To do this, you drop a spherical glass bead (SG = 2.7) with a diameter of 0.5 mm into a large vertical column of the oil and measure its settling velocity. If the measured velocity is 3.5 cm/s, what is the viscosity of the oil?

11. A solid particle with a diameter of 5 mm and an SG of 1.5 is immersed in a liquid with a viscosity of 10 P and an SG of 1. How long will it take for the particle to reach 99% of its terminal velocity after it is released?

12. A hot air popcorn popper operates by blowing air through the popping chamber, which carries the popped corn up through a duct and out of the popper leaving the un-popped grains behind. The un-popped grains weigh 0.15 g, half of which is water, and have an equivalent spherical diameter of 4 mm. The popped corn loses half of the water to steam and has an equivalent diameter of 12 mm. What are the upper and lower limits of the air volumetric flow rate at 200°F over which the popper will operate properly, for a duct diameter of 8 cm?

13. You have a granular solid with SG = 4, which has particle sizes of 300 μm and smaller. You want to separate out all of the particles with a diameter of 20 μm and smaller by pumping water upward through a slurry of the particles in a column with a diameter of 10 cm. What flow rate is required to ensure that all particles less than 20 μm are swept out of the top of the column? If the slurry is fed to the bottom of the column through a vertical tube, what should the diameter of this tube be to ensure that none of the particles settle out in it?

14. You want to reproduce the experiment shown in the text which illustrates the wake behind a sphere falling in water at the point where the boundary layer undergoes transition from laminar to turbulent. If the sphere is made of steel with a density of 500 lb_m/ft^3, what should the diameter be?

15. You have a sample of crushed coal containing a range of particle sizes from 1 to 1000 μm in diameter. You wish to separate the particles according to size by entrainment, in which they are dropped into a vertical column of water that is flowing upward. If the water velocity in the column is 3 cm/s, which particles will be swept out of the top of the column and which will settle at the bottom? (SG of the solid is 2.5)

16. A gravity settling chamber consists of a horizontal rectangular duct that is 6 m long, 3.6 m wide, and 3 m high. The chamber is used to trap sulfuric acid mist droplets entrained in an air stream. The droplets settle out as the air passes through the duct and may be assumed to behave as rigid spheres. If the air stream has a flow rate of 6.5 m^3/s, what is the diameter of the largest particle that will not be trapped in the duct?

    (ρ_{acid} = 1.74 g/cm^3; ρ_{air} = 0.01 g/cm^3; μ_{air} = 0.02 cP; μ_{acid} = 2 cP)

17. A small sample of a coal slurry containing particles with equivalent spherical diameters from 1 to 500 μm is introduced at the top of a water column 30 cm high. The particles that fall to the bottom are continuously collected and weighed to determine the particle size distribution in the slurry. If the solid SG is 1.4 and the water viscosity is 1 cP, over what time range must the data be obtained in order to collect and weigh all of the particles in the sample?

18. Construct a plot of C_D versus N_{ReBP} for a sphere falling in a Bingham plastic fluid over the range of $1 < N_{Re} < 100$ and $10 < N_{Bi} < 1000$ using Equation 12.58. Compare the curves for this relation based upon Equations 12.54, 12.55, and 12.57.

19. The viscosity of applesauce at 80°F was measured to be 24.2 P at a shear rate of 10 1/s and 1.45 P at 500 1/s. The density of the applesauce is 1.5 g/cm^3. Determine the terminal velocity of a solid sphere 1 cm in diameter with a density of 3.0 g/cm^3 falling in the applesauce, if the fluid is described by (a) the power law model and (b) the Bingham plastic model.

20. Determine the size of the smallest sphere of SG = 3 that will settle in the applesauce with properties given in Problem 19, assuming it is best described by the Bingham plastic model. Find the terminal velocity of the sphere that has a diameter twice this size.

NOTATION

A	Cross-sectional area of particle normal to flow direction, $[L^2]$
C_D	Particle drag coefficient, [—]
d	Particle diameter, $[L]$
F_D	Drag force on particle, $[F = M\,L/t^2]$
g	Acceleration due to gravity, $[L/t^2]$
K_{w_o}	Low Reynolds number wall correction factor, [—]
K_{w_∞}	High Reynolds number wall correction factor, [—]
m	Power law consistency coefficient, $[M/(L\,t^{2-n})]$
n	Power law flow index, [—]
N_{Ar}	Archimedes number, Equation (12.13), [—]
N_{ReBP}	Bingham plastic Reynolds number, Equation (12.54), [—]
N_{Bi}	Bingham number = $N_{Re}/N_{He} = (d\tau_o/\mu_\infty V)$, [—]
N_{Re}	Reynolds number, [—]
$N_{Re,pl}$	Power law Reynolds number, [—]
N_λ	Dimensionless time constant, $\lambda V/d$, [—]
V	Relative velocity between fluid and particle, $[L/t]$
X	Correction factor to Stokes law to account for non-Newtonian properties, [—]
Y_G	"Gravity yield" parameter for Bingham plastics, defined by Equation 12.50, [—]

GREEK

κ	μ_i/μ_o, [—]
η_o	Low shear limiting viscosity, $[M/(L\,t)]$
η_∞	High shear limiting viscosity, $[M/(L\,t)]$
$\Delta\rho$	$(\rho_s - \rho)$, $[M/L^3]$
ρ	Density, $[M/L^3]$
λ	Carreau fluid time constant parameter, $[t]$
μ	Viscosity (constant), $[M/(L\,t)]$
μ_∞	Bingham plastic limiting viscosity, $[M/(L\,t)]$
τ_o	Bingham plastic yield stress, $[F/L^2 = M/(L\,t^2)]$

SUBSCRIPTS

i	Distributed ("inside") liquid phase
o	Continuous ("outside") liquid phase
s	Solid
t	Terminal velocity condition

REFERENCES

Ansley, R.W. and T.N. Smith, Motion of spherical particles in a Bingham plastic, *AIChE J.*, 13, 1193, 1967.

Blackery, J. and E. Mitsoulis, Creeping motion of a sphere in tubes filled with a Bingham plastic material, *J. Non-Newton. Fluid Mech.*, 70, 59, 1997.

Bush, M.B. and N. Phan-Thien, Drag force on a sphere in creeping motion through a Carreau model fluid, *J. Non-Newton. Fluid Mech.*, 16, 303, 1984.

Chhabra, R.P., Calculating settling velocities of particles, *Chem. Eng.*, 102, 133, September 1995.

Chhabra, R.P., Wall effects on spheres falling axially in cylindrical tubes, in *Transport Processes in Bubbles, Drops and Particles*, D. Dekee and R.P. Chhabra (Eds.), p. 316, Taylor & Francis, New York, 2002.

Chhabra, R.P., *Bubbles, Drops, and Particles in non-Newtonian Fluids*, 2nd edn., CRC Press, Boca Raton, FL, 2006.

Chhabra, R.P., L. Agarwal, and N.K. Sinha, Drag on non-spherical particles: An evaluation of available methods, *Powder Technol.*, 101, 288, 1999.

Chhabra, R.P. and J.F. Richardson, *Non-Newtonian Flow and Applied Rheology*, 2nd edn., Butterworth-Heinemann, Oxford, 2008.

Chhabra, R.P. and P.H.T. Uhlherr, Creeping motion of spheres though shear-thinning elastic fluids described by the Carreau viscosity equation, *Rheol. Acta*, 19, 187, 1980.

Chhabra, R.P. and P.H.T. Uhlherr, Static equilibrium and motion of spheres in viscoplastic liquids, Chapter 21, in *Encyclopedia of Fluid Mechanics*, N.P. Cheremisinoff (Ed.), Vol. 7, Gulf Publishing Co., Houston, TX, 1988.

Clift, R., J. Grace, and M.E. Weber, *Bubbles, Drops and Particles*, Academic Press, New York, 1978.

Coulson, J.M., J.F. Richardson, J.R. Blackhurst, and J.H. Harker, *Chemical Engineering*, Vol. 2, 4th edn., Pergamon Press, Oxford, 1991.

Dallavalle, J.M., *Micromeritics*, 2nd edn., Pitman, New York, 1948.

Darby, R., Determine settling rates of particles in non-Newtonian fluids, *Chem. Eng.*, 103, 107, December 1996.

Das, P.K., A.K. Gupta, N. Nirmalkar, and R.P. Chhabra, Effect of confinement on forced convection from a heated sphere in Bingham plastic fluids, *Korea-Aust. Rheol. J.*, 27(2) 75–94, 2015.

Francis, A.W., Wall effect in falling ball method for viscosity, *Physics*, 4, 403, 1933.

Ganser, G.H., A rational approach to drag prediction of spherical and non-spherical particles, *Powder Technol.*, 77, 143, 1993.

Grace, J.R., Hydrodynamics of liquid drops in immiscible liquids, Chapter 38, in *Handbook of Fluids in Motion*, N.P. Cheremisinoff and R. Gupta (Eds.), Ann Arbor Science, Ann Arbor, MI, 1983.

Gupta, S.V., *Viscometry for Liquids: Calibration of Viscometers*, Springer, Berlin, 2014.

Haider, A. and O. Levenspiel, Drag coefficient and terminal velocity of spherical and non-spherical particles, *Powder Technol.*, 58, 63, 1989.

Khan, A.R. and J.F. Richardson, The resistance to motion of a solid sphere in a fluid, *Chem. Eng. Commun.*, 62, 135, 1987.

Lamb, H., *Hydrodynamics*, 6th edn., Dover, New York, 1945.

Michaelides, E.E., *Particles, Bubbles & Drops: Their Motion, Heat and Mass Transfer*, World Scientific, Singapore, 2006.

Missirlis, K.A., D. Assimacopolous, E. Mitsoulis, and R.P. Chhabra, Wall effects for motion of spheres in power-law fluids, *J. Non-Newton. Fluid Mech.*, 96, 459, 2001.

Nirmalkar, N. and R.P. Chhabra, Momentum and heat transfer from a heated circular cylinder in Bingham plastic fluids, *Int. J. Heat Mass Transf.*, 70, 564, 2014.

Nirmalkar, N., R.P. Chhabra, and R.J. Poole, Numerical prediction of momentum and heat transfer characteristic from a heated sphere in yield stress fluids, *Ind. Eng. Chem. Res.*, 52, 6848, 2013.

Perry, J.H. (Ed.), *Chemical Engineers' Handbook*, 6th edn., McGraw-Hill, New York, 1984.

Rajitha, P., R.P. Chhabra, N.E. Sabiri, and J. Comiti, Drag on non-spherical particles in power-law non-Newtonian media, *Int. J. Miner. Process.*, 78, 110, 2006.

Song, D., R.K. Gupta, and R.P. Chhabra, Wall effects on a sphere falling in power-law fluids in cylindrical tubes, *Ind. Eng. Chem. Res.*, 48, 5845, 2009.

Tran-Cong, S., M. Gay, and E.E. Michaelides, Drag coefficients of irregularly shaped particles, *Powder Technol.*, 139, 21, 2004.

Tripathi, A. and R.P. Chhabra, Drag on spheroidal particles in dilatant fluids, *AIChE J.*, 41, 728, 1995.

Tripathi, A., R.P. Chhabra, and T. Sundararajan, Power-law fluid flow over spheroidal particles, *Ind. Eng. Chem. Res.*, 33, 403, 1994.

Wadell, H., The coefficient of resistance as a function of Reynolds number for solids of various shapes, *J. Franklin Inst.*, 217, 459–490, 1934.

Wham, R.M., O.A. Basaran, and C.H. Byess, Wall effects on flow past solid spheres at finite Reynolds numbers, *Ind. Eng. Chem. Res.*, 35, 864, 1996.

Zakhem, R., P.D. Weidman, and H.C. deGroh, On the drag of model dendrite fragments at low Reynolds numbers, *Metall. Trans.*, 23A, 2169, 1992.

13 Fluid–Solid Separations by Free Settling

"Scientists study the world as it is, engineers create the world that never has been."

—**Theodore von Karman, 1881–1963, Mathematician**

I. FLUID–SOLID SEPARATIONS

The separation of suspended solids from a carrier fluid is a requirement in many engineering operations, e.g., the extensive listings of wide ranging applications and the equipment available in Besendorfer (1996), Chen (1997), Wakeman and Tarleton (2005) and Fernando Concha (2014). The most appropriate method for achieving this depends upon the specific properties of the system, the most important being the size and density of the solid particles and the solids concentration (e.g., the "solids loading") of the feed stream. For example, for relatively dilute systems (e.g., ~10% or less) of relatively large particles (e.g., ~100 μm or more) of fairly dense solids, a gravity settling tank may be appropriate, whereas for more dilute systems of smaller and/or lighter particles, a centrifuge may be more appropriate. For very fine particles or where very high separation efficiency is required, a "barrier" system may be needed, such as a filter or membrane. For highly concentrated systems, a gravity thickener may be adequate or, for more stringent requirements, a filter may be needed (e.g., Wakeman and Tarleton, 2005; Chen, 1997; Coulson et al., 1991).

In this chapter, we will consider relatively dilute systems, for which the effects of particle–particle interactions are relatively unimportant (e.g., gravity and centrifugal separation). Situations in which particle–particle interactions are negligible are referred to as *free settling*, as opposed to *hindered settling* in which such interactions are important. Figure 13.1 shows the approximate regions of solids concentration and density corresponding to free or hindered settling conditions. In Chapter 14, we will consider systems that are controlled by hindered settling or interparticle interactions (e.g., filtration and sedimentation processes).

II. GRAVITY SETTLING

Solid particles can be removed from a dilute suspension by passing the suspension through a vessel that is large enough such that the vertical component of the fluid velocity is lower than the terminal velocity of the particles and the residence time is sufficiently long to allow the particles to settle out. A typical gravity settler is illustrated in Figure 13.2. If the upward velocity of the liquid (Q/A) is less than the terminal velocity of the particles (V_t), the particles will settle at the bottom; otherwise, they will be carried out with the overflow. If Stokes flow is applicable (i.e., $N_{Re} < 1$), the diameter of the smallest particle that will settle out is

$$d = \left(\frac{18\mu Q}{g\Delta\rho A} \right)^{1/2} \tag{13.1}$$

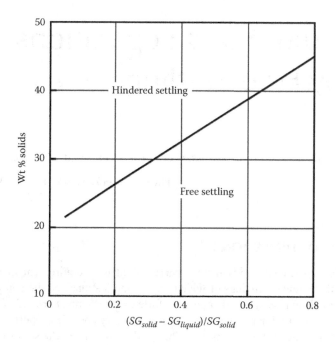

FIGURE 13.1   Approximate regions of hindered and free settling.

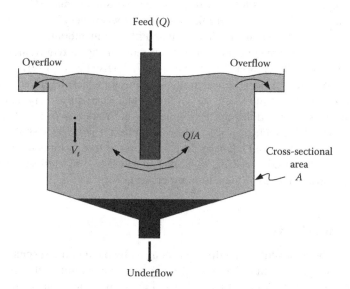

FIGURE 13.2   Schematic of a gravity settling tank.

If Stokes flow is not applicable (or even if it is), the Dallavalle equation in the form of Equation 12.17 can be used to determine the Reynolds number, and hence the diameter, of the smallest settling particle, that is,

$$
\frac{1}{\sqrt{N_{Re}}} = \left(0.00433 + 0.208 \sqrt{\frac{C_D}{N_{Re}}} \right)^{1/2} - 0.0658 \tag{13.2}
$$

where

$$\frac{C_D}{N_{Re}} = \frac{4\mu\Delta\rho g}{3\rho^2 V_t^3}, \quad d = \frac{N_{Re}\mu A}{Q\rho} \tag{13.3}$$

Alternatively, it may be necessary to determine the maximum capacity (e.g., flow rate, Q) at which particles of a given size, d, will (or will not) settle out. This can also be obtained directly from the Dallavalle equation in the form of Equation 12.14, by solving for the unknown flow rate:

$$Q = \left(\frac{\mu A}{d\rho}\right)\left[\left(14.42 + 1.827\sqrt{N_{Ar}}\right)^{1/2} - 3.798\right]^2 \tag{13.4}$$

where

$$N_{Ar} = \frac{d^3\rho g\Delta\rho}{\mu^2} \tag{13.5}$$

III. CENTRIFUGAL SEPARATION

It is not strictly correct to classify centrifugal (and cyclone) separations as "free settling," as the separation in these devices depends on a driving force established by an externally imposed centrifugal force. However, the principles involved in the analysis of these processes are all very similar, so it is reasonable to lump them together.

For very small particles or low-density solids, the terminal velocity may be too low to enable separation by gravity settling in a reasonably sized tank. However, the separation may be carried out in a centrifuge, which operates on the same principle as the gravity settler but employs the (radial) acceleration in a rotating system (e.g., $\omega^2 r$) in place of the vertical gravitational acceleration as the driving force. Centrifuges can be designed to operate at very high rotating speeds, which may be equivalent to many "g's" of acceleration, enhancing the separation.

A simplified schematic of a particle in a centrifuge is illustrated in Figure 13.3. It is assumed that any particle that impacts on the wall of the centrifuge (at r_2) before reaching the outlet will be

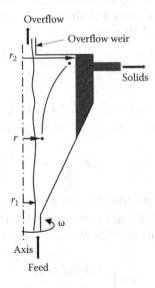

FIGURE 13.3 Schematic of a particle in a centrifuge.

trapped, and all others will not. (It might seem that any particle that impacts the outlet weir barrier would be trapped. However, the fluid circulates around this outlet corner setting up eddies that could sweep these particles out of the centrifuge). It is thus necessary to determine how far the particle will travel in the radial direction while in the centrifuge. To do this, we start with a radial momentum balance on the particle:

$$F_{cf} - F_b - F_D = m_e \frac{dV_r}{dt} \tag{13.6}$$

where
F_{cf} is the centrifugal force on the particle
F_b is the buoyant force (equal to the centrifugal force acting on the displaced fluid)
F_D is the drag force
m_e is the "effective" mass of the particle, which includes the solid particle and the "virtual mass" of the displaced fluid (i.e., half the actual mass of displaced fluid)

Equation 13.6 thus becomes

$$\left(\rho_s - \rho\right)\left(\frac{\pi d^3}{6}\right)\omega^2 r - \rho V_r^2 C_D\left(\frac{\pi d^3}{8}\right) = \left(\rho_s + \frac{\rho}{2}\right)\left(\frac{\pi d^3}{6}\right)\frac{dV_r}{dt} \tag{13.7}$$

When the particle reaches its terminal (radial) velocity, $dV_r/dt = 0$ and Equation 13.7 can be solved for V_{rt} (the radial terminal velocity):

$$V_{rt} = \left(\frac{4\Delta\rho d\omega^2 r}{3\rho C_D}\right)^{1/2} \tag{13.8}$$

If $N_{Re} < 1$, Stokes' law holds and $C_D = 24/N_{Re}$, in which case Equation 13.8 becomes

$$V_{rt} = \frac{dr}{dt} = \frac{\Delta\rho d^2\omega^2 r}{18\mu} \tag{13.9}$$

This shows that the terminal velocity is not a constant but increases with r, because the (centrifugal) driving force increases with r. Assuming that all of the fluid is rotating at the same speed as the centrifuge, integration of Equation 13.9 gives

$$\ln\left(\frac{r_2}{r_1}\right) = \frac{\Delta\rho d^2\omega^2}{18\mu} t \tag{13.10}$$

where t is the time required for the particle to travel a radial distance from r_1 to r_2. The time available for this to occur is the residence time of the particle in the centrifuge, that is,

$$t = \frac{\tilde{V}}{Q}$$

where $\tilde{V}$ is the volume of fluid in the centrifuge. If the region occupied by the fluid is cylindrical, then $\tilde{V} = \pi\left(r_2^2 - r_1^2\right)L$. Combining the above two equations and solving for d gives the smallest particle that will travel from the surface of the fluid (r_1) to the wall (r_2) in time t

$$d = \left(\frac{18\mu Q \ln\left(r_2/r_1\right)}{\Delta\rho\omega^2\tilde{V}}\right)^{1/2} \tag{13.11}$$

Rearranging Equation 13.11 to solve for Q gives

$$Q = \frac{\Delta \rho d^2 \omega^2 \tilde{V}}{18 \mu \ln (r_2/r_1)} = \left(\frac{\Delta \rho g d^2}{18 \mu} \right) \left(\frac{\tilde{V} \omega^2}{g \ln (r_2/r_1)} \right)$$

(13.12)

which can also be written as

$$Q = V_t \Sigma \quad \text{where} \quad \Sigma = \left(\frac{\tilde{V} \omega^2}{g \ln (r_2/r_1)} \right)$$

(13.13)

where

V_t is the terminal velocity (in the radial direction) of the particle in a gravitational field

Σ is the cross-sectional area of the gravity settling tank that would be required to remove the same size particles as in the centrifuge

This driving force can be extremely large, if the centrifuge operates at a speed corresponding to "many g's."

This analysis is based on the assumption that Stokes' law applies, that is, $N_{Re} < 1$. This is frequently a poor assumption, since many industrial centrifuges operate under conditions where $N_{Re} > 1$. If such is the case, an analytical solution to the problem is still possible using the Dallavalle equation for C_D, rearranged to solve for N_{Re} as follows:

$$N_{Re} = \left(\frac{d\rho}{\mu} \right) \frac{dr}{dt} = \left[\left(14.42 + 1.827 \sqrt{N_{Ar}} \right)^{1/2} - 3.797 \right]^2$$

(13.14)

where

$$N_{Ar} = \frac{d^3 \rho \omega^2 r \Delta \rho}{\mu^2}$$

(13.15)

Equation 13.14 can be integrated from r_1 to r_2 and solved for t to give

$$t = N_{12} \frac{\mu}{d^2 \omega^2 \Delta \rho}$$

(13.16)

where

$$N_{12} = 0.599 \left(N_{Re2} - N_{Re1} \right) + 13.65 \left(\sqrt{N_{Re2}} - \sqrt{N_{Re1}} \right)$$

$$+ 17.29 \ln \left(\frac{N_{Re2}}{N_{Re1}} \right) + 48.34 \left(\frac{1}{\sqrt{N_{Re2}}} - \frac{1}{\sqrt{N_{Re1}}} \right)$$

(13.17)

The values of N_{Re2} and N_{Re1} are computed using Equation 13.14 and the values of N_{Ar2} and N_{Ar1} evaluated at r_1 and r_2, respectively. Since $t = \tilde{V}/Q$, Equation 13.16 can be rearranged to solve for Q:

$$Q = \frac{\Delta \rho d^2 \omega^2 \tilde{V}}{\mu N_{12}} = \frac{\Delta \rho d^2 \omega^2 \tilde{V}}{18 \mu \ln (r_2/r_1)} \left[18 \frac{\ln (r_2/r_1)}{N_{12}} \right]$$

(13.18)

where the term in the square brackets is a "correction factor" that can be applied to the Stokes flow solution to account for deviations from the Stokes flow assumption.

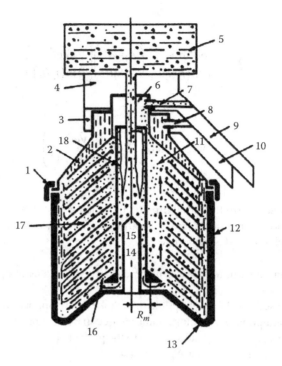

FIGURE 13.4 Schematic of a disk-bowl centrifuge: 1, Ring; 2, bowl; 3, 4, collectors for products; 5, feed tank; 6, tube; 7, 8, discharge nozzles; 9, 10, funnels for collectors; 11, through channels; 12, bowl; 13, bottom; 14, thick-walled tube; 15, hole for guide; 16, disk fixator; 17, disks; 18, central tube. (From Azbel, D.S. and Cheremisinoff, N.P., *Fluid Mechanics and Unit Operations*, Ann Arbor Science, Ann Arbor, MI, 1983.)

For separating very fine solids, emulsions, and immiscible liquids, a disk-bowl centrifuge is frequently used in which the settling occurs in the spaces between a stack of conical disks, as illustrated in Figure 13.4. The advantage of this arrangement is that the particles have a much smaller radial distance to travel before striking a wall and being trapped. The disadvantage is that the carrier fluid circulating between the disks has a higher velocity in the restricted spaces, which can retard the settling motion of the particles. Separation will occur only when $V_{rt} > V_{rf}$, where V_{rt} is the radial terminal velocity of the particle and V_{rf} is the radial velocity component of the carrier fluid in the region where the fluid flow is in the inward radial direction.

A. SEPARATION OF IMMISCIBLE LIQUIDS

The problem of separating immiscible liquids in a centrifuge can best be understood by first considering the static gravity separation, as illustrated in Figure 13.5, where the subscript 1 represents the lighter liquid and 2 represents the heavier liquid. In a continuous system, the static head of the heavier liquid in the overflow pipe must be balanced by the combined head of the lighter and heavier liquids in the separator, that is,

$$\rho_2 z g = \rho_2 z_2 g + \rho_1 z_1 g \tag{13.19}$$

or

$$z = z_2 + z_1 \frac{\rho_1}{\rho_2} \tag{13.20}$$

In a centrifuge, the position of the overflow weir is similarly determined by the relative amounts of the heavier and lighter liquids and their densities, along with the size and speed of the centrifuge. The feed

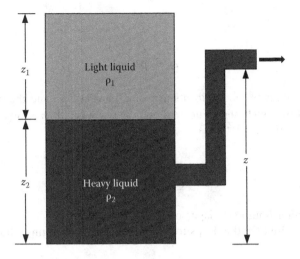

FIGURE 13.5 Gravity separation of immiscible liquids.

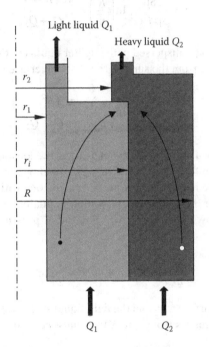

FIGURE 13.6 Centrifugal separation of immiscible liquids.

stream may consist of either the lighter liquid (1) dispersed in the heavier liquid (2) or vice versa. An illustration of the overflow weir positions is shown in Figure 13.6. Because there is no slip at the interface between the liquids, the axial velocity must be the same at that point for both fluids, that is,

$$V_1 = \frac{Q_1}{A_1} = \frac{Q_1}{\pi(r_i^2 - r_1^2)}$$

$$V_2 = \frac{Q_2}{A_2} = \frac{Q_2}{\pi(R^2 - r_i^2)}$$

(13.21)

or

$$\frac{Q_1}{Q_2} = \frac{r_i^2 - r_1^2}{R^2 - r_i^2} \tag{13.22}$$

This provides a relation between the locations of the interface (r_i) and the inner weir (r_1) and the relative feed rates of the two liquids. Furthermore, the residence time for each of the two liquids in the centrifuge must be the same, that is,

$$t = \frac{V_T}{Q_T} = \frac{\pi L \left(R^2 - r_1^2 \right)}{Q_1 + Q_2} \tag{13.23}$$

For drops of the lighter liquid (1) dispersed in the heavier liquid (2), assuming that Stokes flow applies, the time required for the drops to travel from the maximum radius (R) to the interface (r_i) is

$$t = \frac{18\mu_2}{\Delta\rho d^2 \omega^2} \ln\left(\frac{R}{r_i}\right) = \frac{\pi L \left(R^2 - r_1^2 \right)}{Q_1 + Q_2} \tag{13.24}$$

For drops of the heavier liquid (2) dispersed in the lighter liquid (1), the corresponding time required for the maximum radial travel from the surface (r_1) to the interface (r_i) is

$$t = \frac{18\mu_1}{\Delta\rho d^2 \omega^2} \ln\left(\frac{r_i}{r_1}\right) = \frac{\pi L \left(R^2 - r_1^2 \right)}{Q_1 + Q_2} \tag{13.25}$$

Equations 13.22 and 13.24 or Equation 13.25 determines the locations of the light liquid weir (r_1) and the interface (r_i) for given feed rates, centrifuge size, and operating conditions.

The proper location of the heavy liquid weir (r_2) can be determined by a balance of the radial pressure difference through the liquid layers, which is analogous to the gravity head balance in the gravity separator shown in Figure 13.5. The radial pressure gradient due to centrifugal force is

$$\frac{dP}{dr} = \rho\omega^2 r \quad \text{or} \quad \Delta P = \frac{1}{2}\rho\omega^2 \Delta r^2 \tag{13.26}$$

Since both the heavy liquid surface at r_2 and the light liquid surface at r_1 are at atmospheric pressure, the sum of the pressure differences from r_1 to R to r_2 must be zero:

$$\frac{1}{2}\rho_2\omega^2\left(r_2^2 - R^2\right) + \frac{1}{2}\rho_2\omega^2\left(R^2 - r_i^2\right) + \frac{1}{2}\rho_1\omega^2\left(r_i^2 - r_1^2\right) = 0 \tag{13.27}$$

which can be rearranged to give

$$\frac{\rho_2}{\rho_1} = \frac{r_i^2 - r_1^2}{r_i^2 - r_2^2} \tag{13.28}$$

Solving for r_2:

$$r_2^2 = \frac{\rho_1}{\rho_2}\left[r_i^2\left(\frac{\rho_2}{\rho_1} - 1\right) + r_1^2\right] \tag{13.29}$$

Equations 13.22, 13.24, or 13.25 and 13.29 thus determine the three design parameters r_i, r_1, and r_2. These equations can be arranged in the following dimensionless form. From Equation 13.22:

$$\left(\frac{r_1}{R}\right)^2 = \left(\frac{r_i}{R}\right)^2 \left(1 + \frac{Q_1}{Q_2}\right) - \frac{Q_1}{Q_2} \tag{13.30}$$

For drops of the light liquid in the heavy liquid, Equation 13.24 becomes

$$\left(\frac{r_1}{R}\right)^2 = 1 - \frac{18\mu_2 (Q_1 + Q_2)}{\pi L R^2 \Delta\rho d^2 \omega^2} \ln\left(\frac{R}{r_i}\right) \tag{13.31}$$

For drops of the heavy liquid in the light liquid, Equation 13.25 becomes

$$\left(\frac{r_1}{R}\right)^2 = 1 - \frac{18\mu_1 (Q_1 + Q_2)}{\pi L R^2 \Delta\rho d^2 \omega^2} \ln\left[\left(\frac{r_i}{R}\right)\left(\frac{R}{r_1}\right)\right] \tag{13.32}$$

Also, Equation 13.29 is equivalent to

$$\left(\frac{r_2}{R}\right)^2 = \frac{\rho_1}{\rho_2} \left[\frac{r_i^2}{R^2}\left(\frac{\rho_2}{\rho_1} - 1\right) + \frac{r_1^2}{R^2}\right] \tag{13.33}$$

These three equations can be solved simultaneously (by iteration) for (r_1/R), (r_2/R), and (r_i/R). It is assumed that the size of the suspended drops is known, as are the density and viscosity of the liquids and the overall dimensions and speed of the centrifuge.

IV. CYCLONE SEPARATIONS

A. GENERAL CHARACTERISTICS

Centrifugal force can also be used to separate solid particles from fluids by inducing the fluid to undergo a rotating or spiraling flow pattern in a stationary vessel (e.g., a cyclone), which has no moving parts. Cyclones are widely used to remove small particles from gas streams (e.g., "aerocyclones") and suspended solids from liquid streams (e.g., "hydrocyclones").

A typical cyclone is illustrated in Figure 13.7 (this is sometimes referred to as a "reverse flow" cyclone). The suspension enters through a rectangular or circular duct tangential to the cylindrical separator, which usually has a conical bottom. The circulating flow generates a rotating vortex motion that imparts centrifugal force to the particles, which are thrown outward to the walls of the vessel where they fall by gravity to the conical bottom and are removed from the dusty gas. The carrier fluid spirals inward and downward to the cylindrical exit duct (also referred to as the "vortex finder"), where it travels back up and leaves the vessel at the top. The separation is not perfect, and some solid particles leave in the overflow as well as the underflow. The particle size for which 50% leaves in the overflow and 50% leaves in the underflow is called the *cut size*.

The diameter of hydrocyclones can range from 10 mm to 2.5 m, cut sizes from 2 to 250 μm, and flow rates (capacities) from 0.1 to 7200 m³/h. The pressure drop can range from 0.3 to 6 atm (Svarovsky, 1984). For aerocyclones, very little fluid leaves with the solids underflow, although for hydrocyclones the underflow solids content is typically 45%–50% by volume. Aerocyclones can separate particles as small as 2–5 μm.

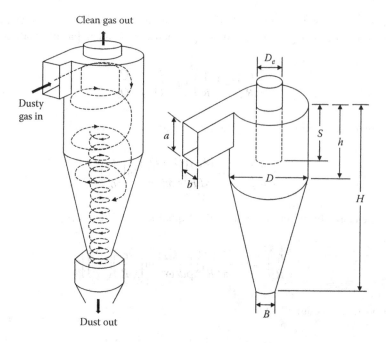

FIGURE 13.7   Typical reverse flow cyclone with geometric dimensions.

The advantages of the cyclone separator include the following (Svarovsky, 2000):

1. *Versatility*—Virtually, any slurry or suspension can be concentrated, liquids degassed, or the solids classified by size, density, or shape.
2. *Simplicity and economy*—They have no moving parts and require little maintenance.
3. *Small*—Low residence times and relatively fast response.
4. *High shear forces*—Can break up agglomerates, clusters of particles, friable solids etc.

The primary disadvantages are as follows:

1. *Inflexibility*—A given design is not easily adapted to a range of conditions. Performance is strongly dependent upon flow rate and feed composition, and the turndown ratio (i.e., range of operation) is small.
2. *Limited separation performance*—In terms of the sharpness of the cut, range of cut size, etc.
3. Susceptibility to erosion for abrasive particles.
4. High shear prevents utilization of flocculants to aid the separation, as can often be done in gravity settlers.

An increase in any one operating parameter generally increases all others, as well. For example, increasing the flow rate will increase both separation efficiency and pressure drop, and vice versa.

B.   AEROCYCLONES

1.   Velocity Distribution

Although the dominant velocity component in the cyclone is in the angular (tangential) direction, the swirling flow field includes significant velocity components in the radial and axial directions as well, which complicate the motion and makes a rigorous analysis impossible. This complex flow field also results in significant particle–particle collisions, which cause some particles of a given size to be carried out in both the overhead and underflow discharge, thus affecting the separation efficiency.

Cyclone analysis and design is not an exact science, and there are a variety of approaches to the analysis of cyclone performance. A critical review of the various methods for analyzing hydrocyclones has been given by Svarovsky (1996), and a review of different approaches to aerocyclone analysis has been given by Leith and Jones (1997). There are a number of different approaches to the analysis of aerocyclones, one of the most comprehensive being that of Bohnet et al. (1997). The presentation here follows that of Leith and Jones (1997), which outlines the basic principles and some of the practical "working relations." The reader is referred to the additional references cited, especially the works of Bohnet (1983) and Bohnet et al. (1997), for more details on specific cyclone design.

The performance of a cyclone is dependent upon the geometry as described by the values of the various dimensionless "length ratios" (see Figure 13.7): a/D, b/D, D_e/D, S/D, h/D, H/D, and B/D. Typical values of these ratios for various "standard designs" are given in Table 13.1.

The complex three-dimensional flow pattern within the cyclone is dominated by the radial (V_r) and tangential (V_θ) velocity components. The vertical component is also significant but plays only an indirect role in the separation. The tangential velocity in the vortex varies with the distance from the axis in a complex manner, which can be described by the following equation:

$$V_\theta r^n = \text{constant} \tag{13.34}$$

For a uniform angular velocity (ω = const., i.e., a "solid body rotation") $n = -1$, whereas for a uniform tangential velocity ("plug flow") $n = 0$, and for inviscid free vortex flow $\omega = c/r^2$, that is, $n = 1$. Empirically, the exponent n has been found to vary typically between 0.5 and 0.9. The maximum value of V_θ occurs in the vicinity of the outlet or exit duct (vortex finder) at $r = D_e/2$. For aerocyclones, the exponent n has been correlated with the cyclone diameter by the following expression

$$n = 0.67 D_m^{0.14} \tag{13.35}$$

where D_m is the cyclone diameter in meters. The exponent also decreases as the temperature increases according to

$$\frac{1-n}{1-n_1} = \left(\frac{T}{T_1}\right)^{0.3} \tag{13.36}$$

There is a "core" of rotating flow below the gas exit duct (vortex finder), in which the velocity decreases as the radius decreases and is nearly zero at the axis.

TABLE 13.1
Standard Designs for Reverse Flow Cyclones

Ref.	Duty	D	a/D	b/D	D_e/D	S/D	h/D	H/D	B/D	K_f	Q/D² (m/h)
1	High η	1	0.5	0.2	0.5	0.5	1.5	4.0	0.375	6.4	3,500
2	High η	1	0.44	0.21	0.4	0.5	1.4	3.9	0.4	9.2	4,940
3	Gen	1	0.5	0.25	0.5	0.625	2.0	4.0	0.25	8.0	6,860
2	Gen	1	0.5	0.25	0.5	0.6	1.75	3.75	0.4	7.6	6,680
1	High Q	1	0.75	0.375	0.75	0.875	1.5	4.0	0.375	7.2	16,500
2	High Q	1	0.8	0.35	0.75	0.85	1.7	3.7	0.4	1.0	12,500

Source: Leith, D. and Jones, D.L., Cyclones, Chapter 15, in *Handbook of Powder Science and Technology*, 2nd edn., M.E. Fayed and L. Otten, (Eds.), Chapman & Hall, New York, 1997. 1. Stairmand (1951). 2. Swift (1969). 3. Lapple (1951).

2. Pressure Drop

The pressure drops throughout the cyclone due to several factors: (1) gas expansion, (2) vortex formation, (3) friction loss, and (4) changes in kinetic energy. The total pressure drop can be expressed in terms of an equivalent loss coefficient, K_f:

$$\Delta P = \frac{K_f \rho_G V_i^2}{2} \tag{13.37}$$

where V_i is the gas inlet velocity, $V_i = Q/ab$. A variety of expressions have been developed for K_f, but one of the simplest that gives reasonable results is

$$K_f = 16 \frac{ab}{D_e^2} \tag{13.38}$$

This (and other) expression may be accurate to only about ±50% or so, and more reliable pressure drop information can only be obtained by experimental testing on a specific geometry. Typical values of K_f for the "standard" designs are also given in Table 13.1.

3. Separation Efficiency

The efficiency of the cyclone (η) is defined as the fraction of particles of a given size, which is separated by the cyclone. The efficiency increases with the following trends:

1. Increasing particle diameter (d) and density
2. Increasing gas velocity
3. Decreasing cyclone diameter
4. Increasing cyclone length
5. Venting some of the gas through the bottom solids exit
6. Wetting the walls

A typical plot of efficiency versus particle diameter is shown in Figure 13.8. This is called a *grade-efficiency* curve. Although the efficiency varies with the particle size, a more easily determined characteristic is the "cut diameter" (d_{50}), which is that particle size collected with 50% efficiency.

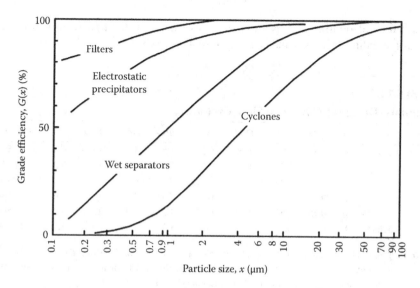

FIGURE 13.8 Typical cyclone grade-efficiency curve.

The particles are subject to centrifugal, inertial, and drag forces as they are carried in the spiraling flow, and it is assumed that the particles that strike the outer wall before the fluid reaches the vortex finder will be collected. It is assumed that the tangential velocity of the particle is the same as that of the fluid, that is, no slip ($V_{p\theta} = V_\theta$), but the radial velocity $V_{pr} \neq V_r$, since the particles move radially toward the wall relative to the fluid. The centrifugal force acting on the particle is

$$F_e = m_p \omega^2 r = \frac{\pi d^3 \rho_s V_\theta^2}{6r} \tag{13.39}$$

Assuming Stokes flow, the drag force is

$$F_d = 3\pi\mu d \left(V_{pr} - V_r\right) \tag{13.40}$$

Equation 13.34 provides a relation between the tangential velocity at any point and that at the wall:

$$V_\theta r^n = V_{\theta w} r_w^n \tag{13.41}$$

Although the velocity right at the wall is zero, the boundary layer at the wall is quite small, so this equation applies up to the boundary layer very near the wall. Setting the sum of the forces equal to the particle accelerating momentum and substituting $V_{pr} = dr/dt$ and $V_\theta = V_{\theta r}(r_w/r)^n$ gives the governing equation for the particle radial position as a function of time:

$$\frac{d^2 r}{dt^2} + \frac{18\mu}{d^2 \rho_s} \frac{dr}{dt} - \left(\frac{V_{\theta w}^2 r_w^{2n}}{r^{(2n+1)}} + \frac{18\mu V_r}{d^2 \rho_s}\right) = 0 \tag{13.42}$$

There is no general solution to this nonlinear differential equation, and various analyses have been based upon specific approximations or simplifications of the equation. One approximation considers the time it takes for the particle to travel from the entrance point, r_i, to the wall, $r_w = D/2$, relative to the residence time of the fluid in the cyclone. By neglecting the acceleration term and the fluid radial velocity and assuming the velocity of the fluid at the entrance to be the same as the tangential velocity at the wall ($V_i = V_{\theta w}$), Equation 13.42 can be integrated to give the time required for the particle to travel from its initial position (r_i) to the wall ($D/2$). If this time is equal to or less than the residence time of the fluid in the cyclone, that particle will be trapped. The result thus gives the size of the theoretical smallest particle that will be trapped completely (in principle):

$$d_{100} = \left[\frac{9\mu D^2 \left(1 - 2r_i/D\right)^{(2n+2)}}{4(n+1)V_i^2 \rho_s t}\right]^{1/2} \tag{13.43}$$

The residence time is related to the "number of turns" (N) that the fluid makes in the vortex, which can vary from 0.3 to 10, with an average value of 5. If the 50% "cut diameter" particle is assumed to enter at $(D - b)/2$, with a residence time of

$$t = \frac{\pi D N}{V_i}\left(1 - \frac{D_e}{2D}\right) \tag{13.44}$$

and it is assumed that $n = 0$, then the cut diameter is

$$d_{50} = \left(\frac{9\mu b}{2\pi \rho_s V_i N_i}\right)^{1/2} \tag{13.45}$$

Another approach is to consider the particle for which the drag force of the gas at the edge of the core (where the velocity is maximum) just balances the centrifugal force. This reduces Equation 13.42 to a "steady state," with no net acceleration or velocity of the particle. The maximum velocity is given by Equation 13.41 applied at the edge of the core: $V_{\theta w}^2 r_w^{2n} = V_\theta^2 r_{core}^{2n}$. When this is introduced into Equation 13.42, the result is

$$d_{100} = \left(\frac{9Q\mu}{\pi(H-S)\rho_s V_{max}^2} \right)^{1/2} \tag{13.46}$$

Although this predicts that all particles larger than this size will be trapped and all smaller particles will escape, the actual grade efficiency depends on the particle size because of the variation of the inward radial velocity of the gas.

Leith and Licht (1972) incorporated the effect of turbulent reentrainment of the solids in a solution of Equation 13.42 to derive the following expression for the grade efficiency:

$$\eta = 1 - \exp\left[-2\left(N_G N_{St}\right)^{1/(2n+2)} \right] \tag{13.47}$$

where

$$N_{St} = \frac{d^2 \rho_s V_i (n+1)}{18\mu D} \tag{13.48}$$

is the Stokes' number and N_G is a dimensionless geometric parameter, defined as

$$N_G = \frac{\pi D^2}{ab} \left\{ 2\left[1 - \left(\frac{D_e}{D} \right)^2 \right]\left(\frac{S}{D} - \frac{a}{2D} \right) + \frac{1}{3}\left(\frac{S + z_c - h}{D} \right) \right.$$

$$\left. \times \left[1 + \frac{d_c}{D} + \left(\frac{d_c}{D} \right)^2 \right] + \frac{h}{D} - \left(\frac{D_e}{D} \right)^2 \frac{z_c}{D} - \frac{S}{D} \right\} \tag{13.49}$$

where z_c is the core length, given by

$$z_c = 2.3 D_e \left(\frac{D^2}{ab} \right)^{1/3} \tag{13.50}$$

and d_c is the core diameter, given by

$$d_c = D - (D-B)\left(\frac{S + z_c - h}{H - h} \right) \tag{13.51}$$

Equation 13.47 implies that the efficiency increases as N_G and/or N_{St} increases.

These equations can serve as a guide for estimating performance but cannot be expected to provide precise prediction of the behavior. However, they can be used effectively to scale experimental results for similar designs of different sizes operating under various conditions. For example, two cyclones of a given design should have the same efficiency when the value of N_{St} is the same for both. That is, if a given cyclone has a known efficiency for particles of diameter d_1, a similar cyclone will have the same efficiency for particles of diameter d_2, where

$$d_2 = d_1 \left[\frac{Q_1}{Q_2}\left(\frac{\rho_{s1}}{\rho_{s2}} \right)\left(\frac{\mu_2}{\mu_1} \right)\left(\frac{D_2}{D_1} \right) \right]^{1/2} \tag{13.52}$$

Thus, the grade efficiency of a similar cyclone can be constructed from the grade efficiency of the known (tested) cyclone.

4. Other Effects

Increasing the solids loading in the feed increases the collection efficiency and decreases the pressure drop. The reduction in pressure drop is attributed to the lowering of the tangential gas velocity due to the increased wall friction at high solids loading. The effect on pressure drop is given approximately by

$$(\Delta P)_c = \frac{(\Delta P)_o}{1 + 0.0086 C_i^{1/2}} \tag{13.53}$$

where C_i is the inlet solids loading in g/m³. On the other hand, the increased collection efficiency can be explained as follows: there is a critical load of particles that can be transported by the gas at a given velocity, and the amount of solids in excess of this load is deposited on the wall and captured by the cyclone, thereby increasing the overall efficiency. The effect of solids loading on overall collection efficiency is given by

$$\frac{100 - \eta}{100 - \eta_1} = \left(\frac{C_{i1}}{C_i}\right)^{0.182} \tag{13.54}$$

If the velocity near the wall is too high, particles will bounce off the wall and become reentrained. The inlet velocity above which this effect occurs is given by the empirical correlation in SI units (Leith and Jones, 1997):

$$V_{ic} = 2400\left(\frac{\mu \rho_s}{\rho_G^2}\right)\left(\frac{D^{0.2}(b/D)^{1.2}}{1 - b/D}\right) \tag{13.55}$$

The cyclone efficiency increases with V_i up to about $1.25V_{ic}$, after which reentrainment results in a decrease in efficiency.

C. Hydrocyclones

A similar approach to the analysis of hydrocyclones has been presented by Svarovsky (2000), Besendorfer (1996), Salcudean et al., (2003), and others. They deduced that the system can be described in terms of three dimensionless groups, in addition to various dimensionless geometric parameters. These groups are the Stokes' number:

$$N_{St50} = \frac{V_{tr}}{V_i} = \frac{\Delta \rho d_{50}^2 Q}{4\pi \mu D^3} \tag{13.56}$$

the Euler number:

$$N_{Eu} = \frac{\Delta P}{\rho V_i^2/2} = \frac{\pi^2 \Delta P D^4}{8\rho Q^2} \tag{13.57}$$

(which is equivalent to the loss coefficient, K_f), and the Reynolds number:

$$N_{Re} = \frac{DV_i \rho}{\mu} = \frac{4Q\rho}{\pi D\mu} \tag{13.58}$$

In each of these groups, the characteristic length is the cyclone diameter, D, and the characteristic velocity is $V_i = 4Q/\pi D^2$. Various empirical models for hydrocyclones indicate that the relationship between these groups can be represented by

$$N_{St50} N_{Eu} = C \tag{13.59}$$

and

$$N_{Eu} = K_p N_{Re}^{n_p} \tag{13.60}$$

The quantities C, K_p, and n_p are empirical constants, with the same values for a given family of geometrically similar cyclones. The value of C ranges from 0.06 to 0.33, the exponent n_p varies from 0 to 0.8, and K_p ranges from 2.6 to 6300. A summary of these parameters corresponding to some known hydrocyclone designs is given in Table 13.2 (Svarovsky, 1984). The references in this table are found in Svarovsky's book, and the notation in the table is as follows: $D_i = 2r_i = (4ab/\pi)^{1/2}$ is the equivalent diameter of the inlet, $D_o = D_e$ is the gas exit diameter, $l = S$ is the length of the vortex finder, and $L = H$ is the total length of the hydrocyclone. These equations can be used to predict the performance of a given cyclone as follows.

TABLE 13.2
Operational Parameters from Some Known Hydrocyclone Designs[a]

Cyclone Type and Size of Hydrocyclone	Geometrical Ratios				Angle	Scale-Up Constants			Operating Cost Criterion
	D_i/D_o	D_o/D	l/D	L/D	θ (°)	$N_{St50}N_{Eu}$	K_p	n_p	$N_{St50}N_{Eu}$
Rietema's design (optimum separation), $D = 0.075$ m	0.28	0.34	0.4	5	20	0.0611	316	0.134	2.12
Bradley's design $D = 0.038$ m	0.133 (1/7.5)	0.2 (1/5)	0.33 (1/3)	6.85	9	0.1111	446.5	0.323	2.17
Mozley cyclone $D = 0.022$ m	0.154 (1/6.5)	0.214 (3/14)	0.57 (4/7)	7.43	6	0.1203	6381	0	3.20
Mozley cyclone $D = 0.044$ m	0.160 (1/6.25)	0.250 (1/4)	0.57 (4/7)	7.71	6	0.1508	4451	0	4.88
Mozley cyclone $D = 0.044$ m	0.197 (1/5)	0.32 (1/3)	0.57 (4/7)	7.71	6	0.2182	3441	0	8.70
Warman 3 in. Model R, $D = 0.076$ m	0.29 (1/3.5)	0.20 (1/5)	0.31	4	15	0.1079	2.618	0.8	2.07
RW 2515 (AKW), $D = 0.125$ m	0.20 (1/5)	0.32 (1/3)	0.80	6.24	15	0.1642	2458	0	6.66
Hi-Klone model 2, $D = 0.097$ m	0.175	0.25	0.92 (0.59)*	5.6	10		873.5	0.2	
Hi-Klone model 3, $D = 0.1125$ m	0.150	0.20	0.80 (0.51)*	5.4	10		815.5	0.2	
Demco, $D = 0.051$ m	0.217	0.50	1	4.7	25				
Demco, $D = 0.102$ m	0.244	0.303	0.833	3.9	20				

Source: Svarovsky, L., Hydrocyclones, Chapter 6, in Solid–Liquid Separation, L. Svarovsky, (Ed.), 4th edn., Butterworth Heinemann, Oxford, U.K., 2000.

[a] Represents a modification; blank spaces in the table are where the figures are not known.

Equation 13.60 can be solved for the capacity, Q, to give

$$Q^{2+n_p} = \frac{\pi^2 \Delta P D^4}{8 \rho K_p} \left(\frac{\pi D \mu}{4 \rho} \right)^{n_p} \tag{13.61}$$

and the cut size obtained from Equation 13.59:

$$d_{50}^2 = \frac{4 \pi N_{St50} N_{Eu} D^3 \mu}{K_p Q \Delta \rho} \left(\frac{\pi D \mu}{4 Q \rho} \right)^{n_p} \tag{13.62}$$

Actually, an extensive database reported by Medronho (Antunes and Medronho, 1992) indicates that the product $N_{St50} N_{Eu}$ is not constant but depends on the underflow to feed ratio (R) and the feed volumetric concentration (C_v) and N_{Eu} also depends on C_v as well as N_{Re}.

The Rietema and Bradley geometries are the two common families of geometrically similar designs, as defined by the geometric parameters in Table 13.3 (Antunes and Medronho, 1992). The Bradley hydrocyclone has a lower capacity than the Rietema geometry but is more efficient. For the Rietema cyclone geometry, the correlations are (Antunes and Medronho, 1992):

$$N_{St50} N_{Eu} = 0.0474 \left[\ln(1/R) \right]^{0.742} \exp\left(8.96 C_v \right) \tag{13.63}$$

and

$$N_{Eu} = 371.5 N_{Re}^{0.116} \exp\left(-2.12 C_v \right) \tag{13.64}$$

where

$$R = 1218 (B/D)^{4.75} N_{Eu}^{0.30} \tag{13.65}$$

For the Bradley geometry, the corresponding correlations are (Antunes and Medronho, 1992)

$$N_{Stk50} N_{Eu} = 0.055 \left[\ln(1/R) \right]^{0.66} \exp\left(12 C_v \right) \tag{13.66}$$

and

$$N_{Eu} = 258 N_{Re}^{0.37} \tag{13.67}$$

where

$$R = 1.21 \times 10^6 (B/D)^{2.63} N_{Eu}^{-1.12} \tag{13.68}$$

TABLE 13.3

Families of Geometric Cyclones

Cyclone	$2r_{in}/D$	D_e/D	$2S/D$	$2h/D$	$2H/D$	Cone Angle (°)
Rietema	0.28	0.34	0.4	—	5.0	10–20
Bradley	1/7	1/5	1/3	1/2	—	9

These equations can be used to either predict the performance of a given cyclone or size a cyclone for given conditions. For example, if the definitions of N_{Eu} and N_{Re} from Equations 13.57 and 13.58 are substituted into Equation 13.67, the resulting expression for D is

$$D = 3.5 \frac{\rho^{0.31} Q^{0.54}}{\mu^{0.085} (\Delta P)^{0.23}} \tag{13.69}$$

which is dimensionally consistent.

SUMMARY

The main points to be retained from this chapter are

- The distinction between "free" and "hindered" settling
- Operation of the gravity settling tank
- Analysis of centrifugal separators
- Analysis of liquid–liquid separators
- General principles and empirical correlations for separation efficiency of hydrocyclones and aerocyclones

PROBLEMS

FREE SETTLING FLUID–PARTICLE SEPARATIONS

1. A slurry containing solid particles having a density of 2.4 g/cm³, ranging in diameter from 0.001 in. to 0.1 in., is fed to a settling tank 10 ft in diameter. Water is pumped into the tank at the bottom and overflows at the top, carrying some of the solid with it. If it is desired to separate out all particles of diameter 0.02 in. and smaller, what flow rate of the water in gpm is required?

2. A handful of sand and gravel is dropped into a tank of water 5 ft deep. The time required for the solids to reach the bottom is measured and found to vary from 3 to 20 s. If the solid particles behave as equivalent spheres and have $SG = 2.4$, what is the range of equivalent particle diameters?

3. It is desired to determine the size of pulverized coal particles by measuring the time it takes them to fall a given distance in a known fluid. It is found that the coal particles ($SG = 1.35$) take a time ranging from 5 s to 1000 min to fall 23 cm through a column of methanol ($SG = 0.785$, $\mu = 0.88$ cP). What is the size range of the particles in terms of their equivalent spherical diameters? Assume the particles are falling at their terminal velocities at all times.

4. A water slurry containing coal particles ($SG = 1.35$) is pumped into the bottom of a large tank (10 ft diameter, 6 ft high), at a rate of 500 gal/h, and overflows at the top. What is the largest coal particle that will be carried over in the overflow? If the flow rate is increased to 5000 gal/h, what size particles would you expect in the overflow? The slurry properties can be taken to be the same as for water.

5. In order to determine the settling characteristics of a sediment, you drop a sample of the material into a column of water. You measure the time it takes for the solids to fall a distance of 2 ft and find that it ranges from 1 to 20 s. If the solid has a $SG = 2.5$, what is the range of particle sizes in the sediment, in terms of the diameters of equivalent spheres?

6. You want to separate all the coal particles having a diameter of 100 µm or larger from a slurry. To do this, the slurry is pumped into the bottom of a large tank. It flows upward and overflows

over the top of the tank, where it is collected in a trough. If the coal has $SG = 1.4$ and the total flow rate is 250 gpm, how big should the tank be?

7. A gravity settling chamber consists of a horizontal rectangular duct 6 m long, 3.6 m wide, 3 m high. The duct is used to trap sulfuric acid mist droplets entrained in an air stream. The droplets settle out as the air passes through the duct and may be assumed to behave as rigid spheres. If the air stream has a flow rate of 6.5 m³/s, what is the diameter of the largest particle that will not be trapped by the duct? (For the *Acid*: $\rho = 1.75$ g/cm³, $\mu = 3$ cP; For *Air*: $\rho = 0.01$ g/cm³, $\mu = 0.02$ cP).

8. Solid particles of diameter 0.1 mm and density 2 g/cm³ are to be separated from air in a horizontal settling chamber. If the air flow rate is 100 ft³/s and the maximum height of the chamber is 4 ft, what should the minimum length and width be for all of the particles to hit the bottom before exiting the chamber? (*Air*: $\rho = 0.075$ lb$_m$/ft³, $\mu = 0.018$ cP).

9. A settling tank contains solid particles that have a wide range of sizes. Water is pumped into the tank from the bottom and overflows on top at a rate of 10,000 gal/h. If the tank diameter is 3 ft, what separation of particle size is achieved (i.e., what size particles are carried out to the top of the tank, assuming the particles to be spherical)? Solid density = 150 lb$_m$/ft³.

10. You want to use a viscous Newtonian fluid to transport small granite particles through a horizontal 1 in. ID 100 ft long pipeline. The granite particles have a diameter of 1.5 mm and $SG = 4.0$. The SG of the fluid can be assumed to be 0.95. The fluid should be pumped as fast as possible, to minimize settling of the particles in the pipe, but must be kept in laminar flow, so you design the system to operate at a pipe Reynolds number of 1000. The flow rate must be fast enough that the particles will not settle at a distance greater than 1/2 the ID of the pipe, from the entrance to the exit. What should the viscosity of the fluid be, and what should the flow rate be (in gpm) at which it is pumped through the pipe?

11. An aqueous slurry containing particles with the size distribution shown in the following table is fed to a 20 ft diameter settling tank (see, e.g., McCabe et al. (1993) or Perry's Handbook for the definition of mesh sizes):

Tyler Mesh Size	% of Total Solids in Feed
8/10	5.0
10/14	12.0
14/20	26.0
20/28	32.0
28/35	21.0
35/48	4.0

The feed enters near the center of the tank, and the liquid flows upward and overflows at the top of the tank. The solids loading of the feed is 0.5 lb$_m$ of solids per gallon of slurry, and the feed rate is 50,000 gpm. What is the total solids concentration and particle size distribution in the overflow? The density of solids is 100 lb$_m$/ft³. Assume (1) the particles are spherical, (2) the particles in the tank are unhindered, and (3) the feed and overflow have the same properties as water.

12. A water stream contacts a bed of particles, which have diameters ranging from 1 to 1000 μm and $SG = 2.5$. The water stream flows upward at a rate of 3 cm/s. What size particles will be carried out by the stream, and what size will be left behind?

13. An aqueous slurry containing particles with $SG = 4$ and a range of sizes up to 300 μm flows upward through a small tube into a larger vertical chamber with a diameter of 10 cm. You want the liquid to carry all of the solids through the small tube, but you want only those particles with diameters less than 20 μm to be carried out to the top of the larger chamber. What should the flow rate of the slurry be (in gpm) and what size should the smaller tube be?

14. A dilute aqueous $CaCO_3$ slurry is pumped into the bottom of a classifier, at a rate of 0.4 m³/s, and overflows at the top. The density of the solids is 2.71 g/cm³.
    (a) What should the diameter of the classifier be if the overflow is to contain no particles larger than 0.2 mm in diameter?
    (b) The same slurry as in (a) is sent to a centrifuge, which operates at 5000 rpm. The centrifuge diameter is 20 cm, its length is 30 cm, and the liquid layer thickness is 20% of the centrifuge radius. What is the maximum flow rate that the centrifuge can handle and achieve the same separation as the classifier?

15. A centrifuge that is 40 cm ID and 30 cm long has an overflow weir that is 5 cm wide. The centrifuge operates at a speed of 3600 rpm.
    (a) What is the maximum capacity of the centrifuge (in gpm) for which particles with a diameter of 25 μm and $SG = 1.4$ can be separated from the suspension?
    (b) What would be the diameter of a settling tank that would do the same job?
    (c) If the centrifuge ID was 30 cm, how fast would it have to rotate to do the same job, everything else being equal?

16. Solid particles with a diameter of 10 μm and $SG = 2.5$ are to be removed from an aqueous suspension in a centrifuge. The centrifuge inner radius is 1 ft, the outer radius is 2 ft, and length is 1 ft. If the required capacity of the centrifuge is 100 gpm, what should the operating speed (in rpm) be?

17. A centrifuge is used to remove solid particles with a diameter of 5 μm and $SG = 1.25$ from a dilute aqueous stream. The centrifuge rotates at 1200 rpm and is 3 ft high, the radial distance to the liquid surface is 10 in., and the radial distance to the wall is 14 in.
    (a) Assuming that the particles must strike the centrifuge wall to be removed, what is the maximum capacity of this centrifuge, in gpm?
    (b) What is the diameter, in feet, of the gravity settling tank that would be required to do the same job?

18. A dilute aqueous slurry containing solids with a diameter of 20 μm and $SG = 1.5$ is fed to a centrifuge rotating at 3000 rpm. The radius of the centrifuge is 18 in., its length is 24 in., and the overflow weir is 12 in. from the centerline.
    (a) If all of the solids are to be removed in the centrifuge, what is the maximum capacity that it can handle (in gpm)?
    (b) What is the diameter of the gravity settling tank that would be required for this separation, at the same flow rate?

19. A centrifuge with a radius of 2 ft and a length of 1 ft has an overflow weir located 1 ft from the centerline. If particles with $SG = 2.5$ and diameters of 10 μm and less are to be removed from an aqueous suspension at a flow rate of 100 gpm, what should the operating speed of the centrifuge be (in rpm)?

20. A centrifuge with a diameter of 20 in. operates at a speed of 1800 rpm. If there is a water layer 3 in. thick on the centrifuge wall, what is the pressure exerted on the wall?

21. A vertical centrifuge, operating at 100 rpm, contains an aqueous suspension of solid particles with $SG = 1.3$ and radius of 1 mm. When the particles are 10 cm from the axis of rotation, determine the direction in which they are moving relative to a horizontal plane.

22. You are required to design an aerocyclone to remove as much dust as possible from the exhaust coming from a rotary drier. The gas is air at 100°C, 1 atm, and a rate of 40,000 m³/h. The effluent from the cyclone will go to a scrubber for final cleanup. The maximum loading to the scrubber should be 10 g/m³, although 8 g/m³ or less is preferable. Measurements on the stack gas indicate that the solids loading from the drier is 50 g/m³. The pressure drop in the cyclone must be less than 2 kPa. Use the Stairmand standard design parameters from Table 13.1 as the basis for your design.

NOTATION

a Cyclone inlet height, [L]

b Cyclone inlet width, [L]

A Cross-sectional area, [L²]

B Cyclone bottom exit diameter, [L]

C_D Drag coefficient, [—]

C_i Inlet solids loading, [M/L³]

d Particle diameter, [L]

d_{50} Diameter of 50% cut particle, [L]

d_{100} Diameter of smallest trapped particle, [L]

D Cyclone diameter, [L]

D_m Cyclone diameter in meters, [L]

D_e Cyclone top exit diameter, [L]

F Force, [F = M L/t²]

F_c Centrifugal force, [F = M L/t²]

g Acceleration due to gravity, [L/t²]

H Total cyclone height, [L]

h Height of cyclone cylindrical section, [L]

K_f Loss coefficient, [—]

m_e Effective mass of moving particle, [M]

n Exponent in Equation 13.34, [—]

N Number of turns in vortex, [—]

N_{Ar} Archimedes number, Equations 13.5 and 13.15, [—]

N_{Eu} Euler number, Equation 13.57, [—]

N_G Dimensionless gravity number, Equation 13.49, [—]

N_{Re} Reynolds number, Equation 13.58, [—]

N_{St} Stokes' number, Equation 13.48, [—]

N_{12} Parameter defined by Equation 13.17, [—]

Q Volumetric flow rate, [L³/t]

r Radial position, [L]

S Cyclone vortex finder height, [L]

T Temperature, [T]

t Time, [t]

V Volume, [L³]

$\tilde{V}$ Volume of fluid in centrifuge, [L³]

V_r Radial velocity of particle, [L/t]

V_{rt} Radial terminal velocity of particle, [L/t]

V_t Terminal velocity, [L/t]

$V*_t$ Gravity settling velocity, [L/t]

V_θ Tangential velocity in cyclone, [L/t]

z Vertical distance measured upward, [L]

GREEK

η   Efficiency, [—]

Δ() $()_2 - ()_1$

ρ   Density, [M/L^3]

μ   Viscosity, [$M/(L\ t)$]

Σ   Equivalent gravity settling area for centrifuge, Equation 13.13, [L^2]

ω   Angular velocity, [$1/t$]

SUBSCRIPTS

1,2   Reference points

c   Core

e   Exit

G   Gas

i   Inlet

o   Solids free

s   Solid

w   Wall

θ   Angular direction

REFERENCES

Antunes, M., and R.A. Medronho, Bradley cyclones: Design and performance analysis, in *Hydrocyclones, Analysis and Applications*, L. Svarovsky and M.T. Thew, (Eds.), Kluwer Academic Publishers, Dordrecht, The Netherlands, 1992.

Azbel, D.S. and N.P. Cheremisinoff, *Fluid Mechanics and Unit Operations*, Ann Arbor Science, Ann Arbor, MI, 1983.

Besendorfer, C., Exert the force of hydrocyclones, *Chem. Eng.*, 103(9), 108–114, 1996.

Bohnet, M., Design methods for aerocyclones and hydrocyclones, Chapter 32, in *Handbook of Fluids in Motion*, N.P Cheremisinoff and R. Gupta, (Eds.), Ann Arbor Science, Ann Arbor, MI, 1983.

Bohnet, M., O. Gottschalk. and M. Morweiser, Modern design of aerocyclones, *Adv. Particle Tech.*, 8(2), 137–161, 1997.

Chen, W., Solid-liquid separation, *Chem. Eng.*, 104(2), 66–72, 1997.

Coulson, J.M., J.F. Richardson, J.R. Blackhurst, and J.H. Harker, *Chemical Engineering*, Vol. 2, 4th Ed., Pergamon Press, Oxford, U.K., 1991.

Fernando Concha, A., *Solid-Liquid Separation in the Mining Industry*, Springer, New York, 2014.

Lapple, C.E., Processes use many collection types, *Chem. Eng.*, 58(5), 144–151, 1951.

Leith, D. and D.L. Jones, Cyclones, Chapter 15, in *Handbook of Powder Science and Technology*, 2nd edn., M.E. Fayed and L. Otten, (Eds.), Chapman & Hall, New York, 1997.

Leith, D. and W. Licht, The collection efficiency of cyclone type particle collectors - a new theoretical approach, *AIChE Sym. Ser.*, 68, 196–206, 1972.

McCabe, W.L., J.C. Smith, and P. Harriott, *Unit Operations of Chemical Engineering*, McGraw-Hill, New York, 1993.

Salcudean, M., I. Gartshore, and E.C. Statie, Test hydrocyclones before they are built, *Chem. Eng.*, 110(4), 66–71, 2003.

Starimand, C.J., Design and performance of cyclone separators, *Trans. Inst. Chem. Engrs.*, 29, 356–372, 1951.

Svarovsky, L., *Hydrocyclones*, Technomic Publishing Co., Lancaster, PA, 1984.

Svarovsky, L., A critical review of hydrocyclone models, in *Hydrocyclones '96*, D. Claxton, L. Svarovsky, and M., Thew, (Eds.), Mechanical Engineering Publications Ltd., London, U.K., 1996.

Svarovsky, L., Hydrocyclones, Chapter 6, in *Solid-Liquid Separation*, L. Svarovsky, (Ed.), 4th edn., Butterworths, Oxford, U.K., 2000.

Swift, P., Dust controls in industry, *Steam Heating Engr.*, 38, 453–456, 1969.

Wakeman, R.J. and E.S. Tarleton, *Solid-Liquid Separation*, Elsevier, London, U.K., 2005.

14 Flow in Porous Media

"If your experiment needs statistics, you ought to have done a better experiment."

—**Ernest Rutherford, 1871–1937, Physicist**

I. DESCRIPTION OF POROUS MEDIA

By a "porous medium" is meant a solid, or a collection of solid particles, with sufficient open space in or around the particles to enable a fluid to pass through or around them. There are various conceptual ways of describing a porous medium.

One concept is a continuous solid body with pores in it, such as a brick or a block of sandstone. Such a medium is referred to as *consolidated*, and the pores may be unconnected ("closed cell" or impermeable) or connected ("open cell" or permeable). Another concept is a collection (or "pile") of solid particles in a packed bed, where the fluid can pass through the voids in between the particles. This is referred to as *unconsolidated* media. A schematic representation is shown in Figure 14.1. Either of these concepts may be valid, depending upon the specific medium under consideration, and both have been used as the basis for developing the equations that describe fluid flow behavior within such a medium. In practice, porous media may range from a "tight" oil-bearing rock formation to a packed column containing relatively large packing elements and large void spaces.

The "pile of solid particles" concept is useful for either consolidated or unconsolidated media as a basis for analyzing the flow phenomena, since many consolidated media are actually made up of individual particles that are just stuck together (e.g., sandstone). One of the key properties of a porous medium is its porosity, ε, or void fraction, which is defined by

$$\varepsilon = \frac{\text{Total volume} - \text{Volume of solids}}{\text{Total volume}} \tag{14.1}$$

For an isotropic and homogeneous medium, the porosity can be written in terms of the corresponding areas as

$$\varepsilon = 1 - \frac{A_{solid}}{A} = \frac{A_{void}}{A}$$

where A_{solid} is the area of the solid phase in a total cross section of area A. For instance, certain rocks and sandstone may have very low values of porosity (~0.15–0.2), whereas fibrous beds, glass wool, metallic or plastic foams, structured packing, etc., may have high values (~≥ 0.9) of porosity.

We also distinguish between the velocity of approach, or the "superficial" velocity of the fluid,

$$V_s = \frac{Q}{A} \tag{14.2}$$

and the "interstitial" velocity, which is the actual velocity within the pores or voids:

$$V_i = \frac{Q}{\varepsilon A} = \frac{V_s}{\varepsilon} \tag{14.3}$$

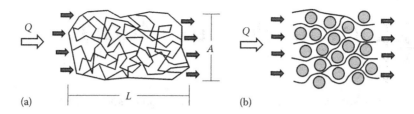

(a) L (b)

FIGURE 14.1   Schematic representation of porous media (a) consolidated and (b) unconsolidated.

A.   HYDRAULIC DIAMETER

Because the fluid in a porous medium follows a tortuous path through channels of varying size and shape, one method of describing the flow behavior in the pores is to consider the flow path as a "noncircular conduit." This requires an appropriate definition of the hydraulic diameter, that is,

$$D_h = 4\frac{A_i}{W_p} = 4\frac{A_i L}{W_p L} = 4\frac{\text{Flow volume}}{\text{Internal wetted surface area}}$$

$$= 4\frac{\varepsilon \times \text{Bed volume}}{(\text{Number of particles})(\text{Surface area/Particle})} \tag{14.4}$$

The medium, with overall dimensions of height L and cross-sectional area A, is assumed to be made up of a collection of individual particles that may be either consolidated or unconsolidated. The number of particles in the medium can be expressed as

$$\text{Number of particles} = \frac{(\text{Bed volume})(\text{Fraction of solids in bed})}{\text{Volume/Particle}}$$

$$= \frac{(\text{Bed volume})(1-\varepsilon)}{\text{Volume/Particle}} \tag{14.5}$$

Substitution of this into Equation (14.4) leads to

$$D_h = 4\left(\frac{\varepsilon}{1-\varepsilon}\right)\frac{1}{a_s} \tag{14.6}$$

where a_s is the specific surface area of each particle, that is, the particle surface area per unit volume of the particle. If the particles are spherical with diameter d, then $a_s = 6/d$. Thus, for a medium composed of uniform spherical particles, the hydraulic diameter is

$$D_h = \frac{2d\varepsilon}{3(1-\varepsilon)} \tag{14.7}$$

For non-spherical particles, the parameter d may be replaced by the equivalent diameter defined as

$$d = \psi d_s = \frac{6}{a_s} \tag{14.8}$$

where ψ is the *sphericity factor*, which is defined by

$$\psi = \frac{\text{Surface area of a sphere with same volume as the particle}}{\text{Surface area of the particle}} \qquad (14.9)$$

and d_s is the diameter of the sphere with the same volume as the particle. This definition of ψ is identical to that given by Equation 12.25 or 12.26.

B. Porous Medium Friction Factor

The expressions for the hydraulic diameter and the superficial velocity can be incorporated into the definition of the friction factor (i.e., Equation 7.5) to give an equivalent expression for the porous medium friction factor, as follows:

$$f \equiv \frac{e_f}{(4L/D_h)(V_i^2/2)} = \frac{e_f d\varepsilon}{3L(1-\varepsilon)V_i^2} = \frac{e_f d\varepsilon^3}{3L(1-\varepsilon)V_s^2} \qquad (14.10)$$

Most references use Equation 14.10 without the numerical factor of 3 in the denominator as the definition of the porous medium friction factor, i.e.

$$f_{PM} \equiv \frac{e_f d\varepsilon^3}{L(1-\varepsilon)V_s^2} \qquad (14.11)$$

C. Porous Medium Reynolds Number

In a like fashion, the hydraulic diameter and the superficial velocity can be introduced into the definition of the Reynolds number to give

$$N_{Re} = \frac{D_h V_i \rho}{\mu} = \frac{2d\varepsilon V_i \rho}{3(1-\varepsilon)\mu} = \frac{2dV_s \rho}{3(1-\varepsilon)\mu} \qquad (14.12)$$

Here again, the usual porous medium Reynolds number is defined by Equation 14.12 without the numerical factor (2/3):

$$N_{RePM} = \frac{dV_s \rho}{(1-\varepsilon)\mu} \qquad (14.13)$$

II. FRICTION LOSS IN POROUS MEDIA

A. Laminar Flow

By analogy with laminar flow in a tube, the friction factor in laminar flow would be

$$f = \frac{16}{N_{Re}} \quad \text{or} \quad f_{PM} = \frac{72}{N_{RePM}} \qquad (14.14)$$

However, this expression assumes that the total resistance to flow is due to the shear deformation of the fluid, as in a uniform pipe. In reality, the resistance is a result of both shear and stretching (extensional) deformation as the fluid moves through the nonuniform converging–diverging flow cross section within the pores. This "stretching resistance" is the product of the extension (stretch) rate and the extensional viscosity. The extension rate in porous media is of the same order as the shear rate, and the extensional viscosity for a Newtonian fluid is three times the shear viscosity. Thus, in practice, a value of 150–180 instead of 72 is in closer agreement with experimental observations at low Reynolds numbers, i.e.:

$$f_{PM} = \frac{180}{N_{RePM}} \quad \text{for} \quad N_{RePM} < 10 \tag{14.15}$$

This is known as the *Blake–Kozeny* equation and, as noted, applies for $N_{RePM} < 10$. Alternately, the use of the numerical value of 150 (or 180) as opposed to the theoretical value of 72 in Equation 14.14 can be partially attributed to the tortuosity of a porous medium, that is, an actual fluid particle travels a longer distance than the length of the bed L. For example, for a fluid particle to travel around a spherical particle, the tortuosity is $\pi/2$ and $72 \times \pi/2 \approx 115$.

B. TURBULENT FLOW

At high Reynolds numbers (i.e., high turbulence levels), the flow is dominated by inertial forces and "wall roughness," as in pipe flow. The porous medium can be considered an "extremely rough" conduit, with $\varepsilon/d \sim 1$. Thus, the flow at a sufficiently high Reynolds numbers should be fully turbulent and the friction factor should be constant. This has been confirmed by observations, with the value of the constant equal to approximately 1.75:

$$f_{PM} = 1.75 \quad \text{for} \quad N_{RePM} > 1000 \tag{14.16}$$

This is known as the *Burke–Plummer* equation and, as noted, applies for $N_{RePM} > 1000$.

C. ALL REYNOLDS NUMBERS

An expression that adequately represents the porous medium friction factor over all values of Reynolds number is

$$f_{PM} = 1.75 + \frac{180}{N_{RePM}} \tag{14.17}$$

This equation (with a value of 150 instead of 180) is called the *Ergun equation* and is simply the sum of Equations 14.15 and 14.16. Obviously, for $N_{RePM} < 10$, the first term is small relative to the second, and the Ergun equation reduces to the Blake–Kozeny equation. Likewise, for $N_{RePM} > 1000$, the first term is much larger than the second, and the equation reduces to the Burke–Plummer equation.

If the definitions of f_{PM} and N_{RePM} are inserted into the Ergun equation, the resulting expression for the frictional energy loss (dissipation) per unit mass of fluid in the medium is

$$e_f = 1.75 \frac{V_s^2}{d} \frac{(1-\varepsilon)}{\varepsilon^3} L + 180 \frac{V_s \mu (1-\varepsilon)^2 L}{d^2 \varepsilon^3 \rho} \tag{14.18}$$

III. PERMEABILITY

The "permeability" of a porous medium (K) is defined as the proportionality constant that relates the flow rate through the medium to the pressure drop, cross-sectional area, fluid viscosity, and net flow length through the medium, that is,

$$Q = K \frac{(-\Delta P) A}{\mu L} \tag{14.19}$$

This equation defines permeability (K) and is known as *Darcy's law*. The most common unit for permeability is the *"darcy,"* which is defined as the flow rate in cm³/s that results when a pressure drop of one atmosphere is applied to a porous medium that is 1 cm² in cross-sectional area and one cm long, for a fluid with a viscosity of one centipoise. It should be evident that the dimensions of the darcy are L^2 and the conversion factors are (approximately) 10^{-8} cm²/darcy = 10^{-11} ft²/darcy. The flow properties of tight crude oil–bearing rock formations are often described in permeability units of *millidarcies*. Typical values of permeability range from ~10^{-14} m² for silica powder to 10^{-11} m² for fiberglass.

If the Blake–Kozeny equation for laminar flow is used to describe the friction loss, which is then equated to $\Delta P/\rho$ from the Bernoulli equation, the resulting expression for the flow rate is

$$Q = \frac{-\Delta PA}{\mu L}\left(\frac{d^2\varepsilon^3}{180(1-\varepsilon)^2}\right) \tag{14.20}$$

By comparison of Equations 14.19 and 14.20, it is evident that permeability is identical to the term in brackets in Equation 14.20. This shows how the permeability is related to the equivalent particle size and porosity of the medium. Since Equation 14.20 applies only to laminar flow, it is evident that permeability has no meaning under turbulent flow conditions. It should be noted that although permeability increases with increasing porosity, one cannot be predicted from the other due to the presence of both "open pores" and "closed pores" (Dullien, 1992).

IV. MULTIDIMENSIONAL FLOW

Flow in a porous medium in two or three dimensions is important in situations such as the production of crude oil from reservoir formations, for example. Thus, it is of interest to consider this situation briefly and to point out some characteristics of the governing equations.

Consider the flow of an incompressible fluid through a 2D porous medium, as illustrated in Figure 14.2. Assuming that the kinetic energy change is negligible and that the flow is laminar as characterized by Darcy's law, the Bernoulli equation becomes

$$-\left(\frac{\Delta P}{\rho} + g\Delta z\right) = e_f = \frac{\mu V_s L}{K\rho} \tag{14.21}$$

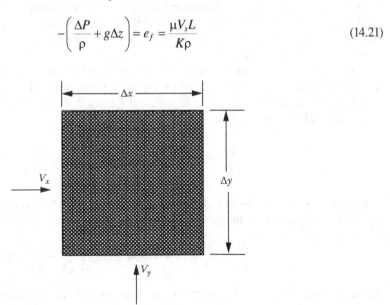

FIGURE 14.2 Schematic of two-dimensional flow in a porous medium.

or

$$\Delta\left(\frac{\Phi}{\rho}\right) = -\frac{\mu V_s L}{K\rho} \tag{14.22}$$

where the density cancels out if the fluid is incompressible. Equation 14.22 can be applied in both the x and y directions, by taking $L = \Delta x$ for the x direction and $L = \Delta y$ for the y direction:

$$\frac{\Delta\Phi}{\Delta x} = -\frac{\mu V_x}{K} = \frac{\partial\Phi}{\partial x} \tag{14.23}$$

and

$$\frac{\Delta\Phi}{\Delta y} = -\frac{\mu V_y}{K} = \frac{\partial\Phi}{\partial y} \tag{14.24}$$

If Equation 14.23 is differentiated with respect to x and Equation 14.24 is differentiated with respect to y and the results are added, assuming μ and K to be constant, we get

$$\frac{\partial^2\Phi}{\partial x^2} + \frac{\partial^2\Phi}{\partial y^2} = -\frac{\mu}{K}\left(\frac{\partial V_x}{\partial x} + \frac{\partial V_y}{\partial y}\right) = 0 \tag{14.25}$$

For an incompressible fluid, the term in brackets is zero as a result of the conservation of mass (e.g., the microscopic continuity equation). In this case, Equation 14.25 can be generalized to three dimensions as

$$\nabla^2\Phi = 0 \tag{14.26}$$

This is called the *Laplace equation*. The solution of this equation, along with appropriate boundary conditions, determines the potential (e.g., pressure) distribution within the medium. The derivatives of this potential then determine the velocity distribution in the medium (e.g., Equations 14.23 and 14.24). The Laplace equation also governs the 3D (potential) flow of an inviscid fluid. Note that the Laplace equation follows from Equation (14.25) for either an incompressible viscous fluid, by virtue of the continuity equation, or for any flow with negligible viscosity effects (e.g., compressible flow outside the boundary layer near a solid boundary). It is interesting that the same equation governs both of these extreme cases.

The Laplace equation also applies to the distribution of electrical potential and current flow in an electrically conducting medium, as well as the temperature distribution and heat flow in a thermally conducting medium. For example, if $\Phi \Rightarrow E$, $V \Rightarrow i$, and $\mu/K \Rightarrow r_e$, where r_e is the electrical resistivity ($r_e = RA/\Delta x$), Equation 14.22 becomes *Ohm's law*:

$$\frac{\partial E}{\partial x} = -r_e i_x, \quad \nabla^2 E = 0, \quad \text{and} \quad \left(\frac{\partial i_x}{\partial x} + \frac{\partial i_y}{\partial y}\right) = 0 \tag{14.27}$$

Furthermore, if $\Phi \Rightarrow T$, $V \Rightarrow q$, and $K/\mu \Rightarrow k$, where q is the heat flux and k is the thermal conductivity, the same equations govern the flow of heat in a thermally conducting medium (i.e., *Fourier's law*):

$$\frac{\partial T}{\partial x} = -\frac{1}{k}q_x, \quad \nabla^2 T = 0, \quad \text{and} \quad \left(\frac{\partial q_x}{\partial x} + \frac{\partial q_y}{\partial y}\right) = 0 \tag{14.28}$$

By making use of these analogies, electrical analog models can be constructed that can be used, for example, to determine the pressure and flow distribution in a porous medium from measurements of the voltage and current distribution in a conducting medium. The process becomes more complex, however, when the local permeability varies with position within the medium (i.e., in anisotropic and/or nonhomogeneous porous media), which is often the case in practice.

V. PACKED COLUMNS

At the other end of the spectrum from a "porous rock" is the unconsolidated medium composed of beds of relatively large packing elements. These elements may include a variety of shapes, such as rings, saddles, grids, and meshes, which are generally used to provide a large gas–liquid interface for promoting mass transfer in operations such as distillation, absorption, or liquid–liquid extraction. A typical application might be the removal of an impurity from a gas stream by selective absorption with a solvent in an absorption column filled with packing. The gas (or lighter liquid, in the case of liquid–liquid extraction) typically enters the bottom of the column, and the heavier liquid enters the top and drains by gravity, the flow being countercurrent as illustrated in Figure 14.3.

For single-phase flow through packed beds, the pressure drop can generally be predicted adequately by the Ergun equation, Equation 14.17. However, because the flow in packed columns is normally countercurrent two-phase flow, this situation is more complex. The effect of increasing the liquid mass flow rate (L) on the pressure drop through the column for a given gas mass flow rate (G), starting with dry packing, is illustrated in Figure 14.4. The pressure drop for wet drained packing is higher than for dry packing, since the liquid occupies some of the void space between packing

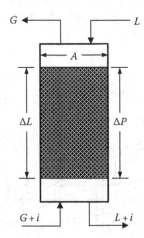

FIGURE 14.3   Schematic of a packed column.

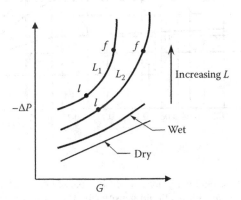

FIGURE 14.4   Effect of liquid rate on pressure drop.

elements even in the "drained" condition. As the liquid flow rate increases, the liquid occupies an increasing portion of the void space, so that the area available to the gas is reduced and therefore the total pressure drop increases. As the liquid flow rate increases, the curve of ΔP versus G becomes increasingly steeper. The points labeled "l" in Figure 14.4 are referred to as the "loading" points and indicate the point where there is a marked increase in the interaction between the liquid and the gas, which is the desired operating point for the column. The points labeled "f" on Figure 14.4 are the "flooding" points. At these points, the pressure drop through the column is equal to the static head of liquid in the column. When this occurs, the pressure drop due to the gas flow balances the static head of liquid so that the liquid can no longer drain through the packing by gravity and the column is said to be "flooded." It is obviously undesirable to operate a packed column at or near the flooding point, since a slight increase in gas flow at this point will carry the liquid out of the top of the column.

The pressure drop through packed columns, and the flooding conditions, can be estimated from the generalized correlation of Leva (1992), as shown in Figure 14.5. The pressure gradient in (mm of water) per (m of packed height) is the parameter on the curves, and interpolation is usually necessary to determine the pressure drop (note that the pressure drop is not linearly proportional to the spacing between the curves). Correction factors for liquid density and viscosity, which are applied to the y axis of this correlation, are also shown. The parameter F in this correlation is called the *packing factor*. Values of F are given in Table 14.1, which shows the dimensions and physical properties of a variety of packings. Note that in Table 14.1, the term S_B is equal to $a_s (1 - \varepsilon)$, where a_s is the specific surface area (i.e., the surface area per unit volume of the packing element). The packing factor F is comparable to the term S_B/ε^3 in the definition of f_{PM} but is an empirical factor that characterizes the packing somewhat better than S_B/ε^3.

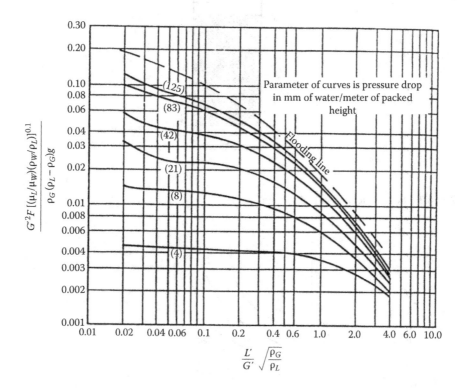

FIGURE 14.5 Generalized correlation for pressure drop in packed columns. L, liquid mass flux [$lb_m/(s\ ft^2)$], kg/(s m²); G, gas mass flux [$lb_m/(sft^2)$]; ρ_L, liquid density (lb_m/ft^3, kg/m³); F, packing factor (Table 14.1); μ_L, liquid viscosity (mN s/m²); g, (9.81 m/s², 32.2 ft/s²); w, water at same T and P as column. (From Coulson, J.M. et al., *Chemical Engineering*, Vol. 2, 5th edn., Butterworth-Heinemann, Oxford, U.K., 2002.)

TABLE 14.1
Design Data for Various Column Packings

	Size		Wall thickness		Number density		Bed density		Contact surface S_B		Free space %	Packing factor	
	in.	mm	in.	mm	ft⁻³	m⁻³	lbₘ/ft³	kg/m³	ft²/ft³	m²/m³	(100 ε)	ft²/ft³	m²/m³
Ceramic Raschig rings	0.25	6.35	0.03	0.76	85,600	2,022,600	60	961	242	794	62	1600	5250
	0.38	9.65	0.05	1.27	24,700	972,175	61	977	157	575	67	1000	3280
	0.50	12.7	0.07	1.78	10,700	377,825	55	881	112	368	64	640	2100
	0.75	19.05	0.09	2.29	3090	109,110	50	801	73	240	72	255	840
	1.0	25.4	0.14	3.56	1350	47,670	42	673	58	190	71	160	525
	1.25	31.75			670	23,660	46	737			71	125	410
	1.5	38.1			387	13,665	43	689			73	95	310
	2.0	50.8	0.25	6.35	164	5790	41	657	29	95	74	65	210
	3.0	76.2			50	1765	35	561			78	36	120
Metal Raschig rings	0.25	6.35	0.03125	0.794	88,000	3,107,345	133	2131			72	700	2300
	0.38	9.65	0.03125	0.794	27,000	953,390	94	1506			81	390	1280
	0.50	12.7	0.03125	0.794	11,400	402,540	75	1201	127	417	85	300	980
	0.75	19.05	0.03125	0.794	3340	117,940	52	833	84	276	89	185	605
	0.75	19.05	0.0625	1.59	3140	110,875	94	1506			80	230	750
	1.0	25.0	0.03125	0.794	1430	50,494	39	625	63	207	92	115	375
	1.0	25.0	0.0625	1.59	1310	46,260	71	1137			86	137	450
	1.25	31.75	0.0625	1.59	725	25,600	62	993			87	110	360
	1.5	38.1	0.0625	1.59	400	14,124	49	785			90	83	270
	2.0	50.8	0.0625	1.59	168	5932	37	593	31	102	92	57	190
	3.0	76.2	0.0625	1.59	51	1800	25	400	22	72	95	32	105
Carbon Raschig rings	0.25	6.35	0.0625	1.59	85,000	3,001,410	46	737	212	696	55	1600	5250
	0.50	12.7	0.0625	1.59	10,600	374,290	27	433	114	374	74	410	1350
	0.75	19.05	0.125	3.175	3140	110,875	34	545	75	246	67	280	920
	1.0	25.0	0.125	3.175	1325	46,787	27	433	57	187	74	160	525
	1.25	31.75			678	23,940	31	496			69	125	410
	1.5	38.1			392	13,842	34	545			67	130	425
	2.0	50.8	0.250	6.35	166	5862	27	433	29	95	74	65	210
	3.0	76.2	0.312	7.92	49	1730	23	368	19	62	78	36	120

(*Note:* Bed densities are for mild steel; multiply by 1.105, 1.12, 1.37, 1.115 for stainless steel, copper, aluminum, and monel, respectively)

(*Continued*)

TABLE 14.1 (Continued)
Design Data for Various Column Packings

	Size		Wall thickness		Number density		Bed density		Contact surface S_B		Free space %	Packing factor	
	in.	mm	in.	mm	ft⁻³	m⁻³	lbm/ft³	kg/m³	ft²/ft³	m²/m³	(100 ε)	ft²/ft³	m²/m³
Metal Pall rings (*Note*: Bed densities are for mild steel)	0.625	15.9	0.018	0.46	5950	210,098	37	593	104	341	93	70	230
	1.0	25.4	0.024	0.61	1400	49,435	30	481	64	210	94	48	160
	1.25	31.75	0.030	0.76	375	13,240	24	385	39	128	95	28	92
	2.0	50.8	0.036	0.915	170	6003	22	353	31	102	96	20	66
	3.5	76.2	0.048	1.219	33	1165	17	273	20	65.6	97	16	52
Plastic Pall Rings (*Note*: Bed densities are for polypropylene)	0.625	15.9	0.03	0.762	6050	213,630	7.0	112	104	341	87	97	320
	1.0	25.4	0.04	1.016	1440	50,848	5.5	88	63	207	90	52	170
	1.5	38.1	0.04	1.016	390	13,770	4.75	76	39	128	91	40	130
	2.0	50.8	0.06	1.524	180	6356	4.25	68	31	102	92	25	82
	3.5	88.9	0.06	1.524	33	1165	4.0	64	26	85	92	16	52
Ceramic Intalox saddles	0.25	6.35			117,500	4,149,010	54	865			65	725	2400
	0.38	9.65			49,800	1,758,475	50	801			67	330	1080
	0.50	12.7			18,300	646,186	46	737			71	200	660
	0.75	19.05			5640	199,150	44	705			73	145	475
	1.0	25.4			2150	75,918	42	673			73	92	300
	1.5	38.1			675	23,835	39	625	59	194	76	52	170
	2.0	50.8			250	8828	38	609			76	40	130
	3.0	76.2			52	1836	36	577			79	22	72
Plastic Super Intalox	No. 1				1620	57,200	6.0	96	63	207	90	33	108
	No. 2				190	6710	3.75	60	33	108	93	21	69
	No. 3				42	1483	3.25	52	27	88.6	94	16	52
Intalox metal		25			4770	168,425					96.7	41	135
		40			1420	50,140					97.3	25	82
		50			416	14,685					97.8	16	52
		70			131	4625					98.1	13	43

(*Continued*)

TABLE 14.1 (Continued)
Design Data for Various Column Packings

	Size		Wall thickness		Number density		Bed density		Contact surface S_B		Free space %	Packing factor	
	in.	mm	in.	mm	ft⁻³	m⁻³	lb_m/ft³	kg/m³	ft²/ft³	m²/m³	(100 ε)	ft²/ft³	m²/m³
Hy-Pak (*Note:* Bed densities are for mild steel)	No. 1				850	30,014	19	304			96	43	140
	No. 2				107	3778	14	224			97	18	59
	No. 3				31	1095	13	208			97	15	49
Plastic Cascade Mini Rings	No. 1											25	82
	No. 2											15	49
	No. 3											12	39
Metal Cascade Mini Rings	No. 0											55	180
	No. 1											34	110
	No. 2											22	72
	No. 3											14	46
	No. 4											10	33
Ceramic Cascade Mini Rings	No. 2											38	125
	No. 3											24	79
	No. 5											18	59

Note: The packing factor F replaces the term S_B/ε^3. Use of the given value of F in Fig. 14.5 permits more predictable performance of designs incorporating packed beds, because the values quoted are derived from operating characteristics of the packings rather than from their physical dimensions.

Source: Coulson et al. (2002).

VI. FILTRATION

For fine suspended solids that are too small to separate from the liquid by gravitational or centrifugal methods, a "barrier" method such as a filter may be used. The liquid is passed through a filter medium (usually a cloth or screen), which provides a support for a cake formed from the solid particles removed from the slurry. In reality, the pores in the filter medium are frequently larger than the particles, which may penetrate some distance into the medium before being trapped. The layer of solids that builds up on the surface of the medium is called the *cake*, and it is the cake that provides the actual filtration. The pressure–flow characteristics of the porous cake primarily determine the performance of the filter. A recent article by Gabelman (2015) provides a good overview of filtration and Branan (1994) and Cheremisinoff (1998) treat filtration in greater detail.

A. GOVERNING EQUATIONS

A schematic of the flow through the cake and the filter medium is shown in Figure 14.6. The slurry flow rate is Q, and the total volume of the filtrate that passes through the filter is $\tilde{V}$. The flow through the cake and the filter medium is invariably laminar, so the resistance can be described by Darcy's law, Equation 14.19, and the permeability of the medium (K):

$$\frac{-\Delta P}{L} = \frac{Q\mu}{KA} \tag{14.29}$$

Applying this relation across both the cake and the filter medium in series gives

$$P_1 - P_2 = \left(\frac{L}{K}\right)_{cake} \frac{Q\mu}{A} \tag{14.30}$$

$$P_2 - P_3 = \left(\frac{L}{K}\right)_{FM} \frac{Q\mu}{A} \tag{14.31}$$

The total pressure drop across the filter is the sum of these:

$$P_1 - P_3 = \frac{Q\mu}{A}\left[\left(\frac{L}{K}\right)_{cake} + \left(\frac{L}{K}\right)_{FM}\right] \tag{14.32}$$

where
$(L/K)_{cake}$ is the resistance of the cake
$(L/K)_{FM}$ is the resistance of the filter medium

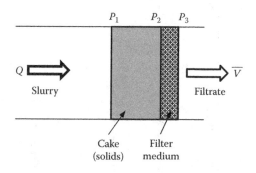

FIGURE 14.6   Schematic of flow through filter cake and medium.

The latter is higher for a "dirty" filter medium than for a clean one, but once the initial particles become imbedded in the medium and the cake starts to build up, it remains relatively constant. The cake resistance, on the other hand, continues to increase with time as the cake thickness increases. The cake thickness is directly proportional to the volume of solids that has been deposited from the slurry and inversely proportional to the area:

$$L_{cake} = \frac{\tilde{V}_{cake}}{A} = \frac{\tilde{V}_{solids}}{A(1-\varepsilon)} = \frac{M_{solids}}{A\rho_s(1-\varepsilon)} \tag{14.33}$$

Now $(M_{solids}/\tilde{V})$ is the mass of solids per unit volume of liquid in the slurry feed (e.g., the "solids loading" of the slurry). $\tilde{V}$ is the volume of liquid (filtrate) that has passed to deposit M_{solids} on the cake

Thus, the cake thickness can be expressed as

$$L_{cake} = \left(\frac{M_{solids}}{\tilde{V}}\right)\left(\frac{\tilde{V}}{A}\right)\frac{1}{\rho_s(1-\varepsilon)} = W\frac{\tilde{V}}{A} \tag{14.34}$$

where $W = (M_{solids}/\tilde{V})/\rho_s(1-\varepsilon)$ is a property of the specific slurry and cake. The density of the cake is given by

$$\rho_c = (1-\varepsilon)\rho_s + \varepsilon\rho_{liq} \tag{14.35}$$

Substituting Equation 14.34 into Equation 14.32 and rearranging the result in the basic equation governing the filter performance

$$\boxed{\frac{Q}{A} = \frac{1}{A}\frac{d\tilde{V}}{dt} = \frac{P_1 - P_3}{\mu\left[(\tilde{V}W/AK) + a\right]}} \tag{14.36}$$

where a is the filter medium resistance, that is, $a = (L/K)_{FM}$.

It should be recognized that the operation of a filter is an unsteady cyclic process. As the cake builds up, its resistance increases with time, and either the flow rate (Q) will drop or the pressure drop (ΔP) will increase with time. The specific behavior depends upon how the filter is operated, that is, constant pressure or constant flow operation, as follows.

B. CONSTANT PRESSURE OPERATION

If the slurry is fed to the filter using a centrifugal pump, which delivers (approximately) a constant head, or if the filter is operated by a controlled pressure or vacuum, the pressure drop across the filter will remain essentially constant during filtering operation and the flow rate will drop as the cake thickness (resistance) increases. In this case, Equation 14.36 can be integrated at constant pressure (i.e., $\Delta P = P_1 - P_3$) to give

$$C_1\left(\frac{\tilde{V}}{A}\right)^2 + C_2\left(\frac{\tilde{V}}{A}\right) = (-\Delta P)t \tag{14.37}$$

where
$$C_1 = \mu W/2K$$
$$C_2 = \mu a$$

Both of these are assumed to be independent of pressure (we will consider compressible cakes later).

The constant of integration has been evaluated using the condition that at $t = 0$, $\tilde{V} = 0$. In Equation 14.37, t is the time required to pass volume $\tilde{V}$ of the filtrate through the filter.

As C_1 and C_2 are unique properties of a specific slurry–cake system, it is usually appropriate to determine their values from laboratory tests using samples of the specific slurry and the filter medium that are to be used in the plant. For this purpose, it is more convenient to rearrange Equation 14.37 in the following form:

$$\left(\frac{-\Delta P t}{\tilde{V}/A}\right) = C_1\left(\frac{\tilde{V}}{A}\right) + C_2 \tag{14.38}$$

If $\tilde{V}$ is measured as a function of t in a lab experiment for given values of ΔP and A, the data can be arranged in the form of Equation 14.38. When the left hand side is plotted versus $\tilde{V}/A$, the result should be a straight line with slope C_1 and intercept C_2 (which are easily determined by linear regression).

C. CONSTANT FLOW OPERATION

If the slurry is fed to the filter by a positive displacement pump, the flow rate will be constant regardless of the pressure drop, which will increase with time. In this case, noting that $\tilde{V} = Qt$, Equation 14.36 can be rearranged to give

$$-\Delta P = 2C_1\left(\frac{Q}{A}\right)^2 t + C_2\left(\frac{Q}{A}\right) \tag{14.39}$$

This shows that for given Q and A, the plot of ΔP versus t should be a straight line and the system constants C_1 and C_2 can be determined from the slope, $2C_1(Q/A)^2$, and intercept, $C_2(Q/A)$.

It is evident that the filter performance is governed by the system constants C_1 and C_2, regardless of whether the operation is at constant pressure or constant flow rate. These constants can be evaluated from laboratory data taken under either type of operation and used to analyze the performance of the full-scale filter for either type of operation.

D. CYCLE TIME

As noted earlier, the operation of a filter is cyclic. As the filtration proceeds, and the pressure either increases or the flow rate drops either a cake will eventually build up to fill the space available for it or the pressure drop will reach the operational limit. At that point, the filtration must cease and the cake removed. There is often a wash cycle prior to removal of the cake in order to remove the slurry carrier liquid from the pores of the cake, using a clean liquid. The pressure–flow behavior during the wash period is a steady-state operation, controlled by the maximum cake and filter medium resistance, since no solids are deposited during this period. The cake can be removed by physically disassembling the filter, removing the filter medium and the cake (as for a plate-and-frame filter), and then reassembling the filter and starting the cycle over. Or in the case of a rotary drum filter, the wash period and cake removal are part of the rotating drum cycle. The drum rotates continuously, although the filtration operation is still cyclic (as discussed in Section F below).

The variable t in the equations given earlier is the actual time (t_{filter}) that is required to pass a volume $\tilde{V}$ of the filtrate through the medium and is only part of the total time of the cycle (t_{cycle}). The rest of the cycle time, which may include wash time, disassembly and assembly time, and cleaning time, we shall call "dead" time (t_{dead}):

$$t_{cycle} = t_{filter} + t_{dead} \tag{14.40}$$

The net (average) filter capacity is determined by the amount of slurry processed during the total cycle time, not just the "filter" time, and represents the average flow rate ($\bar{Q}$):

$$\bar{Q} = \left(\frac{\tilde{V}_{cycle}}{t_{cycle}}\right)_{const\ \Delta P} = \left(\frac{Qt_{filter}}{t_{cycle}}\right)_{const\ Q} \tag{14.41}$$

E. PLATE-AND-FRAME FILTER

A plate-and-frame filter press consists of alternate solid plates and hollow frames in a "sandwich" arrangement. The open frames are covered by a filter medium (e.g., the filter cloth), and the slurry enters through the frames and deposits a cake on the filter medium. The operation is "batch," in that the filter must be disassembled when the cake fills the available frame space and then cleaned and reassembled, after which the entire process is repeated. A schematic of a plate-and-frame press is shown in Figure 14.7. In the arrangement shown, all of the frames are in parallel, and the total filter area (which appears in the equations) is

$$A = 2nA_f \tag{14.42}$$

where
n is the number of frames
A_f is the filter area of (one side) of the frame

The flow rate Q in the equations is the total flow rate, and $Q/A = Q/(2nA_f)$ is the total flow per unit total filtering area, or the flow rate per filter side per unit area of the filter side.

There are a variety of arrangements that operate in the same manner as the plate-and-frame filter, such as the "leaf filter," which may consist of one or more "frames" covered by the filter medium, which are immersed in the slurry. They are often operated by a vacuum that draws the filtrate through the filter, and the cake collects on the filter medium on the outside of the frame.

F. ROTARY DRUM FILTER

The rotary drum filter is a "continuous" process, since it does not have to be shut down during the cycle, although the operation is still cyclic. A schematic is shown in Figure 14.8. The drum rotates at a rate N (rpm), and the filter area is the total drum surface, that is, $A = \pi DL$. However, if the fraction

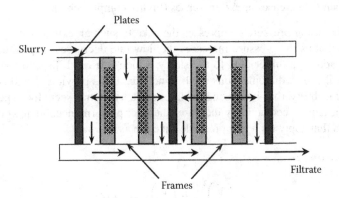

FIGURE 14.7 Schematic of a plate-and-frame filter.

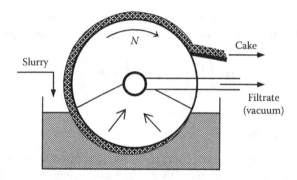

FIGURE 14.8   Schematic of a rotary drum filter.

of the drum that is in contact with the slurry is f, then the total time of the cycle during which any one point on the surface is actually filtering is f/N:

$$t_{cycle} = \frac{1}{N}, \quad t_{filt} = \frac{f}{N} \tag{14.43}$$

The rotary drum filter is commonly operated under constant pressure conditions (i.e. under a vacuum).

G.   COMPRESSIBLE CAKE

The equations presented so far all assume that the cake is incompressible, that is, the porosity, permeability, and density of the cake are constant. For many cakes, this is not so, since the cake properties may vary with pressure due to consolidation (e.g., flocs, gels, fibers, pulp). For such cases, the basic filter equation (Equation 14.36) can be expressed in the following form:

$$\frac{Q}{A} = \frac{(-\Delta P)^{1-s}}{\mu \left[\alpha(\tilde{V}/A)(M_s/\tilde{V}) + a \right]} = \frac{1}{A} \frac{d\tilde{V}}{dt} \tag{14.44}$$

where the pressure dependence is characterized by the parameter s, and α and a are the pressure-independent properties of the cake. There are several modes of performance of the filter, depending upon the value of s:

1. If $s = 0$, then Q proportional ($\propto$) ΔP (the cake is incompressible).
2. If $s < 1$, then Q increases as ΔP increases (slightly compressible).
3. If $s = 1$, then Q is independent of ΔP (compressible).
4. If $s > 1$, then Q decreases as ΔP increases (highly compressible).

In case 4, the increasing pressure compresses the cake to such an extent that it actually "squeezes off" the flow so that as the pressure increases the flow rate decreases. This situation can be compensated for by adding a "filter aid" to the slurry, which is a rigid dispersed solid that forms an incompressible cake (e.g., diatomaceous earth, sand, etc.). This provides "rigidity" to the cake and enhances its permeability, thus increasing the filter capacity (it may seem like a paradox that adding more solids to the slurry feed actually increases the filter performance, but it works).

The equations that apply for a compressible cake are as follows:

Constant pressure drop:

$$C_1 \left(\frac{\tilde{V}}{A} \right)^2 + C_2 \left(\frac{\tilde{V}}{A} \right) = t(-\Delta P)^{1-s} \tag{14.45}$$

Constant flow rate:

$$2C_1\left(\frac{Q}{A}\right)^2 t + C_2\left(\frac{Q}{A}\right) = \left(-\Delta P\right)^{1-s} \qquad (14.46)$$

where
$$C_1 = (\mu\alpha/2)(M_s/\tilde{V})$$
$$C_2 = \mu a$$

There are now three parameters that must be determined empirically from laboratory measurements: C_1, C_2, and s. The easiest way to do this would be to utilize the constant pressure mode in the laboratory (e.g., a Buchner funnel, with a set vacuum pressure difference) and obtain several sets of data for $\tilde{V}$ as a function of t, with each set at a different value of ΔP. For each data set, the plot of $t/\tilde{V}$ versus $\tilde{V}$ should yield a straight line, with a slope of $C_1/[A^2(-\Delta P)^{1-s}]$ and intercept of $C_2/[A(-\Delta P)^{1-s}]$. Thus, a log–log plot of either the slope or the intercept versus ΔP should have a slope of $(s - 1)$, which determines s.

SUMMARY

The following are the major points that should be retained from this chapter:

- The difference between consolidated and unconsolidated porous media
- The definitions of porosity, hydraulic diameter, sphericity factor, and permeability for porous media
- Laminar and turbulent friction loss in porous media
- The pressure drop behavior and the concept of flooding in packed columns
- Filtration under constant pressure or constant flow conditions
- Operation of plate-and-frame and rotary drum filters

PROBLEMS

POROUS MEDIA

1. A packed bed is composed of crushed rock with a density of 175 lb_m/ft^3 having a size and shape such that the average ratio of surface area to volume for the particles is 50 in.^{2}/in.^{3} The bed is 6 ft deep, has a porosity of 0.3, and is covered by a 2 ft deep layer of water that drains by gravity through the bed. Calculate the flow rate of water through the bed in gpm/ft^2, assuming it exits at one atmosphere pressure.

2. An impurity in a water stream at a very small concentration is to be removed in a charcoal trickle bed filter. The filter is in a cylindrical column that is 2 ft in diameter, and the bed is 4 ft deep. The water is stored at a level that is 2 ft above the top of the bed, and it trickles through by gravity flow. If the charcoal particles have a geometric surface area to volume ratio of 48 in.$^{-1}$ and they pack with a porosity of 0.45, what is the flow rate of water through the column, in gpm?

3. A trickle bed filter is composed of a packed bed of broken rock that has a shape such that the average ratio of the surface area to volume for the rock particles is 30 in.$^{-1}$ The bed is 2 ft deep, has a porosity of 0.3, and is covered by a layer of water that is 2 ft. deep and drains by gravity through the bed.
    (a) Determine the volume flow rate of the water through the bed per unit bed area (in gpm/ft^2).
    (b) If the water is pumped upward through the bed (e.g., to flush it out), calculate the flow rate (in gpm/ft^2 of bed area) that would be required to fluidize the bed.
    (c) Calculate the corresponding flow rate that would sweep the rock particles away with the water. The rock density is 120 lb_m/ft^3.

PACKED COLUMNS

4. A packed column that is 3 ft in diameter with a packing height of 25 ft is used to absorb an impurity from a methane gas stream using an amine solution absorbent. The gas flow rate is 2000 scfm, and the liquid has a density of 1.2 g/cm^3 and a viscosity of 2 cP. If the column operates at 1 atm and 80°F, determine the liquid flow rate at which flooding would occur in the column and the pressure drop at 50% of the flooding liquid rate for the following packings:
    (a) 2 in. ceramic Raschig rings
    (b) 2 in. plastic Pall rings

5. A packed column is used to scrub SO$_2$ from air using water. The gas flow rate is 500 scfm/ft^2 and the column operates at 90°F and 1 atm. If the column contains No. 1 plastic Intalox packing, what is the maximum liquid flow rate (per unit cross section of column) that could be used without flooding?

6. A stripping column packed with 2 in. metal Pall rings uses air at 5 psig, 80°C, to strip an impurity from an absorber oil ($SG = 0.9$, viscosity = 5 cP, $T = 20°C$). If the flow rate of the oil is 500 lb$_m$/min and that of the air is 20 lb$_m$/min, determine the following:
    (a) What is the minimum column diameter that can be used without flooding?
    (b) If the column diameter is 50% greater than the minimum size, what is the pressure drop per ft of column height?

7. A packed column, which is 0.6 m in diameter and 4 m high containing 25 mm Raschig rings, is used in a gas absorption process to remove an impurity from the gas stream by absorbing it in a liquid solvent. The liquid, which has a viscosity of 5 cP and $SG = 1.1$, enters the top of the column at a rate of 2.5 kg/s m^2, and the gas, which may be assumed to have the same properties as air, enters the bottom of the column at a rate of 0.6 kg/s m^2. The column operates at atmospheric pressure and 25°C. Determine
    (a) The pressure drop through the column, in inches of water
    (b) How high the liquid rate could be increased before the column would flood

8. A packed column is used to absorb SO$_2$ from flue gas using an ethanol amine solution. The column is 4 ft in diameter and has a packed height of 20 ft and is packed with 2 in. plastic Pall rings. The flue gas is at a temperature of 180°F and has an average molecular weight of 31. The amine solution has a specific gravity of 1.02 and a viscosity at the operating temperature of 1.5 cP. If the gas must leave the column at 25 psig and a flow rate of 10,000 scfm, determine
    (a) The flow rate of the liquid (in gpm) that is 50% of that at which flooding would occur
    (b) The horsepower that would be required for the blower to move the gas through the column, if the blower is 80% efficient

9. A packed absorption tower is used to remove SO$_2$ from an air stream by absorption in a solvent. The tower is 5 ft. in diameter and 60 ft. high and contains 1.5 in. plastic Pall rings. The temperature and pressure in the tower are 90°F and 30 psig. The gas stream flow rate is 6500 scfm, and the liquid SG is 1.25, with a viscosity of 25 cP.
    (a) What is the liquid flow rate (in gpm) at which the column will flood?
    (b) If the column operates at a liquid rate that is 75% of the flooding value, what is the total pressure drop through the tower in psi?

10. A packed absorption column removes an impurity from a gas stream by contact with a liquid solvent. The column is 3 ft in diameter and contains 25 ft of No. 2 plastic Super Intalox packing. The gas has a MW of 28 and enters the column at 120°F and leaves at 10 psig at a rate of 5000 scfm. The liquid has a SG of 1.15 and a viscosity of 0.8 cP. Determine

(a) The flow rate of the liquid in gpm that would be 50% of the flow rate at which the column would flood

(b) The pressure drop through the column in psi

(c) The horsepower of the blower required to move the gas through the column, if it is 60% efficient

FILTRATION

11. A fine aqueous suspension containing 1 lb_m of solids per ft^3 of suspension is to be filtered in a constant pressure filter. It is desired to filter at an average rate of 100 gpm, and the filter cake must be removed when it gets 2 in. thick. What filter area is required? Data: $\Delta P = 10$ psi, ρ (wet cake) = 85 lb_m/ft^3, K (permeability) = 0.118 darcies, $a = 2 \times 10^9$ ft^{-1}

12. An aqueous slurry containing 1.5 lb_m of solid per gallon of liquid is pumped through a filter cloth by a centrifugal pump. If the pump provides a constant pressure drop of 150 psig, how long will it take for the filter cake to build up to a thickness of 2 in.? The density of the filter cake is 30 lb_m/ft^3 and its permeability is 0.01 darcies.

13. A packed bed that consists of the same medium as that in Problem 3 is to be used to filter solids from an aqueous slurry. To determine the filter properties, you test a small section of the bed, which is 6 in. in diameter and 6 in. deep, in the lab. When the slurry is pumped through this test model at a constant flow rate of 30 gpm, the pressure drop across the bed rises to 2 psia after 10 min. How long will it take to filter 100,000 gal of water from the slurry in a full-size bed, which is 10 ft in diameter and 2 ft deep, if the slurry is maintained at a depth of 2 ft over the bed and drains by gravity through the bed?

14. A slurry containing 1 lb_m of solids per gallon of water is to be filtered in a plate-and-frame filter having a total filtering area of 60 ft^2. The slurry is fed to the filter by a centrifugal pump, which develops a head of 20 psig. How long would it take to build up a layer of filter cake 4 in. thick on the filter medium? Laboratory data were taken on the slurry using a positive displacement pump operating at 5 gpm and 1 ft^2 of the filter medium. It was found that the pressure drop increased linearly with time from an initial value of 0.2 psi, reaching a value of 50 psi after 1 min. The density of the dry filter cake was found to be 0.85 g/cm^3.

15. A rotary drum filter that is 6 ft in diameter and 8 ft long is to be used to filter a slurry. The drum rotates at 0.5 rpm, and one-third of the drum's surface is submerged under the slurry. A vacuum is drawn in the drum so that a constant pressure drop of 10 psi is maintained across the drum and filter cake. You test the slurry in the lab by pumping it at a constant filtrate rate of 20 gpm through 1 ft^2 of the drum filter screen and find that after 1 min the pressure drop is 8 psi and after 3 min the pressure drop is 12 psi. How long will it take to filter 100,000 gal of the filtrate from the slurry using the rotary drum?

16. A plate-and-frame filter press contains 16 frames and operates at a constant flow rate of 30 gpm. Each frame has an active filtering area of 4 ft^2, and it takes 15 min to disassemble, clean, and reassemble the press. The press must be shut down for disassembly when the pressure difference builds up to 10 psi. What is the total net filtration rate in gpm for a slurry having properties determined by the following lab test. A sample of the slurry is pumped at a constant pressure differential of 5 psi through 0.25 ft^2 of the filter medium. After 3 min, 1 gal of the filtrate had been collected. The resistance of the filter medium may be neglected.

17. A rotary drum filter is used to filter a slurry. The drum rotates at a rate of 3 min/cycle, and 40% of the drum surface is submerged beneath the slurry. A constant pressure drop at 3 psi is maintained across the filter. If the drum is 5 ft in diameter and 10 ft long, calculate the total net filtration rate in gpm that is possible for a slurry having properties as determined by a lab test,

as follows. A sample of the slurry was pumped at a constant flow rate of 1 gpm through 0.25 ft²
of the filter medium. After 10 min, the pressure difference across the filter had risen to 2.5 psi.
The filter medium resistance may be neglected.

18. You must filter 1000 lb_m/min of an aqueous slurry containing 40% solids by weight using a
rotary drum filter with a diameter of 4 m and a length of 4 m, which operates at a vacuum of
25 in. Hg with 30% of its surface submerged beneath the slurry. A lab test is run on a sample of
the slurry using 200 cm² of the same filter medium, under a vacuum of 25 in. Hg. During the
first minute of operation, 300 cm³ of the filtrate is collected, and during the second minute, an
additional 140 cm³ is collected.
    (a) How fast should the drum be rotated?
    (b) If the drum is rotated at 2 rpm, what would the filter capacity be in lb_m of the slurry filtered
        per minute?

19. A rotary drum filter is to be used to filter a lime slurry. The drum rotates at a rate of 0.2 rpm,
and 30% of the drum surface is submerged beneath the slurry. The filter operates at a constant
ΔP of 10 psi. The slurry properties were determined from a lab test at a constant flow rate of
0.5 gpm using 1/2 ft² of the filter medium. The test results indicated that the pressure drop rose
to 2 psi after 10 s and to 10 psi after 60 s. Calculate the net filtration rate per unit area of the
drum under these conditions, in gpm/ft².

20. A plate-and-frame filter press operating at a constant ΔP of 150 psi is to be used to filter a
sludge containing 2 lb_m of solids per ft³ of water. The filter must be disassembled and cleaned
when the cake thickness builds up to 1 in. The frames have a projected area of 4 ft², and the
downtime for cleaning is 10 min/frame. The properties of the sludge and cake were determined
in a lab test operating at a constant flow rate of 0.2 gpm of the filtrate and a filter area of 1/4 ft².
The test results show that the pressure drop rises to 3 psi after 20 s and to 8 psi after 60 s.
Calculate the overall net filtration rate per frame in the filter in gpm of the filtrate, accounting
for the downtime. The density of the cake was found to be 150 lb_m/ft³.

21. A packed bed composed of crushed rock having a density of 175 lb_m/ft³ is to be used as a filter.
The size and shape of the rock particles is such that the average surface area to volume ratio
is 50 in.²/in.³ and the bed porosity is 0.3. A lab test using the slurry to be filtered is run on a
small bed of the same particles, which is 6 in. deep and 6 in. in diameter. The slurry is pumped
through this bed at a constant filtrate rate of 10 gpm, and it is found that after 5 min the pressure
drop is 5 psi, while after 10 min it is 8 psi. Calculate how long it would take to filter 100,000 gal
of filtrate from the slurry in a full-scale bed that is 10 ft in diameter and 2 ft deep, if the slurry is
maintained at a depth of 2 ft above the bed and drains through it by gravity. Assume the slurry
densities to be the same as water.

22. A rotary drum filter has a diameter of 6 ft and a length of 8 ft and rotates at a rate of 30 s/cycle.
The filter operates at a vacuum of 500 mm Hg, with 30% of its surface submerged. The slurry to
be filtered is tested in the lab using 0.5 ft² of the drum filter medium in a filter funnel operating
at 600 mm Hg vacuum. After 5 min of operation, 250 cm³ of the filtrate was collected through
the funnel, and after 10 min, a total of 400 cm³ is collected. What would be the net (average)
filtration rate of this slurry in the rotary drum filter, in gpm?

23. A rotary drum filter, 10 ft in diameter and 8 ft long, is to be used to filter a slurry of incompressible solids. The drum rotates at 1.2 rpm, and 40% of its surface is submerged beneath the slurry
at all times. A vacuum in the drum maintains a constant pressure drop of 10 psi across the drum
and filter cake. The slurry is tested in the lab by pumping it at a constant rate of 5 gpm through
0.5 ft² of the drum filter screen. After 1 min, the pressure drop is 9 psi, and after 3 min it has
risen to 15 psi. How long will it take to filter 1 million gal of the filtrate from the slurry using
the rotary drum? How long would it take if the drum rotated at 3 rpm?

24. A slurry is being filtered at a net rate of 10,000 gal/day by a plate-and-frame filter with 15 frames with an active filtering area of 1.5 ft² per frame, fed by a positive displacement pump. The pressure drop varies from 2 psi at start-up to 25 psi after 10 min, at which time it is shut down for cleanup. It takes 10 min to disassemble, clean out, and reassemble the filter. Your boss decides that it would be more economical to replace this filter by a rotary drum filter, using the same filter medium. The rotary filter operates at a vacuum of 200 mm Hg with 30% of its surface submerged and rotates at a rate of 5 min/rev. If the drum length is equal to its diameter, how big should it be?

25. You want to select a rotary drum filter to filter a coal slurry at a rate of 100,000 gal of the filtrate per day. The filter operates at a differential pressure of 12 psi, and 30% of the surface is submerged in the slurry at all times. A sample of the slurry is filtered in the lab through a 6 in. diameter sample of the filter medium at a constant rate of 1 gpm. After 1 min, the pressure drop across this filter is 3 psi, and after 5 min it is 10 psi. If the drum rotates at a rate of 3 rpm, what total filter area is required?

26. A slurry containing 40% solids by volume is delivered to a rotary drum filter, which is 4 ft in diameter and 6 ft long and operates at a vacuum of 25 in. Hg. A lab test is run with a 50 cm² sample of the filter medium and the slurry, at a constant flow rate of 200 cm³/min. After 1 min, the pressure across the lab filter is 6 psi and after 3 min it is 16 psi. If 40% of the rotary drum is submerged under the slurry, how fast should it be rotated (rpm) in order to filter the slurry at an average rate of 250 gpm?

27. A slurry is to be filtered with a rotary drum filter that is 5 ft in diameter, 8 ft long, rotates once every 10 s and has 20% of its surface immersed in the slurry. The drum operates with a vacuum of 20 in. Hg. A lab test was run on a sample of the slurry using 1/4 ft² of the filter medium at a constant flow rate of 40 cm³/s. After 20 s, the pressure drop was 30 psi across the lab filter, and after 40 s, it was 35 psi. How many gallons of the filtrate can be filtered per day in the rotary drum?

28. A rotary drum filter is to be installed in your plant. You run a test in the lab on the slurry to be filtered using a 0.1 ft² sample of the filter medium at a constant pressure drop of 10 psi. After 1 min, you find that 500 cc of the filtrate has passed through the filter, and after 2 min, the filtrate volume is 715 cc. If the rotary drum filter operates under a vacuum of 25 in. of Hg with 25% of its surface submerged, determine the following:
    (a) The capacity of the rotary drum filter, in gallons of filtrate per square foot of surface area, if it operates at (1) 2 rpm and (2) 5 rpm.
    (b) If the drum has a diameter of 4 ft and a length of 6 ft, what is the total filter capacity in gallons/day for each of the operating speeds of 2 and 5 rpm?

29. A slurry of $CaCO_3$ in water at 25°C containing 20% solids by weight is to be filtered in a plate-and-frame filter. The slurry and filter medium are tested in a constant pressure lab filter, having an area of 0.0439 m², at a pressure drop of 338 kPa. It is found that 10^{-3} m³ of the filtrate is collected after 9.5 s and 5×10^{-3} m³ is collected after 107.3 s. The plate-and-frame filter has 20 frames, with 0.873 m² of the filter medium per frame, and operates at a constant flow rate of 0.00462 m³ of slurry per second. The filter is operated until the pressure drop reaches 500 kPa, at which time it is shut down for cleaning. The downtime is 15 min per cycle. Determine how much filtrate passes through the filter in each 24 h period of operation (SG of $CaCO_3$ is 1.6).

30. An algal sludge is to be clarified by filtering. A lab test is run on the sludge using an area A of the filter medium. At a constant pressure drop of 40 kN/m², a plot of the time required to collect a volume $\tilde{V}$ of the filtrate times $\Delta P/(\tilde{V}/A)$ versus $\tilde{V}/A$ gives a straight line with a slope of 1.2×10^6 kN s/m⁴ and an intercept of 6.0×10^4 kN s/m³. A repeat of the data at a pressure drop of 200 kN/m² also gave a straight line on the same type of plot, with the same intercept but with

a slope of 2.1×10^6 kN s/m^4. When a filter aid was added to the sludge in an amount equal to 20% of the algae by weight, the lab test gave a straight line with the same intercept but with a slope of 1.4×10^6 kN s/m^4.

(a) What does this tell you about the sludge?

(b) The sludge is to be filtered using a rotary drum filter, with a diameter of 4 ft and a length of 6 ft, operating at a vacuum of 700 mm Hg with 35% of the drum submerged. If the drum is rotated at a rate of 2 rpm, how many gal of the filtrate will be collected in a day, with and without the filter aid?

(c) What would the answer to (b) be if the drum speed was 4 rpm?

31. A slurry containing 0.2 kg of solids per kg water is filtered through a rotary drum filter, operating at a pressure difference of 65 kN/m^2. The drum is 0.6 m in diameter, and 0.6 m long, rotates once every 350 s, and has 20% of its surface submerged below the slurry.

(a) If the overall average filtrate flow rate is 0.125 kg/s, the cake is incompressible with a porosity of 50%, and the solids $SG = 3.0$, determine the maximum thickness of the cake on the drum (you may neglect the filter medium resistance).

(b) The filter breaks down, and you want to replace it with a plate-and-frame filter having the same overall capacity, which operates at a pressure difference of 275 kN/m^2. The frames are 10 cm thick, and the maximum cake thickness at which the filter will still operate properly is 4 cm. It will take 100 s to disassemble the filter, 100 s to clean it out, and 100 s to reassemble it. If the frames are 0.3 m square, how many frames should the filter contain?

32. You want to filter an aqueous slurry using a rotary drum filter, at a total rate (of filtrate) of 10,000 gal/day. The drum rotates at a rate of 0.2 rpm, with 25% of the drum surface submerged in the slurry, at a vacuum of 10 psi. The properties of the slurry are determined from a lab test using a Buchner funnel under a vacuum of 500 mm Hg, using a 100 cm^2 sample of the filter medium and the slurry, which resulted in the data given in the following table. Determine the total filter area of the rotary drum required for this job.

Lab data:

Time (s)	Volume of Filtrate (cc)
50	10
100	18
200	31
400	51

33. You want to use a plate-and-frame filter to filter an aqueous slurry at a rate of 1.8 m^3/8 h day. The filter frames are square, with a length on each side of 0.45 m. The "downtime" for the filter press is 300 s plus an additional 100 s per frame for cleaning. The filter operates with a positive displacement pump, and the maximum operation pressure differential for the filter is 45 psi, which is reached after 200 s of operation.

(a) How many frames must be used in this filter to achieve the required capacity?

(b) At what flow rate (in gpm) should the pump be operated?

The following lab data were taken with the slurry at a constant ΔP of 10 psi and a 0.05 m^2 sample of the filter medium:

After 300 s, the total volume of the filtrate was 400 cc.

After 900 s, the total volume of the filtrate was 800 cc.

34. An aqueous slurry is filtered in a plate-and-frame filter, which operates at a constant ΔP of 100 psi. The filter consists of 20 frames, each of which have a projected area per side of 900 cm^2. A total filtrate volume of 0.7 m^3 is passed through the filter during a filtration time of 1200 s, and the downtime for the filter is 900 s. The resistance of the filter medium is negligible relative

to that of the cake. You want to replace the plate-and-frame filter with a rotary drum filter with the same overall average capacity, using the same filter medium. The drum is 2.2 m in diameter and 1.5 m long and operates at 5 psi vacuum with 25% of the drum surface submerged in the slurry. At what speed (in rpm) should the drum be operated?

35. You must transport a sludge product from an open storage tank to a separations unit at 1 atm, through a 4 in. sch 40 steel pipeline that is 2000 ft long, at a rate of 250 gpm. The sludge is 30% solids by weight in water and has a viscosity of 50 cP with Newtonian properties. The solid particles in the sludge have a density of 3.5 g/cc. The pipeline contains 4 gate valves and 6 elbows.
   (a) Determine the pump head (in feet) required to do this job. You can select any pump with the characteristics given in Appendix H, and you must find the combination of motor speed, motor horsepower, and impeller diameter that should be used.
   (b) You want to install a long radius venturi meter in the line to monitor the flow rate, and you want the maximum pressure drop to be measured to be equal to or less than 40 in. of water. What should the diameter of the venturi throat be?
   (c) At the separations unit, the sludge is fed to a settling tank. The solids settle in the tank, and the water overflows the top. What should the diameter of the tank be if it is desired to limit the size of the particles in the overflow to 100 μm or less?
   (d) If the sludge is fed to a centrifuge instead of the settling tank, at what speed (rpm) should the centrifuge operate to achieve the same separation as the settling tank, if the centrifuge dimensions are $L = D = 1$ ft, $R_1 = R_2/2 = 0.25$ ft.
   (e) Suppose the sludge is fed to a rotary drum filter instead, which removes all of the solids from the stream. The drum operates at a vacuum of 6 psi, has dimensions $L = D = 4$ ft, and operates with 30% of the surface submerged. A lab test is performed on the sludge using 1 ft² of the same filter medium as on the drum, operating at a vacuum of 500 mm Hg. In this test, it is found that 8 gal passes the filter after 2 min, and a total of 20 gal passes through after 10 minutes. What speed (rpm) should the rotary drum filter be operated?

36. Consider a dilute aqueous slurry containing solid particles with diameters from 0.1 to 1000 μm and a density of 2.7 g/cc, flowing at a rate of 500 gpm.
   (a) If the stream is fed to a settling tank in which all particles with a diameter greater than 100 μm are to be removed, what should the tank diameter be?
   (b) The overflow from the settling tank contains almost all of the water plus the fines not removed in the tank. This stream is fed to a centrifuge, which has a diameter of 20 in., a length of 18 in., and an overflow dam that is 6 in. from the centerline. What speed in rpm should the centrifuge rotate in order to separate all particles with a diameter of 1 μm and larger?
   (c) If the centrifuge rotates at 2500 rpm, what size particles will be removed?
   (d) Instead of the tank and centrifuge, the slurry is fed to a rotary drum filter, which has a diameter of 5 ft and a length of 10 ft. The drum operates under a vacuum of 10 in. Hg, with 35% of its surface submerged in the slurry. A lab test is run on the slurry at a constant flow rate of 100 cc/min, using 50 cm² of the filter medium. In the test filter, the pressure drop reached 10 mm Hg after 1 min and 80 mm Hg after 10 min. How fast should the drum rotate (in rpm) to handle the slurry stream?

NOTATION

A	Area, [L²]
a	Filter medium resistance, [1/L]
a_s	Particle surface area/volume, [1/L]
C_1	Filter parameter = $(\mu W)/(2K)$, [M/(L³ t)]
C_2	Filter parameter = μa, [M/(L² t)]
D	Diameter, [L]

d	Particle diameter, [L]
d_s	Diameter of equal volume sphere [L]
D_h	Hydraulic diameter, [L]
e_f	Energy dissipated per unit mass of fluid, [F L/M = L^2/t^2]
F	Correction factor, Fig. 14.5 [—]
f	Friction factor [—]
f_{pm}	Porous media friction factor, Equation 14.11, [—]
G	Gas mass flux, [M/(L^2 s)]
K	Permeability, [L^2]
L	Length, [L]
M_{solids}	Mass of solids [M]
N	Rotation rate, rpm, [1/t]
n	Number of frames, [—]
N_{RePM}	Porous media Reynolds number, Equation 14.13, [—]
P	Pressure, [F/L^2 = M/(L t^2)]
Q	Volumetric flow rate, [L^3/t]
s	Compressibility parameter, Equation, 14.44, [—]
t	Time, [t]
V	Velocity, [L/t]
$\tilde{V}$	Volume of filtrate, [L^3]
W	Slurry/cake solids loading parameter = $(M_{solids}/\tilde{V})/\rho_s(1 - \varepsilon)$, [—]
W_p	Wetted perimeter, [L]
x, y, z	Coordinate directions, [L]

GREEK

$\Delta()$	$()_2 - ()_2$
ε	Porosity or void fraction, [—]
Φ	Potential = $P + \rho g z$
μ	Viscosity, [M/(L t)]
ρ	Density, [M/L^3]
ψ	Sphericity factor, [—]

SUBSCRIPTS

1,2,3	Reference points
f	Filter frame side
G	Gas
i	Interstitial
L	Liquid
s	Superficial or solids

REFERENCES

Branan, C.R., *Rules of Thumb for Chemical Engineers*, Gulf Publishing Co., Houston, TX, 1994.

Cheremisinoff, N.P., *Liquid Filtration*, 2nd edn., Butterworth-Heinemann, Oxford, U.K., 1998.

Coulson, J.M., J.F. Richardson, J.R. Blackhurst, and J.H. Harker, *Chemical Engineering*, Vol. 2, 5th edn., Butterworth-Heinemann, Oxford, U.K., 2002.

Dullien, R.A.L., *Porous Media: Fluid Transport and Pore Structure*, 2nd edn., Academic, New York, 1992.

Gabelman, A., An overview of filtration, *Chem. Eng.*, 122(11), 50–58, 2015.

Leva, M., Reconsider packed tower pressure drop correlations, *Chem. Eng. Prog.*, 88, 65–72, 1992.

15 Fluidization and Sedimentation

I. FLUIDIZATION

When a fluid is passed upward through a bed of particles, as illustrated in Figure 15.1, the pressure drop increases as the fluid velocity increases. The product of the pressure drop and the bed cross-sectional area represents the net upward force acting on the bed, and when this force becomes equal to the weight of the bed (solids plus fluid), the bed becomes suspended by the fluid. In this state, the particles can move freely within the "bed," which thus behaves much like a boiling liquid. Under these conditions, the bed is said to be "fluidized." This "freely flowing" or bubbling behavior results in a high degree of mixing in the bed, which provides a great advantage for heat or mass transfer efficiency as compared to that of a fixed bed. Fluid bed operations are found in refineries (i.e., fluid catalytic crackers), polymerization reactors, fluid bed combustors, drying of cohesive solids, etc. If the fluid velocity within the bed is greater than the terminal velocity of the particles, however, the fluid will tend to entrain the particles and carry them out of the bed. However, if the superficial velocity above the bed, which is less than the interstitial velocity within the bed, is less the terminal velocity of the particles, they will fall back and remain in the bed. Thus, there is a specific range of velocity over which the bed remains in a fluidized state without particles being entrained by the gas fluid (gas or liquid).

The scope of this chapter concerns mainly the sedimentation and fluidization behavior of non-cohesive granular particles. More detailed discussion of these topics as well as the fluidization and sedimentation of fibrous systems (encountered in paper-pulp suspensions) and/or fine cohesive powders are available in other references, such as Yang (2003) and Millan (2013).

A. Governing Equations

The Bernoulli equation relates the pressure drop across the bed to the fluid flow rate and the bed properties:

$$\frac{-\Delta P}{\rho_f} - gh = e_f = \frac{f_{PM} h V_S^2}{d}\left(\frac{1-\varepsilon}{\varepsilon^3}\right) \tag{15.1}$$

where the porous medium friction factor is given by the Ergun equation:

$$f_{PM} = 1.75 + \frac{180}{N_{RePM}} \tag{15.2}$$

and the porous medium Reynolds number is defined as:

$$N_{RePM} = \frac{d V_S \rho}{(1-\varepsilon)\mu} \tag{15.3}$$

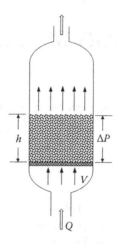

FIGURE 15.1   Schematic representation of a fluidized bed.

Now the criterion for incipient fluidization is that the force due to the pressure drop must balance the buoyant weight of the bed, that is,

$$-\Delta P = \frac{Bed\ Wt.}{A} = \rho_s (1-\varepsilon) gh + \rho \varepsilon gh \tag{15.4}$$

where the first term on the right-hand side is the pressure due to the weight of the solids and the second is that due to the weight of the fluid in the bed. When the pressure drop is eliminated from Equations 15.1 and 15.4, an equation for the "minimum fluidization velocity" (V_{mf}) results:

$$(\rho_s - \rho)(1-\varepsilon) g = \frac{\rho e_f}{h} = 1.75 \frac{\rho V_{mf}^2 (1-\varepsilon)}{d\varepsilon^3} + 180 \frac{V_{mf}\mu (1-\varepsilon)^2}{d^2\varepsilon^3} \tag{15.5}$$

which can be written in dimensionless form:

$$N_{Ar} = 1.75 \frac{\hat{N}_{Re}^2}{\varepsilon^3} + 180 \left(\frac{1-\varepsilon}{\varepsilon^3} \right) \hat{N}_{Re} \tag{15.6}$$

where

$$N_{Ar} = \frac{\rho g \Delta \rho d^3}{\mu^2}, \quad \hat{N}_{Re} = \frac{d V_S \rho}{\mu} \tag{15.7}$$

Equation 15.6 can be solved for the Reynolds number to give

$$\hat{N}_{Re} = \left(C_1^2 + C_2 N_{Ar} \right)^{1/2} - C_1 \tag{15.8}$$

where

$$C_1 = \frac{180(1-\varepsilon)}{3.5}, \quad C_2 = \frac{\varepsilon^3}{1.75} \tag{15.9}$$

Equation 15.8 gives the dimensionless superficial velocity (V_S) for incipient fluidization in terms of the Archimedes number and the bed porosity, ε.

B. Minimum Bed Voidage

Before the bed can become fluidized, however, the particles must dislodge from their "packed" state, after which the bed can expand. Thus, the porosity (ε) in Equations 15.5 and 15.9 is not the initial "packed bed" porosity but is the "expanded bed" porosity at the point of incipient fluidization (ε_{mf}), that is, the "minimum bed voidage" in the bed just prior to fluidization. Actually, the values of C_1 and C_2 in Equation 15.8 that give the best results for fluidized beds of uniform spherical particles have been found from empirical observations to be

$$C_1 = 27.2, \quad C_2 = 0.0408 \tag{15.10}$$

By comparing these empirical values of C_1 and C_2 with Equation 15.9, the C_1 value of 27.2 is seen to be equivalent to $\varepsilon_{mf} = 0.471$, while the C_2 value of 0.0408 is equivalent to $\varepsilon_{mf} = 0.415$. In practice, the value of ε_{mf} may vary considerably with the nature of the solid particles, as shown in Figure 15.2.

C. Nonspherical Particles

Many particles are not spherical and so will not have the same drag properties as spherical particles. The effective diameter for such particles is often characterized by the equivalent *Stokes diameter*, which is the diameter of the sphere that has the same terminal velocity as the nonspherical particle. This can be determined from a direct measurement of the settling rate of the particles and provides the best value of equivalent diameter for use in applications involving fluid drag on the particles.

An alternative description of nonspherical particles is often represented by the "sphericity factor" (ψ), which is the number that, when multiplied by the diameter of a sphere with the same volume as the particle (d_s), gives the particle effective diameter (d_p):

$$d_p = \psi d_s \tag{15.11}$$

The sphericity factor is defined as

$$\psi = \frac{\text{Surface area of the sphere with the same volume as the particle}}{\text{Surface area of the particle}} \tag{15.12}$$

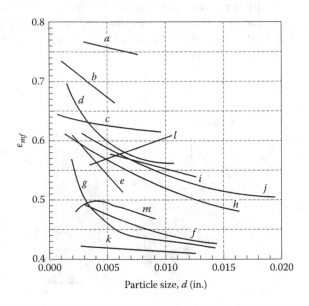

FIGURE 15.2 Values of ε_{mf} for various solids. (From Azbel, D.S. and Cheremisinoff, N.P., *Fluid Mechanics and Unit Operations*, Ann Arbor Science, Ann Arbor, MI, 1983.)

Thus

$$\psi = \frac{A_s}{A_p} = \frac{A_s/V_S}{A_p/V_p} = \frac{6/d_s}{a_s} \tag{15.13}$$

Equations 15.11 and 15.13 show that $d_p = 6/a_s$, where a_s is the surface-to-volume ratio for the particle (A_p/V_p), as deduced in Chapter 14. Since $V_p = V_S$ (by definition), an equivalent definition of ψ is

$$\psi = \frac{6}{d_s}\left(\frac{V_p}{A_p}\right) = \left(6^2\pi\right)^{1/3}\frac{V_p^{2/3}}{A_p} = \frac{4.84 V_p^{2/3}}{A_p} \tag{15.14}$$

The minimum bed porosity at incipient fluidization for nonspherical particles can be estimated from

$$\varepsilon_{mf} \cong \left(14\psi\right)^{-1/3} \tag{15.15}$$

For spherical particles ($\psi = 1$), Equation 15.15 reduces to $\varepsilon_{mf} = 0.415$.

II. SEDIMENTATION

Sedimentation, or thickening, involves increasing the solids content of a slurry or a suspension by gravity settling in order to effect separation (or partial separation) of the solids and the fluid. It differs from the gravity settling process that was previously considered in that the solids fraction is relatively high in these systems, so that particle settling rates are strongly influenced by the presence of the surrounding particles. This is referred to as *hindered settling*. Fine particles (10 μm or less) tend to behave differently than larger or coarse particles (100 μm or more), because fine particles may exhibit a high degree of flocculation due to the presence of surface forces and high specific surface area. Figure 13.1 shows a rough illustration of the effect of solids concentration and particle/fluid density ratio on the free and hindered settling regimes.

A. Hindered Settling

A mixture of particles of different sizes can settle in different ways, according to Coulson et al. (2002), as illustrated in Figure 15.3. Case (a) corresponds to a suspension with a range of particle sizes less than about 6:1. In this case, all of the particles settle at about the same velocity in the

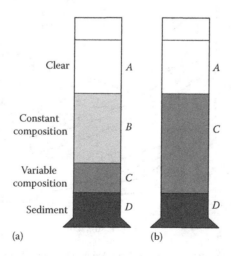

FIGURE 15.3   Modes of settling. (a) Narrow particle size range; (b) broad particle size range.

"constant composition zone" (*B*), leaving a layer of clear liquid above. As the sediment (*D*) builds up, however, the liquid that is "squeezed out" of this layer serves to further retard the particles just above it, resulting in a zone of variable composition (*C*). Case (b) in Figure 15.3 is less common and corresponds to a broad particle size range, in which the larger particles settle at a rate significantly greater than that of the smaller ones, and consequently there is no constant composition zone.

The sedimentation characteristics of hindered settling systems differ significantly from those of freely settling particles in several ways:

1. The large particles are hindered by the small particles, which increase the effective resistance of the suspending medium for the large particles. At the same time, however, the small particles tend to be "dragged down" by the large particles, so that all particles tend to fall at about the same rate (unless the size range is very large, i.e., greater than 6:1 or so).
2. The upward velocity of the displaced fluid flowing in the interstices between the particles is significant, so that the apparent settling velocity (relative to a fixed point) is significantly lower than the particle velocity relative to the fluid.
3. The velocity *gradient* in the suspending fluid flowing upward between the particles is increased, resulting in greater shear forces.
4. Because of the high surface area-to-volume ratio for small particles, surface forces are important resulting in flocculation or "clumping" of the smaller particles into larger effective particle groups. This effect is more pronounced in a highly ionic (conducting) fluid, because the electrostatic surface forces that would cause the particles to be repelled are "shorted out" by the conductivity of the surrounding fluid.

There are essentially three different approaches to describing hindered settling. One approach is to define a "correction factor" to the Stokes free settling velocity in an infinite Newtonian fluid (which we will designate V_o), as a function of the solids loading. A second approach is to consider the suspending fluid properties (e.g., viscosity and density) to be modified by the presence of fine particles. A third approach is to consider the collection or "swarm" of particles equivalent to a moving porous bed, the resistance to flow through the bed being determined by an equivalent of the Kozeny equation. There is insufficient evidence to say that any one of these approaches is any better or worse than the others. For many systems, they may all give comparable results, whereas for others one of these methods may be better or worse than the others.

If all of the solids are relatively fine and/or the slurry is sufficiently concentrated so that settling is extremely slow, the slurry can usually be approximated as a uniform continuous "pseudo-homogeneous" medium with properties (viscosity and density) that depend upon the solids loading, particle size and density, and interparticle forces (surface charges, conductivity, etc.). Such systems are generally quite non-Newtonian, with properties that can be described by the Bingham plastic or power law models. If the particle size distribution is broad and a significant fraction of the particles are fines (e.g., less than about 30 μm or so), the suspending fluid plus fines can be considered to be a continuous medium with a characteristic viscosity and density through which the larger particles must move. Such systems may or may not be non-Newtonian, depending on solids loading, particle size distribution, etc., but are most commonly non-Newtonian. If the solids loading is relatively low (e.g., below about 10% solids by volume) and/or the particle size and/or density are relatively large, the system will be "heterogeneous" and the larger particles will settle readily. Such systems are usually Newtonian. A summary of the flow behavior of these various systems has been presented by Darby (1986), Shook and Roco (1991), and Wilson et al. (2008).

B. FINE PARTICLES

For suspensions of fine particles, or systems containing a significant amount of fines, the suspending fluid can be considered to be homogeneous with the density and viscosity modified by the presence of the fines. These properties depend primarily upon the solids loading of the suspension, which may be described in terms of either the porosity or void fraction (ε) or, more commonly, the volume fraction of solids, φ ($\varphi = 1 - \varepsilon$). The buoyant force on the particles is due to the difference in density between the solid (ρ_s) and the surrounding suspension (ρ_φ), which is

$$\rho_s - \rho_\varphi = \rho_s - \left[\rho_s(1-\varepsilon) + \rho\varepsilon\right] = \varepsilon(\rho_s - \rho) = (1-\varphi)(\rho_s - \rho) \tag{15.16}$$

where ρ is the density of the clear fluid phase.

The viscosity of the suspension (μ_φ) is also modified by the presence of the solids. For uniform spheres at a volumetric fraction of 2% or less, Einstein (1906) showed that

$$\mu_\varphi = \mu(1 + 2.5\varphi) \tag{15.17}$$

where μ is the viscosity of the suspending fluid phase. For more concentrated suspensions, a wide variety of expressions have been proposed in the literature (see, e.g., Darby [1986], Mewis and Wagner [2012], and Chhabra [2016]). For example, Vand (1948) proposed the expression

$$\mu_\varphi = \mu\exp\left(\frac{2.5\varphi}{1 - 0.609\varphi}\right) \tag{15.18}$$

However, Mooney (1951) concluded that the numerical constant 0.609 in Equation 15.18 varies from 0.75 to 1.5, depending upon the system. Equation 15.17 or 15.18 (or equivalent) may be used to modify the viscosity and density in Stokes' law, that is,

$$V_o = \frac{(\rho_s - \rho)gd^2}{18\mu} \tag{15.19}$$

In this equation, V_o is the relative velocity between the unhindered particle and the fluid. However, in a hindered suspension, this velocity is increased by the velocity of the displaced fluid, which flows back up through the suspension in the void space between the particles. Thus, if V_S is the (superficial) settling velocity of the suspension (e.g., "swarm") and V_L is the velocity of the fluid, the total flux of solids and liquid is $[\varphi V_S + (1 - \varphi)V_L]$. The relative velocity between the fluid and solids in the swarm is $V_r = V_S - V_L$. If the total net flux is zero (e.g., "batch" settling in a closed-bottom container with no outflow), elimination of V_L gives

$$V_r = \frac{V_S}{1 - \varphi} \tag{15.20}$$

This also shows that $V_L = -\varphi V_S/(1 - \varphi)$, that is, V_L is negative relative to V_S in batch settling.

From Equations 15.16, 15.18, and 15.20, the ratio of the settling velocity of the suspension (V_S) to the terminal velocity of a single freely settling sphere (V_o) can be determined to be

$$\frac{V_S}{V_o} = \frac{(1-\varphi)^2}{\exp\left(\dfrac{2.5\varphi}{1 - k_2\varphi}\right)} \tag{15.21}$$

where the value of the constant k_2 can vary from 0.61 to 1.5, depending upon the system. However, Coulson et al. (2002) remark that the use of a modified viscosity for the suspending fluid is more appropriate for the settling of large particles through a suspension of fines than for the uniform settling of a "swarm" of uniform particles with a narrow size distribution. They state that in the latter case, the increased resistance is due to the higher-velocity gradients in the interstices rather than to an increased viscosity. However, the net effect is essentially the same for either mechanism. This approach and the other two mentioned earlier all result in expressions of the general form:

$$\frac{V_S}{V_o} = \varepsilon^2 fn(\varepsilon) \quad \text{where} \quad \varepsilon = (1-\varphi) \tag{15.22}$$

which is consistent with Equation 15.21.

A widely quoted empirical expression for the function in Equation 15.22 is that of Richardson and Zaki (1954):

$$fn(\varepsilon) = \varepsilon^n \tag{15.23}$$

where

$$n = \begin{cases} 4.65 & \text{for} & N_{Rep} < 0.2 \\ 4.35 N_{Rep}^{-0.03} & \text{for} & 0.2 < N_{Rep} < 1 \\ 4.45 N_{Rep}^{0.1} & \text{for} & 1 < N_{Rep} < 500 \\ 2.39 & \text{for} & N_{Rep} > 500 \end{cases}$$

and N_{Rep} is the single-particle Reynolds number in an "infinite" fluid. An alternative expression due to Davies et al. (1977) for the ratio of the two velocities is

$$\frac{V_S}{V_o} = \exp(-k_1\varphi) \tag{15.24}$$

which agrees well with Equation 15.23 for $k_1 = 5.5$. Yet another expression for $fn(\varepsilon)$, deduced by Steinour (1944) from settling data on tapioca in oil, is

$$fn(\varepsilon) = 10^{-1.82(1-\varepsilon)} \tag{15.25}$$

Barnea and Mizrahi (1973) considered the effects of the modified density and viscosity of the suspending fluid, as represented by Equation 15.21, as well as a "crowding" or hindrance effect that decreases the effective space around the particles and increases the drag. This additional "crowding factor" is given by $(1 + k_2\varphi^{1/3})$, which, when included in Equation 15.21, gives

$$\frac{V_S}{V_o} = \frac{(1-\varphi)^2}{(1+\varphi^{1/3})\exp[5\varphi/3(1-\varphi)]} \tag{15.26}$$

for the modified Stokes velocity, where the constant 2.5 in Equation 15.21 has been replaced by 5/3, and the constant k_2 set equal to unity, based upon experimental observations in a range of systems.

C. COARSE PARTICLES

Coarser particles (e.g., ~100 μm or larger) have a relatively small specific surface, so that flocculation is not common. Also, the suspending fluid surrounding the particles is the liquid phase rather than a "pseudo continuous" phase of fines in suspension which modify the fluid viscosity and density properties. Thus, the properties of the continuous phase can be taken to be those of the pure fluid unaltered by the presence of fine particles. In this case, it can be shown by dimensional analysis that the dimensionless settling velocity V_S/V_o must be a function of the particle drag coefficient, which in turn is a unique function of the particle Reynolds number, N_{Rep}, the void fraction (porosity), $\varepsilon = (1 - \varphi)$, and the ratio of the particle diameter to container diameter, d/D. Since there is a unique relationship between the drag coefficient, the Reynolds number, and the Archimedes number for settling particles, the result can be expressed in functional form as

$$\frac{V_S}{V_o} = fn\left(N_{Ar}, d/D, \varepsilon\right) \tag{15.27}$$

It has been found that this relationship can be represented by the following empirical expression (Coulson et al., 2002):

$$\frac{V_S}{V_o} = \varepsilon^n \left(1 + 2.4\frac{d}{D}\right)^{-1} \tag{15.28}$$

where the exponent n is given by

$$n = \frac{4.8 + 2.4X}{X + 1} \tag{15.29}$$

and

$$X = 0.043 N_{Ar}^{0.57} \left[1 - 2.4\left(\frac{d}{D}\right)^{2.7}\right] \tag{15.30}$$

D. ALL FLOW REGIMES

The above expressions give the suspension velocity (V_S) relative to the single-particle free settling velocity, V_o, that is, the Stokes velocity. However, it is not necessary that the particle settling conditions correspond to the Stokes regime to use these equations. As shown in Chapter 12, the Dallavalle equation can be used to calculate the single-particle terminal velocity V_o under any flow conditions from a known value of the Archimedes number, as follows:

$$V_o = \frac{\mu}{\rho d}\left[\left(14.42 + 1.827\sqrt{N_{Ar}}\right)^{1/2} - 3.798\right]^2 \tag{15.31}$$

where

$$N_{Ar} = \frac{d^3 \rho g \Delta \rho}{\mu^2} \tag{15.32}$$

This result can also be applied directly to coarse particle "swarms." For fine particle systems, the suspending fluid properties are assumed to be modified by the fines in suspension, which necessitates modifying the fluid properties in the definitions of the Reynolds and Archimedes numbers accordingly. Furthermore, since the particle drag is a direct function of the local relative velocity between the fluid and the solid (i.e., the interstitial relative velocity, V_r), it is this velocity that must be used in the drag equations (e.g., the modified Dallavalle equation). Knowing that $V_r = V_S/(1 - \varphi) = V_S/\varepsilon$, the appropriate definitions for the Reynolds number and drag coefficient for the suspension (e.g., the particle "swarm") are (after Barnea and Mizrahi, 1973)

$$N_{Re\varphi} = \frac{dV_r \rho}{\mu_\varphi} = N_{Re0} \left(\frac{V_S}{V_o} \right) \frac{1}{(1-\varphi)\exp\left[5\varphi/3(1-\varphi)\right]} \tag{15.33}$$

and

$$C_{D\varphi} = C_{Do} \left(\frac{V_o}{V_S} \right)^2 \frac{(1-\varphi)^2}{\left(1+\varphi^{1/3}\right)} \tag{15.34}$$

where $N_{Reo} = dV_o\rho/\mu$ and $C_{Do} = 4gd(\rho_S - \rho)/(3\rho V_o^2)$ are the Reynolds number and drag coefficient, respectively, for a single particle in an infinite fluid. Data presented by Barnea and Mizrahi (1973) show that the "swarm" dimensionless groups $N_{Re\varphi}$ and $C_{D\varphi}$ are related by the same expression as the corresponding groups for single particles, for example, by the Dallavalle equation:

$$C_{D\varphi} = \left(0.6324 + \frac{4.8}{N_{Re\varphi}^{1/2}} \right)^2 \tag{15.35}$$

The settling velocity or the terminal velocity of the "swarm" may therefore be determined from

$$N_{Re\varphi} = \left[\left(14.42 + 1.827 N_{Ar\varphi}^{1/2} \right)^{1/2} - 3.798 \right]^2 \tag{15.36}$$

where

$$N_{Ar\varphi} = \frac{3}{4} C_{D\varphi} N_{Re\varphi}^2 = \left(\frac{d^3 \rho g (\rho_s - \rho)}{\mu^2 (1 + \varphi^{1/3})} \right) \exp \left(\frac{-10\varphi}{3(1-\varphi)} \right) \tag{15.37}$$

III. GENERALIZED SEDIMENTATION/FLUIDIZATION

The above relations all apply to hindered settling of a suspension (or "swarm") of particles (in most cases of uniform size) in a stagnant suspending medium. Barnea and Mizrahi (1973) showed that these generalized relations may be applied to fluidization as well, since a fluidized bed may be considered a particle "swarm" suspended by the fluid flowing upward at the terminal velocity of the swarm. In this case, the above equations apply with V_S replaced by the velocity V_f, that is, the superficial velocity of the fluidizing medium. Once $N_{Re\varphi}$ is found from Equations 15.36 and 15.37, the settling velocity (V_S) is determined from Equation 15.33. Barnea and Mizrahi (1973) presented data for both settling and fluidization, which cover a very wide range of the dimensionless parameters, as shown in Figure 15.4. In this figure, the x-coordinate is independent of velocity and the y-coordinate is independent of particle size.

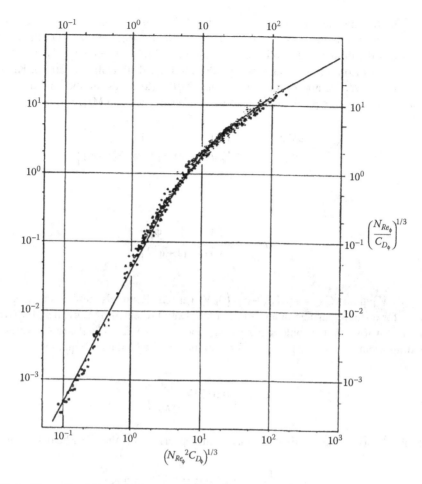

FIGURE 15.4   Generalized dimensionless correlation of settling and fluidizing velocities. (From Barnea, E. and Mizrahi, J., A generalized approach to the fluid dynamics of particulate systems, Part 1: General correlation for fluidization and sedimentation in solid multi-particle systems, *Chem. Eng. J.*, 5, 171, 1973.)

IV.   THICKENING

The process of thickening involves the concentration of a slurry, suspension, or sludge usually by gravity settling. Since concentrated suspensions and/or fine particle dispersions are often involved, the result is usually not a complete separation of the solids from the liquid but is instead a separation into a more concentrated (underflow) stream and a diluted (overflow) stream. Thickeners and clarifiers are essentially identical. The only difference is that the clarifier is designed to produce a clean liquid overflow with a specified purity, whereas the thickener is designed to produce a concentrated underflow product with a specified concentration (McCabe et al., 1993; Christian, 1994; Tiller and Tang, 1995).

A schematic of a thickener/clarifier is shown in Figure 15.5. As indicated in Figure 15.3, several settling regions or zones can be identified, depending upon the solids concentration and interparticle interactions. For simplicity, we consider three primary zones, as indicated in Figure 15.5 (with the understanding that there are transition zones in between). The top, or clarifying, zone contains relatively clear liquid from which most of the particles have settled out. Any particles remaining in this zone will settle by free settling. The middle zone is a region of varying composition through which the particles move by hindered settling. The size of this region and the settling rate depend upon the local solids concentration. The bottom zone is a

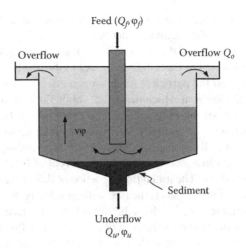

Feed (Q_f, φ_f)

Overflow

Overflow Q_o

$v\varphi$

Sediment

Underflow
Q_u, φ_u

FIGURE 15.5 Schematic of a thickener.

highly concentrated settled or compressed region containing the settled particles. The particle settling rate in this zone is very slow.

In the top (clarifying) zone, the relatively clear liquid moves upward and overflows the top. In the middle zone, the solid particles settle as the displaced liquid moves upward, and both the local solids concentration and the settling velocity vary from point to point. In the bottom (compressed) zone, the solids and liquid both move downward at a rate that is determined mainly by the underflow draw-off rate from the thickener. For a given feed rate and solids loading, the objective is to determine the area of the thickener and the optimum underflow (draw-off) rate to achieve a specified underflow concentration (φ_u), or the underflow rate and underflow concentration for a stable steady-state operation.

The solids concentration can be expressed in terms of either the solids volume fraction (φ) or the mass ratio of solids to fluid (R). If φ_f is the volume fraction of solids in the feed stream (flow rate Q_f) and φ_u is the volume fraction of solids in the underflow (flow rate Q_u), then the solids ratio in the feed, $R_f = [(mass\ of\ solids)/(mass\ of\ fluid)]_{feed}$, and in the underflow, $R_u = [(mass\ of\ solids)/(mass\ of\ liquid)]_u$, are given by

$$R_f = \frac{\varphi_f \rho_s}{(1-\varphi_f)\rho}, \quad R_u = \frac{\varphi_u \rho_s}{(1-\varphi_u)\rho} \tag{15.38}$$

These relations can be rearranged to give the solids volume fractions in terms of the solids ratio:

$$\varphi_f = \frac{R_f}{R_f + (\rho_s/\rho)}, \quad \varphi_u = \frac{R_u}{R_u + (\rho_s/\rho)} \tag{15.39}$$

Now the total (net) flux of the solids plus liquid moving through the thickener at any point is given by

$$q = \frac{Q}{A} = q_S + q_L = \varphi V_S + (1-\varphi)V_L \tag{15.40}$$

where

$q_s = \varphi V_S$ is the local solids flux, defined as the volumetric settling rate of the solids per unit cross-sectional area of the settler

$q_L = (1-\varphi)V_L$ is the local liquid flux

The solids flux depends upon the local concentration of solids, the settling velocity of the solids at this concentration *relative to the liquid*, and the net velocity of the liquid. Thus, the local solids flux will vary within the thickener because the concentration of solids increases with depth. The amount of liquid that is displaced (upward) by the solids decreases as the solids concentration increases, thus affecting the "upward drag" on the particles. As these two effects act in opposite directions, there will be some point in the thickener at which the actual solids flux is a minimum. This point determines the conditions for stable steady-state operation, as explained in the following.

The settling behavior of a slurry is normally determined by measuring the velocity of the interface between the top (clear) and middle suspension zones in a *batch settling test* using a closed system (e.g., a graduated cylinder) as illustrated in Figure 15.3. A typical batch settling curve is shown in Figure 15.6 (e.g., Foust et al., 1980). The initial linear portion of this curve corresponds to free (unhindered) settling, and the slope of this region is the free settling velocity, V_o. The nonlinear region of the curve corresponds to hindered settling in which the solids flux depends upon the local solids concentration, which can be determined from the batch settling curve, as follows (Kynch, 1952).

If the initial height of the suspension with a solids fraction of φ_o is Z_o, at some later time the height of the interface between the clear layer and the hindered settling zone will be $Z(t)$, where the average solids fraction in this zone is $\varphi(t)$. Since the total amount of solids in the system is constant, and assuming the amount of solids in the clear layer to be negligible, it follows that

$$Z(t)\varphi(t) = Z_o\varphi_o \quad \text{or} \quad \varphi(t) = \frac{\varphi_o Z_o}{Z(t)} \tag{15.41}$$

Thus, given the initial height and concentration (Z_o, φ_o), the average solids concentration $\varphi(t)$ corresponding to any point on the $Z(t)$ curve can be determined using Equation 15.41. Furthermore, the hindered settling velocity and batch solids flux at this point can be determined from the slope of the curve at that point, that is, $V_{Sb} = -(dZ/dt)$ and $q_{sb} = \varphi V_{Sb}$. Thus, the batch settling curve can be converted to a *batch flux curve*, as shown in Figure 15.7. The batch flux curve exhibits a maximum and a minimum, since the settling velocity is nearly constant in the free settling region and the flux is directly

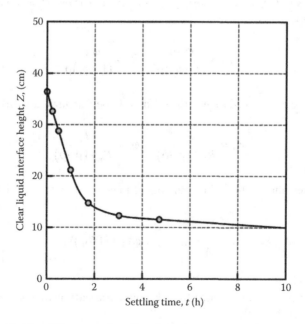

FIGURE 15.6 Typical batch settling curve for a limestone slurry.

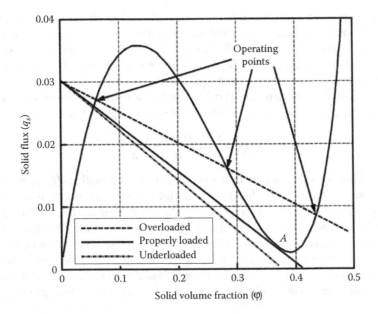

FIGURE 15.7 Typical batch flux curve with operating lines.

proportional to the solids concentration, whereas the settling velocity and the flux drop rapidly with the increasing solids concentration in the hindered settling region, as explained above. However, the solids flux in the bottom (compressed) zone is much higher because of the high concentration of solids in this zone. The minimum in this curve represents a "pinch" or "critical" condition in the thickener, which limits the total solids flux that can be obtained under steady (stable) operation.

Because the batch flux data are obtained in a closed system with no outflow, the net solids flux is zero in the batch system and Equation 15.40 reduces to $V_L = -\varphi V_S/(1 - \varphi)$. Note that V_L and V_S are of opposite sign, since the displaced liquid moves upward as the solids settle downward. The relative velocity between the solids and liquid is $V_r = V_S - V_L$, which, from Equation 15.20, is $V_r = V_S/(1 - \varphi)$. It is this relative velocity that controls the dynamics in the thickener. If the underflow draw-off rate from the thickener is Q_u, the additional solids flux in the thickener due to superimposition of this underflow is $q_u = Q_u/A = V_u$. Thus, the total solids flux at any point in the thickener (q_S) is equal to the settling flux relative to the suspension (i.e., the batch flux q_{sb}) at that point, plus the bulk flux due to the underflow draw-off rate, φV_u, that is, $q_S = q_{sb} + \varphi q_u$. Furthermore, at steady state, the net local solids flux in the settling zone (q_S) must be equal to that in the underflow, that is, $q_S = q_u \varphi_u$. Eliminating q_u and rearranging leads to

$$q_{sb} = q_S\left(1 - \frac{\varphi}{\varphi_u}\right) \tag{15.42}$$

This equation represents a straight line on the batch flux curve (q_{sb} vs. φ) that passes through the points (q_S, 0) and (0, φ_u), that is, the line intersects the φ axis at φ_u and the q_{sb} axis at q_S, which is the net local solids flux in the thickener at the point where the solids fraction is φ. This line is called the "operating line" for the thickener, and its intersection with the batch flux curve determines the stable operating point for the thickener, as shown in Figure 15.7. The "properly loaded" operating line is tangent to the batch flux curve (point A). At the tangent point, called the critical (or "pinch") point, the local solids flux corresponds to the steady-state value at which the net critical (minimum) settling rate in the thickener equals the total underflow solids rate. The "underloaded" line represents a condition for which the underflow draw-off rate is higher than the critical settling rate, so that no solids layer can build up and excess clear liquid will eventually be drawn out of the bottom (i.e., the draw-off rate is too high). The "overloaded" line represents the condition at which the underflow draw-off rate is lower than the critical settling rate, so that the bottom solids layer will build up and eventually rise to the overflow (i.e., the underflow rate is too low).

Once the operating line is set, the equations that govern the thickener operation are determined from a solids mass balance as follows. At steady-state (stable) operating conditions, the net solids flux is

$$q_S = \frac{Q_s}{A} = \frac{Q_f \varphi_f}{A} = \frac{Q_u \varphi_u}{A} \tag{15.43}$$

This equation relates the thickener area (A) and the feed rate and loading (Q_f, φ_f) to the solids underflow rate (Q_u) and the underflow loading (φ_u), assuming that there are no solids in the overflow.

The area of a thickener required for a specified underflow loading can be determined as follows. For a given underflow solids loading (φ_u), the operating line is drawn on the batch flux curve from φ_u on the φ axis tangent to the batch flux curve at the critical point, (q_c, φ_c). The intersection of this line with the vertical axis ($\varphi = 0$) gives the local solids flux (q_S) in the thickener, which results in stable or steady (properly loaded) conditions. This value is determined from the intersection of the operating line on the q_{sb} axis or from the equation of the operating line that is tangent to the critical point (q_c, φ_c):

$$q_S = \frac{q_c}{1 - \varphi_c/\varphi_u} \tag{15.44}$$

If the feed rate (Q_f) and solids loading (φ_f) are specified, the thickener area A is determined from Equation 15.43. If it is assumed that none of the solids are carried over with the overflow, the overflow rate Q_o is given by

$$Q_o = Q_f \left(1 - \varphi_f\right) - Q_f \varphi_f \frac{\left(1 - \varphi_u\right)}{\varphi_u} \tag{15.45}$$

or

$$\frac{Q_o}{Q_f} = 1 - \frac{\varphi_f}{\varphi_u} \tag{15.46}$$

Likewise, the underflow rate Q_u is given by

$$Q_u = Q_f - Q_o = Q_f - Q_f \left(1 - \frac{\varphi_f}{\varphi_u}\right) \tag{15.47}$$

or

$$\frac{Q_u}{Q_f} = \frac{\varphi_f}{\varphi_u} \tag{15.48}$$

SUMMARY

The following are some of the major points covered in this chapter:

- The principles governing the fluidization of solid particles, the minimum fluidization velocity, and the minimum bed voidage for both spherical and nonspherical particles
- The difference in the settling characteristics of suspensions of fine particles and coarse particles, and the method of prediction of the settling rate in both cases

- The use of the Archimedes number in predicting the settling velocity of particle "swarms" and the similarity to the fluidization velocity of the "swarm"
- The use of batch settling curves and batch flux curves to determine the properly loaded stable operating point for a thickener

PROBLEMS

1. Calculate the flow rate of air (in scfm) required to fluidize a bed of sand ($SG = 2.4$), if the air exits the bed at 1 atm, 70°F. The sand grains have an equivalent diameter of 500 μm and the bed is 2 ft in diameter and 1 ft deep, with a porosity of 0.35. What flow rate of air would be required to blow the sand away?

2. Calculate the flow rate of water (in gpm) required to fluidize a bed of 1/16 in. diameter lead shot ($SG = 11.3$). The bed is 1 ft in diameter, 1 ft deep, and has a porosity of 0.38. What water flow rate would be required to sweep the bed away?

3. Calculate the range of water velocities that will fluidize a bed of glass spheres ($SG = 2.1$) if the sphere diameter is (a) 2 mm, (b) 1 mm, and (c) 0.1 mm.

4. A coal gasification reactor operates with particles of 500 μm diameter and a density of 1.4 g/cm³. The gas may be assumed to have properties of air at 1000°F and 30 atm. Determine the range of superficial gas velocity over which the bed is in a fluidized state.

5. A bed of coal particles, 2 ft in diameter and 6 ft deep, is fluidized using a hydrocarbon liquid with a viscosity 15 cP and a density of 0.9 g/cm³. The coal particles have a density of 1.4 g/cm³ and an equivalent spherical diameter of 1/8 in. If the bed porosity is 0.4,
   (a) Determine the range of liquid superficial velocities over which the bed is fluidized.
   (b) Repeat the problem using the "particle swarm" (Barnea and Mizrahi) "swarm terminal velocity" approach, assuming (1) $\varphi = 1 - \varepsilon$; (2) $\varphi = 1 - \varepsilon_{mf}$.

6. A catalyst having spherical particles with $d_p = 50$ μm and $\rho_s = 1.65$ g/cm³ is used to contact a hydrocarbon vapor in a fluidized reactor at 900°F, 1 atm. At operating conditions, the fluid viscosity is 0.02 cP and its density is 0.21 lb$_m$/ft³. Determine the range of fluidized bed operation, that is, calculate
   (a) Minimum fluidization velocity for $\varepsilon_{mf} = 0.42$
   (b) The particle terminal velocity

7. A fluid bed reactor contains catalyst particles with a mean diameter of 500 μm and a density of 2.5 g/cm³. The reactor feed has properties equivalent to 35° API distillate at 400°F. Determine the range of superficial velocities over which the bed will be in a fluidized state.

8. Water is pumped upward through a bed of 1 mm diameter iron oxide particles ($SG = 5.3$). If the bed porosity is 0.45, over what range of superficial water velocity will the bed be fluidized?

9. A fluidized bed combustor is 2 m in diameter and is fed with air at 250°F, 10 psig, at a rate of 2000 scfm. The coal has a density of 1.6 g/cm³, and a shape factor of 0.85. The flue gas from the combustor has an average MW of 35 and leaves the combustor at a rate of 2100 scfm at 2500°F and 1 atm. What is the size range of the coal particles that can be fluidized in this system.

10. A fluid bed incinerator, 3 m in diameter and 0.56 m high, operates at 850°C using a sand bed. The sand density is 2.5 g/cm³, and the average sand grain has a mass of 0.16 mg and a sphericity of 0.85. In the stationary (packed) state, the bed porosity is 35%. Find

(a) The range of air velocities that will fluidize the bed.

(b) The compressor power required, if the bed is operated at 10 times the minimum fluidizing velocity and the compressor efficiency is 70%. The compressor takes air in from the atmosphere at 20°C, and the gases leave the bed at 1 atm.

11. Determine the range of flow rates (in gpm) that will fluidize a bed of 1 mm cubic silica particles ($SG = 2.5$) with water. The bed is 10 in. in diameter, 15 in. deep.

12. Determine the range of velocities over which a bed of granite particles ($SG = 3.5$, $a_s = 0.012 \, \mu m^{-1}$, $\psi = 0.8$, $d = 0.6/a_s$ would be fluidized using the following fluids:
    (a) Water at 70°F
    (b) Air at 70°F and 20 psig

13. Calculate the velocity of water that would be required to fluidize spherical particles with $SG = 1.6$ and a diameter of 1.5 mm, in a tube with a diameter of 10 mm. Also, determine the water velocity that would sweep the particles out of the tube. Use each of the two following methods, and compare the results:
    (a) The bed starts as a packed bed and is fluidized when the pressure drop due to friction through the bed balances the weight of the bed.
    (b) The bed is considered to be a "swarm" of particles, falling at the terminal velocity of the "swarm." (Assume $\varepsilon = 0.45$.)
    Comment on any uncertainties or limitations in your results.

14. You want to fluidize a bed of solid particles using water. The particles are cubical, with a length on each side of 1/8 in., and a SG of 1.2.
    (a) What is the sphericity factor for these particles, and their equivalent diameter?
    (b) What is the approximate bed porosity at the point of fluidization of the bed?
    (c) What velocity of water would be required to fluidize the bed?
    (d) What velocity of water would sweep the particles out of the bed?

15. Solid particles with a density of 1.4 g/cm³ and a diameter of 0.01 cm are fed from a hopper into a line where they are mixed with water, which is draining by gravity from an open tank, to form a slurry having 0.4 lb$_m$ of solids/lb$_m$ of water. The slurry is transported by a centrifugal pump, through a 6 in. sch 40 pipeline that is 0.5 mile long, at a rate of 1000 gpm. The slurry can be described as a Bingham plastic, with a yield stress of 120 dyn/cm² and a limiting viscosity of 50 cP.
    (a) If the pipeline is at 60°F, and the pump is 60% efficient with a required NPSH (net positive suction head) of 15 ft, what horsepower motor would be required to drive the pump?
    (b) If the pump is 6 ft below the bottom of the water storage tank, and the water in the line upstream of the pump is at 90°C ($P_v = 526$ mmHg), what depth of water in the tank would be required to prevent the pump from cavitating?
    (c) A venturi meter is installed in the line to measure the slurry flow rate. If the maximum pressure drop reading for the venturi is 29 in. of water, what diameter should the venturi throat be?
    (d) The slurry is discharged from the pipeline to a settling tank, where it is desired to concentrate the slurry to 1 lb$_m$ of solids/lb$_m$ of water (in the underflow). Determine the required diameter of the settling tank, and the volumetric flow rates of the overflow (Q_o) and underflow (Q_u), in gpm.
    (e) If the slurry were to be sent to a rotary drum filter instead, to remove all of the solids, determine the required size of the drum (assuming the drum length and diameter are equal). The drum rotates at 3 rpm, with 25% of its surface submerged in the slurry, and operates at a vacuum of 20 in. of mercury. Lab test data taken on the slurry with 0.5 ft² of the filter

medium, at a constant flow rate of 3 gpm, indicated a pressure drop of 1.5 psi after 1 min of filtration and 2.3 psi after 2 min of operation.

16. A sludge is clarified in a thickener, which is 50 ft in diameter. The sludge contains 35% solids by volume ($SG = 1.8$) in water, with an average particle size of 25 μm. The sludge is pumped into the center of the tank, where the solids are allowed to settle and the clarified liquid overflows the top. Estimate the maximum flow rate of the sludge (in gpm) that this thickener can handle. Assume that the solids are uniformly distributed across the tank and that all particle motion is vertical.

17. In a batch thickener, an aqueous sludge containing 35% by volume of solids ($SG = 1.6$) with an average particle size of 50 μm is allowed to settle. The sludge is fed to the settler at a rate of 1000 gpm, and the clear liquid overflows the top. Estimate the minimum tank diameter required for this separation.

18. Ground coal is slurried with water in a pit, and the slurry is pumped out of the pit at a rate of 500 gpm with a centrifugal pump and into a classifier. The classifier inlet is 50 ft above the slurry level in the pit. The piping system consists of an equivalent length of 350 ft of 5 in. sch 40 pipe and discharges into the classifier at 2 psig. The slurry may be assumed to be a Newtonian fluid, with a viscosity of 30 cP, a density of 75 lb_m/ft^3, and a vapor pressure of 30 mmHg. The solid coal has a $SG = 1.5$.

(a) How much power would be required to pump the slurry?

(b) Using the pump characteristic charts in Appendix H, select the best one of these for this job. Specify the pump size, motor speed (rpm), and impeller diameter that you would use. Also determine the pump efficiency and NPSH requirement.

(c) What is the maximum height above the level of the slurry in the pit that the pump could be located without cavitating?

(d) A venturi meter is located in a vertical section of the line to monitor the slurry flow rate. The meter has a 4 in. diameter throat, and the pressure taps are 1 ft apart. If a DP cell (transducer) is used to measure the pressure difference between the taps, what would it read (in inches of water)?

(e) A 90° flanged elbow is located in the line at a point where the pressure (upstream of the elbow) is 10 psig. What are the forces transmitted to the pipe by the elbow from the fluid inside the elbow (neglect the weight of the fluid)?

(f) The classifier consists of three collection tanks in series that are full of water. The slurry enters at the top on the side of the first tank and leaves at the top on the opposite side, which is 5 ft from the entrance. The solids settle into the tank as the slurry flows into it and then overflows into the next tank. The space through which the slurry flows above the tank is 2 ft wide and 3 ft high. All particles for which the settling time in the space above the collection tank is less than the residence time of the fluid flowing in the space over the collection tank will be trapped in that tank. Determine the diameter of the largest particle that will not settle into each of the three collection tanks. Assume that the particles are equivalent spheres and that they fall at their terminal velocity.

(g) The suspension leaving the classifier is transferred to a rotary drum filter to remove the remaining solids. The drum operates at a constant pressure difference of 5 psi and rotates at a rate of 2 rpm with 20% of the surface submerged. Lab tests on a sample of the suspension through the same filter medium were conducted at a constant flow rate of 1 gpm through 0.25 ft^2 of the medium. It was found that the pressure drop increased to 2.5 psi after 10 min, and the resistance of the medium was negligible. How much filter area would be required to filter the liquid?

19. You want to concentrate a slurry from 5% (by vol.) solids to 30% (by vol.) in a thickener. The solids density is 200 lb_m/ft^3 and that of the liquid is 62.4 lb_m/ft^3. A batch settling test was run on the slurry, and the analysis of the tests yielded the following information:

φ (Vol. Fraction Solids)	Settling Rate (lb_m/h ft^2)
0.05	73.6
0.075	82.6
0.1	79.8
0.125	70.7
0.15	66
0.2	78
0.25	120
0.3	200

If the feed flow rate of the slurry is 500 gpm, what should the cross-sectional area of the thickener tank be? What are the overflow and underflow rates?

20. You must determine the maximum feed rate that a thickener can handle to concentrate a waste suspension from 5% solids by volume to 40% solids by volume. The thickener has a diameter of 40 ft. A batch flux test in the laboratory for the settled height versus time was analyzed to give the following data for the solids flux versus solids volume fraction. Determine (a) the proper feed rate of liquid in gpm, (b) the overflow liquid rate in gpm, and (c) the underflow liquid rate in gpm.

φ (Solids Volume Fraction)	Solids Flux (ft^3/h ft^2)
0.03	0.15
0.05	0.38
0.075	0.46
0.10	0.40
0.13	0.33
0.15	0.31
0.20	0.38
0.25	0.60
0.30	0.80

NOTATION

A	Area, $[L^2]$
A_p	Surface area of particle, $[L^2]$
A_s	Surface area of equal volume sphere, $[L^2]$
a_s	Particle surface area/volume, $[1/L]$
C_1, C_2	Constants, Equation 15.9, [—]
C_D	Drag coefficient, [—]
C_{DO}	Single-particle drag coefficient, [—]
$C_{D\varphi}$	Swarm drag coefficient, [—]
D	Container diameter, $[L]$
d	Particle diameter, $[L]$
d_p	Particle effective diameter, $[L]$
d_s	Equal volume sphere diameter, $[L]$
e_f	Energy dissipated per unit mass of fluid, $[F\,L/M = L^2/t^2]$
f_{PM}	Porous media friction factor, [—]

g	Acceleration due to gravity, $[L/t^2]$
h	Height of bed, $[L]$
N_{Ar}	Archimedes number, Equation 15.32, $[—]$
$N_{Ar\varphi}$	Swarm Archimedes number, Equation 15.37, $[—]$
N_{Rep}	Unconfined single-particle Reynolds number, $[—]$
$\hat{N}_{Re}$	Reynolds number defined by Equation 15.7, $[—]$
$N_{Re\varphi}$	Swarm Reynolds number, Equation 15.33, $[—]$
$N_{Re,PM}$	Porous media Reynolds number, Equation 15.3, $[—]$
n	Richardson–Zaki index, Equation 15.23, $[—]$
P	Pressure, $[F/L^2 = M/(L\ t^2)]$
Q_f	Slurry feed rate, $[L^3/t]$
Q_o	Overflow rate, $[L^3/t]$
Q_u	Solids underflow rate (thickener), $[L^3/t]$
q_s	Solids flux, $[L/t]$
t	Time, $[t]$
V	Velocity, $[L/t]$
V_o	Stokes velocity, $[L/t]$
V_p	Particle volume, $[L^3]$
V_r	Relative velocity between solid and fluid, Equation 15.20, $[L/t]$
$Z(t)$	Instantaneous height of liquid/suspension interface, $[L]$

GREEK

$\Delta(\ )$	$(\ )_2 - (\ )_1$
ε	Porosity or void fraction, $[—]$
μ	Viscosity, $[M/(L\ t)]$
φ	Volume fraction of solids, $[—]$
ρ	Density, $[M/L^3]$
ψ	Sphericity factor, $[—]$

SUBSCRIPTS

c	Critical point
i	Inlet
s	Superficial
f	Fluid, feed
L	Liquid
mf	Minimum fluidization condition
o	Infinitely dilute condition, overflow
p	Particle
s	Solid
u	Underflow
φ	Solid suspension of volume fraction φ

REFERENCES

Azbel, D.S. and N.P. Cheremisinoff, *Fluid Mechanics and Unit Operations*, Ann Arbor Science, Ann Arbor, MI, 1983.

Barnea, E. and J. Mizrahi, A generalized approach to the fluid dynamics of particulate systems, Part I, General correlation for fluidization and sedimentation in solid multi-particle systems, *Chem. Eng. J.*, 5, 171–189, 1973.

Chhabra, R.P., Rheology: From simple fluids to complex suspensions, in *Lignocellulosic Fibers and Wood Handbook*, N. Belgacem and A. Pizzi, (Eds.), pp. 407–438, Scrivener, New York, 2016.

Christian, J.B., Improve clarifier and thickener design and operation, *Chem. Eng. Prog.*, 90(7), 50–56, 1994.

Coulson, J.M., J.F. Richardson, J.R. Blackhurst, and J.H. Harker, *Chemical Engineering*, Vol. 2, 5th edn., Butterworth-Heinemann, Oxford, U.K., 2002.

Darby, R., Hydrodynamics of slurries and suspensions, Chapter 2, in *Encyclopedia of Fluid Mechanics*, Vol. 5, N.P. Cheremisinoff, (Ed.), Gulf, Houston, TX, 1986, pp. 49–92.

Davies, L., D. Dollimore, and G.B. McBride, Sedimentation of suspensions: simple methods of calculating sedimentation parameters, *Powder Technol.*, 16, 45–49, 1977.

Einstein, A., Eine neue Bestimmung der Molekul-dimensionen, *Ann. Phys.*, 19, 289–306, 1906.

Foust, A.S., L.A. Wenzel, C.W. Clump, L. Maus, and L.B. Anderson, *Principles of Unit Operations*, 2nd edn., Wiley, New York, 1980.

Kynch, G.J., A theory of sedimentation, *Trans. Faraday Soc.*, 48, 166, 1952.

McCabe, W.L., J.C. Smith, and P. Harriott, *Unit Operations of Chemical Engineering*, 5th edn., McGraw-Hill, New York, 1993.

Mewis, J. and N.J. Wagner, *Colloid Suspension Rheology*, Cambridge University Press, New York, 2012.

Millan, J.M.V., *Fluidization of Fine Powders: Cohesive versus Dynamical Aggregation*, Springer, New York, 2013.

Mooney, M., The viscosity of a concentrated suspension of spherical particles, *J. Colloid Sci.*, 6, 162–170, 1951.

Richardson, J.F. and W.N. Zaki, Sedimentation and fluidization, *Trans. Inst. Chem. Engrs.*, 32, 35, 1954.

Shook, C.A. and M.C. Roco, *Slurry Flow: Principles and Applications*, Butterworth-Heinemann, Oxford, U.K., 1991.

Steinour, H.H., Rate of sedimentation, *Ind. Eng. Chem.*, 36, 618, 840, 901, 1944.

Tiller, F.M. and D. Tarng, Try deep thickeners and clarifiers, *Chem. Eng. Prog.*, 91, 75–80, March 1995.

Vand, V., Viscosity of solutions and suspensions, *J. Phys. Colloid Chem.*, 52, 277–299, 1948.

Wilson, K.C., G.R. Addie, A. Sellgren, and R. Clift, *Slurry Transport Using Centrifugal Pumps*, 3rd edn., Springer, New York, 2008.

Yang, W.-C., *Handbook of Fluidization and Fluid-Particle Systems*, Taylor & Francis, New York, 2003.

16 Two-Phase Flow

"For every simple problem, there is a complex solution that is wrong."

—Anonymous

I. SCOPE

The term "two-phase flow" covers an extremely broad range of situations, and it is possible to address only a small portion of this spectrum in one book, let alone one chapter. A "two-phase flow" includes any combination of two of the three phases: solid, liquid, and gas, that is, solid–liquid, gas–liquid, solid–gas, or liquid–liquid. Also, if both phases are fluids (combinations of liquid and/or gas), one of the phases may be continuous and the other distributed (e.g., gas in liquid or liquid in gas). Furthermore, the mass ratio of the two phases may be fixed or variable throughout the system. Examples of the former are nonvolatile liquids with entrained solids or noncondensable gases, whereas examples of the latter are flashing liquids, soluble solids in liquids, and partly miscible liquids in liquids. Additional situations are encountered in boilers and condensers in which phase change occurs. One can also encounter three phases like in slurry reactors, three phase fluidized beds, etc.

In addition, in pipe flows, the two phases may be highly mixed and uniformly distributed over the cross section (i.e., "homogeneous") or they may be "separated" into two distinct regions, and the conditions under which these states prevail are different for horizontal than for vertical flow and depend upon the type of system, for example, gas–liquid, gas–solid, solid–liquid, liquid–liquid, or even gas–liquid–solid. For uniformly distributed homogeneous flows, the fluid can be described as a "pseudo-single-phase fluid," with properties of the mixture being average values of the two phases over the flow cross section. Such flows can be described as "1-D," as opposed to "separated" or heterogeneous, in which the phase distribution varies over the cross section.

In this chapter, we will focus on two-phase flows in pipes, which includes the transport of solids as slurries and suspensions in a continuous liquid phase, pneumatic transport of solid particles in a continuous gas phase, and mixtures of gas or vapor with liquids in which either phase may be continuous. Although it may appear that only one additional "variable" is added to the single-phase problems previously considered, the complexity of two-phase flows is indeed greater by orders of magnitude. It is emphasized that this is only an introduction and literally thousands of articles abound in the literature on various aspects of two-phase flows. It should also be realized that if these problems were simple or straightforward, the number of papers required to describe them would be orders of magnitude smaller. Detailed treatments can also be found in several excellent books including Wallis (1969), Govier and Aziz (1983), Hetsroni (1982), Levy (1999) and Michaelides et al. (2016).

II. DEFINITIONS

Before proceeding, it is appropriate to define the terminology for the various flow rates, velocities, and concentrations for two-phase flows. There is a bewildering variety of notation in the literature relative to two-phase flow, and we will attempt to use a notation that is consistent with the following definitions for solid–liquid, solid–gas, and liquid–gas systems.

The subscripts m, L, S, and G represent the local two-phase mixture, liquid phase, solid phase, and gas phase, respectively. The definitions in the following text are given in terms of solid–liquid (S-L)

mixtures, where the solid is the more dense distributed phase and the liquid the less dense continuous phase. The same definitions can be applied to gas–liquid (*G-L*) flows in which the gas is the continuous phase and the liquid the distributed phase if the subscript S is replaced by L (the more dense distributed phase) and L by G (the less dense continuous phase). The symbol φ is used for the volume fraction of the distributed (dense) phase and ε is the volume fraction of the continuous phase (obviously $\varphi = 1 - \varepsilon$). (For gas–liquid flows, the volume fraction of gas is often denoted by α (i.e., $\alpha = \varepsilon$).

An important distinction is made between (φ, ε) and (φ_m, ε_m). The former (φ, ε) refers to the overall flow-average (equilibrium) values entering the pipe, that is,

$$\varphi = \frac{Q_S}{Q_S + Q_L} = 1 - \varepsilon \tag{16.1}$$

whereas the latter (φ_m, ε_m) refer to the local values at a given position in the pipe. These are different ($\varphi_m \neq \varphi$, $\varepsilon_m \neq \varepsilon$) when the local velocities of the two phases are not the same (i.e., *slip* is significant), as will be shown later.

Mass flow rate ($\dot{m}$) and volume flow rate (Q):

$$\dot{m}_m = \dot{m}_S + \dot{m}_L = \rho_S Q_S + \rho_L Q_L = \rho_m Q_m \tag{16.2}$$

Mass flux (G):

$$G_m = G_S + G_L = \frac{\dot{m}_S + \dot{m}_L}{A} \tag{16.3}$$

where A is the conduit cross-sectional area.

Volume flux:

$$J_m = J_S + J_L = \frac{G_m}{\rho_m} = \frac{G_S}{\rho_S} + \frac{G_L}{\rho_L} = \frac{Q_S + Q_L}{A} = V_m \tag{16.4}$$

Phase velocity:

$$V_S = \frac{J_S}{\varphi} = \frac{J_S}{1 - \varepsilon}, \quad V_L = \frac{J_L}{\varepsilon} = \frac{J_L}{1 - \varphi} \tag{16.5}$$

Relative (slip) velocity and slip ratio:

$$V_r = V_L - V_S, \quad S = \frac{V_L}{V_S} \tag{16.6}$$

Note that the total *volume flux (J_m)* of the mixture is the same as the *superficial velocity, V_m* (the volumetric flow rate divided by the total flow area). However, the local velocity of each phase (V_i) is greater than the volume flux of that phase (J_i), since each phase occupies only a fraction of the total flow area. This is akin to the interstitial velocity versus the superficial velocity in the context of flow in packed beds and porous media (Chapter 14). The volume flux of each phase is the total volume flow rate of that phase divided by the total flow area.

The relative slip velocity (or slip ratio) is an extremely important variable. This occurs primarily when the distributed phase density is greater than that of the continuous phase and the heavier distributed phase tends to lag behind the lighter phase for various reasons (explained later). The resulting relative velocity (slip) between the phases determines the drag exerted by the continuous (lighter) phase on the distributed (heavier) phase. One consequence of slip (as shown later) is that the in situ concentration or "holdup" of the more dense phase within the pipe (φ_m) is greater than that entering or leaving the pipe, since its residence time is longer than that of the continuous phase. Consequently, the concentrations and local phase velocities within a pipe under slip conditions depend upon the properties and degree of interaction of the phases and cannot be determined solely from the knowledge of the entering and leaving concentrations and flow rates. Slip can only be determined indirectly by measurement of some local flow property within the pipe, such as the holdup, concentration profiles, local phase velocity, or the local mixture density.

For example, for transport of solid particles by a liquid, if φ is the solids volume fraction entering the pipe at velocity V and φ_m is the local volume fraction in the pipe where the solid velocity is V_S, a component balance shows

$$\frac{V_S}{V} = \frac{\varphi}{\varphi_m} \quad \text{and} \quad \frac{V_L}{V} = \frac{1-\varphi}{1-\varphi_m} \tag{16.7}$$

Substituting these expressions in the definition of the slip velocity, Equation 16.6, and dividing by the inlet velocity (V) to make the results dimensionless gives

$$\bar{V}_r = \frac{V_r}{V} = \frac{V_L - V_S}{V_L + V_S} = \frac{S-1}{S+1} = \frac{\varphi_m - \varphi}{\varphi_m(1-\varphi_m)} \tag{16.8}$$

This can be solved for φ_m in terms of $\bar{V}_r$ and φ:

$$\varphi_m = \frac{1}{2}\left\{1 - \frac{1}{\bar{V}_r} + \left[\left(\frac{1}{\bar{V}_r} - 1\right)^2 + \frac{4\varphi}{\bar{V}_r}\right]^{1/2}\right\} \tag{16.9}$$

As an example, if the entering solids fraction φ is 0.4, the corresponding values of the local solids fraction φ_m for relative slip velocities $(\bar{V}_r)$ of 0.01, 0.1, and 0.5 are 0.403, 0.424, and 0.525, respectively. There are many "theoretical" expressions for predicting the slip velocity, but practical applications depend on experimental observations and correlations (which will be presented later). In gas–liquid or gas–solid flows, φ_m will vary along the pipe since the gas expands as the pressure drops and speeds up as it expands, which increases the slip. This, in turn, increases the holdup of the denser phase (more on slip and holdup later).

The mass fraction (x) of the less dense phase (which, for gas–liquid flows, is called the *quality*) is $x = \dot{m}_L/(\dot{m}_S + \dot{m}_L)$, so that the mass flow ratio can be written as

$$\frac{\dot{m}_L}{\dot{m}_S} = \frac{x}{1-x} = \frac{\rho_L V_L A \varepsilon_m}{\rho_S V_S A(1-\varepsilon_m)} = S\left(\frac{\rho_L}{\rho_S}\right)\left(\frac{\varepsilon_m}{1-\varepsilon_m}\right) \tag{16.10}$$

This can be rearranged to give the less dense phase volume fraction (ε_m or α) in terms of the mass fraction and slip ratio:

$$\varepsilon_m = \frac{x}{x + S(1-x)\rho_L/\rho_S} \tag{16.11}$$

The local density of the mixture is given by

$$\rho_m = \varepsilon_m \rho_L + (1-\varepsilon_m)\rho_S \tag{16.12}$$

which depends on the slip ratio S through Equation 16.11. The corresponding expression for the local in-situ holdup of the denser phase is

$$\varphi_m = (1-\varepsilon_m) = \frac{S(1-x)(\rho_L/\rho_S)}{x + S(1-x)(\rho_L/\rho_S)} \tag{16.13}$$

Note that both the local mixture density and holdup increase as the slip ratio (S) increases. The "no-slip" $(S = 1)$ density or volume fraction is identical to the equilibrium value entering (or leaving) the pipe.

III.   FLUID–SOLID TWO-PHASE PIPE FLOWS

The conveying of solids by a fluid in a pipe involves a wide range of flow conditions and phase distributions, depending on the density, viscosity, and velocity of the fluid; the density, size, shape, and concentration of the solid particles; or the orientation of the pipe (vertical, horizontal, or inclined). The flow regime can vary from essentially uniformly distributed solids in a "pseudohomogeneous" (symmetric) flow regime for sufficiently small and/or light particles and/or a very high mixture velocity, to an almost completely segregated or stratified (asymmetric) transport of a bed of particles along the pipe wall. The demarcation between the "homogeneous" and "heterogeneous" flow regimes depends in a complex manner on the size and density of the solids, the fluid density and viscosity, the velocity of the mixture, and the volume fraction of solids. Furthermore, the transition from one flow regime to another is far from being sharp. Figure 16.1 illustrates the approximate effect of particle size, density, and solids loading on these regimes.

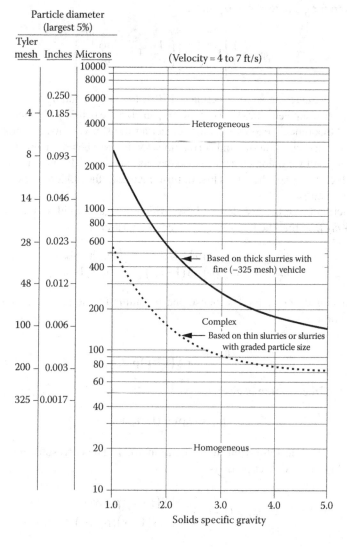

FIGURE 16.1   Approximate slurry flow regimes. (From Aude, T.C. et al., Slurry piping systems: Trends, design methods, guidelines, *Chem. Eng.*, 74, June 28, 1971.)

Either a liquid or a gas can be used as the carrier fluid, depending on the size and properties of the particles. However, there are important differences between hydraulic (liquid) and pneumatic (gas) transport. For example, in hydraulic (liquid) transport, fluid–particle and particle–particle interactions dominate over the particle–wall interactions, whereas in gas (pneumatic) transport, the particle–particle and particle–wall interactions tend to dominate over the fluid–particle interactions. A typical "practical" approach, which gives reasonable results for a wide variety of flow conditions in both cases, is to determine the "fluid only" pressure drop and then apply a correction to account for the effect of the particles from the fluid–particle, particle–particle, and/or particle–wall interactions. A great number of studies have been devoted to this subject, and summaries of much of this work have been given by Darby (1986), Govier and Aziz (1983), Klinzing et al. (2010), Molerus (1993), Wasp et al. (1977), Brown and Heywood (1991), Shook and Roco (1991) and Wilson et al. (2008). This approach will be described later.

A. Pseudohomogeneous Flows

If the solid particles are very small (e.g., typically less than 100 μm), and/or not greatly denser than the fluid, and/or the flow is highly turbulent, the mixture may behave as a uniform suspension with essentially continuous properties with no slip between the two phases. In this case, the mixture can be described as a "pseudo-single-phase" uniform fluid, and the effect of the presence of the particles can be accounted for by appropriate modification of the fluid properties (density and viscosity). For relatively dilute suspensions (e.g., 5% by volume or less), the mixture will behave as a Newtonian fluid with a viscosity given by the Einstein equation:

$$\mu = \mu_L \left(1 + 2.5\varphi\right) \tag{16.14}$$

where

μ_L is the viscosity of the suspending (continuous) Newtonian fluid
$\varphi = (1 - \varepsilon)$ is the volume fraction of solids

and the density of the mixture is given by

$$\rho_m = \rho_L \left(1 - \varphi\right) + \rho_S \varphi \tag{16.15}$$

For greater concentrations of fine particles, the suspension is more likely to be non-Newtonian, in which case the viscous properties can probably be adequately described by the power law or Bingham plastic (preferred) models. The pressure drop–flow relationship for pipe flow under these conditions can be determined by the methods presented in Chapters 6 and 7.

B. Heterogeneous Liquid–Solid Flows

Figure 16.2 shows how the pressure gradient and flow regimes in a horizontal pipe depend upon the velocity for a typical heterogeneous suspension. It is seen that the pressure gradient exhibits a minimum at the "minimum deposit velocity," which is the velocity at which a significant amount of solids begin to settle in the pipe. Under these conditions, most of the particles are transported in the form of a sliding or moving bed along the pipe wall. This not only leads to a high pressure gradient but it can also cause severe erosion and wear of the pipe. A variety of correlations have been proposed in the literature for the prediction of the minimum deposit velocity, one of the more useful being that of Hanks (1980):

$$V_{md} = 1.32\varphi^{0.186} \left[2gD\left(s-1\right)\right]^{1/2} \left(\frac{d}{D}\right)^{1.6} \tag{16.16}$$

where $s = \rho_S/\rho_L$. At velocities below this value, the solids settle out and form a bed along the bottom of the pipe. This bed can build up and eventually plug the pipe if the velocity is too low, or it can be swept along the pipe wall if the velocity is near the minimum deposit velocity.

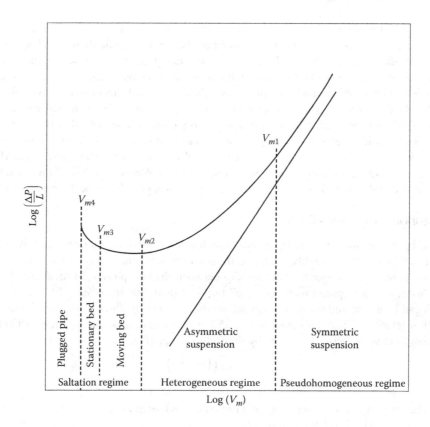

FIGURE 16.2   Pressure gradient and flow regimes for slurry flow in a horizontal pipe.

Above the minimum deposit velocity, the particles are suspended but are not uniformly distributed ("symmetric") until the turbulent mixing is high enough to overcome the settling forces and/or can resuspend at least some of the settled particles. One criterion for a nonsettling suspension has been given by Wasp et al. (1977):

$$\frac{V_t}{V^*} \le 0.022 \tag{16.17}$$

where
V_t is the particle terminal velocity in a quiescent fluid
V^* is the *friction velocity*, defined as

$$V^* = \sqrt{\frac{\tau_w}{\rho}} = \sqrt{\frac{\Delta P D}{4 \rho L}} \tag{16.18}$$

For heterogeneous flow, one approach to determining the pressure drop in a pipe is to compute separately the contributions of the fluid and the solids at the same velocity:

$$\Delta P_m = \Delta P_L + \Delta P_S \tag{16.19}$$

where
ΔP_L is the "fluid only" pressure drop
ΔP_S is an additional pressure drop due to the presence of the solids

For uniform sized particles in a Newtonian liquid, ΔP_L is determined as for any Newtonian fluid in a pipe at the same flow rate as that of the mixture. For a broad particle size distribution, the suspension may behave more like a heterogeneous suspension of the larger particles in a carrier vehicle composed of a homogeneous suspension of the finer particles. In this case, the homogeneous carrier will likely be non-Newtonian, and the methods given in Chapter 6 for such fluids should be used to determine ΔP_L.

The procedure for determining ΔP_S that will be presented here is that of Molerus (1993). The basis of the method is a consideration of the extra energy dissipated in the flow as a result of the fluid–particle interaction. This is characterized by the particle terminal settling velocity in an infinite fluid in terms of the drag coefficient, C_d:

$$C_d = \frac{4g(s-1)d}{3V_t^2} \tag{16.20}$$

where $s = \rho_S/\rho_L$. Molerus (1993) applied dimensional analysis to the variables in this system, along with energy dissipation considerations, to arrive at the following dimensionless groups:

$$\frac{V_r}{V}\left(\frac{1}{s}\right)^{1/2} = \frac{\bar{V}_r}{\sqrt{s}} \tag{16.21}$$

$$N_{Frp}^2 = \frac{V^2}{(s-1)dg} \tag{16.22}$$

$$N_{Frt}^2 = \frac{V_t^2}{(s-1)Dg} \tag{16.23}$$

where
  V is the overall average velocity in the pipe
  $V_r = V_L - V_S$ is the relative ("slip") velocity between the fluid and the solid
  V_t is the terminal velocity of the solid particle
  d is the particle diameter
  D is the tube diameter
  N_{Frp} is the particle Froude number
  N_{Frt} is the tube Froude number

The slip velocity is the key parameter in the mechanism of transport and energy dissipation because the drag force exerted by the fluid on the particle depends on the *relative* velocity between the fluid and particles. That is, the fluid must move faster than the particles in order to carry them along the pipe. The particle terminal velocity is related to the particle drag coefficient and Reynolds number, as discussed in Chapter 12 (e.g., unknown velocity), for either a Newtonian or non-Newtonian carrier medium.

Molerus (1993) developed a "state diagram" that shows a correlation between these dimensionless groups based on an extremely wide range of data covering $25 < D < 315$ mm, $12 < d < 5200$ μm, and $1270 < \rho_S < 5250$ kg/m³ for both hydraulic and pneumatic transport. This state diagram is shown in Figure 16.3 in the following form:

$$\frac{\bar{V}_{ro}}{\sqrt{s}} = fn\left(\sqrt{s}N_{Frp}, N_{Frt}^2\right) \tag{16.24}$$

where $\bar{V}_{ro}$ is the dimensionless "single-particle" slip velocity as determined from the diagram,

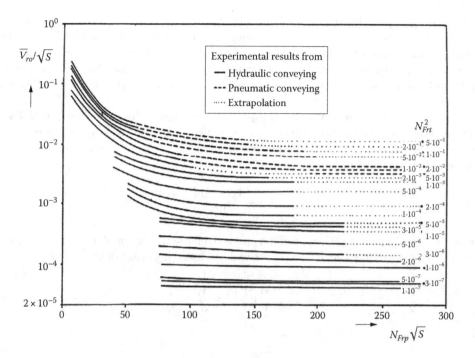

FIGURE 16.3   State diagram for suspension transport. (From Molerus, O., *Principles of Flow in Disperse Systems*, Chapman & Hall, London, U.K., 1993.)

which, in turn, is used to define the parameter

$$X_o = \frac{\bar{V}_{ro}^2}{1 - \bar{V}_{ro}}$$ (16.25)

Using this value of X_o and the entering solids volume fraction (φ), a value of X is determined as follows:

For $0 < \varphi < 0.25$	$X = X_o$
For $\varphi > 0.25$	$X = X_o + 0.1 N_{Frt}^2 (\varphi - 0.25)$

The parameter X is the dimensionless solids contribution to the pressure drop:

$$X = \frac{\Delta P_S}{\varphi \rho_L g L (s - 1)} \left(\frac{V_t}{V} \right)^2$$ (16.26)

Knowing X determines ΔP_S, which is added to ΔP_L to get the total pressure drop in the pipe. This procedure is straightforward if all the particles are of the same diameter (d). However, if the solid particles cover a broad range of sizes, the procedure must be applied for each particle size (diameter d_i, with concentration φ_i) to determine the corresponding contribution of that particle size to the pressure drop ΔP_{Si}. The total solids contribution to the pressure drop is then $\Sigma \Delta P_{Si}$. If the carrier vehicle exhibits non-Newtonian properties with a yield stress, particles for which $d \leq 5 \, \tau_o/(g \Delta \rho)$ (approximately) will not fall at all.

For vertical transport, the major difference is that no "bed" can form on the pipe wall but, instead, the pressure gradient must overcome the weight of the solids as well as the fluid/particle drag. Thus, the solids holdup and hence the fluid velocity are significantly higher for vertical transport conditions than for horizontal transport. However, vertical flow of slurries and suspensions is generally avoided where possible due to the much greater possibility of plugging if the velocity falls.

Example 16.1

Determine the pressure gradient (in *psi/ft*) required to transport a slurry at 300 gpm through a 4 in. sch 40 pipeline. The slurry contains 50% (by weight) solids (*SG* of 2.5) in water. The slurry contains a "bimodal" particular size distribution, with half of the particles below 100 μm and the other half about 2000 μm. The suspension of fines is stable and constitutes a pseudohomogeneous non-Newtonian vehicle in which the larger particles are suspended. The vehicle can be described as a Bingham plastic, with a limiting viscosity of 30 cP and a yield stress of 55 dyn/cm².

Solution

First, convert the mass fraction of solids to a volume fraction using $x = 0.5$ and $s = 2.5$:

$$\varphi = \frac{x}{[s - (s-1)x]} = 0.286$$

where $s = \rho_S/\rho_L$. Half of the solids are in the non-Newtonian "vehicle" and half will be "settling," with a volume fraction of 0.143. Thus, the density of the "vehicle" is

$$\rho_m = \rho_S \, \varphi + \rho_L(1 - \varphi) = 1.215 \text{ g/cc}$$

Then calculate the contribution to the pressure gradient due to the continuous Bingham plastic vehicle, as well as the contribution from the "nonhomogeneous" solids. For the first part, we use the method presented in Chapter 6, Section V.C, for Bingham plastics. From the given data, we can calculate $N_{Re,BP} = 9,540$ and $N_{He} = 77,600$. From Equation 6.62, this gives a friction factor of $f = 0.0629$ and a corresponding pressure gradient of $(\Delta P/L)_f = 2fpV^2/D = 1.105$ psi/ft.

The pressure gradient due to the heterogeneous component is determined by the Molerus method. This first requires the determination of the terminal velocity of the settling particles, which is done using the method given in Chapter 12, Section IV.D, for the larger particles settling in a Bingham plastic. This requires determining $N_{Re,BP}$, N_{Bi}, and C_d for the particle, all of which depend on V_t. This can be done using an iterative procedure to find V_t, such as the "solve" function on a calculator or spreadsheet. The result is $V_t = 19.5$ cm/s. This is used to calculate the particle and tube Froude numbers, $N_{Frp}^2 = 25.6$ and $N_{Frt}^2 = 0.0358$. These values are used with Figure 16.3 to find $(\bar{V}_r/\sqrt{s}) = 0.05$, which corresponds to a value of $X = 0.00279$. From the definition of X, this gives $(\Delta P/L)_s = 0.0312$ psi/ft and thus a total pressure gradient of $(\Delta P/L)_t = 1.14$ psi/ft. In this case, the pressure drop due to the Bingham plastic "vehicle" is much greater than that due to the heterogeneous particle contribution, which may be due to the relatively low concentration of coarse particles.

C. PNEUMATIC SOLIDS TRANSPORT

The transport of solid particles by a gaseous medium provides a considerable challenge, since the solid is typically about three orders of magnitude more dense than the fluid (as compared with hydraulic transport, in which the solid and liquid densities normally differ by less than an order of magnitude). Hence, the problems that may be associated with instability in hydraulic conveying are greatly magnified in the case of pneumatic conveying. The complete design of a pneumatic conveying system requires proper attention to the prime mover (i.e., a fan, blower, or compressor); the feeding, mixing, and accelerating conditions and equipment; and the downstream separation equipment as well as the conveying system. A complete description of such a system is beyond the scope of this book, and the interested reader should consult the more specialized literature in the field, such as the extensive treatise of Fan and Zhu (1998) or Klinzing et al. (2010).

One major difference between pneumatic transport and hydraulic transport, as noted earlier, is that the gas–solid interaction for pneumatic transport is generally much weaker than the particle–particle and particle–wall interactions. There are two primary modes of pneumatic transport: *dense phase* and *dilute phase*. In the former, the transport occurs below the *saltation velocity* (which is roughly equivalent to the minimum deposit velocity) in plug flow, dune flow, or sliding bed flow. Dilute phase transport occurs above the saltation velocity in suspended flow. The saltation velocity is not the same as the entrainment or "pick up" velocity, however, which is approximately 50% greater than the saltation velocity. The pressure gradient–velocity relation is similar to that for hydraulic transport, as shown in Figure 16.4, except that transport is possible in the dense phase in which the pressure gradient, though quite large, is still usually not as large as for hydraulic transport. The entire curve shifts up and to the right as the solids mass flux increases. A comparison of typical operating conditions for dilute phase and dense phase pneumatic transport is shown in Table 16.1.

Although a lot of information is available on dilute phase transport, which is useful for designing such systems, transport in the dense phase is much more difficult and more sensitive to detailed properties of

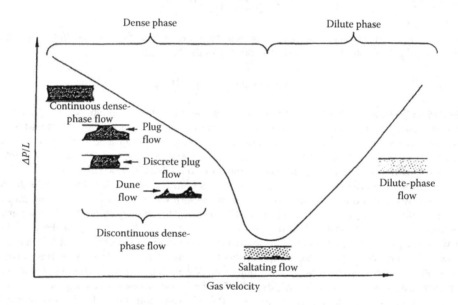

FIGURE 16.4 Pressure gradient–velocity relation for horizontal pneumatic flow.

TABLE 16.1
Dilute Phase versus Dense Phase Pneumatic Transport

Conveying Mode	Solids Loading (e.g., kg$_s$/kg$_g$)	Conveying Velocity ft/s (m/s)	ΔP psi (kPa)	Solids Volume Fraction
Dilute phase	<15	>35 (10)	<15 (100)	<1%
Dense phase	>15	<35 (10)	>15 (100)	>30%

the specific solids. Thus, because operating experimental data on the particular materials of interest are usually needed for dense phase transport, we will limit our treatment here to the dilute phase.

There are a variety of correlations for the saltation velocity, one of the most popular being that of Rizk (1973):

$$\mu_s = \frac{\dot{m}_S}{\dot{m}_G} = 10^{-\delta} N_{Frs}^{\chi} \tag{16.27}$$

where

$$\delta = 1.44d + 1.96$$

$$\chi = 1.1d + 2.5$$

and

$$N_{Frs} = \frac{V_{gs}}{\sqrt{gD}} \tag{16.28}$$

where
   μ_s is the "solids loading" (mass of solids/mass of gas)
   V_{gs} is the saltation gas velocity
   d is the particle diameter (mm)
   D is the pipe diameter

It should be pointed out that correlations such as this are based, of necessity, on a finite range of conditions and have a relatively broad range of uncertainty, for example, $\pm 50\%$–60% is not unusual (see, e.g., the recent work of Wilms and Dhodapkar, 2014).

1. Horizontal Transport

The two major effects that contribute to the pressure drop in horizontal flow are acceleration of the particles and friction loss. Initially, the inertia of the particles must be overcome as they are accelerated up to speed, and then the friction loss in the mixture must be overcome. If V_S is the solid particle velocity and $\dot{m}_S = \rho_S V_S A(1 - \varepsilon_m)$ is the solids mass flow rate, the acceleration component of the pressure drop is

$$\Delta P_{acS} + \Delta P_{acG} = V_S \frac{\dot{m}_S}{A} + \frac{\rho_G V_G^2}{2} = \frac{\rho_G V_G^2}{2} \left(1 + 2\frac{\dot{m}_S V_S}{\dot{m}_G V_G} \right) \tag{16.29}$$

The slip ratio $V_G/V_S = S$ can be estimated, for example, from the IGT correlation (e.g., Klinzing et al., 2010):

$$\frac{1}{S} = \frac{V_S}{V_G} = 1 - \frac{0.68 d^{0.92} \rho_S^{0.5}}{\rho_G^{0.2} D^{0.54}} \tag{16.30}$$

in which
   d and D are the particle and pipe diameters, respectively, in meters
   ρ_S and ρ_G are in kg/m^3

For vertical transport, the major differences are that no "bed" on the pipe wall is possible but, instead, the pressure gradient must overcome the weight of the solids as well as the fluid and the fluid/particle drag, so that the solids holdup and hence the fluid velocity must be significantly higher under transport conditions. The steady flow pressure drop in the pipe can be deduced from a momentum balance on a differential slice of the fluid–particle mixture in a constant diameter pipe, as was done in Chapter 5 for single-phase flow (see Figure E5.4). For steady uniform flow through area A_x:

$$\sum F_x = 0 = dF_{xp} + dF_{xg} + dF_{xw}$$

$$= -A_x dP - \left[\rho_S(1-\varepsilon_m) + \rho_G \varepsilon_m\right] g A_x dz - (\tau_{wS} + \tau_{wG}) W_p dX \tag{16.31}$$

where
   τ_{wS} and τ_{wG} are the effective wall stresses resulting from energy dissipation due to the particle–particle and particle–wall and gas–wall interactions
   W_p is the wetted perimeter

Dividing by A_x, integrating, and solving for the pressure drop, $-\Delta P = P_1 - P_2$, gives

$$-\Delta P = \left[\rho_S(1-\varepsilon_m) + \rho_G \varepsilon_m\right] g \Delta z + (\tau_{wS} + \tau_{wG}) 4L/D_h \tag{16.32}$$

where $D_h = 4A_x/W_p$ is the hydraulic diameter. The void fraction ε_m is the volume fraction of the gas in the pipe, i.e.,

$$\varepsilon_m = 1 - \frac{\dot{m}_S}{\rho_S V_S A} = \frac{x}{x + S(1-x)\rho_G/\rho_S} \tag{16.33}$$

The wall stresses are related to the corresponding friction factors by

$$\tau_{wS} = \frac{\rho_S f_S}{2}(1-\varepsilon_m) V_S^2 = \frac{\Delta P_{fS}}{4L/D_h}, \quad \tau_{wG} = \frac{\rho_G f_G}{2}\varepsilon_m V_G^2 = \frac{\Delta P_{fG}}{4L/D_h} \tag{16.34}$$

Here ΔP_{fG} is the pressure drop due to "gas only" flow (i.e., the gas flowing alone in the full pipe cross section). Note that if the pressure drop is less than about 30% of P_1, the incompressible flow equations can be used to determine ΔP_{fG} using the average gas density. Otherwise, the compressibility must be considered and the methods outlined in Chapter 9 must be used to determine ΔP_{fG}. The pressure drop is related to the pressure ratio P_1/P_2 by

$$P_1 - P_2 = \left(1 - \frac{P_2}{P_1}\right) P_1 \tag{16.35}$$

   The solids contribution to the pressure drop ΔP_{fS} includes contributions from both the particle–wall and the particle–particle interactions. The latter is reflected in the dependence of the friction factor f_S on the particle diameter, along with the drag coefficient, density, and the relative (slip) velocity (Hinkle, 1953):

$$f_S = \frac{3}{8}\left(\frac{\rho_G}{\rho_S}\right)\left(\frac{D}{d}\right) C_d \left(\frac{V_G - V_S}{V_S}\right)^2 \tag{16.36}$$

A variety of other expressions for f_S have been proposed by various authors (see, e.g., Klinzing et al., 2010) and that of Yang (1983) for horizontal flow

$$f_S = 0.117\left(\frac{1-\varepsilon}{\varepsilon^3}\right)\left[(1-\varepsilon)\frac{N_{Ret}}{N_{Rep}}\frac{V_G}{\varepsilon\sqrt{gD}}\right]^{-1.15} \tag{16.37}$$

and for vertical flow

$$f_S = 0.0206\left(\frac{1-\varepsilon}{\varepsilon^3}\right)\left[(1-\varepsilon)\frac{N_{Ret}}{N_{Rep}}\right]^{-0.869}$$ (16.38)

where

$$N_{Ret} = \frac{dV_t\rho_G}{\mu_G}, \quad N_{Rep} = \frac{d(V_G/\varepsilon - V_S)\rho_G}{\mu_G}$$ (16.39)

V_t is the particle terminal velocity.

2. Vertical Transport

The principles governing vertical pneumatic transport are the same as those given earlier, and the method for determining the pressure drop is identical (with an appropriate expression for f_S). However, there is one major difference in vertical transport, which occurs as the gas velocity is decreased. As the velocity falls, the frictional pressure drop decreases but the slip increases since the drag force exerted by the gas entraining the particles also decreases. The result is an increase in the solids holdup, with a corresponding increase in the static head opposing the flow, which in turn causes an increase in the pressure drop. A point will be reached at which the gas can no longer entrain all of the solids and a slugging fluidized bed results with large pressure fluctuations. This condition is known as *choking* (not to be confused with choking that occurs when the gas velocity reaches the speed of sound) and represents the lowest gas velocity at which vertical pneumatic transport can be attained at a specified solids mass flow rate. The choking velocity, V_c, and the corresponding void fraction, ε_c, are related by the following two equations (Yang, 1983):

$$\frac{V_c}{V_t} = 1 + \frac{V_S}{V_t(1-\varepsilon_c)}$$ (16.40)

and

$$\frac{2gD(\varepsilon_c^{-4.7}-1)}{(V_c-V_t)^2} = 6.81\times10^5\left(\frac{\rho_o}{\rho_S}\right)^{2.2}$$ (16.41)

where ρ_o is the gas phase density at upstream conditions. These two equations must be solved simultaneously for V_c and ε_c.

IV. GAS–LIQUID TWO-PHASE PIPE FLOW

The two-phase flow of gases and liquids has been the subject of literally thousands of publications in the literature, and it is clear that we can only provide a brief introduction to the subject here. Although the single-phase flow of liquids and gases is relatively straightforward, the two-phase combined flow can be significantly more complex. Two-phase gas–liquid flows are also more complex than fluid–solid flows because of the wider variety of possible flow regimes and the possibility that the liquid may be volatile and/or the gas a condensable vapor, with the result that the mass ratio of the two phases may change throughout the system.

A. FLOW REGIMES

The configuration or distribution of the two phases in a pipe depends upon the phase ratio and the relative velocities of the two phases. These regimes can be described qualitatively as illustrated in Figure 16.5a for horizontal flow and Figure 16.5b for vertical flow. The patterns for horizontal flow are seen to be more complex than for vertical flow because of the asymmetric effect of gravity. The boundaries or transitions between these regimes have been mapped by various investigators based on observations in terms of various flow and property parameters. A number of these maps have been compared by Rouhani and Sohal (1983). Typical flow regime maps for horizontal and vertical flow are shown in Figure 16.6a and b. In Figures 16.5 and 16.6,

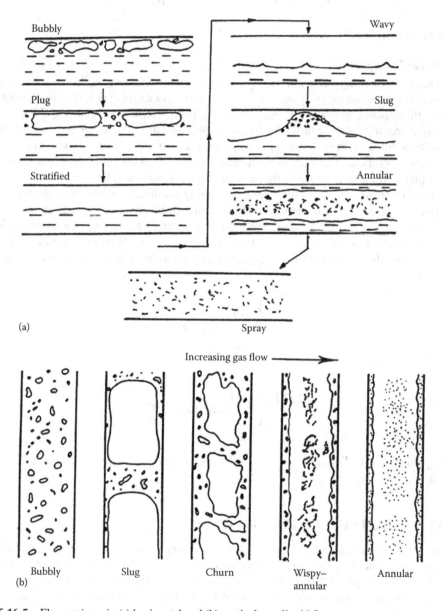

FIGURE 16.5 Flow regimes in (a) horizontal and (b) vertical gas–liquid flow.

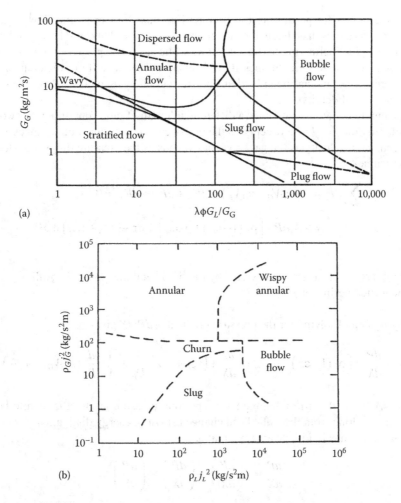

FIGURE 16.6 Flow regimes map for (a) horizontal and (b) vertical gas–liquid flow. (a: From Baker, O., *Oil Gas J.*, 53, 185, 1954; b: From Hewitt, G.F. and Roberts, D.N., Studies of two-phase flow patterns by x-ray and flash photography, Report AERE-M 2159, HMSO, London, U.K., 1969.)

$G_G = \dot{m}_G/A$ is the mass flux of the gas, $G_L = \dot{m}_L/A$ is the mass flux of the liquid, and λ and Φ are fluid property correction factors:

$$\lambda = \left(\frac{\rho_G}{\rho_A} \frac{\rho_L}{\rho_W} \right)^{1/2} \tag{16.42}$$

$$\Phi = \frac{\sigma_W}{\sigma_L} \frac{\rho_W}{\rho_L} \left(\frac{\mu_L}{\mu_W} \right)^{1/2} \tag{16.43}$$

where
  σ is surface tension
  W and A subscripts refer to water and air, respectively, at 20°C

A quantitative model for predicting the flow regime map for horizontal flow in terms of five dimensionless variables has been developed by Taitel and Dukler (1976) and was further extended to any pipe inclination by Barnea (1987). The approach proposed in this work, referred to as "mechanistic models," combines fluid flow equations describing different flow patterns, with closing relations, and has been further improved upon by many researchers (see, e.g., Zhang (2003) for additional details).

The momentum equation written for a differential length of pipe containing the two-phase mixture is similar to Equation 16.31, except that the rate of change of momentum changes along the tube due to the change in velocity as the gas or vapor expands as the pressure drops. For steady uniform flow through area A_x:

$$\sum dF_x = d\left(\dot{m}_G V_G + \dot{m}_L V_L\right) = dF_{xp} + dF_{xg} + dF_{xw} = 0$$

$$= -A_x dP - \left[\rho_L\left(1-\varepsilon_m\right)+\rho_G\varepsilon_m\right]A_x dz - \left(\tau_{wL}+\tau_{wG}\right)W_p dX \qquad (16.44)$$

where

τ_{wL} and τ_{wG} are the shear stresses exerted by the liquid and the gas on the wall
W_p is the wetted perimeter

Dividing by $A_x\, dx$ and solving for the pressure gradient, $-dP/dX$ gives

$$-\frac{dP}{dX} = \left[\rho_L\left(1-\varepsilon_m\right)+\rho_G\varepsilon_m\right]g\frac{dz}{dX} + \left(\tau_{wL}+\tau_{wG}\right)\frac{4}{D_h} + \frac{1}{A_x}\frac{d}{dX}\left(\dot{m}_G V_G + \dot{m}_L V_L\right) \qquad (16.45)$$

where $D_h = 4A_x/W_p$ is the hydraulic diameter. The total pressure gradient is seen to be composed of three terms resulting from the static head change (gravity), energy dissipation (friction loss), and acceleration (the change in kinetic energy), that is,

$$-\frac{dP}{dX} = -\left(\frac{dP}{dX}\right)_g - \left(\frac{dP}{dX}\right)_f - \left(\frac{dP}{dX}\right)_{acc} \qquad (16.46)$$

which is equivalent to Equation 9.14 for pure gas flow.

B. Homogeneous Gas–Liquid Models

In principle, the energy dissipation (friction loss) associated with the gas–liquid, gas–wall, and liquid–wall interactions can be evaluated separately and summed to get the total energy dissipation for the system. However, even for distributed (nonhomogeneous) flows, it is common practice to evaluate the friction loss as a single term which, however, depends in a complex manner upon the nature of the flow pattern and fluid properties in both phases. This is referred to as the "homogeneous" model:

$$-\left(\frac{dP}{dX}\right)_f = \frac{4f_m}{D_h}\left(\frac{\rho_m V_m^2}{2}\right) = \frac{2f_m G_m^2}{\rho_m D_h} \qquad (16.47)$$

The homogeneous model also assumes both phases to be moving at the same velocity, that is, no slip. Since the total mass flux is constant, the acceleration (or kinetic energy change) term can be written:

$$-\left(\frac{dP}{dX}\right)_{acc} = \rho_m V_m \frac{dV_m}{dX} = G_m^2 \frac{d\nu_m}{dX} \qquad (16.48)$$

where $v_m = 1/\rho_m$ is the average specific volume of the homogeneous two-phase mixture

$$v_m = \frac{1}{\rho_m} = \frac{x}{\rho_G} + \frac{1-x}{\rho_L} = v_G x + v_L(1-x) \tag{16.49}$$

and x is the quality (i.e., the mass fraction of gas). For "frozen" flows in which there is no phase change (e.g., air and cold water), the acceleration term is often negligible in steady pipe flow (although it can be appreciable in entrance flows and in non-uniform cross-section channels). However, if a phase change occurs (e.g., flashing of hot water or other volatile liquids), this term can be very significant. Evaluating the derivative of v_m from the previous equation gives

$$\frac{dv_m}{dX} = x\frac{dv_G}{dX} + (v_G - v_L)\frac{dx}{dX} = x\frac{dv_G}{dP}\frac{dP}{dX} + v_{GL}\frac{dx}{dX} \tag{16.50}$$

where $v_{GL} = v_G - v_L$. The first term on the right describes the effect of the gas expansion on the acceleration for constant mass fraction and the last term represents the additional acceleration resulting from a phase change from liquid to gas (e.g., a flashing liquid).

Substituting the expressions for the acceleration and friction loss pressure gradients into the total pressure gradient equation and rearranging gives

$$-\frac{dP}{dX} = \frac{\dfrac{2f_m G_m^2}{\rho_m D} + G_m^2 v_{GL}\dfrac{dx}{dX} + \rho_m g\dfrac{dz}{dX}}{1 + G_m^2 x\dfrac{dv_G}{dP}} \tag{16.51}$$

Finding the pressure drop corresponding to a total mass flux G_m from this equation requires a stepwise procedure using physical property data from which the density of both the gas phase and the mixture can be determined as a function of pressure. For example, if the upstream pressure P_1 and the mass flux G_m are known, the equation is used to evaluate the pressure gradient at point 1 and hence the change in pressure ΔP over a finite length ΔL, and then the pressure $P_{1+i} = P_1 - \Delta P$. The densities are then determined at pressure P_{1+i}, and the process is repeated at successive increments until the end of the pipe is reached. If the flow is choked, the pressure gradient will tend toward infinity at the choke point.

There are a number of special cases that permit simplification of Equation 16.51. For example, if the pressure is high and the pressure gradient moderate, the term in the denominator that represents the acceleration due to gas expansion can be neglected. Likewise for "frozen" flow for which there is no phase change (e.g., air and cold water), the quality (x) is constant and the second term in the numerator is zero. For flashing flows, the change in quality with length (dx/dX) must be determined from a total energy balance from the pipe inlet (or stagnation) conditions, along with the appropriate vapor–liquid equilibrium data for the flashing liquid. If the Clausius–Clapeyron equation is used, this becomes

$$\left(\frac{\partial v_G}{\partial P}\right)_T = -\frac{v_{GL}^2 C_p T}{\lambda_{GL}^2} \tag{16.52}$$

where
   λ_{GL} is the heat of vaporization
   v_{GL} is the change in specific volume due to vaporization

For an ideal gas,

$$\left(\frac{\partial v}{\partial P}\right)_T = -\frac{1}{\rho P}, \quad \left(\frac{\partial v}{\partial P}\right)_s = -\frac{P_1^{1/k}}{\rho_1 k P^{(1+k)/k}} \tag{16.53}$$

It should be noted that the derivative is negative, so that if conditions are such that if the denominator of Equation 16.51 is zero, an infinite pressure gradient would result. This condition corresponds to the speed of sound, that is, choked flow. For a nonflashing liquid and an ideal gas mixture, the corresponding maximum (choked) mass flux, G_m^*, follows directly from the definition of the speed of sound

$$G_m^* = c_m \rho_m = \rho_m \sqrt{k \left(\frac{\partial P}{\partial \rho_m} \right)_T} = \sqrt{\frac{\rho_m kP}{\varepsilon}} \tag{16.54}$$

The ratio of the sonic velocity in a homogeneous two-phase mixture to that in a gas alone is $c_m/c = \sqrt{\rho_G/(\varepsilon_m \rho_m)} = \sqrt{\rho_L/[\rho_L \varepsilon_m (1-\varepsilon_m)]}$. This ratio can be much smaller than unity, so that choking can occur in a two-phase mixture at a higher downstream pressure than that for a single-phase gas flow (i.e., at a lower pressure drop and a corresponding lower mass flux).

Evaluation of each term in Equation 16.51 is straightforward, except for the friction factor. One approach is to treat the two-phase mixture as a "pseudo-single-phase" fluid with appropriate properties. The friction factor is found assuming the usual Newtonian methods (i.e., Moody diagram, Churchill equation) using a suitable Reynolds number, for example,

$$N_{Re,TP} = \frac{DG_m}{\mu_m} \tag{16.55}$$

where μ_m is an appropriate viscosity for the two-phase mixture. A wide variety of methods have been proposed for estimating this viscosity, but the one that seems most logical is the local volume-weighted average (Dukler et al., 1964b):

$$\mu_m = \varepsilon \mu_G + (1-\varepsilon) \mu_L \tag{16.56}$$

The corresponding density ρ_m is the "no-slip" or equilibrium density of the mixture:

$$\rho_m = \varepsilon \rho_G + (1-\varepsilon) \rho_L = \frac{1}{x/\rho_G + (1-x)/\rho_L} \tag{16.57}$$

Note that the frictional pressure gradient is inversely proportional to the fluid density for a given mass flux:

$$\left(-\frac{\partial P}{\partial X} \right)_f = \frac{2 f_m G_m^2}{\rho_m D} \tag{16.58}$$

The corresponding pressure gradient for purely liquid flow is

$$\left(-\frac{\partial P}{\partial X} \right)_{fL} = \frac{2 f_L G_L^2}{\rho_L D} \tag{16.59}$$

Taking the reference liquid mass flux to be the same as that for the two-phase flow ($G_L = G_m$) and the friction factors to be the same ($f_L = f_m$), then

$$\left(-\frac{\partial P}{\partial X} \right)_{fm} = \frac{\rho_L}{\rho_G} \left(-\frac{\partial P}{\partial X} \right)_{fL} = \left(x \frac{\rho_L}{\rho_G} + 1 - x \right) \left(-\frac{\partial P}{\partial X} \right)_{fL} \tag{16.60}$$

A similar relation could be written by taking the single-phase gas flow as the reference instead of the liquid, that is, $G_G = G_m$. This is the basis for the *two-phase multiplier* method:

$$\left(-\frac{\partial P}{\partial X}\right)_{fm} = \Phi_R^2 \left(-\frac{\partial P}{\partial X}\right)_{fR} \tag{16.61}$$

where

R represents a reference single-phase flow

Φ_R^2 is the *two-phase multiplier*

There are four possible reference flows:

1. $R = L$: The total mass flow is liquid ($G_m = G_L$).
2. $R = G$: The total mass flow is gas ($G_m = G_G$).
3. $R = L_{Lm}$: The total mass flow is that of the "liquid only" in the mixture ($G_{Lm} = (1 - x)G_m$).
4. $R = G_{Gm}$: The total mass flow is that of the "gas only" in the mixture ($G_{Gm} = xG_m$).

The two-phase multiplier method is utilized primarily for separated flows, which will be discussed later.

1. Omega Method for Homogeneous Equilibrium Flow

For homogeneous equilibrium (no slip) flow in a uniform pipe, the governing (Bernoulli) equation can be written (equivalent to Equation 16.45) as

$$\frac{dP}{dX} + G_m^2 \frac{dv_m}{dX} + \frac{2f_m v_m G_m^2}{D} + \frac{g\Delta z}{v_m L} = 0 \tag{16.62}$$

where $v_m = 1/\rho_m$. By integrating over the pipe length, L, and assuming the friction factor to be constant, this can be rearranged as follows:

$$\frac{4f_m L}{D} = K_f = \int \frac{-v_m \left(1 + G_m^2 dv_m / dX\right)}{G_m^2 v_m^2 / 2 + \left(gD/4f_m\right)\Delta z/L} dP \tag{16.63}$$

where X is the distance along the pipe. Leung (1996) utilized a linearized two-phase equation of state to evaluate $v_m = fn(P)$:

$$\frac{v_m}{v_o} = \omega\left(\frac{P_o}{P_m} - 1\right) + 1 = \frac{\rho_o}{\rho_m} \tag{16.64}$$

where ρ_o is the two-phase density at the upstream (stagnation) pressure P_o. The parameter ω represents the compressibility of the fluid and can be determined from property data for $\rho = fn(P)$ at two pressure values or estimated from the physical properties at the upstream (stagnation) state. This equation is based on the fact that isentropic lines for most single-component fluids in the vapor–liquid two-phase region are almost linear in the coordinates ($1/P, v$). The parameter ω is the inverse value of the isentropic exponent in the two-phase region ($\omega = 1/k_{P_o}$, see Chapter 9) and can be calculated using the following equations:

For any flashing system

$$\omega = \frac{C_{pmo}T_o P_o}{v_o}\left(\frac{v_{GLo}}{\lambda_{GLo}}\right)^2 - 2\frac{P_o v_{GLo}}{\lambda_{GLo}}\left[\varepsilon_o\hat{\beta}_G + \left(1 - \varepsilon_o\right)\hat{\beta}_L\right] + \frac{\varepsilon_o}{\hat{\kappa}_G} + \frac{1 - \varepsilon_o}{\hat{\kappa}_L} \tag{16.65}$$

where $\hat{\beta} = -\left(\partial \ln \rho / \partial \ln T\right)_P$ and $\hat{\kappa} = \left(\partial \ln P / \partial \ln \rho\right)_T$ are the dimensionless thermal expansion coefficient and isothermal bulk modulus, respectively. For an ideal gas, $\hat{\beta}_G = \hat{\kappa}_G = 1$, $\hat{\beta}_L = 0$, $\hat{\kappa}_L = \infty$, and

$$\omega = \varepsilon_o \left(1 - 2\frac{P_o v_{GLo}}{\lambda_{GLo}}\right) + \frac{C_{pmo}T_o P_o}{v_o}\left(\frac{v_{GLo}}{\lambda_{GLo}}\right)^2 \tag{16.66}$$

For nonflashing (frozen) flows of an ideal gas and liquid:

$$\omega = \varepsilon_o \frac{x_o C_{vGo} + \left(1 - x_o\right)C_{pLo}}{x_o C_{pGo} + \left(1 - x_o\right)C_{pLo}} \tag{16.67}$$

Using Equation 16.64, Equation 16.63 can be written as

$$\frac{4 f_m L}{D} = K_f = -\int_{\eta_1}^{\eta_2} \frac{\left[\left(1-\omega\right)\eta^2 + \omega\eta\right]\left(1 - G^{*2}\omega/\eta^2\right)}{G^{*2}\left[\left(1-\omega\right)\eta + \omega\right]^2/2 + \eta^2 N_{Fi}} d\eta \tag{16.68}$$

where $\eta = P/P_o$, $G^* = G_m/(P_o \rho_o)^{1/2}$ and

$$N_{Fi} = \frac{\rho_o g \Delta z}{P_o\left(4 f_m L/D\right)} \tag{16.69}$$

is the "flow inclination number." From the definition of the speed of sound, it follows that the exit pressure ratio at which choking occurs is given by

$$\eta_{2c} = G_m^* \sqrt{\omega} \tag{16.70}$$

For horizontal flow, Equation 16.68 can be evaluated analytically to give

$$\frac{4 f_m L}{D} = \frac{2}{G^{*2}}\left[\frac{\eta_1 - \eta_2}{1-\omega} + \frac{\omega}{\left(1-\omega\right)^2} \ln\left\{\frac{\left(1-\omega\right)\eta_2 + \omega}{\left(1-\omega\right)\eta_1 + \omega}\right\}\right]$$
$$- 2 \ln\left[\frac{\left(1-\omega\right)\eta_2 + \omega}{\left(1-\omega\right)\eta_1 + \omega}\left(\frac{\eta_1}{\eta_2}\right)\right] \tag{16.71}$$

As $\omega \to 1$ (i.e., setting $\omega = 1.001$ in Equation 16.71), this reduces to the solution for ideal isothermal gas flow (Equation 9.17), and for $\omega = 0$, it reduces to the incompressible flow solution. For inclined pipes, Leung (1996) has given the solution of Equation 16.68 in graphical form for various values of N_{Fi}.

A so-called universal equation for designing pipelines for two-phase flow has been presented by Kim et al. (2015). This method is basically an extension of the Leung omega method, which utilizes average values of the density and specific volume over the pressure range of interest in terms of two empirical parameters determined by a fit of ρ versus P data. This method is not applicable to subcooled flashing flows.

It is also possible to use Equations 9.97 and 9.98 for two-phase homogeneous equilibrium flow in the region where isentropic exponents and coefficients in these equations change slowly (which is usually the case for values that are not too small, that is, where $\omega = 1/k_{P_\rho} \leq 2$).

For single-component flashing flow, $k_{P_\rho} = \omega^{-1}$ so that, in this case, ω can be calculated by Equation 16.65 and $k_{TP} = P_o v_{GLo}/\lambda_{GLo}$. Equation 9.96 also can be used for two-phase flow with a simple approximation $\omega = 1/k_{P_\rho} \leq 2$.

2. Generalized (Homogeneous Direct Integration) Method for All Homogeneous Flow Conditions

The omega method is limited to systems for which the linearized two-phase equation of state (Equation 16.64) is a good approximation to the two-phase density (i.e., single-component systems and multicomponent mixtures of similar compounds). Fluid property data for ρ versus P for at least two values of P are required to evaluate the parameter ω.

For any system (including single- or two-phase and frozen or flashing flows), the governing equations for homogeneous flow can be evaluated numerically using either experimental or theoretical (EOS) thermodynamic properties for the P–ρ relation.

For example, the basic Bernoulli equation for pipe flow can be integrated and rearranged to give

$$G^2 = \frac{-\int_{P_1}^{P_2} \rho \, dP}{ln\left(\dfrac{\rho_1}{\rho_2}\right) + \dfrac{1}{2}\left(\sum_1^2 K_f + 1\right)} \cong \frac{\sum_{P_1}^{P_2}\left(\rho_{i+1} + \rho_{i+2}\right)\left(P_{i+1} - P_{i+2}\right)}{2ln\left(\dfrac{\rho_1}{\rho_2}\right) + \left(\sum_1^2 K_f + 1\right)} \tag{16.72}$$

where the right-hand side of Equation 16.72 incorporates the finite difference version of the integral. This method is referred to as the homogeneous direct integration (HDI) method and is similar to that for sizing relief valves for homogeneous flow. The term $\sum_1^2 K_f$ includes the pipe loss coefficient (i.e., $K_{pipe} = 4fL/D$) as well as any fittings in the line.

A thermodynamic property database is used to evaluate the mixture density as a function of pressure at suitable pressure intervals along an adiabatic path from P_1 to P_2. An adiabatic path is more appropriate for pipe flow, as opposed to an isentropic path as used for relief valve nozzles. However, as the pipe length increases, adiabatic flow approaches isothermal flow (see Figure 9.3). Alternately, an isentropic path can be used for the numerical approximation, as the friction loss term, $\sum_1^2 K_f$, represents the irreversible nonisentropic contribution to the integral. By starting at $P_2 = P_1$, values of G are calculated from Equation 16.72 as P_2 is reduced until the exit pressure is reached. If the value of G reaches a maximum before the exit pressure is reached, then the flow is choked at that value of G and the corresponding value of P_2 is the choke pressure, P_c. This method can be applied directly to single- or multiphase fluids and frozen or flashing flows as long as suitable property data are available. Multicomponent fluids require an adiabatic flash routine to determine the required $P(\rho)$ function. A good inexpensive thermodynamic database for many single-component and some multicomponent fluids that can be used for this purpose is the NIST Database (NIST, 2015).

C. Separated Flow Models

Homogeneous flow models apply when the velocity of the two-phase mixture is large enough that turbulence will mix the two phases sufficiently that the mixture can be described as a "pseudo-single-phase" fluid, with appropriate "average" physical properties of the mixture.

At lower flow rates, the fluids will flow as separate phases, each occupying a specific portion of the flow field. Separated flow models consider each phase to occupy a defined fraction of the flow cross section and account for possible differences in the phase velocities (i.e., slip). There are a variety of such models in the literature, and many of these have been compared against data for various horizontal flow regimes by Chisholm (1983), Dukler et al. (1964a) and later by Ferguson and Spedding (1995).

The "classic" Lockhart–Martinelli (1949) method is based upon the two-phase multiplier defined previously for either "liquid only" (L_m) or "gas only" (G_m) reference flows, that is,

$$\left(-\frac{\partial P}{\partial X}\right)_{fm} = \Phi_{Lm}^2 \left(-\frac{\partial P}{\partial X}\right)_{fLm} \tag{16.73}$$

or

$$\left(-\frac{\partial P}{\partial X}\right)_{fm} = \Phi_{Gm}^2 \left(-\frac{\partial P}{\partial X}\right)_{fGm} \tag{16.74}$$

where

$$\left(-\frac{\partial P}{\partial X}\right)_{fLm} = \frac{2f_{Lm}\left(1-x\right)^2 G_m^2}{\rho_L D} \tag{16.75}$$

and

$$\left(-\frac{\partial P}{\partial X}\right)_{fGm} = \frac{2f_{Gm}x^2 G_m^2}{\rho_G D} \tag{16.76}$$

Here, Φ is the two-phase multiplier, which is correlated as a function of the Lockhart–Martinelli correlating parameter χ^2, defined as

$$\chi^2 = \left(-\frac{\partial P}{\partial X}\right)_{fLm} \left(-\frac{\partial P}{\partial X}\right)_{fGm}^{-1} \tag{16.77}$$

The correlation is shown in Figure 16.7. There are four curves for each multiplier, depending upon the flow regime in each phase, that is, both turbulent (*tt*), both laminar (vv), liquid turbulent and gas laminar (*tv*), or liquid laminar and gas turbulent (v*t*). The curves can also be represented by the following equations:

$$\Phi_{Lm}^2 = 1 + \frac{C}{\chi} + \frac{1}{\chi^2} \tag{16.78}$$

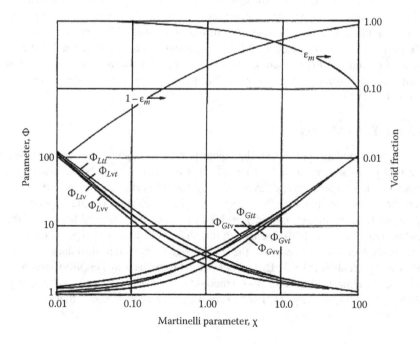

FIGURE 16.7 Lockhart–Martinelli two-phase correlating parameters.

TABLE 16.2

Values of Constant C in Two-Phase Multiplier Equations

Flow State	Liquid	Gas	C
tt	Turbulent	Turbulent	20
vt	Laminar	Turbulent	12
tv	Turbulent	Laminar	10
vv	Laminar	Laminar	5

and

$$\Phi_{Gm}^2 = 1 + C\chi + \chi^2 \tag{16.79}$$

where the values of C for the various flow combinations are shown in Table 16.2.

Here f_{Lm} is the pipe friction factor based on the "liquid only" Reynolds number $N_{ReLm} = (1 - x) G_m D/\mu_L$ and f_{Gm} is the friction factor based on the "gas only" Reynolds number $N_{ReGm} = xG_m D/\mu_G$. The curves cross at $\chi = 1$, and it is best to use the "G" reference curves if $\chi < 1$ and the "L" curves for $\chi > 1$.

Using similar analysis, Duckler et al. (1964a,b) deduced that

$$\left(-\frac{\partial P}{\partial X}\right)_{fm} = \frac{2f_L G_m^2}{\rho_L D} \frac{\rho_L}{\rho_m} \alpha(\varphi)\beta \tag{16.80}$$

or

$$\left(-\frac{\partial P}{\partial X}\right)_{fm} = \frac{2f_G G_m^2}{\rho_G D} \frac{\rho_G}{\rho_m} \alpha(\varphi)\beta \tag{16.81}$$

which is equivalent to the following Martinelli parameters:

$$\Phi_{Lm}^2 = \frac{\rho_L}{\rho_m} \alpha(\varphi)\beta \quad \text{and} \quad \Phi_{Gm}^2 = \frac{\rho_G}{\rho_m} \alpha(\varphi)\beta \tag{16.82}$$

where

$$\alpha(\varphi) = 1.0 + \frac{-\ln\varphi}{1.281 - 0.478(-\ln\varphi) + 0.444(-\ln\varphi)^2 - 0.094(-\ln\varphi)^3 + 0.00843(-\ln\varphi)^4} \tag{16.83}$$

and

$$\beta = \frac{\rho_L}{\rho_m} \frac{\varphi^2}{\varphi_m} + \frac{\rho_G}{\rho_m} \frac{(1-\varphi)^2}{(1-\varphi_m)}$$

Here φ and ρ are the equilibrium ("no slip") properties. Another major difference is that Duckler deduced that the friction factors f_L and f_G should both be evaluated at the following *mixture* Reynolds number:

$$N_{Rem} = \frac{DG_m}{\mu_m} \beta \tag{16.84}$$

D. Slip and Holdup

A major complication, especially for separated flows, arises from the effect of slip in gas/liquid flows. Slip occurs because the less dense and less viscous phase exhibits a lower resistance to flow, as well as expansion and acceleration of the gas phase as the pressure drops along the pipe length. The net result is an increase in the local holdup of the more dense phase within the pipe (φ_m) (or the corresponding two-phase density, ρ_m), as given by Equation 16.11. A large number of expressions and correlations for the holdup or (equivalent) slip ratio have appeared in the literature, and that deduced by Lockhart and Martinelli is shown in Figure 16.7. Many of these slip models can be summarized in terms of a general correlation of the form

$$S = a_o \left(\frac{1-x}{x} \right)^{(a_1 - 1)} \left(\frac{\rho_G}{\rho_L} \right)^{(a_2 - 1)} \left(\frac{\mu_L}{\mu_G} \right)^{a_3} \tag{16.85}$$

where the values of the parameters are shown in Table 16.3.

Although many additional slip models have been proposed in the literature, it is not clear which of these should be used under any given set of circumstances. In some cases, a constant slip ratio (S) may give satisfactory results. For example, a comparison of calculated and experimental mass flux data for high velocity air–water flows through nozzles (Jamerson and Fisher, 1999) found that $S = 1.1$–1.8 accurately represents the data over a range of $x = 0.02$–0.2, with the value of S increasing as the quality (x) increases.

A general correlation for slip has been given by Butterworth and Hewitt (1977):

$$S = 1 + a \sqrt{\frac{Y}{1 + bY} - bY} \tag{16.86}$$

where

$$Y = \left(\frac{x}{1-x} \right) \left(\frac{\rho_L}{\rho_G} \right) \tag{16.87}$$

$$a = 1.578 N_{ReL}^{-0.19} \left(\frac{\rho_L}{\rho_G} \right)^{0.22} \tag{16.88}$$

$$b = 0.0273 N_{We} N_{ReL}^{-0.51} \left(\frac{\rho_L}{\rho_G} \right)^{0.08} \tag{16.89}$$

$$N_{ReL} = \frac{G_m D}{\mu_L}, \quad N_{We} = \frac{G_m^2 D}{\sigma \rho_L} \tag{16.90}$$

TABLE 16.3
Parameters for Slip Model Equation (16.85)

Model	a_o	a_1	a_2	a_3
Homogeneous	1	1	1	0
$S = (\rho_L/\rho_G)^{1/2}$ (Fauske, 1962)	1	1	1/2	0
$S = (\rho_L/\rho_G)^{1/3}$ (Moody, 1965)	1	1	2/3	0
Thom (1964)	1	1	0.89	0.18
Baroczy (1966)	1	0.74	0.75	0.13
Lockhart–Martinelli (t-t) (1949)	1	0.75	0.417	0.083

An empirical correlation of holdup was developed by Mukherjee and Brill (1983) based on more than 1500 measurements of air with oil and kerosene in horizontal, inclined, and vertical flow (inclination of $\pm 90°$). Their results for the holdup were correlated by an empirical equation of the form

$$\varepsilon_m = \exp\left[\left(c_1 + c_2 \sin\theta + c_3 \sin^2\theta + c_4 N_L \frac{N_{GV}^{c_5}}{N_{LV}^{c_6}}\right)\right] \tag{16.91}$$

where

$$N_L = \mu_L \left(\frac{g}{\rho_L \sigma^3}\right)^{0.25}$$

$$N_{LV} = J_L \left(\frac{\rho_L}{g\sigma}\right)^{0.25}$$

$$N_{GV} = J_G \left(\frac{\rho_L}{g\sigma}\right)^{0.25}$$

and σ is the liquid surface tension. The constants in Equation 16.91 are given in Table 16.4 for the various flow inclinations.

A correlation for holdup by Hughmark (1962) was found to represent data quite well for both horizontal and vertical gas–liquid flow over a wide range of conditions. This was found by Dukler et al. (1964a) to be superior to a number of other relations that were checked against a variety of data sets. The Hughmark (1962) correlation is equivalent to the following expression for slip:

$$S = \frac{1 - K + \left[(1-x)/x\right]\rho_G/\rho_L}{K\left[(1-x)/x\right]\rho_G/\rho_L} = \left(\frac{x}{1-x}\right)\left(\frac{1-\varepsilon_m}{\varepsilon_m}\right), \quad K = \frac{\varepsilon_m}{\varepsilon} \tag{16.92}$$

The parameter K was found to correlate well with the dimensionless parameter Z:

$$Z = N_{Re}^{1/6} N_{Fr}^{1/8} / (1-\varepsilon)^{1/4} \tag{16.93}$$

where ε is the "no-slip" volume fraction of gas. The volume average viscosity of the two phases is used in the Reynolds number, and $N_{Fr} = V^2/gD$, where V is the average velocity of the two-phase mixture. Hughmark (1962) presented the correlation between K and Z in graphical form, which can be represented quite well by the expression:

$$K = \left(\frac{1}{1 + 0.12/Z^{0.95}}\right)^{19} \tag{16.94}$$

TABLE 16.4
Coefficients for Equation 16.91

Flow Direction	Flow Pattern	c_1	c_2	c_3	c_4	c_5	c_6
Uphill and horizontal	All	−0.3801	0.12988	−0.1198	2.3432	0.47569	0.28866
Downhill	Stratified	−1.3303	4.8081	4.17584	56.262	0.07995	0.50489
	Other	−0.5166	0.78981	0.55163	15.519	0.37177	0.39395

Slip is related to the holdup by Equation (16.13). The presence of slip also means that the acceleration term in the governing equation (Equation 16.45) cannot be evaluated in the same manner as for homogeneous flow conditions. When the acceleration term is expanded to account for the difference in phase velocities, the momentum equation when solved for the total pressure gradient becomes

$$-\frac{dP}{dX} = \frac{\left[\left(-\frac{\partial P}{\partial X}\right)_{fm} + G_m^2 \frac{dx}{dX} A(\varepsilon_m, x) + \rho_m g \frac{dz}{dX}\right]}{1 + G_m^2 \left\{\frac{x^2}{\varepsilon_m} \frac{dv}{dP} + \left(\frac{\partial \varepsilon_m}{\partial P}\right)_x \left[\frac{(1-x)^2}{\rho_L (1-\varepsilon_m)^2} - \frac{x^2}{\rho_G \varepsilon_m^2}\right]\right\}}$$ (16.95)

where

$$A(\varepsilon_m, x) = \left[\frac{2x}{\rho_G \varepsilon_m} - \frac{2(1-x)}{\rho_L (1-\varepsilon_m)}\right] + \left(\frac{\partial \varepsilon_m}{\partial x}\right)_P \left[\frac{(1-x)^2}{\rho_L (1-\varepsilon_m)^2} - \frac{x^2}{\rho_G \varepsilon_m^2}\right]$$ (16.96)

The pressure drop over a given length of pipe must be determined by a stepwise procedure, as described for homogeneous flow. The major additional complication in this case is the evaluation of the holdup (φ_m, or the equivalent slip ratio S) using one of the aforementioned correlations.

In some special cases, simplifications are possible, which makes the process easier, such as

1. If the denominator of Equation 16.95 is close to unity
2. If f_m, ρ_L, and ρ_G are nearly constant over the length of pipe

---

Example 16.2

Estimate the pressure gradient (in psi/ft) for a two-phase mixture of air and water entering a horizontal 6 in. sch 40 pipe at a total mass flow rate of 6500 lb_m/min at 150 psia, 60°F, with a quality (x) of 0.1 (mass fraction of air). Compare your answer using the (a) Omega, (b) Lockhart–Martinelli, (c) Dukler, and (d) HDI methods.

Solution

At the entering temperature and pressure, the density of air is 0.7799 lb_m/ft^3, its viscosity is 0.02 cP, the density of water is 62.4 lb_m/ft^3, and its viscosity is 1 cP. The no-slip volume fraction corresponding to the given quality is (by Equation 16.11) 0.899, and the corresponding density of the mixture (by Equation 16.12) is 7.01 lb_m/ft^3. The viscosity of the mixture by Equation 16.56 is 0.119 cP. The slip ratio can be estimated from Equation 16.85 using the Lockhart–Martinelli constants from Table 16.3 to be $S = 10.28$. Using this value in Equation 16.13 gives the in situ holdup $\varphi_m = 0.6027$. From the given mass flow rate and diameter, the total mass flux $G_m = 540$ lb_m/ft^2 s.

(a) *Omega method*: Since this is a "frozen" flow (no phase changes), the value of ω is given by Equation 16.67, with $k = 1.4$ for air, which gives $\omega = 0.642$. From the given data, $N_{Rem} = D G_m/\mu_m = 3.41 \times 10^6$, which, assuming a pipe roughness of 0.005 in., gives $f = 0.00412$ and a value of $4fL/D = 0.0326$. The pressure gradient is determined from Equation 16.70, with $G^* = G_m/(P_o \rho_o)^{1/2} = 0.245$ and $\eta_1 = 1$. The equation is solved by iteration for $\eta_2 = 0.999$ or $P_2 = 149.9$ psia. The pressure gradient is thus $(P_1 - P_2)/L = \mathbf{0.0969}$ **psi/ft**. This pressure gradient will apply until the pressure drops to the choke pressure, which from Equation 16.70 is 7.19 psia.

(b) *Lockhart–Martinelli method*: Using the "liquid only" basis, the corresponding Reynolds number is $N_{ReLm} = (1 - x)DG_m/\mu_L = 3.66 \times 10^5$, which gives a value of $f_{Lm} = 0.00419$. Likewise, using the "gas only" basis gives $N_{ReGm} = xDG_m/\mu_G = 2.27 \times 10^6$, which gives $f_{Gm} = 0.00383$. These values give the corresponding pressure gradients from Equations 16.76 and 16.77 as 0.0135 and 0.012 psi/ft, respectively. The square root of the ratio of these values gives the Lockhart–Martinelli parameter $\chi = 1.0527$, which, from Equation 16.73, gives $\Phi_{Lm}^2 = 20.9$. The pressure gradient is then calculated from Equation 16.76 to be **0.283 psi/ft**.

(c) *Duckler method*: This method requires determining values for β and α from Equation 16.83. The in-situ holdup determined earlier, $\varphi_m = 0.6027$, is used in the equation for β to give a value of 0.377, and the "no-slip" holdup value of $\varphi = 0.101$ is used in the equation for α to give a value of 2.416. These values are used in Equation 16.80, with a value of $f = 0.00378$, to determine the **pressure gradient of 0.122 psi/ft**.

(d) *HDI method*: For this problem, Equation 16.51 can be used. For horizontal flow ($dz = 0$) and constant quality (x), this equation reduces to

$$-\frac{dP}{dX} = \frac{2f_m G_m^2}{\rho_m D\left(1 + G_m^2 x \dfrac{dv_G}{dP}\right)} \tag{16.97}$$

Finding the pressure drop corresponding to a total mass flux G_m from this equation requires a stepwise procedure using physical property data from which the density of both the gas phase and the mixture can be determined as a function of pressure. For example, if the upstream pressure P_1 and the mass flux G_m are known, the equation is used to evaluate the pressure gradient at point 1 and hence the change over a finite pipe length ΔL for the pressure increment ΔP and then at the pressure $P_{1+i} = P_1 - \Delta P$. The densities are then determined at pressure P_{1+i}, and the process is repeated at successive increments until the end of the pipe is reached. Assuming isothermal flow, the density (and specific gravity) of the two-phase mixture will vary down the pipe as the pressure drops and the air expands. The properties of air can be accessed by the NIST Database, over the pressure range of 150–14.7 psia at 2 psi increments, and used to calculate the specific volume of the mixture from

$$v_m = v_G x + v_L (1 - x)$$

Alternately, the air can be assumed to be an ideal gas and its density calculated accordingly (in this example, the ideal gas law is an excellent approximation). A pressure increment of 2 psi is selected, and the density (and/or specific volume) of the mixture is tabulated in the spreadsheet as a function of pressure from 150 to 14.7 psia. This requires a determination of the pipe friction factor that depends on the Reynolds number, which, in turn, depends on the mixture viscosity, which is unknown, that is, $N_{Re} = DG/\mu$.

Air Density from NIST Database: Same Values as Calculated for Ideal Gas (Can Assume Ideal Gas)

psia	lb$_m$/ft³	lb$_m$/ft³	ft³/lb$_m$	ft³/lb$_m$	(ft²s/lb$_m$)²	psi/ft	ft	ft
P_2	ρ_G	ρ_m	v_G	v_m	dv_G/dP	$-dP/dX$	dx	L
150	0.7820	7.0274	1.279E+00	0.142300315	−1.79E−06			0
148	0.7720	6.9465	1.295E+00	0.143956756	−1.79E−06	1.36E−01	1.47E+01	14.69
146	0.7610	6.8573	1.314E+00	0.145829122	−2.02E−06	1.38E−01	1.45E+01	29.20
144	0.7510	6.7760	1.332E+00	0.147578869	−1.89E−06	1.40E−01	1.43E+01	43.53
142	0.7400	6.6864	1.351E+00	0.149558212	−2.13E−06	1.41E−01	1.41E+01	57.67
140	0.7300	6.6046	1.370E+00	0.151409378	−2.00E−06	1.43E−01	1.40E+01	71.64
138	0.7196	6.5191	1.390E+00	0.153394687	−2.14E−06	1.45E−01	1.38E+01	85.43
136	0.7090	6.4322	1.410E+00	0.1554668	−2.23E−06	1.47E−01	1.36E+01	99.03
134	0.6980	6.3416	1.433E+00	0.157689553	−2.40E−06	1.49E−01	1.34E+01	112.45
132	0.6880	6.2589	1.453E+00	0.159771914	−2.25E−06	1.51E−01	1.32E+01	125.69

(Continued)

Air Density from NIST Database: Same Values as Calculated for Ideal Gas (Can Assume Ideal Gas)

psia	lb$_m$/ft^3	lb$_m$/ft^3	ft^3/lb$_m$	ft^3/lb$_m$	(ft^{2}s/lb$_m$)^2	psi/ft	ft	ft
P_2	ρ_G	ρ_m	ν_G	ν_m	$d\nu_G/dP$	$-dP/dX$	dx	L
130	0.6770	6.1678	1.477E+00	0.162133564	−2.55E−06	1.53E−01	1.30E+01	138.73
128	0.6670	6.0846	1.499E+00	0.164348114	−2.39E−06	1.55E−01	1.29E+01	151.60
126	0.6570	6.0013	1.522E+00	0.166630078	−2.46E−06	1.58E−01	1.27E+01	164.29
124	0.6460	5.9094	1.548E+00	0.169221839	−2.79E−06	1.60E−01	1.25E+01	176.79
122	0.6356	5.8221	1.573E+00	0.171759523	−2.74E−06	1.62E−01	1.23E+01	189.11
120	0.6252	5.7345	1.600E+00	0.174381797	−2.83E−06	1.65E−01	1.21E+01	201.24
118	0.6147	5.6468	1.627E+00	0.177092962	−2.92E−06	1.67E−01	1.19E+01	213.18
116	0.6043	5.5587	1.655E+00	0.179897615	−3.02E−06	1.70E−01	1.18E+01	224.94
114	0.5939	5.4704	1.684E+00	0.182800677	−3.13E−06	1.73E−01	1.16E+01	236.51
112	0.5835	5.3819	1.714E+00	0.18580742	−3.24E−06	1.76E−01	1.14E+01	247.89
110	0.5730	5.2926	1.745E+00	0.188943147	−3.38E−06	1.79E−01	1.12E+01	259.09
108	0.5626	5.2036	1.778E+00	0.192175	−3.49E−06	1.82E−01	1.10E+01	270.09
106	0.5522	5.1143	1.811E+00	0.19552881	−3.62E−06	1.85E−01	1.08E+01	280.91
104	0.5417	5.0248	1.846E+00	0.199011612	−3.76E−06	1.88E−01	1.06E+01	291.54
102	0.5313	4.9351	1.882E+00	0.202630995	−3.90E−06	1.92E−01	1.04E+01	301.97
100	0.5207	4.8433	1.920E+00	0.206472242	−4.14E−06	1.95E−01	1.02E+01	312.22
98	0.5103	4.7530	1.960E+00	0.210391612	−4.23E−06	1.99E−01	1.01E+01	322.27
96	0.4999	4.6626	2.001E+00	0.21447429	−4.40E−06	2.03E−01	9.86E+00	332.13
94	0.4895	4.5718	2.043E+00	0.218730699	−4.59E−06	2.07E−01	9.67E+00	341.80
92	0.4790	4.4808	2.087E+00	0.223172169	−4.79E−06	2.11E−01	9.48E+00	351.28
90	0.4686	4.3896	2.134E+00	0.227811038	−5.00E−06	2.15E−01	9.28E+00	360.56
88	0.4582	4.2981	2.182E+00	0.232660764	−5.23E−06	2.20E−01	9.09E+00	369.66
86	0.4478	4.2063	2.233E+00	0.237736059	−5.47E−06	2.25E−01	8.90E+00	378.55
84	0.4374	4.1143	2.286E+00	0.243053035	−5.73E−06	2.30E−01	8.70E+00	387.25
82	0.4268	4.0205	2.343E+00	0.248724858	−6.12E−06	2.35E−01	8.50E+00	395.76
80	0.4164	3.9280	2.402E+00	0.254582402	−6.32E−06	2.41E−01	8.31E+00	404.07
78	0.4060	3.8352	2.463E+00	0.260740334	−6.64E−06	2.47E−01	8.11E+00	412.18
76	0.3956	3.7422	2.528E+00	0.267222367	−6.99E−06	2.53E−01	7.91E+00	420.09
74	0.3850	3.6475	2.597E+00	0.274163337	−7.48E−06	2.59E−01	7.71E+00	427.81
72	0.3746	3.5539	2.670E+00	0.281378344	−7.78E−06	2.66E−01	7.52E+00	435.32
70	0.3642	3.4601	2.746E+00	0.289005637	−8.22E−06	2.73E−01	7.32E+00	442.64
68	0.3538	3.3661	2.827E+00	0.297081595	−8.71E−06	2.81E−01	7.12E+00	449.76
66	0.3434	3.2717	2.912E+00	0.305647005	−9.24E−06	2.89E−01	6.92E+00	456.68
64	0.3330	3.1771	3.003E+00	0.314747752	−9.81E−06	2.98E−01	6.72E+00	463.40
62	0.3226	3.0823	3.100E+00	0.324435645	−1.04E−05	3.07E−01	6.52E+00	469.92
60	0.3121	2.9866	3.204E+00	0.334833202	−1.12E−05	3.17E−01	6.32E+00	476.24
58	0.3017	2.8912	3.315E+00	0.345881827	−1.19E−05	3.27E−01	6.11E+00	482.35
56	0.2913	2.7955	3.433E+00	0.357719639	−1.28E−05	3.38E−01	5.91E+00	488.26
54	0.2809	2.6995	3.560E+00	0.370434327	−1.37E−05	3.50E−01	5.71E+00	493.97
52	0.2705	2.6033	3.697E+00	0.384127067	−1.48E−05	3.63E−01	5.51E+00	499.48
50	0.2601	2.5068	3.845E+00	0.398915227	−1.59E−05	3.77E−01	5.30E+00	504.78
48	0.2497	2.4100	4.005E+00	0.414935733	−1.73E−05	3.92E−01	5.10E+00	509.88
46	0.2393	2.3129	4.179E+00	0.432349327	−1.88E−05	4.09E−01	4.89E+00	514.77
44	0.2289	2.2156	4.369E+00	0.451345975	−2.05E−05	4.27E−01	4.69E+00	519.45
42	0.2185	2.1180	4.577E+00	0.472151827	−2.24E−05	4.47E−01	4.48E+00	523.93
40	0.2081	2.0200	4.806E+00	0.495038264	−2.47E−05	4.68E−01	4.27E+00	528.21
38	0.1977	1.9218	5.059E+00	0.520333801	−2.73E−05	4.92E−01	4.06E+00	532.27

(Continued)

Air Density from NIST Database: Same Values as Calculated for Ideal Gas (Can Assume Ideal Gas)								
psia	lb$_m$/ft^3	lb$_m$/ft^3	ft^3/lb$_m$	ft^3/lb$_m$	(ft^{2}s/lb$_m$)^2	psi/ft	ft	ft
P_2	ρ_G	ρ_m	v_G	v_m	dv_G/dP	$-dP/dX$	dx	L
36	0.1871	1.8218	5.345E+00	0.54889662	−3.08E−05	5.19E−01	3.85E+00	536.12
34	0.1767	1.7231	5.659E+00	0.580336241	−3.39E−05	5.49E−01	3.64E+00	539.77
32	0.1663	1.6242	6.013E+00	0.615705813	−3.81E−05	5.82E−01	3.43E+00	543.20
30	0.1559	1.5249	6.414E+00	0.655791329	−4.32E−05	6.20E−01	3.22E+00	546.43
28	0.1455	1.4253	6.872E+00	0.701603347	−4.94E−05	6.64E−01	3.01E+00	549.44
26	0.1351	1.3254	7.400E+00	0.754463368	−5.70E−05	7.14E−01	2.80E+00	552.24
24	0.1247	1.2253	8.017E+00	0.816133392	−6.65E−05	7.72E−01	2.59E+00	554.83
22	0.1143	1.1248	8.746E+00	0.889016148	−7.86E−05	8.41E−01	2.38E+00	557.21
20	0.1039	1.0241	9.621E+00	0.976475455	−9.43E−05	9.24E−01	2.16E+00	559.37
18	0.0936	0.9230	1.069E+01	1.083370164	−1.15E−04	1.03E+00	1.95E+00	561.33
16	0.0832	0.8217	1.203E+01	1.21698855	−1.44E−04	1.15E+00	1.74E+00	563.06
14.7	0.0764	0.7557	1.309E+01	1.323337877	−1.15E−04	**1.25E+00**	1.60E+00	564.66
14	0.0728	0.7201	1.374E+01	1.388783618	−7.06E−05	1.31E+00	1.52E+00	566.18
12	0.0623	0.6175	1.605E+01	1.619559514	−2.49E−04	1.53E+00	1.30E+00	567.49

If a reasonable "guesstimate" is made for the viscosity, for example, 20 cP, and the pipe roughness is assumed to be 0.0018 in., the resulting value of the friction factor is found to be about 0.0038. The pressure gradient is found as a function of P from Equation (16.51), which can then be used to determine ΔX (and L) versus P. As L increases, the pressure gradient increases, and as the flow approaches choked, the pressure gradient approaches infinity. The spreadsheet output for this problem is shown above, which indicates that at the end of the pipe ($L = 565$ ft, $P_2 = 14.7$ psia), the pressure gradient is 1.25 psi/ft. **Note the pressure gradient at the pipe entrance is 0.136 psi/ft, and at the end of the pipe, it is 1.25 psi/ft.** An increase of almost a factor of 10 indicates that the flow is almost at the choke pressure at the pipe exit.

The flow regime can be determined from Figure 16.6a using an ordinate of 1 and an abscissa of 2635 kg/m²s, to be well in the dispersed flow regime, so each of these methods should be applicable.

SUMMARY

The following points should be retained from this chapter:

- Understand the definitions of the various measures of mass flux, volume flux, phase velocity and slip concentrations of the heavier and lighter phases in two-phase solid–liquid and gas/vapor–liquid flows.
- Know how to determine the pressure drop in a pipe for homogeneous and heterogeneous (separated) (gas/vapor)/liquid and solid–liquid flows.
- Understand the difference in the mechanisms for dense phase and dilute phase pneumatic transport, as well as horizontal versus vertical transport.
- Recognize the difference in flow regimes for horizontal and vertical gas–liquid pipe flows.

PROBLEMS

1. An aqueous slurry contains 45% (vol.) solids, with SG = 4 and psd given below. The slurry is a Bingham plastic, with properties given below. Calculate the pressure gradient (psi/ft) for the slurry at 8 ft/s in a 10 in. ID pipe.

   The procedure is the same as that outlined in Example 15.1, using the Molerus state diagram for the heterogeneous component and the Bingham plastic pipe flow equations for the homogeneous component. The terminal velocity of the largest particles is found using the correlation for Bingham plastic in Chapter 11. A "no solution" is obtained from this correlation for conditions corresponding to particles which are too small to fall in the Bingham fluid, which sets the division between the heterogeneous and homogeneous regions, and the respective solids size and concentration. The concentration of solids in the homogeneous fraction is used to calculate the density of this phase.

US Screen Mesh	Mesh Opening (μm)	Fraction Passing
400	37	0.05
325	44	0.05
200	74	0.05
140	105	0.05
100	149	0.05
60	250	0.1
35	500	0.1
18	1000	0.15
10	2000	0.2
5	4000	0.2

Slurry pipe flow

Given: Total solids fraction:

$C_v = 0.45 \quad C_{vcont} = 0.369$

Bingham properties:

$\tau_0 = 5 \text{ dyn/cm}^2$

$\mu_\infty = 0.3 \text{ P}$

$\rho_s = 3 \text{ g/cm}^3, \rho_L = 1 \text{ g/cm}^3$

$\rho_m = 1.738 \text{ g/cm}^3$

$S = 1.726122 \ \rho_s / \rho_m$

Pipe $D = 18$ in.

$V = 8$ ft/s

$\rho_m = \rho_s C_v + \rho_L(1 - C_v)$

2.   An aqueous slurry contains 45% (vol.) solids, with SG = 4 and psd the same as in Problem 1. The slurry is a power law fluid, with properties given. Calculate the pressure gradient (psi/ft) for the slurry at 8 ft/s in a 10 in. ID pipe.

   The procedure is the same as that outlined in Example 16.1, using the Molerus state diagram for the heterogeneous component and the power law pipe flow equations for the homogeneous component. The terminal velocity of the largest particles is found using the correlation for power law fluid in Chapter 11.

Slurry pipe flow

Given: Total solids fraction:

$C_v = 0.45 \quad C_{vcont} = 0.4095$

Power law properties:

$m = 60$ poise

$n = 0.18$

$\rho_s = 4 \text{ g/cm}^3 \ \rho_L = 1 \text{ g/cm}^3$

$\rho_m = 2.2285 \text{ g/cm}^3$

$S = 1.795 \ \rho_s / \rho_m$

Pipe $D = 18$ in.

$V = 8$ ft/s

$\rho_m = \rho_s C_v + \rho_L(1 - C_v)$

3. An aqueous slurry is composed of 45% solids (by volume). The solids have an SG of 4 and a particle size distribution shown here:

US Screen Mesh	Mesh Opening (µm)	Fraction Passing
400	37	0.02
325	44	0.06
200	74	0.08
140	105	0.10
100	149	0.15
60	250	0.18
35	500	0.20
18	1000	0.12
10	2000	0.08
5	4000	0.01

The slurry behaves as a non-Newtonian fluid, which can be described as a Bingham plastic with a yield stress of 40 dyn/cm^2 and a limiting viscosity of 100 cP. Calculate the pressure gradient (in psi/ft) for this slurry flowing at a velocity of 8 ft/s in a 10 in. ID pipe.

4. Spherical polymer pellets with a diameter of 1/8 in. and an SG of 0.96 are to be transported pneumatically using air at 80°F. The pipeline is horizontal, 6 in. ID and 100 ft long, and discharges at an atm pressure. It is desired to transport 15% by volume of solids at a velocity that is 1 ft/s above the minimum deposit velocity.
    (a) What is the pressure of the air that is required at the entrance to the pipe to overcome the friction loss in the pipe? (*Note*: There is an additional pressure gradient required to accelerate the particles after contacting with the air, but your answer should address only the friction loss).
    (b) If a section of this pipe is vertical, what would the choking velocity be in this line and the pressure gradient (in psi/ft) at a velocity of 1 ft/s above this velocity?

5. Saturated ethylene enters a 4 in. sch 40 pipe at 400 psia. The ethylene flashes as the pressure drops through the pipe, and the quality at any pressure can be estimated by assuming constant enthalpy along the pipe. If the pipe is 80 ft long and discharges at a pressure of 100 psia, what is the mass flow rate through the pipe? Use 50 psi pressure increments in the stepwise calculation procedure.

6. A two-phase mixture of natural gas (methane) and a 40°API liquid are being pumped through a 6 in. sch 40 pipeline at 80°F. The mixture enters the pipe at 500 psia at a total rate of 6000 lb$_m$/min and 6% quality. What is the total pressure gradient in the pipe at this point (in psi/ft)?

7. Repeat Example 16.2 using the HDI method, and compare the results with the other methods used in this example.

NOTATION

A_x	x-component of area, [L^2]
a	Slip correlating parameter, Equation 16.86, [—]
b	Slip correlating parameter, Equation 16.88, [—]
c	Speed of sound, [L/t]
C_d	Particle drag coefficient, [—]
c_p	Specific heat, [F L/(M T) = L^2/(t^2 T)]
D	Pipe diameter, [L]
D_h	Hydraulic diameter, [L]

d	Particle diameter, [L]
F	Force, $[F = M\,L/t^2]$
f	Fanning friction factor, [—]
G	Mass flux, $[M/(L^2\,t)]$
G^*	Dimensionless mass flux, [—]
g	Acceleration due to gravity, $[L/t^2]$
J	Volume flux (superficial velocity), [L/t]
k	Isentropic exponent (specific heat ratio for ideal gas), [—]
K	$\varepsilon_m/\varepsilon$, parameter in Hughmark correlation, Equation 16.92, [—]
L	Pipe length, [L]
$\dot{m}$	Mass flow rate, [M/t]
N_{Fi}	Flow inclination number, Equation 16.69 [—]
N_{Fr}	Froude number, [—]
N_{Frp}	Particle Froude number, Equation 16.22, [—]
N_{Frs}	Solids Froude number, Equation 16.28, [—]
N_{Frt}	Pipe Froude number, Equation 16.23, [—]
N_{Re}	Reynolds number, [—]
N_{Ret}	Particle terminal velocity Reynolds number, Equation 16.39, [—]
N_{Rep}	Particle relative velocity Reynolds number, Equation 16.39, [—]
$N_{Re,TP}$	Two-phase Reynolds number, [—]
N_{We}	Weber number, Equation 16.90, [—]
P	Pressure, $[F/L^2 = m/(L\,t^2)]$
Q	Volumetric flow rate, $[L^3/t]$
S	Velocity slip ratio, [—]
s	ρ_S/ρ_L, [—]
V	Velocity, [L/t]
V_c	Choke velocity for vertical pneumatic transport, [L/t]
V^*	Friction velocity, Equation 16.18, [L/t]
V_{gs}	Saltation gas velocity, [L/t]
V_m	Velocity of mixture, [L/t]
V_{md}	Minimum deposit velocity, [L/t]
V_r	Relative (slip) velocity, [L/t]
$\bar{V}_r$	Dimensionless slip velocity, V_r/V_m, [—]
$\bar{V}_{ro}$	Dimensionless single-particle slip velocity, Equation 16.24
V_t	Particle terminal velocity, [L/t]
W_p	Wetted perimeter, [L]
X	Dimensionless solids contribution to pressure drop, Equation 16.26, [—]
X	Horizontal coordinate direction, [L]
X_o	Dimensionless parameter defined by Equation 16.25, [—]
x	Mass fraction of less dense phase (quality, for gas–liquid flow), [—]
Y	Slip correlating parameter, Equation 16.87, [—]
z	Vertical direction measured upward, [L]
Z	Hughmark dimensionless parameter, defined by Equation 16.93

GREEK

α	Volume fraction of gas phase, [—]
$\hat{\beta}$	Dimensionless thermal expansion coefficient, [—]
δ	Parameter defined in Equation 16.27, [—]
χ^2	Lockhart–Martinelli two-phase correlating parameter, Equation 16.77 [—]
η	Pressure ratio P/P_o, [—]

ΔP_{Si}	Contribution to total pressure drop by particle size fraction Si, $[F/L^2]$
Δz	Increase in vertical position, $[L]$
ε	Volume fraction of the continuous phase, $[—]$
ε_c	Void fraction at choke point for vertical pneumatic transport, $[—]$
ε_m	Local value of ε at a specific point in pipe, $[—]$
φ_m	Local value of φ at a specific point in pipe, $[—]$
Φ	Property correction factor for gas/liquid flow regimes, Equation 16.43, $[—]$
Φ_R^2	Two-phase multiplier with reference to single-phase R, Equation 16.61, $[—]$
φ	Volume fraction of the distributed phase, $[—]$
$\hat{\kappa}$	Dimensionless isothermal bulk module, $[—]$
λ	Latent heat, $[F\,L/M = L^2/t^2]$
λ	Density correction factor, Equation 16.42, $[—]$
λ	Parameter for liquid/gas flow regimes, Equation 16.42
η	Pressure ratio, $[—]$
μ	Viscosity, $[M/(L\,t)]$
μ_s	Solids loading (mass of solids/mass of gas) in pneumatic transport, $[—]$
μ_s	Ratio of mass of solids to mass of gas, $[—]$
v	Specific volume, $[L^3/M]$
ρ	Density, $[M/L^3]$
σ	Surface tension, $[F/L = M/t^2]$
τ	Shear stress, $[F/L^2 = m/(L\,t^2)]$
ω	Two-phase equation of state parameter, Equation 16.64, $[—]$

SUBSCRIPTS

1,2	Reference points
C	Choking condition
f	Friction loss
G	Gas
L	Liquid
m	Mixture

REFERENCES

Aude, T.C., N.T. Cowper, T.L. Thompson, and E. Wasp, Slurry piping systems: Trends, design methods, guidelines, *Chem. Eng.*, 78, 74–90, June 28, 1971.

Baker, O., Simultaneous flow of oil and gas, *Oil Gas J.*, 53, 185–195, 1954.

Barnea, D., A unified model for prediction flow-pattern transitions for the whole range of pipe inclinations, *Int. J. Multiphase Flow*, 13(1), 1–12, 1987.

Baroczy, C.J., A systematic correction for two-phase pressure drop, *CEP Symp. Ser.* 62(44), 232–249, 1966.

Brown, N.P. and N.I. Heywood, (Eds.), *Slurry Handling: Design of Solid–Liquid Systems*, Elsevier, London, U.K., 1991.

Butterworth, D. and G.F. Hewitt, (Eds.), *Two-Phase Flow and Heat Transfer*, Oxford University Press, Oxford, U.K., 1977.

Chisholm, D., *Two-Phase Flow in Pipelines and Heat Exchangers*, The Institution of Chemical Engineers, George Goodwin, London, U.K., 1983.

Center for Chemical Process Safety (CCPS), *Guidelines for Pressure Relief and Effluent Handling Systems*, AIChE, New York, 1998.

Darby, R., Hydrodynamics of slurries and suspensions, Chapter 2, in *Encyclopedia of Fluid Mechanics*, Vol. 5, N.P. Cheremisinoff, (Ed.), Gulf Publishing Co., Houston, TX, 1985.

Dukler, A.E., M. Wicks, III, and R.G. Cleveland, Frictional pressure drop in two-phase flow: A. A comparison of existing correlations for pressure loss and holdup, *AIChE J.*, 10, 38–43, 1964a.

Dukler, A.E., M. Wicks, III, and R.G. Cleveland, Frictional pressure drop in two-phase flow: B. An approach through similarity analysis, *AIChE J.*, 10, 44–51, 1964b.

Fan, L.-S. and C. Zhu, *Principles of Gas-Solid Flows*, Cambridge University Press, Cambridge, U.K., 1998.

Fauske, H.K., Contribution to the theory of two-phase, one-component critical flow, Argonne National Laboratory Report SNL-6673, October 1962.

Ferguson, M.E.G. and P.L. Spedding, Measurement and prediction of pressure drop in two-phase flow, *J. Chem. Tech. Biotechnol.*, 62, 262–278, 1995.

Govier, G.W. and K. Aziz, *The Flow of Complex Mixtures in Pipes*, Van Nostrand Reinhold, 1983

Hanks, R.W., ASME Paper No. 80-PET-45, *ASME Energy Sources Technology Conference*, New Orleans, LA, February 3–7, 1980.

Hetsroni, G., (Ed.), *Handbook of Multiphase Systems*, McGraw Hill, New York, 1982.

Hewitt, G.F. and D.N. Roberts, Studies of two-phase flow patterns by x-ray and flash photography, Report AERE-M 2159, HMSO, London, U.K., 1969.

Hinkle, B.L., Acceleration of particles and pressure drops encountered in horizontal pneumatic conveying, PhD Thesis, Georgia Institute of Technology, Atlanta, GA, June 1953.

Hughmark, G.A., Holdup in gas-liquid flow, *Chem. Eng. Prog.*, 58(4), 62–65, 1962.

Jamerson, S.C. and H.G. Fisher, Using constant slip ratios to model non-flashing (frozen) two-phase flow through nozzles, *Process Safety Progress*, 18(2), 89–98, 1999.

Kim, J.S., T. Oh, and H.J. Dunsheath, A universal equation for designing pipelines, *Chem. Eng.*, 122, 66–71, September 10, 2015.

Klinzing, G.E., R.D. Marcus, F. Rizk, and L.S. Leung, *Pneumatic Conveying of Solids*, 3rd edn., Springer, Berlin, Germany, 2010.

Leung, J.C., Easily size relief devices and piping for two-phase flow, *Chem. Eng. Prog.*, 92, 28–50, December 1996.

Levy, S., *Two-Phase Flow in Complex Systems*, Wiley, New York, 1999.

Lockhart, R.W. and R.C. Martinelli, Proposed correlation of data for isothermal two-phase, two-component flow in pipes, *Chem. Eng. Prog.*, 45(1), 39–48, 1949.

Michaelides, E.E., J.D. Schwarzkopf and C. Crowe, *Multiphase Flow Handbook*, 2nd edn., CRC Press, Boca Raton, FL, 2016.

Molerus, O., *Principles of Flow in Disperse Systems*, Chapman & Hall, London, U.K., 1993.

Moody, F.J., Maximum flow rate of a single component two-phase mixture, *Trans. ASME, J. Heat Trans.*, 87, 134–142, February, 1965.

Mukherjee, H. and J.P. Brill, Liquid holdup correlations for inclined two-phase flow, *J. Petrol. Technol.*, 35, 1003–1008, May, 1983.

NIST (National Institute of Standards and Technology), REFPROP, version 9.5, Washington, DC, 2015.

Rizk, F., Dissertation, University of Karlsruhe, Karlsruhe, Germany, 1973.

Rouhani, S.Z. and M.S. Sohal, Two-phase flow patterns: A review of research results, *Prog. Nucl. Energy*, 11(3), 219–259, 1983.

Shook, C.A. and M.C. Roco, *Slurry Flow—Principles and Practice*, Butterworth-Heinemann, Boston, MA, 1991.

Taitel, Y. and A.E. Dukler, A model for predicting flow regime transitions in horizontal and near horizontal gas-liquid flow, *AIChE J.*, 22(1), 47–55, 1976.

Thom, J.R.S, Prediction of pressure drop during forced circulation boiling of water, *Intl. J. Heat Mass Transfer*, 7, 709–724, 1964.

Wallis, G.B., *One-Dimensional Two-Phase Flow*, McGraw-Hill, New York, 1969.

Wasp, E.J., J.P. Kenny, and R.L. Gandhi, *Solid-Liquid Flow in Slurry Pipeline Transportation*, Trans-Tech Publications, Clausthal, Germany, 1977.

Wilms, H. and S. Dhodapkar, Pneumatic conveying: Optimal system design, operation and control, *Chem. Eng.*, 121, 59–67, October, 2014.

Wilson, K.C., G.R. Addie, A. Sellgren, and R. Clift, *Slurry Transport using Centrifugal Pumps*, 3rd edn., Springer, New York, 2008.

Yang, W.C., Criteria for choking in vertical pneumatic conveying lines, *Powder Technol.*, 35, 143–150, 1983.

Zhang, Y.-Q., Q. Wang, C. Sarica, and J.P. Brill, Unified model for gas-liquid pipe flow via slug dynamics—Part 1: Model development, *Trans. ASME*, 125, 266–273, 2003.

"It is better to know how to learn, than it is to know."

—Dr. Seuss

Appendix A: Viscosities and Other Properties of Gases and Liquids

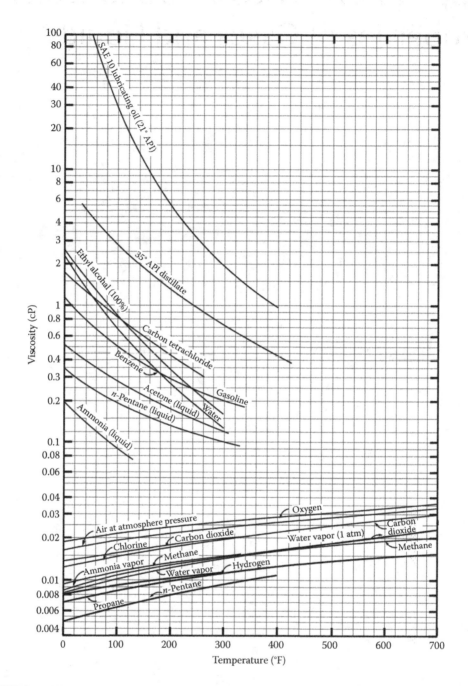

FIGURE A.1 Viscosities of various fluids at 1 atm pressure. 1 cP = 0.01 g/(cm s) = 6.72 × 10⁻⁴ lbₘ/(ft s) = 2.42 lbₘ/(ft h) = 2.09 × 10⁻⁵ lb_f s/ft². (Reproduced from Brown, G.G. et al., *Unit Operations*, Wiley, New York, 1951, p. 586. With permission.)

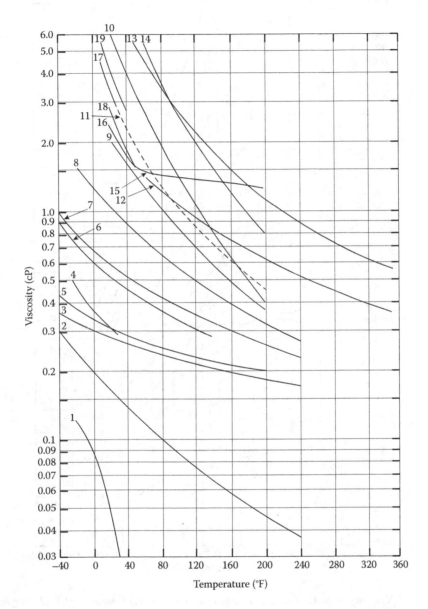

FIGURE A.2 Viscosity of various liquids: (1) Carbon dioxide, (2) ammonia, (3) methyl chloride, (4) sulfur dioxide, (5) Freon 12, (6) Freon 114, (7) Freon 11, (8) Freon 113, (9) ethyl alcohol, (10) isopropyl alcohol, (11) 20% sulfuric acid—20% H_2SO_4, (12) Dowtherm E, (13) Dowtherm A, (14) 20% sodium hydroxide—20% NaOH, (15) mercury, (16) 10% sodium chloride brine—10% NaCl, (17) 20% sodium chloride brine—20% NaCl, (18) 10% calcium chloride brine—10% $CaCl_2$, and (19) 20% calcium chloride brine—20% $CaCl_2$. (From Crane Technical Paper 4-10, Crane Co., Chicago, IL, 1991.)

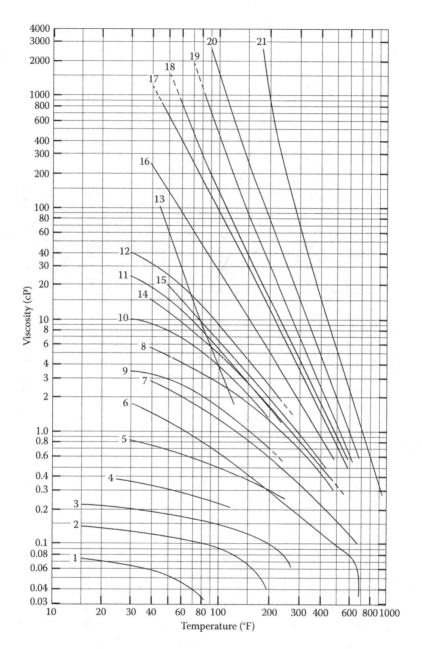

FIGURE A.3 Viscosity of water and liquid petroleum products: (1) Ethane (C_2H_6), (2) propane (C_3H_8), (3) butane (C4H10), (4) natural gasoline, (5) gasoline, (6) water, (7) kerosene, (8) distillate, (9) 48° API crude, (10) 40° API crude, (11) 35.6° API crude, (12) 32.6° API crude, (13) salt creek crude, (14) Fuel 3 (Max.), (15) Fuel 5 (Min.), (16) SAE 10 lube (100 V.I.), (17) SAE 30 lube (100 V.I.), (18) Fuel 5 (Min.) or Fuel 6 (Min.), (19) SAE 70 lube (100 V.I.), (20) Bunker C fuel (Max.) and M.C. Residuum, and (21) asphalt. (From Crane Company, Crane Technical Paper 410, Crane Co., Chicago, IL, 1991.)

TABLE A.1

Viscosities of Liquids (Coordinates Apply to Figure A.4)

Liquid	X	Y
Acetaldehyde	15.2	4.8
Acetic acid, 100%	12.1	14.2
Acetic acid, 70%	9.5	17.0
Acetic anhydride	12.7	12.8
Acetone, 100%	14.5	7.2
Acetone, 35%	7.9	15.0
Acetonitrile	14.4	7.4
Acrylic acid	12.3	13.9
Allyl alcohol	10.2	14.3
Allyl bromide	14.4	9.6
Allyl iodide	14.0	11.7
Ammonia, 100%	12.6	2.0
Ammonia, 26%	10.1	13.9
Amyl acetate	11.8	12.5
Amyl alcohol	7.5	18.4
Aniline	8.1	18.7
Anisole	12.3	13.5
Arsenic trichloride	13.9	14.5
Benzene	12.5	10.9
Brine, CaCl(S)_2(S), 25%	6.6	15.9
Brine, NaCl, 25%	10.2	16.6
Bromine	14.2	13.2
Bromotoluene	20.0	15.9
Butyl acetate	12.3	11.0
Butyl acrylate	11.5	12.6
Butyl alcohol	8.6	17.2
Butyric acid	12.1	15.3
Carbon dioxide	11.6	0.3
Carbon disulfide	16.1	7.5
Carbon tetrachloride	12.7	13.1
Chlorobenzene	12.3	12.4
Chloroform	14.4	10.2
Chlorosulfonic acid	11.2	18.1
Chlorotoluene, ortho	13.0	13.3
Chlorotoluene, meta	13.3	12.5
Chlorotoluene, para	13.3	12.5
Cresol, meta	2.5	20.8
Cyclohexanol	2.9	24.3
Cyclohexane	9.8	12.9
Dibromomethane	12.7	15.8
Dichloroethane	13.2	12.2
Dichloromethane	14.6	8.9

(Continued)

TABLE A.1 (*Continued*)

Viscosities of Liquids (Coordinates Apply to Figure A.4)

Liquid	X	Y
Diethyl ketone	13.5	9.2
Diethyl oxalate	11.0	16.4
Diethylene glycol	5.0	24.7
Diphenyl	12.0	18.3
Dipropyl ether	13.2	8.6
Dipropyl oxalate	10.3	17.7
Ethyl acetate	13.7	9.1
Ethyl acrylate	12.7	10.4
Ethyl alcohol, 100%	10.5	13.8
Ethyl alcohol, 95%	9.8	14.3
Ethyl alcohol, 40%	6.5	16.6
Ethyl benzene	13.2	11.5
Ethyl bromide	14.5	8.1
2-Ethyl butyl acrylate	11.2	14.0
Ethyl chloride	14.8	6.0
Ethyl ether	14.5	5.3
Ethyl formate	14.2	8.4
2-Ethyl hexyl acrylate	9.0	15.0
Ethyl iodide	14.7	10.3
Ethyl propionate	13.2	9.9
Ethyl propyl ether	14.0	7.0
Ethyl sulfide	13.8	8.9
Ethylene bromide	11.9	15.7
Ethylene chloride	12.7	12.2
Ethylene glycol	6.0	23.6
Ethylidene chloride	14.1	8.7
Fluorobenzene	13.7	10.4
Formic acid	10.7	15.8
Freon-11	14.4	9.0
Freon-12	16.8	5.6
Freon-21	15.7	7.5
Freon-22	17.2	4.7
Freon-113	12.5	11.4
Glycerol, 100%	2.0	30.0
Glycerol, 50%	6.9	19.6
Heptane	14.1	8.4
Hexane	14.7	7.0
Hydrochloric acid, 31.5%	13.0	16.6
Iodobenzene	12.8	15.9
Isobutyl alcohol	7.1	18.0
Isobutyric acid	12.2	14.4
Isopropyl alcohol	8.2	16.0
Isopropyl bromide	14.1	9.2
Isopropyl chloride	13.9	7.1

(Continued)

TABLE A.1 (*Continued*)
Viscosities of Liquids (Coordinates Apply to Figure A.4)

Liquid	X	Y
Isopropyl iodide	13.7	11.2
Kerosene	10.2	16.9
Linseed oil, raw	7.5	27.2
Mercury	18.4	16.4
Methanol, 100%	12.4	10.5
Methanol, 90%	12.3	11.8
Methanol, 40%	7.8	15.5
Methyl acetate	14.2	8.2
Methyl acrylate	13.0	9.5
Methyl *r*-butyrate	12.3	9.7
Methyl *n*-butyrate	13.2	10.3
Methyl chloride	15.0	3.8
Methyl ethyl ketone	13.9	8.6
Methyl formate	14.2	7.5
Methyl iodide	14.3	9.3
Methyl propionate	13.5	9.0
Methyl propyl ketone	14.3	9.5
Methyl sulfide	15.3	6.4
Naphthalene	7.9	18.1
Nitric acid	12.8	13.8
Nitric acid, 60%	10.8	17.0
Nitrobenzene	10.6	16.2
Nitrogen dioxide	12.9	8.6
Nitrotoluene	11.0	17.0
Octane	13.7	10.0
Octyl alcohol	6.6	21.1
Pentachloroethane	10.9	17.3
Pentane	14.9	5.2
Phenol	6.9	20.8
Phosphorus tribromide	13.8	16.7
Phosphorus trichloride	16.2	10.9
Propionic acid	12.8	13.8
Propyl acetate	13.1	10.3
Propyl alcohol	9.1	16.5
Propyl bromide	14.5	7.5
Propyl chloride	14.4	7.5
Propyl formate	13.1	9.7
Propyl iodide	14.1	11.6
Sodium	16.4	13.9
Sodium hydroxide, 50%	3.2	25.8
Stannic chloride	13.5	12.8
Succinonitrile	10.1	20.8
Sulfur dioxide	15.2	7.1
Sulfuric acid, 110%	7.2	27.4

(*Continued*)

TABLE A.1 (*Continued*)
Viscosities of Liquids (Coordinates Apply to Figure A.4)

Liquid	X	Y
Sulfuric acid, 100%	8.0	25.1
Sulfuric acid, 98%	7.0	24.8
Sulfuric acid, 60%	10.2	21.3
Sulfuryl chloride	15.2	12.4
Tetrachloroethane	11.9	15.7
Thiophene	13.2	11.0
Titanium tetrachloride	14.4	12.3
Toluene	13.7	10.4
Trichloroethylene	14.8	10.5
Triethylene glycol	4.7	24.8
Turpentine	11.5	14.9
Vinyl acetate	14.0	8.8
Vinyl toluene	13.4	12.0
Water	10.2	13.0
Xylene, ortho	13.5	12.1
Xylene, meta	13.9	10.6
Xylene, para	13.9	10.9

Source: Perry, R.H. and Green, D.W. (eds.), *Perry's Chemical Engineers' Handbook*, 7th edn., McGraw-Hill, New York, 1997. With permission.

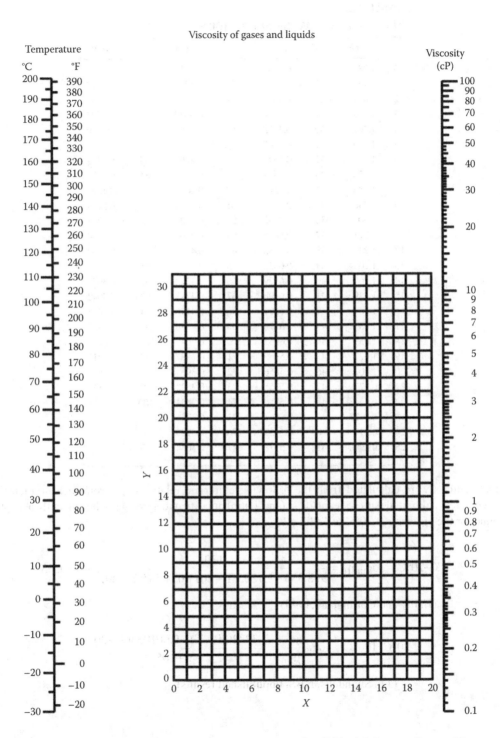

FIGURE A.4 Nomograph for viscosities of liquids at 1 atm. See Table A.1 for coordinates. (To convert centipoise to pascal-seconds, multiply by 0.001.)

TABLE A.2
The Viscosity of Water at 0°C–100°C

°C	μ (cP)	°C	μ (cP)	°C	μ (cP)	°C	μ (cP)
0	1.787	26	0.8705	52	0.5290	78	0.3638
1	1.728	27	0.8513	53	0.5204	79	0.3592
2	1.671	28	0.8327	54	0.5121	80	0.3547
3	1.618	29	0.8148	55	0.5040	81	0.3503
4	1.567	30	0.7975	56	0.4961	82	0.3460
5	1.519	31	0.7808	57	0.4884	83	0.3418
6	1.472	32	0.7647	58	0.4809	84	0.3377
7	1.428	33	0.7491	59	0.4736	85	0.3337
8	1.386	34	0.7340	60	0.4665	86	0.3297
9	1.346	35	0.7194	61	0.4596	87	0.3259
10	1.307	36	0.7052	62	0.4528	88	0.3221
11	1.271	37	0.6915	63	0.4462	89	0.3184
12	1.235	38	0.6783	64	0.4398	90	0.3147
13	1.202	39	0.6654	65	0.4335	91	0.3111
14	1.169	40	0.6529	66	0.4273	92	0.3076
15	1.139	41	0.6408	67	0.4213	93	0.3042
16	1.109	42	0.6391	68	0.4155	94	0.3008
17	1.081	43	0.6178	69	0.4098	95	0.2975
18	1.053	44	0.6067	70	0.4042	96	0.2942
19	1.027	45	0.5960	71	0.3987	97	0.2911
20	1.002	46	0.5856	72	0.3934	98	0.2879
21	0.9779	47	0.5755	73	0.3882	99	0.2848
22	0.9548	48	0.5656	74	0.3831	100	0.2818
23	0.9325	49	0.5561	75	0.3781		
24	0.9111	50	0.5468	76	0.3732		
25	0.8904	51	0.5378	77	0.3684		

Table entries were calculated from the following empirical relationships from measurements in viscometers calibrated with water at 20°C (and 1 atm), modified to agree with the currently accepted value for the viscosity at 20°C of 1.002 cP:

$$0°C–20°C: \quad \log_{10} \eta_T = \frac{1301}{998.333 + 8.1855(T-20) + 0.00585(T-20)^2} - 3.30233$$

(Hardy and Cottingham, 1949)

$$20°C \ 100°C: \quad \log_{10} \frac{\mu_T}{\mu_{20}} = \frac{1.3272(20-T) - 0.001053(T-20)^2}{T+105}$$

(J.F. Swindells, NBS, unpublished results)

TABLE A.3
Physical Properties of Ordinary Water and Common Liquids (SI Units)

Liquid	Temp T (°C)	Density ρ (kg/m³)	Specific Gravity S	Absolute Viscosity μ (N s/m²)	Kinematic Viscosity ν (m²/s)	Surface Tension σ (N/m)	Isothermal Bulk Modulus of Elasticity E_p (N/m²)	Coefficient of Thermal Expansion α_T (K⁻¹)
Water	0	1,000	1.000	1.79E–3	1.79E–6	7.56E–2	1.99E9	6.80E–5
	3.98	1,000	1.000	1.57	1.57	—	—	—
	10	1,000	1.000	1.31	1.31	7.42	2.12	8.80
	20	998	0.998	1.00	1.00	7.28	2.21	2.07E–4
	30	996	0.996	7.98E–4	7.12	7.26	2.94	—
	40	992	0.992	6.53	6.58	6.96	2.29	3.85
	50	988	0.988	5.47	5.48	6.79	2.29	4.58
	60	983	0.983	4.67	4.75	6.62	2.28	5.23
	70	978	0.978	4.04	4.13	6.64	2.24	5.84
	80	972	0.972	3.55	3.65	6.26	2.20	6.41
	90	965	0.965	3.15	3.26	—	2.14	6.96
	100	958	0.958	2.82	2.94	5.89	2.07	7.50
Mercury	0	13,600	13.60	1.68E–3	1.24E–7	—	2.50E10	—
	4	13,590	13.59	—	—	—	—	—
	20	13,550	13.55	1.55	1.14	37.5	2.50E10	1.82E–4
	40	13,500	13.50	1.45	1.07	—	—	1.82
	60	13,450	13.45	1.37	1.02	—	—	1.82
	80	13,400	13.40	1.30	9.70E–8	—	—	1.82
	100	13,350	13.35	1.24	9.29	—	—	—
Ethylene glycol	0	—	—	5.70E–2	—	—	—	—
	20	1,110	1.11	1.99	1.79E–5	—	—	—
	40	1,110	1.10	9.13E–3	8.30E–6	—	—	—
	60	1,090	1.09	4.95	4.54	—	—	—
	80	1,070	1.07	3.02	2.82	—	—	—
	100	1,060	1.06	1.99	1.88	—	—	—
Methyl alcohol (methanol)	0	810	0.810	8.17E–4	1.01E–6	2.45E–2	9.35E8	—
	10	801	0.801	—	—	2.26	8.78	—
	20	792	0.792	5.84	7.37E–7	—	8.23	—
	30	783	0.783	5.10	6.51	—	7.72	—
	40	774	0.774	4.50	5.81	—	7.23	—
	50	765	0.765	3.96	5.18	—	6.78	—
Ethyl alcohol (ethanol)	0	806	0.806	1.77E–3	2.20E–6	2.41E–2	1.02E9	—
	20	789	0.789	1.20	1.52	—	9.02E8	—
	40	772	0.772	8.34E–4	1.08	—	7.89	—
	60	754	0.754	5.92	7.85E–7	—	6.78	—
Normal octane	0	718	0.718	7.06E–7	9.83E–7	—	1.00E9	—
	16	—	—	5.74	—	—	—	—
	20	702	0.702	5.42	7.72	—	—	—
	25	—	—	—	—	—	8.35E8	—
	40	686	0.686	4.33	6.31	—	7.48	—
Benzene	0	900	0.900	9.12E–4	1.01E–6	3.02E–2	1.23E9	—
	20	879	0.879	6.52	7.42E–7	2.76	1.06	—
	40	858	0.857	5.03	5.86	—	9.10E8	—

(*Continued*)

TABLE A.3 (*Continued*)
Physical Properties of Ordinary Water and Common Liquids (SI Units)

Liquid	Temp T (°C)	Density ρ (kg/m³)	Specific Gravity S	Absolute Viscosity μ (N s/m²)	Kinematic Viscosity ν (m²/s)	Surface Tension σ (N/m)	Isothermal Bulk Modulus of Elasticity E_ρ (N/m²)	Coefficient of Thermal Expansion α_T (K⁻¹)
	60	836	0.836	3.92	4.69	—	7.78	—
	80	815	0.815	3.29	4.04	—	6.48	—
Kerosene	−18	841	0.841	7.06E−3	8.40E−6	—	—	—
	20	814	0.814	1.9	2.37	2.9E−2	—	—
Lubricating oil	20	871	0.871	1.31E−6	1.50E−9	—	—	—
	40	858	0.858	6.81E−5	7.94E−8	—	—	—
	60	845	0.845	4.18	4.95	—	—	—
	80	832	0.832	2.83	3.40	—	—	—
	100	820	0.820	2.00	2.44	—	—	—
	120	809	0.809	1.54	1.90	—	—	—

TABLE A.4
Physical Properties of Ordinary Water and Common Liquids (EE Units[a])

Liquid	Temp T (°F)	Density ρ (lb$_m$/ft³)	Specific Gravity S	Absolute Viscosity μ (lb$_f$ s/ft²)	Kinematic Viscosity v (ft²/s)	Surface Tension σ (lb$_f$/ft)	Isothermal Bulk Modulus of Elasticity E_ρ (lb$_f$/in.²)	Coefficient of Thermal Expansion α_T (°R^{-1})
Water	32	62.4	1.00	3.75E–5	1.93E–5	5.18E–3	2.93E–5	2.03E–3
	40	62.4	1.00	3.23	1.66	5.14	2.94	—
	60	62.4	0.999	2.36	1.22	5.04	3.11	—
	80	62.2	0.997	1.80	9.30E–6	4.92	3.22	—
	100	62.0	0.993	1.42	7.39	4.80	3.27	1.7
	120	61.7	0.988	1.17	6.09	4.65	3.33	—
	140	61.4	0.983	9.81E–6	5.14	4.54	3.30	—
	160	61.0	0.977	8.38	4.42	4.41	3.26	—
	180	60.6	0.970	7.26	3.85	4.26	3.13	—
	200	60.1	0.963	6.37	3.41	4.12	3.08	1.52
	212	59.8	0.958	5.93	3.19	4.04	3.00	—
Mercury	50	847	13.6	1.07E–3	1.2E–6	—	—	1.0E–4
	200	834	13.4	8.4E–3	1.0	—	—	1.0E–4
	300	826	13.2	7.4	9.0E–7	—	—	—
	400	817	13.1	6.7	8.0	—	—	—
	600	802	12.8	5.8	7.0	—	—	—
Ethylene glycol	68	69.3	1.11	4.16E–4	1.93E–4	—	—	—
	104	68.7	1.10	1.91	8.93E–5	—	—	—
	140	68.0	1.09	1.03	4.89	—	—	—
	176	66.8	1.07	6.31E–5	3.04	—	—	—
	212	66.2	1.06	4.12	2.02	—	—	—
Methyl alcohol (methanol)	32	50.6	0.810	1.71E–5	1.09E–5	1.68E–3	1.36E–5	—
	68	50.0	0.801	—	—	1.55	1.9	—
	104	49.4	0.792	1.22	7.93E–6	—	1.05	—
	140	48.9	0.783	1.07	7.01	—	—	—
	176	48.3	0.774	9.40E–6	6.25	—	—	—
	212	47.8	0.765	8.27	5.58	—	—	—
Ethyl alcohol (ethanol)	32	50.3	0.806	3.70E–5	2.37E–5	1.65E–3	1.48E–5	—
	68	49.8	0.789	3.03	1.96	—	1.31	—
	104	49.3	0.789	2.51	1.64	—	1.14	—
	140	48.2	0.772	1.74	1.16	—	9.83E–4	—
	176	47.7	0.754	1.24	8.45E–6	—	—	—
	212	47.1	0.745	—	—	—	—	—
Normal octane	32	44.8	0.718	1.47E–5	1.06E–5	—	1.45E–5	—
	68	43.8	0.702	1.13	8.31E–6	—	—	—
	104	42.8	0.686	9.04E–6	6.79	—	1.08	—
Benzene	32	56.2	0.900	1.90E–5	1.09E–5	2.07E–3	1.78E–5	—
	68	54.9	0.879	1.36	7.99E–6	1.89	1.53	—
	104	53.6	0.858	1.05	6.31	—	1.32	—
	140	52.2	0.836	8.19E–6	5.05	—	1.13	—
	176	50.9	0.815	6.87	4.35	—	9.40E–4	—
Kerosene	0	52.5	0.841	1.48E–4	9.05E–5	—	—	—
	77	50.8	0.814	3.97E–5	2.55E–5	—	—	—
Lubricating oil	68	54.5	0.871	2.74E–8	1.61E–8	—	—	—
	104	53.6	0.858	1.42E–7	8.55E–7	—	—	—
	140	52.6	0.845	8.73	5.33	—	—	—
	176	51.9	0.832	5.91	3.66	—	—	—
	212	51.2	0.820	4.18	2.63	—	—	—
	248	50.5	0.809	3.22	2.05	—	—	—

[a] EE, English engineering.

TABLE A.5
Physical Properties of SAE Oils and Lubricants

			SI Units				EE Units[a]	
			Kinematic Viscosity v (m²/s)				Kinematic Viscosity v (ft²/s)	
Fluid	Temp. (°C)	Specific Gravity	Minimum	Maximum	Temp. (°F)	Specific Gravity	Minimum	Maximum
Oil								
SAE 50	99	—	1.68E–5	2.27E–5	210	—	1.81E–4	2.44E–4
	99	—	1.29	1.68	210	—	1.08	1.81
	99	—	9.6E–4	1.29	210	—	1.03E–2	1.08
	99	—	—	5.7E–4	210	—	—	6.14E–3
	–18	0.92	2.60E–3	1.05E–2	0	0.92	2.80E–2	1.13E–1
	–18	0.92	1.30	2.60E–2	0	0.92	1.40	2.80E–2
	–18	0.92	—	1.30	0	0.92	—	1.40
Lubricants								
SAE 250	99	—	4.3E–5	—	210	—	4.6E–4	—
140	99	—	2.5	4.3E–5	210	—	2.7	4.6E–4
90	99	—	1.4	2.5	210	—	1.5	2.7
85W	99	—	1.1	—	210	—	1.2	—
80W	99	—	7.0E–6	—	210	—	7.5E–5	—
75W	99	—	4.2	—	210	—	4.5E–5	—

[a] EE, English engineering.

TABLE A.6
Viscosity of Steam and Water[a]

Viscosity of Steam and Water, μ (cP)

Temp. (°F)	1 (psia)	2 (psia)	5 (psia)	10 (psia)	20 (psia)	50 (psia)	100 (psia)	200 (psia)	500 (psia)	1,000 (psia)	2,000 (psia)	5,000 (psia)	7,500 (psia)	10,000 (psia)	12,000 (psia)
Sat. water	0.667	0.524	0.388	0.313	0.255	0.197	0.164	0.138	0.111	0.094	0.078	—	—	—	—
Sat. steam	0.010	0.010	0.011	0.012	0.012	0.013	0.014	0.015	0.017	0.019	0.023	—	—	—	—
1500	0.041	0.041	0.041	0.041	0.041	0.041	0.041	0.041	0.042	0.042	0.042	0.044	0.046	0.048	0.050
1450	0.040	0.040	0.040	0.040	0.040	0.040	0.040	0.040	0.040	0.041	0.041	0.043	0.045	0.047	0.049
1400	0.039	0.039	0.039	0.039	0.039	0.039	0.039	0.039	0.039	0.040	0.040	0.042	0.044	0.047	0.049
1350	0.038	0.038	0.038	0.038	0.038	0.038	0.038	0.038	0.038	0.038	0.039	0.041	0.044	0.046	0.049
1300	0.037	0.037	0.037	0.037	0.037	0.037	0.037	0.037	0.038	0.037	0.038	0.040	0.043	0.045	0.048
1250	0.035	0.035	0.035	0.035	0.035	0.035	0.035	0.036	0.036	0.036	0.037	0.039	0.042	0.045	0.048
1200	0.034	0.034	0.034	0.034	0.034	0.034	0.034	0.034	0.035	0.035	0.036	0.038	0.041	0.045	0.048
1150	0.034	0.034	0.034	0.034	0.034	0.034	0.034	0.034	0.034	0.034	0.034	0.037	0.041	0.045	0.049
1100	0.032	0.032	0.032	0.032	0.032	0.032	0.032	0.032	0.033	0.033	0.034	0.037	0.040	0.045	0.050
1050	0.031	0.031	0.031	0.031	0.031	0.031	0.031	0.031	0.032	0.032	0.033	0.036	0.040	0.047	0.052
1000	0.030	0.030	0.030	0.030	0.030	0.030	0.030	0.030	0.030	0.031	0.032	0.035	0.041	0.049	0.055
950	0.029	0.029	0.029	0.029	0.029	0.029	0.029	0.029	0.029	0.030	0.031	0.035	0.042	0.052	0.059
900	0.028	0.028	0.028	0.028	0.028	0.028	0.028	0.028	0.028	0.028	0.029	0.035	0.045	0.057	0.064
850	0.026	0.026	0.026	0.026	0.026	0.026	0.027	0.027	0.027	0.027	0.028	0.035	0.052	0.064	0.070
800	0.025	0.025	0.025	0.025	0.025	0.025	0.025	0.025	0.026	0.026	0.027	0.040	0.062	0.071	0.075
750	0.024	0.024	0.024	0.024	0.024	0.024	0.024	0.024	0.025	0.025	0.026	0.057	0.071	0.078	0.081
700	0.023	0.023	0.023	0.023	0.023	0.023	0.023	0.023	0.023	0.024	0.026[b]	0.071	0.079	0.085	0.086
650	0.022	0.022	0.022	0.022	0.022	0.022	0.022	0.022	0.023	0.023	0.023	0.082	0.088	0.092	0.096
600	0.021	0.021	0.021	0.021	0.021	0.021	0.021	0.021	0.021	0.021	0.087	0.091	0.096	0.101	0.104
550	0.020	0.020	0.020	0.020	0.020	0.020	0.020	0.020	0.020	0.019	0.095	0.101	0.105	0.109	0.113
500	0.019	0.019	0.019	0.019	0.019	0.019	0.019	0.018	0.018	0.103	0.105	0.111	0.114	0.119	0.122

(*Continued*)

TABLE A.6 (Continued)
Viscosity of Steam and Water[a]

Viscosity of Steam and Water, μ (cP)

Temp. (°F)	1 (psia)	2 (psia)	5 (psia)	10 (psia)	20 (psia)	50 (psia)	100 (psia)	200 (psia)	500 (psia)	1,000 (psia)	2,000 (psia)	5,000 (psia)	7,500 (psia)	10,000 (psia0	12,000 (psia)
450	0.018	0.018	0.018	0.018	0.017	0.017	0.017	0.017	0.115	0.116	0.118	0.123	0.127	0.131	0.135
400	0.106	0.106	0.106	0.106	0.106	0.106	0.106	0.106	0.131	0.132	0.134	0.138	0.143	0.147	0.150
350	0.015	0.015	0.015	0.015	0.015	0.015	0.015	0.152	0.153	0.154	0.155	0.160	0.164	0.168	0.171
300	0.014	0.014	0.014	0.014	0.014	0.014	0.182	0.183	0.183	0.184	0.185	0.190	0.194	0.198	0.201
250	0.013	0.013	0.013	0.013	0.013	0.228	0.228	0.228	0.228	0.229	0.231	0.235	0.238	0.242	0.245
200	0.012	0.012	0.012	0.012	0.300	0.300	0.300	0.300	0.301	0.301	0.303	0.306	0.310	0.313	0.316
150	0.011	0.011	0.427	0.427	0.427	0.427	0.427	0.427	0.427	0.428	0.429	0.431	0.434	0.437	0.439
100	0.680	0.680	0.680	0.680	0.680	0.680	0.680	0.680	0.680	0.680	0.680	0.681	0.682	0.683	0.683
50	1.299	1.299	1.299	1.299	1.299	1.299	1.299	1.299	1.299	1.298	1.296	1.289	1.284	1.279	1.275
32	1.753	1.753	1.753	1.753	1.753	1.753	1.753	1.752	1.751	1.749	1.745	1.733	1.723	1.713	1.705

[a] Values directly below underscored viscosities are for water.
[b] Critical point.

TABLE A.7
Viscosities of Gases[a] (Coordinates Apply to Figure A.5)

Gas	X	Y	$\mu \times 10^7$ (P)
Acetic acid	7.0	14.6	825 (50°C)
Acetone	8.4	13.2	735
Acetylene	9.3	15.5	1017
Air	10.4	20.4	1812
Ammonia	8.4	16.0	1000
Amylene (β)	8.6	12.2	676
Argon	9.7	22.6	2215
Arsine	8.6	20.0	1575
Benzene	8.7	13.2	746
Bromine	8.8	19.4	1495
Butane (η)	8.6	13.2	735
Butane (iso)	8.6	13.2	744
Butyl acetate (iso)	5.7	16.3	778
Butylene (α)	8.4	13.5	761
Butylene (β)	8.7	13.1	746
Butylene (iso)	8.3	13.9	786
Butyl formate (iso)	6.6	16.0	840
Cadmium	7.8	22.5	5690 (500)
Carbon dioxide	8.9	19.1	1463
Carbon disulfide	8.5	15.8	990
Carbon monoxide	10.5	20.0	1749
Carbon oxysulfide	8.2	17.9	1220
Carbon tetrachloride	8.0	15.3	966
Chlorine	8.8	18.3	1335
Chloroform	8.8	15.7	1000
Cyanogen	8.2	16.2	1002
Cyclohexane	9.0	12.2	701
Cyclopropane	8.3	14.7	870
Deuterium	11.0	16.2	1240
Diethyl ether	8.8	12.7	730
Dimethyl ether	9.0	15.0	925
Diphenyl ether	8.6	10.4	610 (50)
Diphenyl methane	8.0	10.3	605 (50)
Ethane	9.0	14.5	915
Ethanol	8.2	14.5	835
Ethyl acetate	8.4	13.4	743
Ethyl chloride	8.5	15.6	987
Ethylene	9.5	15.2	1010
Ethyl propionate	12.0	12.4	890
Fluoride	7.3	23.8	2250
Freon-11	8.6	16.2	1298 (93)
Freon-12	9.0	17.4	1496 (93)
Freon-14	9.5	20.4	1716
Freon-21	9.0	16.7	1389 (93)

(*Continued*)

TABLE A.7 (*Continued*)
Viscosities of Gases[a] (Coordinates Apply to Figure A.5)

Gas	X	Y	$\mu \times 10^7$ (P)
Freon-22	9.0	17.7	1554 (93)
Freon-113	11.0	14.0	1166 (93)
Freon-114	9.4	16.4	1364 (93)
Helium	11.3	20.8	1946
Heptane (*n*)	9.6	10.6	618 (50)
Hexane (*n*)	8.4	12.0	644
Hydrogen	11.3	12.4	880
Hydrogen–helium			
10% H_2, 90% He	11.0	20.5	1780 (0)
25% H_2, 75% He	11.0	19.4	1603 (0)
40% H_2, 60% He	10.7	18.4	1431 (0)
60% H_2, 40% He	10.8	16.7	1227 (0)
81% H_2, 19% He	10.5	15.0	1016 (0)
Hydrogen–sulfur dioxide	8.7	18.1	1259 (17)
10% H_2, 90% SO_2	8.7	18.1	1259 (17)
20% H_2, 80% SO_2	8.6	18.2	1277 (17)
50% H_2, 50% SO_2	8.9	18.3	1332 (17)
80% H_2, 20% SO_2	9.7	17.7	1306 (17)
Hydrogen bromide	8.4	21.6	1843
Hydrogen chloride	8.5	19.2	1425
Hydrogen cyanide	7.1	14.5	737
Hydrogen iodide	8.5	21.5	1830
Hydrogen sulfide	8.4	18.0	1265
Iodine	8.7	18.7	1730 (100)
Krypton	9.4	24.0	2480
Mercury	7.4	24.9	4500 (200)
Mercury bromide	8.5	19.0	2253
Mercuric chloride	7.7	18.7	2200 (200)
Mercuric iodide	8.4	18.0	2045 (200)
Mesitylene	9.5	10.2	660 (50)
Methane	9.5	15.8	1092
Methane (deuterated)	9.5	17.6	1290
Methanol	8.3	15.6	935
Methyl acetate	8.4	14.0	870 (50)
Methyl acetylene	8.9	14.3	867
3-Methyl-1-butene	8.0	13.3	716
Methyl butyrate (iso)	6.6	15.8	824
Methyl bromide	8.1	18.7	1327
Methyl bromide	8.1	18.7	1327
Methyl chloride	8.5	16.5	1062
3-Methylene-1-butene	8.0	13.3	716
Methylene chloride	8.5	15.8	989
Methyl formate	5.1	18.0	923
Neon	11.1	25.8	3113
Nitric oxide	10.4	20.8	1899

(*Continued*)

TABLE A.7 (*Continued*)

Viscosities of Gases[a] (Coordinates Apply to Figure A.5)

Gas	X	Y	$\mu \times 10^7$ (P)
Nitrogen	10.6	20.0	1766
Nitrous oxide	9.0	19.0	1460
Nonane (*n*)	9.2	8.9	554 (50)
Octane (*n*)	8.8	9.8	586 (50)
Oxygen	10.2	21.6	2026
Pentene (*n*)	8.5	12.3	668
Pentane (iso)	8.9	12.1	685
Phosphene	8.8	17.0	1150
Propane	8.9	13.5	800
Propanol (*n*)	8.4	13.5	770
Propanol (iso)	8.4	13.6	774
Propyl acetate	8.0	14.3	797
Propylene	8.5	14.4	840
Pyridine	8.6	13.3	830 (50)
Silane	9.0	16.8	1148
Stannic chloride	9.1	16.0	1330 (100)
Stannic bromide	9.0	16.7	142 (100)
Sulfur dioxide	8.4	18.2	1250
Thiazole	10.0	14.4	958
Thiophene	8.3	14.2	901 (50)
Toluene	8.6	12.5	686
2,2,3-Trimethylbutane	10.0	10.4	691 (50)
Trimethylethane	8.0	13.0	686
Water	8.0	16.0	1250 (100)
Xenon	9.3	23.0	2255
Zinc	8.0	22.0	5250 (500)

Source: Perry, R.H. and Green, D.W. (eds.), *Perry's Chemical Engineers' Handbook*, 7th edn., McGraw-Hill, New York, 1997. With permission.

[a] Viscosity at 20°C unless otherwise indicated.

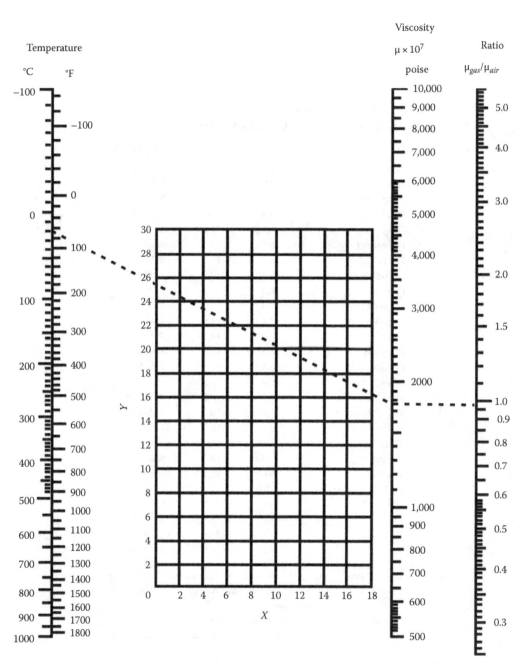

FIGURE A.5 Nomograph for determining the absolute viscosity of a gas near ambient pressure and the relative viscosity of a gas compared with air. (Coordinates from Table A.7.) To convert from poise to pascal-seconds, multiply by 0.1. (From Beerman, *Meas. Control.*, June 1982, 154–157.)

A.1 VISCOSITIES OF GASES AND LIQUIDS

The curves for hydrocarbon vapors and natural gases in the chart at the upper right are taken from Maxwell, and the curves for all other gases (except helium) in the chart are based upon Sutherland's formula, as follows:

$$\mu = \mu_0 \left(\frac{0.555T_0 C}{0.555T + C} \right) \left(\frac{T}{T_0} \right)^{3/2}$$

where
   μ is the viscosity in cP at temperature T
   μ_0 is the viscosity in cP at temperature T_0
   T is the absolute temperature, in °R (460 + °F), for which viscosity is desired
   T_0 is the absolute temperature, in °R, for which viscosity is known
   C is Sutherland's constant

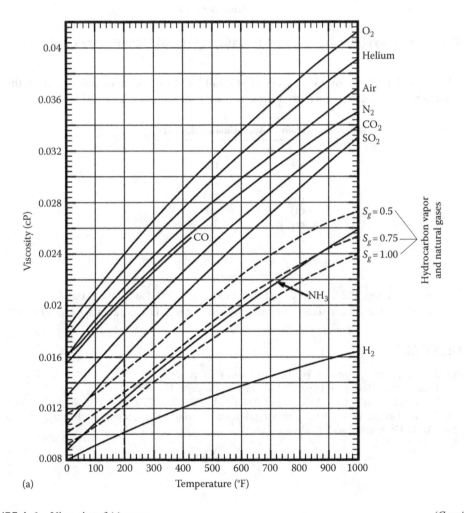

(a)

FIGURE A.6 Viscosity of (a) gases.   (*Continued*)

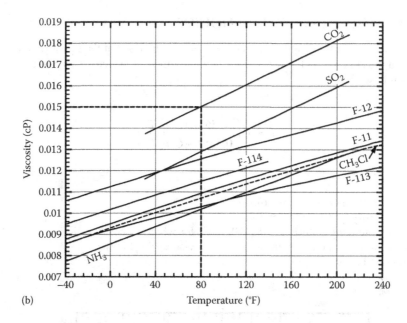

FIGURE A.6 Viscosity of (b) refrigerant vapors. (From Crane Technical Paper 410, Crane Co., Chicago, IL, 1991.)

Note: The variation of viscosity with pressure is small for most gases. For gases given on this page, the correction of viscosity for pressure is less than 10% for pressures up to 500 psi.

Fluid	Approximate Values of C
O_2	127
Air	120
N_2	111
CO_2	240
CO	118
SO_2	416
NH_3	370
H_2	72

Example (*Figure A.6a*): The viscosity of sulfur dioxide gas (SO_2) at 200°F is 0.016 cP.
Example (*Figure A.6b*): The viscosity of carbon dioxide gas (CO_2) at about 80°F is 0.015 cP.

REFERENCES

Brown, G.G. et al., *Unit Operations*, Wiley, New York, 1951, p. 586.
Crane Company, Crane Technical Paper 410, Crane Co., Chicago, IL, 1991.
Hardy, R.C. and R.L. Cottingham, Viscosity of Water at T −20 to 150°C, *J. Res. NBS*, 42, 573, 1949.
Perry, R.H. and Green, D.W. (eds.), *Perry's Chemical Engineers' Handbook*, 7th edn., McGraw-Hill, New York, 1997.

Appendix B: Generalized Viscosity Plot

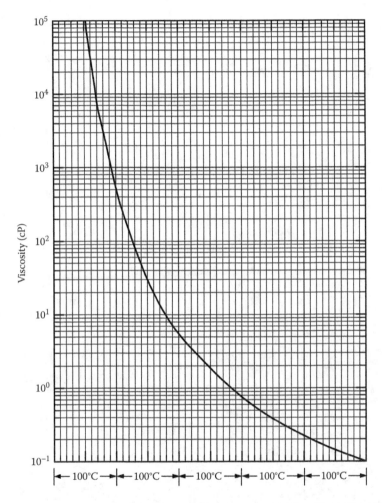

FIGURE B.1 Curves for estimating viscosity from a single measurement value. (From Gambill, W.R., *Chem. Eng.*, 66, 127, 1959.)

REFERENCE

Gambill, W.R., How P & T change liquid viscosity, *Chem. Eng.*, 66, 127, 1959.

Appendix C: Properties of Gases

TABLE C.1
Physical Properties of Gases (Approximate Values at 68°F and 14.8 psia)

Name of Gas	Chemical Formula or Symbol	Approx. Molecular Weight, M	Weight Density ρ (lb$_m$/ft³)	Specific Gravity Relative to Air, S_0	Individual Gas Constant, R (ft.lb$_f$/lb$_m$·°R)	Specific Heat at Room Temperature (Btu/lb°F)		Heat Capacity (BTU/ft³) [°R]		k (Btu/ft³°F)
						c_p^a	c_v^b (lb$_m$)	c_p	c_v	c_p/c_v
Acetylene	C_2H_2	26.0	0.0682	0.907	59.4	0.350	0.269	0.0239	0.0184	1.30
Air	—	29.0	0.0752	1.000	53.3	0.241	0.172	0.0181	0.0129	1.40
Ammonia	NH_3	17.0	0.0448	0.596	91.0	0.523	0.396	0.0234	0.0178	1.32
Argon	Ar	39.9	0.1037	1.379	38.7	0.124	0.074	0.0129	0.0077	1.67
Butane	C_4H_{10}	58.1	0.1554	2.067	26.5	0.395	0.356	0.0614	0.0553	1.11
Carbon dioxide	CO_2	44.0	0.1150	1.529	35.1	0.205	0.158	0.0236	0.0181	1.30
Carbon monoxide	CO	28.0	0.0727	0.967	55.2	0.243	0.173	0.0177	0.0126	1.40
Chlorine	Cl_2	70.9	0.1869	2.486	21.8	0.115	0.086	0.0215	0.0162	1.33
Ethane	C_2H_6	30.0	0.0789	1.049	51.5	0.386	0.316	0.0305	0.0250	1.22
Ethylene	C_2H_4	28.0	0.0733	0.975	55.1	0.400	0.329	0.0293	0.0240	1.22
Helium	He	4.0	0.01039	0.1381	386.3	1.250	0.754	0.130	0.0078	1.66
Hydrogen chloride	HCl	36.5	0.0954	1.268	42.4	0.191	0.135	0.0182	0.0129	1.41
Hydrogen	H_2	2.0	0.00523	0.0695	766.8	3.420	2.426	0.0179	0.0127	1.41
Hydrogen sulfide	H_2S	34.1	0.0895	1.190	45.2	0.243	0.187	0.0217	0.0167	1.30
Methane	CH_4	16.0	0.0417	0.554	96.4	0.593	0.449	0.0247	0.0187	1.32
Methyl chloride	CH_3Cl	50.5	0.1342	1.785	30.6	0.240	0.200	0.0322	0.0268	1.20
Natural gas	—	19.5	0.0502	0.667	79.1	0.560	0.441	0.0281	0.0221	1.27
Nitric oxide	NO	30.0	0.0780	1.037	51.5	0.231	0.165	0.0180	0.0129	1.40
Nitrogen	N_2	28.0	0.0727	0.967	55.2	0.247	0.176	0.0180	0.0127	1.41
Nitrous oxide	N_2O	44.0	0.1151	1.530	35.1	0.221	0.169	0.0254	0.0194	1.31
Oxygen	O_2	32.0	0.0831	1.105	48.3	0.217	0.155	0.0180	0.0129	1.40
Propane	C_3H_8	44.1	0.1175	1.562	35.0	0.393	0.342	0.0462	0.0402	1.15
Propene (propylene)	C_3H_6	42.1	0.1091	1.451	36.8	0.358	0.314	0.0391	0.0343	1.14
Sulfur dioxide	SO_2	64.1	0.1703	2.264	24.0	0.154	0.122	0.0262	0.0208	1.26

Source: Molecular weight, specific gravity, individual gas constant, and specific heat values were based on data in Avallone, E. et al., Mark's *Standard Handbook for Mechanical Engineers*, 11th edn., McGraw Hill, New York, 2006.

Weight density values were obtained by multiplying the density of air by the specific gravity of gas. For values of 60°F, multiply by 1.0154. Natural gas values were representative only. Exact characteristics require knowledge of specific constituents.

[a] Specific heat at constant pressure.
[b] Specific heat at constant volume.

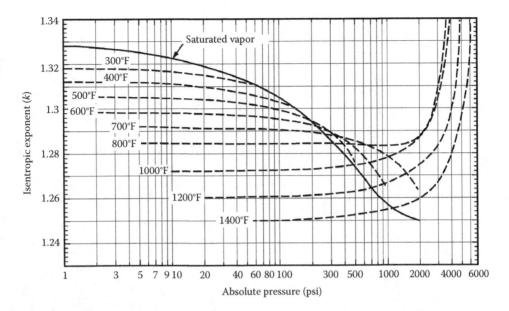

FIGURE C.1 Steam values of isentropic exponent, k (for small changes in pressure (or volume) along an isentrope, pV^k = constant). (From Avallone, E. et al., *Mark's Standard Handbook for Mechanical Engineers*, 11th edn., McGraw Hill, New York, 2006.)

REFERENCE

Avallone, E. et al., Mark's Standard Handbook for Mechanical Engineers, 11th edn., McGraw Hill, New York, 2006.

Appendix D: Pressure–Enthalpy Diagrams for Various Compounds

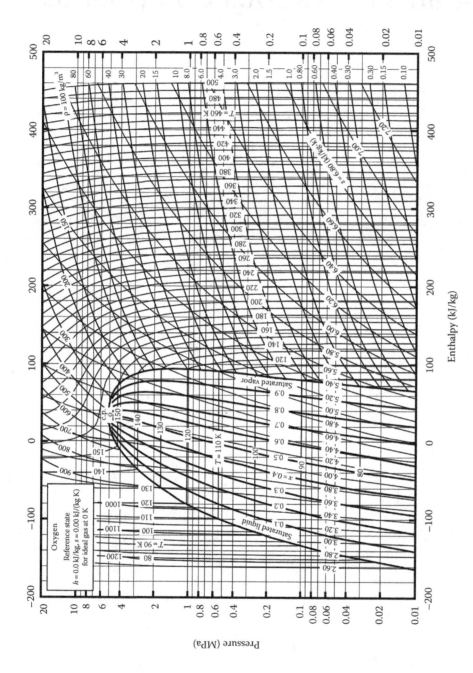

FIGURE D.1   Oxygen pressure–enthalpy diagram. (Adapted from Canjar, L.N. and Manning, F.S., *Thermodynamic Properties and Reduced Correlations for Gases*, Gulf Publishing, Houston, TX, 1967. With permission.)

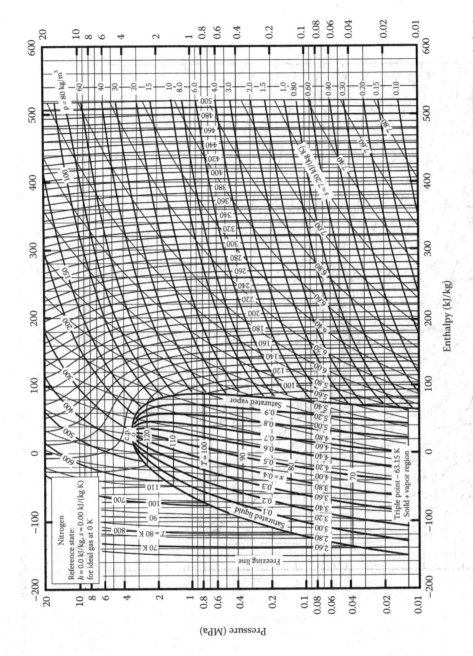

FIGURE D.2 Nitrogen pressure–enthalpy diagram. (Adapted from Tejada, V.M. et al., *Hydrocarbon Process.*, 45(3), 137, 1966. With permission.)

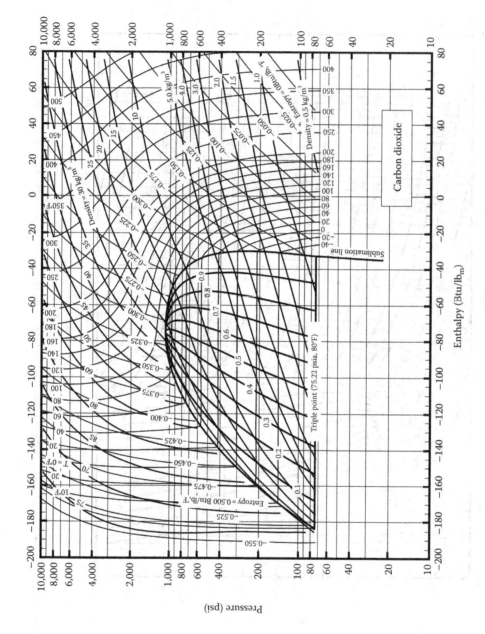

FIGURE D.3   Pressure–enthalpy chart for carbon dioxide. (Adapted from *ASHRAE Handbook of Fundamentals*, 1967.)

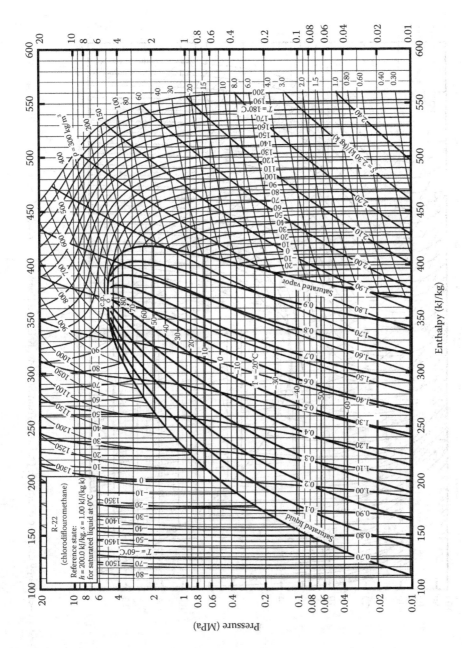

FIGURE D.4 Enthalpy–log pressure diagram for Refrigerant 22. Temperature in °C, entropy in kJ/kg°C, quality in weight percent. (Reprinted by permission of El du Pont de Nemours & Company, 1967.)

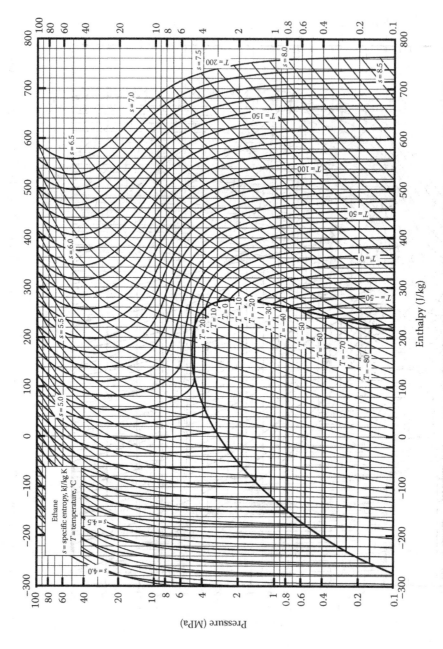

FIGURE D.5   Pressure–enthalpy diagram of ethane. (Adapted from Starling, K.E. and Kwok, Y.C., *Hydrocarbon Process.*, 50(4), 140, 1971.)

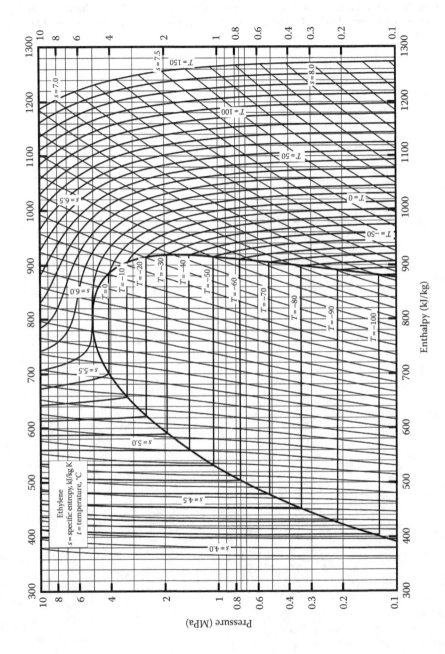

FIGURE D.6 Pressure–enthalpy diagram for ethylene. (Adapted from Starling, K.E., *Fluid Thermodynamic Properties for Light Petroleum Systems*, Gulf Publishing, Houston, TX , 1973. With permission.)

REFERENCES

Canjar, L.N. and F.S. Manning, *Thermodynamic Properties and Reduced Correlations for Gases*, Gulf
   Publishing, Houston, TX, 1967.
Starling, K.E., *Fluid Thermodynamic Properties for Light Petroleum Systems*, Gulf Publishing, Houston, TX,
   1973.
Tejada, V.M., B. Santaorez, L. Lema, A. Lozano, J. Penoranda, and L.M. Bahalla, Thermodynamic properties of non-hydrocarbons, Part 5: Thermodynamic properties of nitrogen. *Hydrocarbon Process*, 45(3),
   137–140, 1966.

Appendix E: Microscopic Conservation Equations in Rectangular, Cylindrical, and Spherical Coordinates

E.1 CONTINUITY EQUATION

Rectangular coordinates (x, y, z):

$$\frac{\partial \rho}{\partial t} + \frac{\partial}{\partial x}(\rho v_x) + \frac{\partial}{\partial y}(\rho v_y) + \frac{\partial}{\partial z}(\rho v_z) = 0$$

Cylindrical coordinates (r, θ, z):

$$\frac{\partial \rho}{\partial t} + \frac{1}{r}\frac{\partial}{\partial r}(\rho r v_r) + \frac{1}{r}\frac{\partial}{\partial \theta}(\rho v_\theta) + \frac{\partial}{\partial z}(\rho v_z) = 0$$

Spherical coordinates (r, θ, ϕ):

$$\frac{\partial \rho}{\partial t} + \frac{1}{r^2}\frac{\partial}{\partial r}(\rho r^2 v_r) + \frac{1}{r\sin\theta}\frac{\partial}{\partial \theta}(\rho v_\theta \sin\theta) + \frac{1}{r\sin\theta}\frac{\partial}{\partial \phi}(\rho v_\phi) = 0$$

E.2 MOMENTUM EQUATION IN RECTANGULAR COORDINATES

x-component:

$$\rho\left(\frac{\partial v_x}{\partial t} + v_x\frac{\partial v_x}{\partial x} + v_y\frac{\partial v_x}{\partial y} + v_z\frac{\partial v_x}{\partial z}\right)$$

$$= -\frac{\partial P}{\partial x} + \left(\frac{\partial \tau_{xx}}{\partial x} + \frac{\partial \tau_{yx}}{\partial y} + \frac{\partial \tau_{zx}}{\partial z}\right) + \rho g_x$$

y-component:

$$\rho\left(\frac{\partial v_y}{\partial t} + v_x\frac{\partial v_y}{\partial x} + v_y\frac{\partial v_y}{\partial y} + v_z\frac{\partial v_y}{\partial z}\right)$$

$$= -\frac{\partial P}{\partial y} + \left(\frac{\partial \tau_{xy}}{\partial x} + \frac{\partial \tau_{yy}}{\partial y} + \frac{\partial \tau_{zy}}{\partial z}\right) + \rho g_y$$

z-component:

$$\rho\left(\frac{\partial v_z}{\partial t} + v_x \frac{\partial v_z}{\partial x} + v_y \frac{\partial v_z}{\partial y} + v_z \frac{\partial v_z}{\partial z}\right)$$

$$= -\frac{\partial P}{\partial y} + \left(\frac{\partial \tau_{xz}}{\partial x} + \frac{\partial \tau_{yz}}{\partial y} + \frac{\partial \tau_{zz}}{\partial z}\right) + \rho g_z$$

E.3   MOMENTUM EQUATION IN CYLINDRICAL COORDINATES

r-component:

$$\rho\left(\frac{\partial v_r}{\partial t} + v_r \frac{\partial v_r}{\partial r} + \frac{v_\theta}{r} \frac{\partial v_r}{\partial \theta} - \frac{v_\theta^2}{r} + v_z \frac{\partial v_r}{\partial z}\right)$$

$$= -\frac{\partial P}{\partial r} + \left(\frac{1}{r}\frac{\partial}{\partial r}\left(r\tau_{rr}\right) + \frac{1}{r}\frac{\partial \tau_{r\theta}}{\partial \theta} - \frac{\tau_{\theta\theta}}{r} + \frac{\partial \tau_{rz}}{\partial z}\right) + \rho g_r$$

θ-*component*:

$$\rho\left(\frac{\partial v_\theta}{\partial t} + v_r \frac{\partial v_\theta}{\partial r} + \frac{v_\theta}{r} \frac{\partial v_\theta}{\partial \theta} - \frac{v_r v_\theta}{r} + v_z \frac{\partial v_\theta}{\partial z}\right)$$

$$= -\frac{1}{r}\frac{\partial P}{\partial \theta} + \left(\frac{1}{r^2}\frac{\partial}{\partial r}\left(r^2\tau_{r\theta}\right) + \frac{1}{r}\frac{\partial \tau_{\theta\theta}}{\partial \theta} + \frac{\partial \tau_{\theta z}}{\partial z}\right) + \rho g_\theta$$

z-component:

$$\rho\left(\frac{\partial v_z}{\partial t} + v_r \frac{\partial v_z}{\partial r} + \frac{v_\theta}{r} \frac{\partial v_z}{\partial \theta} + v_z \frac{\partial v_z}{\partial z}\right)$$

$$= -\frac{\partial P}{\partial z} + \left(\frac{1}{r}\frac{\partial}{\partial r}\left(r\tau_{rz}\right) + \frac{1}{r}\frac{\partial \tau_{\theta z}}{\partial \theta} + \frac{\partial \tau_{zz}}{\partial z}\right) + \rho g_z$$

E.4   MOMENTUM EQUATION IN SPHERICAL COORDINATES

r-component:

$$\rho\left(\frac{\partial v_r}{\partial t} + v_r \frac{\partial v_r}{\partial r} + \frac{v_\theta}{r} \frac{\partial v_r}{\partial \theta} + \frac{v_\phi}{r\sin\theta} \frac{\partial v_r}{\partial \phi} - \frac{v_\theta^2 + v_\phi^2}{r}\right)$$

$$= -\frac{\partial P}{\partial r} + \left(\frac{1}{r^2}\frac{\partial}{\partial r}(r^2\tau_{rr}) + \frac{1}{r\sin\theta}\frac{\partial}{\partial \theta}(\tau_{r\theta}\sin\theta)\right.$$

$$\left. + \frac{1}{r\sin\theta}\frac{\partial \tau_{r\phi}}{\partial \phi} - \frac{\tau_{\theta\theta} + \tau_{\phi\phi}}{r}\right) + \rho g_r$$

θ-*component*:

$$\rho\left(\frac{\partial v_\theta}{\partial t}+v_r\frac{\partial v_\theta}{\partial r}+\frac{v_\theta}{r}\frac{\partial v_\theta}{\partial \theta}+\frac{v_\phi}{r\sin\theta}\frac{\partial v_\theta}{\partial \phi}+\frac{v_r v_\theta}{r}-\frac{v_\phi^2\cot\theta}{r}\right)$$

$$=-\frac{1}{r}\frac{\partial P}{\partial \theta}+\left(\frac{1}{r^2}\frac{\partial}{\partial r}(r^2\tau_{r\theta})+\frac{1}{r\sin\theta}\frac{\partial}{\partial \theta}(\tau_{\theta\theta}\sin\theta)+\frac{1}{r\sin\theta}\frac{\partial\tau_{\theta\phi}}{\partial \phi}\right.$$

$$\left.+\frac{\tau_{r\theta}}{r}-\frac{\cot\theta}{r}\tau_{\phi\phi}\right)+\rho g_\theta$$

φ-*component*:

$$\rho\left(\frac{\partial v_\phi}{\partial t}+v_r\frac{\partial v_\phi}{\partial r}+\frac{v_\theta}{r}\frac{\partial v_\phi}{\partial \theta}+\frac{v_\phi}{r\sin\theta}\frac{\partial v_\phi}{\partial \phi}+\frac{v_\phi v_r}{r}+\frac{v_\theta v_\phi}{r}\cot\theta\right)$$

$$=-\frac{1}{r\sin\theta}\frac{\partial P}{\partial \phi}+\left(\frac{1}{r^2}\frac{\partial}{\partial r}(r^2\tau_{r\phi})+\frac{1}{r}\frac{\partial\tau_{\theta\phi}}{\partial \theta}+\frac{1}{r\sin\theta}\frac{\partial\tau_{\phi\phi}}{\partial \phi}\right.$$

$$\left.+\frac{\tau_{r\phi}}{r}+\frac{2\cot\theta}{r}\tau_{\theta\phi}\right)+\rho g_\phi$$

E.5 COMPONENTS OF THE STRESS TENSOR τ

Rectangular coordinates:

$$\tau_{xx}=+\mu\left[2\frac{\partial v_x}{\partial x}-\frac{2}{3}(\nabla\cdot v)\right]+\kappa(\nabla\cdot v)$$

$$\tau_{yy}=+\mu\left[2\frac{\partial v_y}{\partial y}-\frac{2}{3}(\nabla\cdot v)\right]+\kappa(\nabla\cdot v)$$

$$\tau_{zz}=+\mu\left[2\frac{\partial v_z}{\partial z}-\frac{2}{3}(\nabla\cdot v)\right]+\kappa(\nabla\cdot v)$$

$$\tau_{xy}=\tau_{yx}=+\mu\left[\frac{\partial v_x}{\partial y}+\frac{\partial v_y}{\partial x}\right]$$

$$\tau_{yz}=\tau_{zy}=+\mu\left[\frac{\partial v_y}{\partial z}+\frac{\partial v_z}{\partial y}\right]$$

$$\tau_{zx}=\tau_{xz}=+\mu\left[\frac{\partial v_z}{\partial x}+\frac{\partial v_x}{\partial z}\right]$$

$$(\nabla\cdot v)=\frac{\partial v_x}{\partial x}+\frac{\partial v_y}{\partial y}+\frac{\partial v_z}{\partial z}$$

Cylindrical coordinates:

$$\tau_{rr} = +\mu\left[2\frac{\partial v_r}{\partial r} - \frac{2}{3}(\nabla \cdot v)\right] + \kappa(\nabla \cdot v)$$

$$\tau_{\theta\theta} = +\mu\left[2\left(\frac{1}{r}\frac{\partial v_\theta}{\partial\theta} + \frac{v_r}{r}\right) - \frac{2}{3}(\nabla \cdot v)\right] + \kappa(\nabla \cdot v)$$

$$\tau_{zz} = +\mu\left[2\frac{\partial v_z}{\partial z} - \frac{2}{3}(\nabla \cdot v)\right] + \kappa(\nabla \cdot v)$$

$$\tau_{r\theta} = \tau_{\theta r} = +\mu\left[r\frac{\partial}{\partial r}\left(\frac{v_\theta}{r}\right) + \frac{1}{r}\frac{\partial v_r}{\partial\theta}\right]$$

$$\tau_{\theta z} = \tau_{z\theta} = +\mu\left[\frac{\partial v_\theta}{\partial z} + \frac{1}{r}\frac{\partial v_z}{\partial\theta}\right]$$

$$\tau_{zr} = \tau_{rz} = +\mu\left[\frac{\partial v_z}{\partial r} + \frac{\partial v_r}{\partial z}\right]$$

$$(\nabla \cdot v) = \frac{1}{r}\frac{\partial}{\partial r}(rv_r) + \frac{1}{r}\frac{\partial v_\theta}{\partial\theta} + \frac{\partial v_z}{\partial z}$$

Spherical coordinates:

$$\tau_{rr} = +\mu\left[2\frac{\partial v_r}{\partial r} - \frac{2}{3}(\nabla \cdot v)\right] + \kappa(\nabla \cdot v)$$

$$\tau_{\theta\theta} = +\mu\left[2\left(\frac{1}{r}\frac{\partial v_\theta}{\partial\theta} + \frac{v_r}{r}\right) - \frac{2}{3}(\nabla \cdot v)\right] + \kappa(\nabla \cdot v)$$

$$\tau_{\phi\phi} = +\mu\left[2\left(\frac{1}{r\sin\theta}\frac{\partial v_\phi}{\partial\phi} + \frac{v_r}{r} + \frac{v_\theta\cot\theta}{r}\right) - \frac{2}{3}(\nabla \cdot v)\right] + \kappa(\nabla \cdot v)$$

$$\tau_{r\theta} = \tau_{\theta r} = +\mu\left[r\frac{\partial}{\partial r}\left(\frac{v_\theta}{r}\right) + \frac{1}{r}\frac{\partial v_r}{\partial\theta}\right]$$

$$\tau_{\theta\phi} = \tau_{\phi\theta} = +\mu\left[\frac{\sin\theta}{r}\frac{\partial}{\partial\theta}\left(\frac{v_\phi}{\sin\theta}\right) + \frac{1}{r\sin\theta}\frac{\partial v_\phi}{\partial\theta}\right]$$

$$\tau_{\phi r} = \tau_{r\phi} = +\mu\left[\frac{1}{r\sin\theta}\frac{\partial v_r}{\partial\phi} + r\frac{\partial}{\partial r}\left(\frac{v_\phi}{r}\right)\right]$$

$$(\nabla \cdot v) = \frac{1}{r^2}\frac{\partial}{\partial r}(r^2 v_r) + \frac{1}{r\sin\theta}\frac{\partial}{\partial\theta}(v_\theta\sin\theta) + \frac{1}{r\sin\theta}\frac{\partial v_\phi}{\partial\phi}$$

Appendix F: Standard Steel Pipe Dimensions and Capacities

TABLE F.1

Standard Steel Pipe Dimensions and Capacities[a]

Nominal Pipe Size (in.)	Outside Diameter (in.)	Schedule No.	Wall Thickness (in.)	Inside Diameter (in.)	Cross-Sectional Area Metal (in.²)	Cross-Sectional Area Flow (ft²)	Circumference (ft) or Surface (ft/ft of Length) Outside	Circumference (ft) or Surface (ft/ft of Length) Inside	Capacity at 1 ft/s Velocity U.S. gpm	Capacity at 1 ft/s Velocity lb$_m$/h Water
$\frac{1}{8}$	0.405	10S	0.949	0.307	0.055	0.00051	0.106	0.0804	0.231	115.5
		40ST, 40S	1.068	0.269	0.072	0.00040	0.106	0.0705	0.179	89.5
		80X, 80S	0.095	0.215	0.093	0.00025	0.106	0.0563	0.113	56.5
$\frac{1}{4}$	0.540	10S	0.065	0.410	0.097	0.00092	0.141	0.107	0.412	206.5
		40ST, 40S	0.088	0.364	0.125	0.00072	0.141	0.095	0.323	161.5
		80XS, 80S	0.119	0.302	0.157	0.00050	0.141	0.079	0.224	112.0
$\frac{3}{8}$	0.675	10S	0.065	0.545	0.125	0.00162	0.177	0.143	0.727	363.5
		40ST, 40S	0.091	0.493	0.167	0.00133	0.177	0.129	0.596	298.0
		80XS, 80S	0.126	0.423	0.217	0.00098	0.177	0.111	0.440	220.0
$\frac{1}{2}$	0.840	5S	0.065	0.710	0.158	0.00275	0.220	0.186	1.234	617.0
		10S	0.083	0.674	0.197	0.00248	0.220	0.176	1.112	556.0
		40ST, 40S	0.109	0.622	0.250	0.00211	0.220	0.163	0.945	472.0
		80XS, 80S	0.147	0.546	0.320	0.00163	0.220	0.143	0.730	365.0
		160	0.188	0.464	0.385	0.00117	0.220	0.122	0.527	263.5
		XX	0.294	0.252	0.504	0.00035	0.220	0.066	0.155	77.5
$\frac{3}{4}$	1.050	5S	0.065	0.920	0.201	0.00461	0.275	0.241	2.072	1036.0
		10S	0.083	0.884	0.252	0.00426	0.275	0.231	1.903	951.5
		40ST, 40S	0.113	0.824	0.333	0.00371	0.275	0.216	1.665	832.5
		80XS, 80S	0.154	0.742	0.433	0.00300	0.275	0.194	1.345	672.5
		160	0.219	0.612	0.572	0.00204	0.275	0.160	0.917	458.5
		XX	0.308	0.434	0.718	0.00103	0.275	0.114	0.461	230.5
1	1.315	5S	0.065	1.185	0.255	0.00768	0.344	0.310	3.449	1,725
		10S	0.109	1.097	0.413	0.00656	0.344	0.287	2.946	1,473
		40ST, 40S	0.133	1.049	0.494	0.00600	0.344	0.275	2.690	1,345
		80XS, 80S	0.179	0.957	0.639	0.00499	0.344	0.250	2.240	1,120
		160	0.250	0.815	0.836	0.00362	0.344	0.213	1.625	812.5
		XX	0.358	0.599	1.076	0.00196	0.344	0.157	0.878	439.0
$1\frac{1}{4}$	1.660	5S	0.065	1.530	0.326	0.01277	0.435	0.401	5.73	2,865
		10S	0.109	1.442	0.531	0.01134	0.435	0.378	5.09	2,545
		40ST, 40S	0.140	1.380	0.668	0.01040	0.435	0.361	4.57	2,285
		80XS, 80S	0.191	1.278	0.881	0.0891	0.435	0.335	3.99	1,995
		160	0.250	1.160	1.107	0.00734	0.435	0.304	3.29	1,645
		XX	0.382	0.896	1.534	0.00438	0.435	0.235	1.97	985

(Continued)

TABLE F.1 (*Continued*)

Standard Steel Pipe Dimensions and Capacities[a]

Nominal Pipe Size (in.)	Outside Diameter (in.)	Schedule No.	Wall Thickness (in.)	Inside Diameter (in.)	Cross-Sectional Area Metal (in.²)	Cross-Sectional Area Flow (ft²)	Circumference (ft) or Surface (ft/ft of Length) Outside	Circumference (ft) or Surface (ft/ft of Length) Inside	Capacity at 1 ft/s Velocity U.S. gpm	Capacity at 1 ft/s Velocity lb_m/h Water
$1\frac{1}{2}$	1.900	5S	0.065	1.770	0.375	0.01709	0.497	0.463	7.67	3,835
		10S	0.109	1.682	0.614	0.01543	0.497	0.440	6.94	3,465
		40ST, 40S	0.145	1.610	0.800	0.01414	0.497	0.421	6.34	3,170
		80XS, 80S	0.200	1.500	1.069	0.01225	0.497	0.393	5.49	2,745
		160	0.281	1.338	1.439	0.00976	0.497	0.350	4.38	2,190
		XX	0.400	1.100	1.885	0.00660	0.497	0.288	2.96	1,480
2	2.375	5S	0.065	2.245	0.472	0.02749	0.622	0.588	12.34	6,170
		10S	0.109	2.157	0.776	0.02538	0.622	0.565	11.39	5,695
		40ST, 40S	0.154	2.067	1.075	0.02330	0.622	0.541	10.45	5,225
		80ST, 80S	0.218	1.939	1.477	0.02050	0.622	0.508	9.20	4,600
		160	0.344	1.687	2.195	0.01552	0.622	0.436	6.97	3,485
		XX	0.436	1.503	2.656	0.01232	0.622	0.393	5.53	2,765
$2\frac{1}{2}$	2.875	5S	0.083	2.709	0.728	0.04003	0.753	0.709	17.97	8,985
		10S	0.120	2.635	1.039	0.03787	0.753	0.690	17.00	8,500
		40ST, 40S	0.203	2.469	1.704	0.03322	0.753	0.647	14.92	7,460
		80XS, 80S	0.276	2.323	2.254	0.2942	0.753	0.608	13.20	6,600
		160	0.375	2.125	2.945	0.2463	0.753	0.556	11.07	5,535
		XX	0.552	1.771	4.028	0.01711	0.753	0.464	7.68	3,840
3	3.500	5S	0.083	3.334	0.891	0.06063	0.916	0.873	27.21	13,605
		10S	0.120	3.260	1.274	0.05796	0.916	0.853	26.02	13,010
		40ST, 40S	0.216	3.068	2.228	0.05130	0.916	0.803	23.00	11,500
		80XS, 80S	0.300	2.900	3.016	0.04587	0.916	0.759	20.55	10,275
		160	0.438	2.624	4.213	0.03755	0.916	0.687	16.86	8,430
		XX	0.600	2.300	5.466	0.02885	0.916	0.602	12.95	6,475
$3\frac{1}{2}$	4.0	5S	0.083	3.834	1.021	0.08017	1.047	1.004	35.98	17,990
		10S	0.120	3.760	1.463	0.07711	1.047	0.984	34.61	17,305
		40ST, 40S	0.226	3.548	2.680	0.06870	1.047	0.929	30.80	15,400
		80XS, 80S	0.318	3.364	3.678	0.06170	1.047	0.881	27.70	13,850
4	4.5	5S	0.083	4.334	1.152	0.10245	1.178	1.135	46.0	23,000
		10S	0.120	4.260	1.651	0.09898	1.178	1.115	44.4	22,200
		40ST, 40S	0.237	4.026	3.17	0.08840	1.178	1.054	39.6	19,800
		80XS, 80S	0.337	3.826	4.41	0.07986	1.178	1.002	35.8	17,900
		120	0.438	3.624	5.58	0.07170	1.178	0.949	32.2	16,100
		160	0.531	3.438	6.62	0.06647	1.178	0.900	28.9	14,450
		XX	0.674	3.152	8.10	0.05419	1.178	0.825	24.3	12,150
5	5.563	5S	0.109	5.345	1.87	0.1558	1.456	1.399	69.9	34.950
		10S	0.134	5.295	2.29	0.1529	1.456	1.386	68.6	34,300
		40ST, 40S	0.258	5.047	4.30	0.1390	1.456	1.321	62.3	31,150
		80XS, 80S	0.375	4.813	6.11	0.1263	1.456	1.260	57.7	28,850

(Continued)

TABLE F.1 (*Continued*)

Standard Steel Pipe Dimensions and Capacities[a]

Nominal Pipe Size (in.)	Outside Diameter (in.)	Schedule No.	Wall Thickness (in.)	Inside Diameter (in.)	Cross-Sectional Area		Circumference (ft) or Surface (ft/ft of Length)		Capacity at 1 ft/s Velocity	
					Metal (in.²)	Flow (ft²)	Outside	Inside	U.S. gpm	lb$_m$/h Water
		120	0.500	4.563	7.95	0.1136	1.456	1.195	51.0	25,500
		160	0.625	4.313	9.70	0.1015	1.456	1.129	45.5	22,750
		XX	0.750	4.063	11.34	0.0900	1.456	1.064	40.4	20,200
6	6.625	5S	0.109	6.407	2.23	0.2239	1.734	1.677	100.5	50,250
		10S	0.134	6.357	2.73	0.2204	1.734	1.664	98.9	49,450
		40ST, 40S	0.280	6.065	5.58	0.2006	1.734	1.588	90.0	45,000
		80XS, 80S	0.432	5.761	8.40	0.1810	1.734	1.508	81.1	40,550
		120	0.562	5.501	10.70	0.1650	1.734	1.440	73.9	36,950
		160	0.719	5.187	13.34	0.1467	1.734	1.358	65.9	32,950
		XX	0.864	4.897	15.64	0.1308	1.734	1.282	58.7	29,350
8	8.625	5S	0.109	8.407	2.915	0.3855	2.258	2.201	173.0	86,500
		10S	0.148	8.329	3.941	0.3784	2.258	2.180	169.8	84,900
		20	0.250	8.125	6.578	0.3601	2.258	2.127	161.5	80,750
		30	0.277	8.071	7.265	0.3553	2.258	2.113	159.4	79,700
		40ST, 40S	0.322	7.981	8.399	0.3474	2.258	2.089	155.7	77,850
		60	0.406	7.813	10.48	0.3329	2.258	2.045	149.4	74,700
		80XS, 80S	0.500	7.625	12.76	0.3171	2.258	1.996	142.3	71,150
		100	0.594	7.437	14.99	0.3017	2.258	1.947	135.4	67,700
		120	0.719	7.187	17.86	0.2817	2.258	1.882	126.4	63,200
		140	0.812	7.001	19.93	0.2673	2.258	1.833	120.0	60,000
		XX	0.875	6.875	21.30	0.2578	2.258	1.800	115.7	57,850
		160	0.906	6.813	21.97	0.2532	2.258	1.784	113.5	56,750
10	10.75	5S	0.134	10.482	4.47	0.5993	2.814	2.744	269.0	134,500
		10S	0.165	10.420	5.49	0.5922	2.814	2.728	265.8	132,900
		20	0.250	10.250	8.25	0.5731	2.814	2.685	257.0	128,500
		30	0.307	10.136	10.07	0.5603	2.814	2.655	252.0	126,000
		40ST, 40S	0.365	10.020	11.91	0.5745	2.814	2.620	246.0	123,000
		80S, 60XS	0.500	9.750	16.10	0.5185	2.814	2.550	233.0	116,500
		80	0.594	9.562	18.95	0.4987	2.814	2.503	223.4	111,700
		100	0.719	9.312	22.66	0.4728	2.814	2.438	212.3	106,150
		120	0.844	9.062	26.27	0.4479	2.814	2.372	201.0	100,500
		140, XX	1.000	8.750	30.63	0.4176	2.814	2.291	188.0	94,000
		160	1.125	8.500	34.02	0.3941	2.814	2.225	177.0	88,500
12	12.75	5S	0.156	12.428	6.17	0.8438	3.338	3.26	378.7	189,350
		10S	0.180	12.390	7.11	0.8373	3.338	3.24	375.8	187,900
		20	0.250	12.250	9.82	0.8185	3.338	3.21	367.0	183,500
		30	0.330	12.090	12.88	0.7972	3.338	3.17	358.0	179,000
		ST, 40S	0.375	12.000	14.58	0.7854	3.338	3.14	352.5	176,250
		40	0.406	11.938	15.74	0.7773	3.338	3.13	349.0	174,500
		XS, 80S	0.500	11.750	19.24	0.7530	3.338	3.08	338.0	169,000
		60	0.562	11.626	21.52	0.7372	3.338	3.04	331.0	165,500

(Continued)

TABLE F.1 (Continued)
Standard Steel Pipe Dimensions and Capacities[a]

Nominal Pipe Size (in.)	Outside Diameter (in.)	Schedule No.	Wall Thickness (in.)	Inside Diameter (in.)	Cross-Sectional Area Metal (in.²)	Cross-Sectional Area Flow (ft²)	Circumference (ft) or Surface (ft/ft of Length) Outside	Circumference (ft) or Surface (ft/ft of Length) Inside	Capacity at 1 ft/s Velocity U.S. gpm	Capacity at 1 ft/s Velocity lb$_m$/h Water
		80	0.688	11.374	26.07	0.7056	3.338	2.98	316.7	158,350
		100	0.844	11.062	31.57	0.6674	3.338	2.90	299.6	149,800
		120, XX	1.000	10.750	36.91	0.6303	3.338	2.81	283.0	141,500
		140	1.125	10.500	41.09	0.6013	3.338	2.75	270.0	135,000
		160	1.312	10.136	47.14	0.5592	3.338	2.65	251.0	125,500
14	14	5S	0.156	13.688	6.78	1.0219	3.665	3.58	459	229,500
		10S	0.188	13.624	8.16	1.0125	3.665	3.57	454	227,000
		10	0.250	13.500	10.80	0.9940	3.665	3.53	446	223,000
		20	0.312	13.376	13.42	0.9750	3.665	3.50	438	219,000
		30, ST	0.375	13.250	16.05	0.9575	3.665	3.47	430	215,000
		40	0.438	13.124	18.66	0.9397	3.665	3.44	422	211,000
		XS	0.500	13.000	21.21	0.9218	3.665	3.40	414	207,000
		60	0.594	12.812	25.02	0.8957	3.665	3.35	402	201,000
		80	0.750	12.500	31.22	0.8522	3.665	3.27	382	191,000
		100	0.938	12.124	38.49	0.8017	3.665	3.17	360	180,000
		120	1.094	11.812	44.36	0.7610	3.665	3.09	342	171,000
		140	1.250	11.500	50.07	0.7213	3.665	3.01	324	162,000
		160	1.406	11.188	55.63	0.6827	3.665	2.93	306	153,000
16	16	5S	0.165	15.670	8.21	1.3393	4.189	4.10	601	300,500
		10S	0.188	15.624	9.34	1.3314	4.189	4.09	598	299,000
		10	0.250	15.500	12.37	1.3104	4.189	4.06	587	293,500
		20	0.312	15.376	15.38	1.2985	4.189	4.03	578	289,000
		30, ST	0.375	15.250	18.41	1.2680	4.189	3.99	568	284,000
		40, XS	0.500	15.000	24.35	1.2272	4.189	3.93	550	275,000
		60	0.656	14.688	31.62	1.766	4.189	3.85	528	264,000
		80	0.844	14.312	40.19	1.1171	4.189	3.75	501	250,500
		100	1.031	13.939	48.48	1.0596	4.189	3.65	474	237,000
		120	1.219	13.562	56.61	1.0032	4.189	3.55	450	225,000
		140	1.438	13.124	65.79	0.9394	4.189	3.44	422	211,000
		160	1.594	12.812	72.14	0.8953	4.189	3.35	402	201,000
18	18	5S	0.165	17.760	9.25	1.8029	4.712	4.63	764	382,000
		10S	0.188	17.624	10.52	1.6941	4.712	4.61	760	379,400
		10	0.250	17.500	13.94	1.6703	4.712	4.58	750	375,000
		20	0.312	17.376	17.34	1.6468	4.712	4.55	739	369,500
		ST	0.375	17.250	20.76	1.6230	4.712	4.52	728	364,000
		30	0.438	17.124	24.16	1.5993	4.712	4.48	718	359,000
		XS	0.500	17.000	27.49	1.5763	4.712	4.45	707	353,500
		40	0.562	16.876	30.79	1.5533	4.712	4.42	697	348,500
		60	0.750	16.500	40.54	1.4849	4.712	4.32	666	333,000
		80	0.938	16.124	50.28	1.4180	4.712	4.2	636	318,000
		100	1.156	15.688	61.17	1.3423	4.712	4.11	602	301,000

(Continued)

TABLE F.1 (*Continued*)
Standard Steel Pipe Dimensions and Capacities[a]

Nominal Pipe Size (in.)	Outside Diameter (in.)	Schedule No.	Wall Thickness (in.)	Inside Diameter (in.)	Cross-Sectional Area Metal (in.²)	Cross-Sectional Area Flow (ft²)	Circumference (ft) or Surface (ft/ft of Length) Outside	Circumference (ft) or Surface (ft/ft of Length) Inside	Capacity at 1 ft/s Velocity U.S. gpm	Capacity at 1 ft/s Velocity lb$_m$/h Water
		120	1.375	15.250	71.82	1.2684	4.712	3.99	569	284,500
		140	1.562	14.876	80.66	1.2070	4.712	3.89	540	270,000
		160	1.781	14.438	90.75	1.1370	4.712	3.78	510	255,000
20	20	5S	0.188	19.624	11.70	2.1004	5.236	5.14	943	471,500
		10S	0.218	19.564	13.55	2.0878	5.236	5.12	937	467,500
		10	0.250	19.500	5.51	2.0740	5.236	5.11	930	465,000
		20, ST	0.375	19.250	23.12	2.0211	5.236	5.04	902	451,000
		30, XS	0.500	19.000	30.63	1.9689	5.236	4.97	883	441,500
		40	0.594	18.812	36.21	1.9302	5.236	4.92	866	433,000
		60	0.812	18.376	48.95	1.8417	5.236	4.81	826	413,000
		80	1.031	17.938	61.44	1.7550	5.236	4.70	787	393,500
		100	1.281	17.438	75.33	1.6585	5.236	4.57	744	372,000
		120	1.500	17.000	87.18	1.5763	5.236	4.45	707	353,500
		140	1.750	16.500	100.3	1.4849	5.236	4.32	665	332,500
		160	1.969	16.062	111.5	1.4071	5.236	4.21	632	316,000
24	24	5S	0.218	23.564	16.29	3.0285	6.283	6.17	1359	579,500
		20, 10S	0.250	23.500	18.65	3.012	6.283	6.15	1350	675,000
		20, ST	0.375	23.250	27.83	2.948	6.283	6.09	1325	662,500
		XS	0.500	23.000	36.90	2.885	6.283	6.02	1295	642,500
		30	0.562	22.876	41.39	2.854	6.283	5.99	1281	640,500
		40	0.688	22.624	50.39	2.792	6.283	5.92	1253	626,500
		60	0.969	22.062	70.11	2.655	6.283	5.78	1192	596,000
		80	1.219	21.562	87.24	2.536	6.283	5.64	1138	569,000
		100	1.531	20.938	108.1	2.391	6.283	5.48	1073	536,500
		120	1.812	20.376	126.3	2.264	6.283	5.33	1016	508,000
		140	2.062	19.876	142.1	2.155	6.283	5.20	965	482,500
		160	2.344	19.312	159.5	2.034	6.283	5.06	913	456,500
30	30	5S	0.250	29.500	23.37	4.746	7.854	7.72	2130	1,065,000
		10, 10S	0.312	29.376	29.10	4.707	7.854	7.69	2110	1,055,000
		ST	0.375	29.250	34.90	4.666	7.854	7.66	2094	1,048,000
		20, XS	0.500	29.000	46.34	4.587	7.854	7.59	2055	1,027,500
		30	0.625	28.750	57.68	4.508	7.854	7.53	2020	1,010,000

[a] 5S, 10S, and 40S are extracted from stainless steel pipe, ANSI B36.19-1976, with permission from the publisher, the American Society of Mechanical Engineers, New York. ST, standard wall; XS, extra strong wall; XX, double extra strong wall. Schedules 10–160 are extracted from wrought-steel and wrought-iron pipe, ANSI B36.10-1975, with permission from the same publisher. Decimal thicknesses for respective pipe sizes represent their nominal or average wall dimensions. Mill tolerances as high as ±12.5% are permitted.

A plain-end pipe is produced by a square cut. The pipe is also shipped from the mills threaded, with a threaded coupling on one end, or with the ends beveled for welding or grooved or sized for patented couplings. Weights per foot for threaded and coupled pipe are slightly greater because of the weight of the coupling, but it is not available larger than 12 in. or lighter than Schedule 30 sizes 8 -12 in., or Schedule 40, 6 in. and smaller.

To convert inches to millimeters, multiply by 25.4; to convert square inches to square millimeters, multiply by 645; to convert feet to meters, multiply by 0.3048; to convert square feet to square meters, multiply by 0.0929; to convert pounds per foot to kilograms per meter, multiply by 1.49; to convert gallons to cubic meters, multiply by 3.7854×10^{-3}; and to convert pounds to kilograms, multiply by 0.4536.

Appendix G: Flow of Water/ Air through Schedule 40 Pipe

TABLE G.1
Flow of Water through Schedule 40 Steel Pipe

Pressure Drop per 100 ft and Velocity in Schedule 40 Pipe for Water at 60°F

Discharge		1/8 in.		1/4 in.		3/8 in.		1/2 in.		3/4 in.		1 in.		1¼ in.		1½ in.		2 in.		2½ in.		3 in.		3½ in.		4 in.	
(gpm)	(ft³/s)	Velocity (ft/s)	Pressure Drop (psi)	Velocity (ft/s)	Pressure Drop (psi)	Velocity (ft/s)	Pressure Drop (psi)	Velocity (ft/s)	Pressure Drop (psi)	Velocity (ft/s)	Pressure Drop (psi)	Velocity (ft/s)	Pressure Drop (psi)	Velocity (ft/s)	Pressure Drop (psi)	Velocity (ft/s)	Pressure Drop (psi)	Velocity (ft/s)	Pressure Drop (psi)	Velocity (ft/s)	Pressure Drop (psi)	Velocity (ft/s)	Pressure Drop (psi)	Velocity (ft/s)	Pressure Drop (psi)	Velocity (ft/s)	Pressure Drop (psi)
0.2	0.000446	1.13	1.86	0.616	0.359	...	...																				
0.3	0.000668	1.69	4.22	0.924	0.903	0.504	0.159	0.317	0.061																		
0.4	0.000891	2.26	6.98	1.23	1.61	0.672	0.345	0.422	0.086	...	...																
0.5	0.00111	2.82	10.5	1.54	2.39	0.840	0.539	0.528	0.167	0.301	0.033																
0.6	0.00134	3.39	14.7	1.85	3.29	1.01	0.751	0.633	0.240	0.361	0.041																
0.8	0.00178	4.52	25.0	2.46	5.44	1.34	1.25	0.844	0.408	0.481	0.102																
1	0.00223	5.65	37.2	3.08	8.28	1.68	1.85	1.06	0.600	0.602	0.155	0.371	0.048	...	...												
2	0.00446	11.29	134.4	6.16	30.1	3.36	6.58	2.11	2.10	1.20	0.526	0.741	0.164	0.429	0.044												
3	0.00668	...	...	9.25	64.1	5.04	13.9	3.17	4.33	1.81	1.09	1.114	0.336	0.644	0.090	0.473	0.043										
4	0.00891	...	...	12.33	111.2	6.72	23.9	4.22	7.42	2.41	1.83	1.49	0.565	0.858	0.150	0.630	0.071										
5	0.01114					8.40	36.7	5.28	11.2	3.01	2.75	1.86	0.835	1.073	0.223	0.788	0.104										
6	0.01337					10.08	51.9	6.3	15.8	3.61	3.84	2.23	1.17	1.29	0.309	0.946	0.145	0.574	0.044	...	...						
8	0.01782					13.44	91.1	8.45	27.7	4.81	6.60	2.97	1.99	1.72	0.518	1.26	0.241	0.765	0.073	...	...						
10	0.02228							10.56	42.4	6.02	9.99	3.71	2.99	2.15	0.774	1.58	0.361	0.956	0.108	0.670	0.046						
15	0.03342									9.03	21.6	5.57	6.36	3.22	1.63	2.37	0.755	1.43	0.224	1.01	0.094						
20	0.04456									12.03	37.8	7.43	10.9	4.29	2.78	3.16	1.28	1.91	0.375	1.34	0.158	0.868	0.056				
25	0.05570											9.28	16.7	5.37	4.22	3.94	1.93	2.39	0.561	1.68	0.234	1.09	0.083	0.812	0.041		
30	0.06684											11.14	23.8	6.44	5.92	4.73	2.72	2.87	0.786	2.01	0.327	1.30	0.114	0.974	0.056	...	...
35	0.07798											12.99	32.2	7.51	7.90	5.52	3.64	3.35	1.05	2.35	0.436	1.52	0.151	1.14	0.071	0.882	0.041
40	0.08912											14.85	41.5	8.59	10.24	6.30	4.65	3.83	1.35	2.68	0.556	1.74	0.192	1.30	0.095	1.01	0.052
45	0.1003											...	...	9.67	12.80	7.09	5.85	4.30	1.67	3.02	0.668	1.95	0.239	1.46	0.117	1.13	0.064

(Continued)

TABLE G.1 (Continued)
Flow of Water through Schedule 40 Steel Pipe

Pressure Drop per 100 ft and Velocity in Schedule 40 Pipe for Water at 60°F

Discharge		Velocity (ft/s)	Pressure Drop (psi)	Velocity (ft/s)	Pressure Drop (psi)	Velocity (ft/s)	Pressure Drop (psi)	Velocity (ft/s)	Pressure Drop (psi)	Velocity (ft/s)	Pressure Drop (psi)	Velocity (ft/s)	Pressure Drop (psi)	Velocity (ft/s)	Pressure Drop (psi)	Velocity (ft/s)	Pressure Drop (psi)
(gpm)	(ft³/s)																
50	0.1114	4.78	2.03	3.35	0.839	2.17	0.288	1.62	0.142	1.26	0.076	5 in.	...	10.74	15.66	7.88	7.15
60	0.1337	5.74	2.87	4.02	1.18	2.60	0.406	1.95	0.204	1.51	0.107	...	...	12.89	22.2	9.47	10.21
70	0.1560	6.70	3.84	4.69	1.59	3.04	0.540	2.27	0.261	1.76	0.143	1.12	0.047	...	...	11.05	13.71
80	0.1782	7.65	4.97	5.36	2.03	3.47	0.687	2.60	0.334	2.02	0.180	1.28	0.060	6 in.	...	12.62	17.59
90	0.2005	8.60	6.20	6.03	2.53	3.91	0.861	2.92	0.416	2.27	0.222	1.44	0.074	...	...	14.20	22.0
100	0.2228	9.56	7.59	6.70	3.09	4.34	1.05	3.25	0.509	2.52	0.272	1.60	0.090	1.11	0.036	15.78	26.9
125	0.2785	11.97	11.76	8.38	4.71	5.43	1.61	4.06	0.769	3.15	0.415	2.01	0.135	1.39	0.055	19.72	41.4
150	0.3342	14.36	16.70	10.05	6.69	6.51	2.24	4.87	1.08	3.78	0.580	2.41	0.190	1.67	0.077	...	...
175	0.3899	16.75	22.3	11.73	8.97	7.60	3.00	5.68	1.44	4.41	0.774	2.81	0.253	1.94	0.102	8 in.	...
200	0.4456	19.14	28.8	13.42	11.68	8.68	3.87	6.49	1.85	5.04	0.985	3.21	0.323	2.22	0.130	...	...
225	0.5013	...	...	15.09	14.63	9.77	4.83	7.30	2.32	5.67	1.23	3.61	0.401	2.50	0.162	1.44	0.043
250	0.557	...	...	...	...	10.85	5.93	8.12	2.84	6.30	1.46	4.01	0.495	2.78	0.195	1.60	0.051
275	0.6127	...	...	...	...	11.94	7.14	8.93	3.40	6.93	1.79	4.41	0.583	3.05	0.234	1.76	0.061
300	0.6684	...	...	...	...	13.00	8.36	9.74	4.02	7.56	2.11	4.81	0.683	3.33	0.275	1.92	0.072
325	0.7241	...	...	...	...	14.12	9.89	10.53	4.69	8.19	2.47	5.21	0.797	3.61	0.320	2.08	0.083
350	0.7798	...	...	...	...	...	...	11.36	5.41	8.82	2.84	5.62	0.919	3.89	0.367	2.24	0.095
375	0.8355	...	...	...	...	...	...	12.17	6.18	9.45	3.25	6.02	1.05	4.16	0.416	2.40	0.108
400	0.8912	...	...	...	...	...	...	12.98	7.03	10.08	3.68	6.42	1.19	4.44	0.471	2.56	0.121
425	0.9469	...	...	...	...	...	...	13.80	7.89	10.71	4.12	6.82	1.33	4.72	0.529	2.73	0.136
450	1.003	10 in.	...	...	...	...	...	14.61	8.80	11.34	4.60	7.22	1.48	5.00	0.590	2.89	0.151
475	1.059	1.93	0.054	...	...	...	...	...	...	11.97	5.12	7.62	1.64	5.27	0.653	3.04	0.166
500	1.114	2.03	0.059	...	...	...	...	...	...	12.60	5.65	8.02	1.81	5.55	0.720	3.21	0.182
550	1.225	2.24	0.071	...	...	...	...	...	...	13.85	6.79	8.82	2.17	6.11	0.861	3.53	0.219
600	1.337	2.44	0.083	...	...	...	...	...	...	15.12	8.04	9.63	2.55	6.66	1.02	3.85	0.258
650	1.448	2.64	0.097	12 in.	...	...	...	...	...	...	...	10.43	2.98	7.22	1.18	4.17	0.301

(Continued)

TABLE G.1 (Continued)
Flow of Water through Schedule 40 Steel Pipe

Pressure Drop per 100 ft and Velocity in Schedule 40 Pipe for Water at 60°F

Discharge (gpm)	Discharge (ft³/s)	5 in. Velocity (ft/s)	5 in. Pressure Drop (psi)	6 in. Velocity (ft/s)	6 in. Pressure Drop (psi)	8 in. Velocity (ft/s)	8 in. Pressure Drop (psi)	10 in. Velocity (ft/s)	10 in. Pressure Drop (psi)	12 in. Velocity (ft/s)	12 in. Pressure Drop (psi)	14 in. Velocity (ft/s)	14 in. Pressure Drop (psi)	16 in. Velocity (ft/s)	16 in. Pressure Drop (psi)	18 in. Velocity (ft/s)	18 in. Pressure Drop (psi)	20 in. Velocity (ft/s)	20 in. Pressure Drop (psi)	24 in. Velocity (ft/s)	24 in. Pressure Drop (psi)
700	1.560	11.23	3.43	7.78	1.35	4.49	0.343	2.85	0.112	2.01	0.047	...	...	...	...	...	...	...	...	...	...
750	1.671	12.03	3.92	8.33	1.55	4.81	0.392	3.05	0.127	2.15	0.054	...	...	...	...	...	...	...	...	...	...
800	1.782	12.83	4.43	8.88	1.75	5.13	0.443	3.25	0.143	2.29	0.061	...	...	...	...	...	...	...	...	...	...
850	1.894	13.64	5.00	9.44	1.96	5.45	0.497	3.46	0.160	2.44	0.068	2.02	0.042	...	...	...	...	...	...	...	...
900	2.005	14.44	5.58	9.99	2.18	5.77	0.554	3.66	0.179	2.58	0.075	2.13	0.047	...	...	...	...	...	...	...	...
950	2.117	15.24	6.21	10.55	2.42	6.09	0.613	3.86	0.198	2.72	0.083	2.25	0.052	...	...	...	...	...	...	...	...
1,000	2.228	16.04	6.84	11.10	2.68	6.41	0.675	4.07	0.218	2.87	0.091	2.37	0.057	...	...	...	...	...	...	...	...
1,100	2.451	17.65	8.23	12.22	3.22	7.05	0.807	4.48	0.260	3.15	0.110	2.61	0.068	...	...	...	...	...	...	...	...
1,200	2.674	...	...	13.33	3.81	7.70	0.948	4.88	0.306	3.44	0.128	2.85	0.080	2.18	0.042	...	...	...	...	...	...
1,300	2.896	...	...	14.43	4.45	8.11	1.11	5.29	0.355	3.73	0.150	3.08	0.093	2.36	0.048	...	...	...	...	...	...
1,400	3.119	...	...	15.55	5.13	8.98	1.28	5.70	0.409	4.01	0.171	3.32	0.107	2.54	0.055	...	...	...	...	...	...
1,500	3.342	...	...	16.66	5.85	9.62	1.46	6.10	0.466	4.30	0.195	3.56	0.122	2.72	0.063	...	...	...	...	...	...
1,600	3.565	...	...	17.77	6.61	10.26	1.65	6.51	0.527	4.59	0.219	3.79	0.138	2.90	0.071	...	...	...	...	...	...
1,800	4.010	...	...	19.99	8.37	11.54	2.08	7.32	0.663	5.16	0.276	4.27	0.172	3.27	0.088	2.58	0.050	...	...	...	...
2,000	4.456	...	...	22.21	10.3	12.82	2.55	8.14	0.808	5.73	0.339	4.74	0.209	3.63	0.107	2.87	0.060	...	...	...	...
2,500	5.570	...	...	...	...	16.03	3.94	10.17	1.24	7.17	0.515	5.93	0.321	4.54	0.163	3.59	0.091	...	...	...	...
3,000	6.684	...	...	...	...	19.24	5.59	12.20	1.76	8.60	0.731	7.11	0.451	5.45	0.232	4.10	0.129	3.46	0.075	...	...
3,500	7.798	...	...	...	...	22.44	7.56	14.24	2.38	10.01	0.982	8.30	0.607	6.35	0.312	5.02	0.173	5.05	0.101	...	...
4,000	8.912	...	...	...	...	25.65	9.80	16.27	3.08	11.47	1.27	8.48	0.787	7.26	0.401	5.74	0.222	4.62	0.129	3.19	0.052
4,500	10.03	...	...	...	...	28.87	12.2	18.31	3.87	12.90	1.60	10.67	0.990	8.17	0.503	6.46	0.280	5.20	0.162	3.59	0.065
5,000	11.14	...	...	...	...	...	...	20.35	4.71	14.33	1.95	11.85	1.21	9.08	0.617	7.17	0.340	5.77	0.199	3.99	0.079
6,000	13.37	...	...	...	...	...	...	24.41	6.47	17.20	2.77	14.21	1.71	10.89	0.877	8.61	0.483	6.91	0.280	4.79	0.111

(Continued)

TABLE G.1 (Continued)
Flow of Water through Schedule 40 Steel Pipe

Pressure Drop per 100 ft and Velocity in Schedule 40 Pipe for Water at 60°F

Discharge		Velocity (ft/s)	Pressure Drop (psi)	Velocity (ft/s)	Pressure Drop (psi)	Velocity (ft/s)	Pressure Drop (psi)	Velocity (ft/s)	Pressure Drop (psi)	Velocity (ft/s)	Pressure Drop (psi)	Velocity (ft/s)	Pressure Drop (psi)	Velocity (ft/s)	Pressure Drop (psi)	Velocity (ft/s)	Pressure Drop (psi)
(gpm)	(ft³/s)																
7,000	15.60	28.49	9.11	20.07	3.74	16.60	2.31	12.71	1.18	10.04	0.652	8.08	0.376	5.59	0.150	...	...
8,000	17.82	...		22.93	4.48	18.96	2.99	14.52	1.51	11.47	0.839	9.23	0.488	6.18	0.192	...	...
9,000	20.05	...		25.79	6.09	21.14	3.76	16.34	1.90	12.91	1.05	10.39	0.608	7.18	0.242	...	...
10,000	22.28	...		28.66	7.46	23.71	4.61	18.15	2.34	14.14	1.28	11.54	0.739	7.98	0.294	...	...
12,000	26.74	...		34.40	10.7	28.45	6.59	21.79	3.33	17.21	1.83	13.85	1.06	9.58	0.416	...	...
14,000	31.19	...		...		33.19	8.89	25.42	4.49	20.08	2.45	16.16	1.43	11.17	0.562	...	...
16,000	35.65	...		...		...		29.05	5.83	22.95	3.18	18.47	1.85	12.77	0.723	...	...
18,000	40.10	...		...		...		32.68	7.31	25.82	4.03	20.77	2.32	14.36	0.907	...	...
20,000	44.56	...		...		...		36.31	9.03	28.69	4.93	23.08	2.86	15.96	1.12	...	...

Note: For pipe lengths other than 100 ft, the pressure drop is proportional to the length. Thus, for 50 ft of the pipe, the pressure drop is approximately one-half the value given in the table; for 300 ft, three times the given value, *etc.*

TABLE G.2
Flow of Air through Schedule 40 Steel Pipe

Free Air q'_m ft³/min at 60°F and 14.7 psia	Compressed Air ft³/min at 60°F and 100 psig	Pressure Drop of Air in psi per 100 ft of Schedule 40 Pipe for Air at 100 psi Gauge Pressure and 60°F Temperature								
		⅛ in.	¼ in.	⅜ in.	½ in.	¾ in.	1 in.	1¼ in.	1½ in.	2 in.
1	0.128	0.361	0.083	0.018						
2	0.256	1.31	0.285	0.064	0.020	¾ in.				
3	0.384	3.06	0.605	0.133	0.042					
4	0.513	4.83	1.04	0.226	0.071					
5	0.641	7.45	1.58	0.343	0.106	0.027				
							1 in.			
6	0.769	10.6	2.23	0.408	0.148	0.037				
8	1.025	18.6	3.89	0.848	0.255	0.062	0.019			
10	1.282	28.7	5.96	1.26	0.356	0.094	0.029	1¼ in.	1½ in.	
15	1.922	...	13.0	2.73	0.834	0.201	0.062			
20	2.563	...	22.8	4.76	1.43	0.345	0.102	0.026		
25	3.204	...	35.6	7.34	2.21	0.526	0.156	0.039	0.019	
30	3.845	...	...	10.6	3.15	0.748	0.219	0.055	0.026	
35	4.486	...	...	14.2	4.24	1.00	0.293	0.073	0.035	
40	5.126	...	...	18.4	5.49	1.30	0.379	0.095	0.044	
45	5.767	...	...	23.1	6.90	1.62	0.474	0.116	0.055	2 in.
50	6.408			28.5	8.49	1.99	0.578	0.149	0.067	0.019
60	7.690	2½ in.		40.7	12.2	2.85	0.619	0.200	0.094	0.027
70	8.971			...	16.5	3.83	1.10	0.270	0.126	0.036
80	10.25	0.019		...	21.4	4.96	1.43	0.350	0.162	0.046
90	11.63	0.023		...	27.0	6.25	1.80	0.437	0.203	0.058
100	12.82	0.029	3 in.	...	33.2	7.69	2.21	0.634	0.247	0.070
125	16.02	0.044			...	11.9	3.39	0.825	0.380	0.107
150	19.22	0.062	0.021		...	17.0	4.87	1.17	0.537	0.151
175	22.43	0.083	0.028		...	23.1	6.60	1.58	0.727	0.205
200	25.63	0.107	0.036	3½ in.	...	30.0	8.64	2.05	0.937	0.264
225	28.84	0.134	0.045	0.022	...	37.9	10.8	2.59	1.19	0.331
250	32.04	0.164	0.055	0.027		...	13.3	3.18	1.45	0.404
275	35.24	0.191	0.066	0.032		...	16.0	3.83	1.76	0.484
300	38.45	0.232	0.078	0.037		...	19.0	4.56	2.07	0.573
325	41.65	0.270	0.090	0.043		...	22.3	5.32	2.42	0.673
					4 in.					
350	44.87	0.313	0.104	0.050		...	25.8	6.17	2.80	0.776
375	48.06	0.356	0.119	0.057	0.030	...	29.6	7.05	3.20	0.887
400	51.26	0.402	0.134	0.064	0.034	...	33.6	8.02	3.64	1.00
425	54.47	0.452	0.151	0.072	0.038	...	37.9	9.01	4.09	1.13
450	57.67	0.507	0.168	0.081	0.042	...	...	10.2	4.59	1.26

(Continued)

TABLE G.2 (*Continued*)
Flow of Air through Schedule 40 Steel Pipe

Free Air q'_m ft³/min at 60°F and 14.7 psia	Compressed Air ft³/min at 60°F and 100 psig	Pressure Drop of Air in psi per 100 ft of Schedule 40 Pipe for Air at 100 psi Gauge Pressure and 60°F Temperature								
		1/8 in.	1/4 in.	3/8 in.	1/2 in.					
475	60.88	0.562	0.187	0.089	0.047		...	11.3	5.09	1.40
500	64.08	0.623	0.206	0.099	0.052		...	12.5	5.61	1.55
550	70.49	0.749	0.248	0.118	0.062		...	15.1	6.79	1.87
600	76.90	0.887	0.293	0.139	0.073		...	18.0	8.04	2.21
650	83.30	1.04	0.342	0.163	0.086	5 in.	...	21.1	9.43	2.60
700	89.71	1.19	0.395	0.188	0.099	0.032		24.3	10.9	3.00
750	96.12	1.36	0.451	0.214	0.113	0.036		27.9	12.6	3.44
800	102.5	1.55	0.513	0.244	0.127	0.041		31.8	14.2	3.90
850	108.9	1.74	0.576	0.274	0.144	0.046		35.9	16.0	4.40
900	115.3	1.95	0.642	0.305	0.160	0.051	6 in.	40.2	18.0	4.91
950	121.8	2.18	0.715	0.340	0.178	0.057	0.023	...	20.0	5.47
1,000	128.2	2.40	0.788	0.375	0.197	0.063	0.025	...	22.1	6.06
1,100	141.0	2.89	0.948	0.451	0.236	0.075	0.030	...	26.7	7.29
1,200	153.8	3.44	1.13	0.533	0.279	0.089	0.035	...	31.8	8.63
1,300	166.6	4.01	1.32	0.626	0.327	0.103	0.041	...	37.3	10.1
1,400	179.4	4.65	1.52	0.718	0.377	0.119	0.047			11.8
1,500	192.2	5.31	1.74	0.824	0.431	0.136	0.054	8 in.		13.5
1,600	205.1	6.04	1.97	0.932	0.490	0.154	0.061			15.3
1,800	230.7	7.65	2.50	1.18	0.616	0.193	0.075			19.3
2,000	256.3	9.44	3.06	1.45	0.757	0.237	0.094	0.023		23.9
									10 in.	
2,500	320.4	14.7	4.76	2.15	1.17	0.366	0.143	0.035		37.3
3,000	384.5	21.1	6.82	3.20	1.67	0.524	0.204	0.051	0.016	
3,500	448.6	28.8	9.23	4.33	2.26	0.709	0.276	0.068	0.022	
4,000	512.6	37.6	12.1	5.66	2.94	0.919	0.358	0.088	0.028	
4,500	576.7	47.6	15.3	7.16	3.69	1.16	0.450	0.111	0.035	12 in.
5,000	640.8	...	18.8	8.85	4.56	1.42	0.652	0.136	0.043	0.018
6,000	769.0	...	27.1	12.7	6.57	2.03	0.794	0.195	0.061	0.025
7,000	897.1	...	36.9	17.2	8.94	2.76	1.07	0.262	0.082	0.034
8,000	1025	...	...	22.5	11.7	3.59	1.39	0.339	0.107	0.044
9,000	1153	...	...	28.5	14.9	4.54	1.76	0.427	0.134	0.055
10,000	1282	...	...	35.2	18.4	5.60	2.16	0.526	0.164	0.067
11,000	1410	...	...	...	22.2	6.78	2.62	0.633	0.197	0.081
12,000	1538	...	...	...	26.4	8.07	3.09	0.753	0.234	0.096
13,000	1666	...	...	...	31.0	9.47	3.63	0.884	0.273	0.112
14,000	1794	...	...	...	36.0	11.0	4.21	1.02	0.316	0.129

(Continued)

TABLE G.2 (*Continued*)
Flow of Air through Schedule 40 Steel Pipe

Free Air q'_m	Compressed Air								
ft³/min at 60°F and 14.7 psia	ft³/min at 60°F and 100 psig	Pressure Drop of Air in psi per 100 ft of Schedule 40 Pipe for Air at 100 psi Gauge Pressure and 60°F Temperature							
		⅛ in.	¼ in.	⅜ in.	½ in.				
15,000	1922	...	...	...	...	12.6	4.84	1.17	0.364	0.148
16,000	2051	...	...	...	...	14.3	5.50	1.33	0.411	0.167
18,000	2307	...	...	...	...	18.2	6.96	1.68	0.520	0.213
20,000	2563	...	...	...	...	22.4	8.60	2.01	0.642	0.260
22,000	2820	...	...	...	...	27.1	10.4	2.50	0.771	0.314
24,000	3076	...	...	...	...	32.3	12.4	2.97	0.918	0.371
26,000	3332	...	...	...	...	37.9	14.5	3.49	1.12	0.435
28,000	3588	...	...	...	...	...	16.9	4.04	1.25	0.505
30,000	3545	...	...	...	...	...	19.3	4.64	1.42	0.520

For pipe lengths other than 100 ft, the pressure drop is proportional to the length. Thus, for 50 ft of the pipe, the pressure drop is approximately one-half the value given in the table; for 300 ft, three times the given value, etc.

The pressure drop is also inversely proportional to the absolute pressure and directly proportional to the absolute temperature.

Therefore, to determine the pressure drop for inlet or average pressures other than 100 psi and at temperatures other than 60°F, multiply the values given in the table by the following ratio:

$$\left(\frac{100+147}{P+14.7}\right)\left(\frac{460+T}{520}\right)$$

where
P is the inlet or average gauge pressure in psi
T is the temperature in °F under consideration

The cubic feet per minute of compressed air at any pressure is inversely proportional to the absolute pressure and directly proportional to the absolute temperature.

To determine the cubic feet per minute of compressed air at any temperature and pressure other than standard conditions, multiply the value of cubic feet per minute of free air by the following ratio:

$$\left(\frac{14.7}{14.7+P}\right)\left(\frac{460+T}{520}\right)$$

G.1　Calculations for Pipe Other Than Schedule 40

To determine the velocity of water, or the pressure drop of water or air, through a pipe other than Schedule 40, use the following formulas:

$$V_a = V_{40} \left(\frac{d_{40}}{d_a} \right)^2$$

$$\Delta P_a = \Delta P_{40} \left(\frac{d_{40}}{d_a} \right)^5$$

Subscript *"a"* refers to the schedule of pipe through which velocity or pressure drop is desired.

Subscript "40" refers to the velocity or pressure drop through a Schedule 40 pipe, as given in Table G.1.

Appendix H: Typical Pump Head Capacity Range Charts

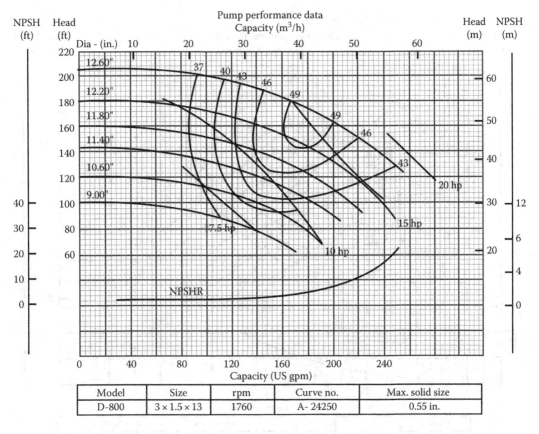

FIGURE H.1 Typical pump characteristic curves.

Model	Size	rpm	Curve no.	Max. solid size
D-800	3 × 1.5 × 13	1760	A- 24250	0.55 in.

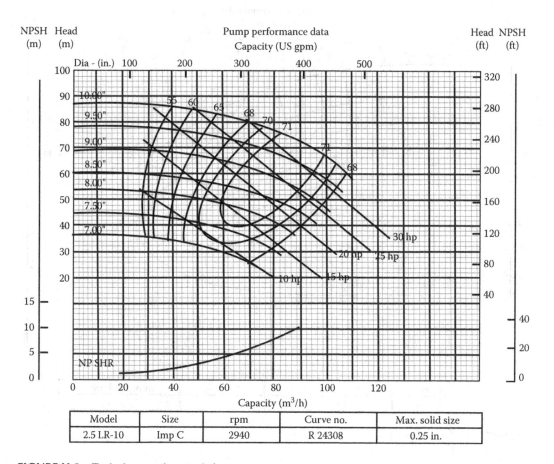

Model	Size	rpm	Curve no.	Max. solid size
2.5 LR-10	Imp C	2940	R 24308	0.25 in.

FIGURE H.2 Typical pump characteristic curves.

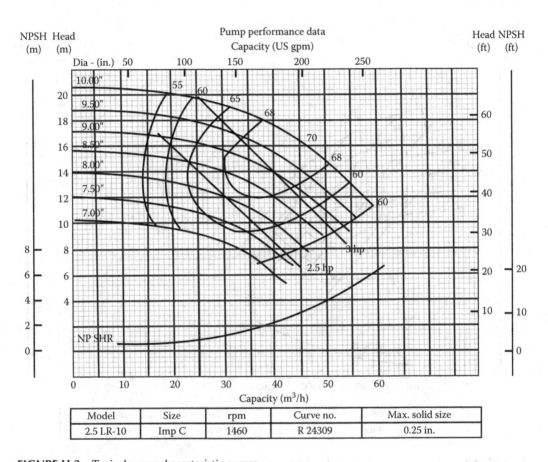

Model	Size	rpm	Curve no.	Max. solid size
2.5 LR-10	Imp C	1460	R 24309	0.25 in.

FIGURE H.3   Typical pump characteristic curves.

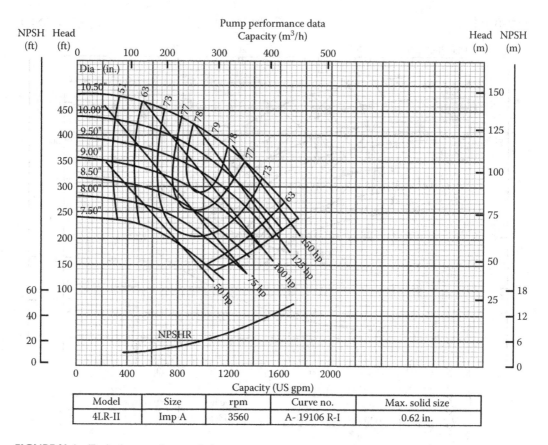

FIGURE H.4 Typical pump characteristic curves.

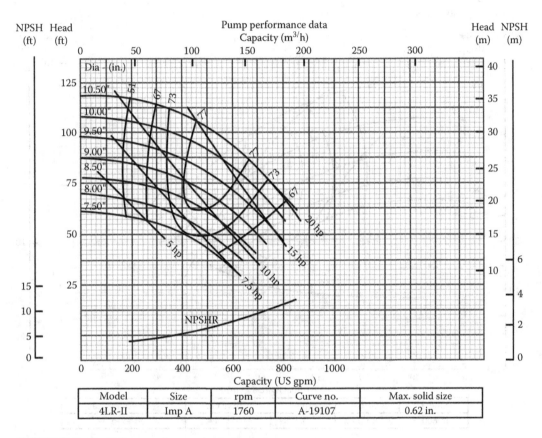

FIGURE H.5 Typical pump characteristic curves.

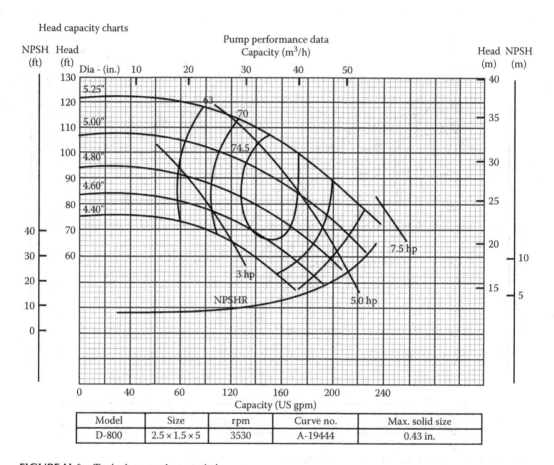

Head capacity charts

Model	Size	rpm	Curve no.	Max. solid size
D-800	2.5 × 1.5 × 5	3530	A-19444	0.43 in.

FIGURE H.6 Typical pump characteristic curves.

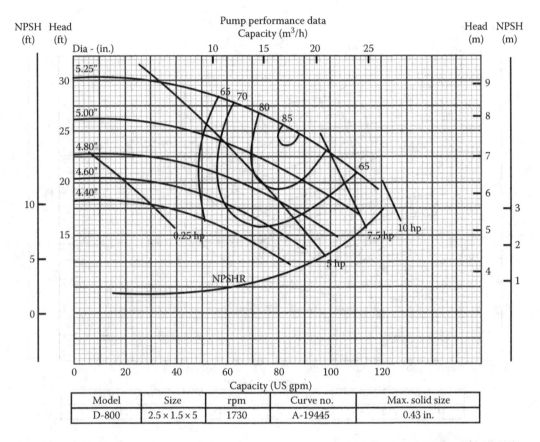

FIGURE H.7   Typical pump characteristic curves.

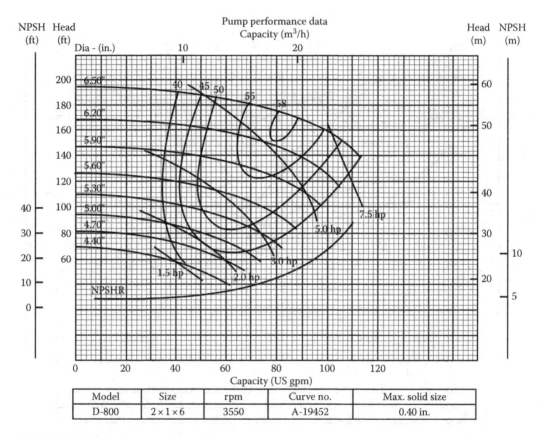

FIGURE H.8 Typical pump characteristic curves.

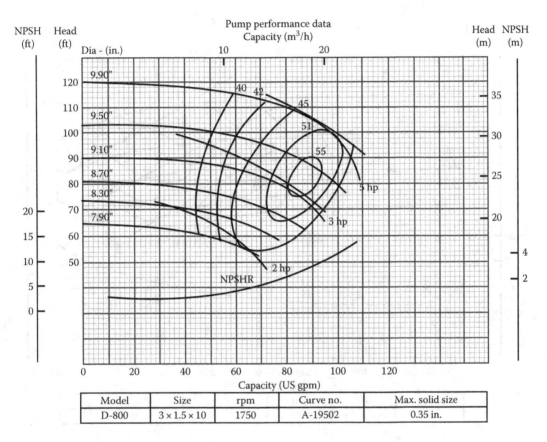

FIGURE H.9   Typical pump characteristic curves.

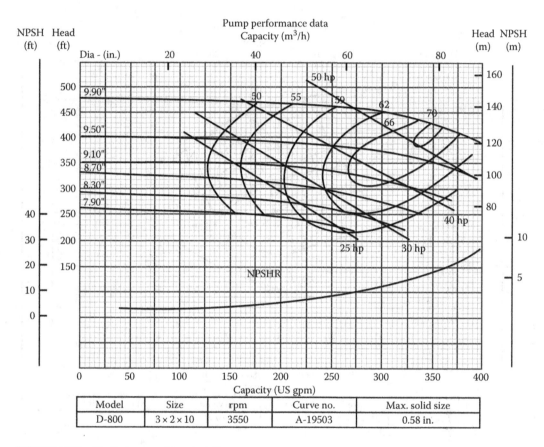

FIGURE H.10 Typical pump characteristic curves.

Model	Size	rpm	Curve no.	Max. solid size
D-800	3 × 2 × 10	3550	A-19503	0.58 in.

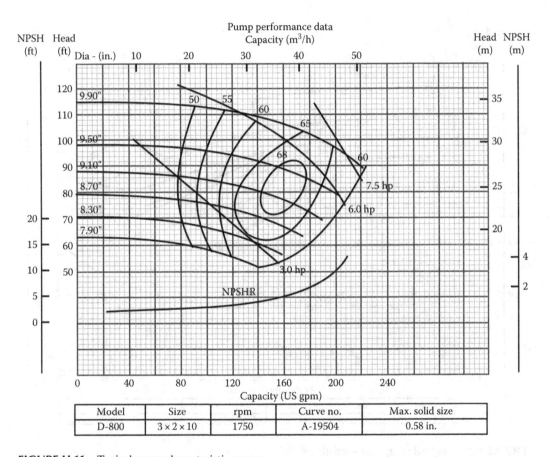

Model	Size	rpm	Curve no.	Max. solid size
D-800	3 × 2 × 10	1750	A-19504	0.58 in.

FIGURE H.11 Typical pump characteristic curves.

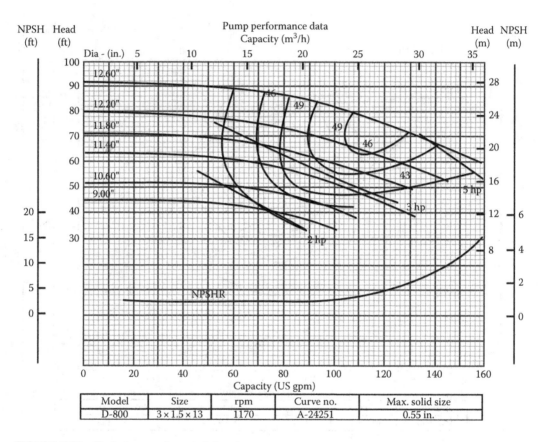

Model	Size	rpm	Curve no.	Max. solid size
D-800	3 × 1.5 × 13	1170	A-24251	0.55 in.

FIGURE H.12 Typical pump characteristic curves.

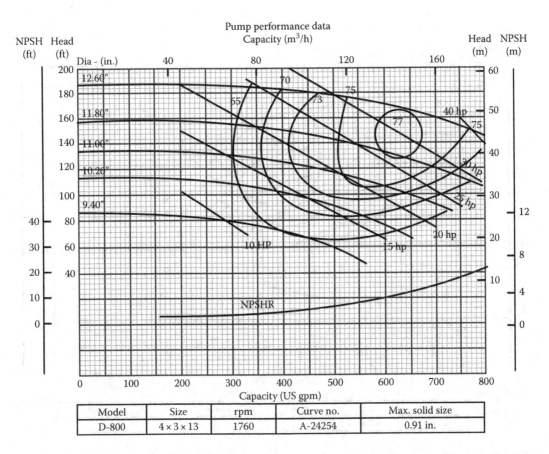

FIGURE H.13 Typical pump characteristic curves.

Model	Size	rpm	Curve no.	Max. solid size
D-800	4 × 3 × 13	1760	A-24254	0.91 in.

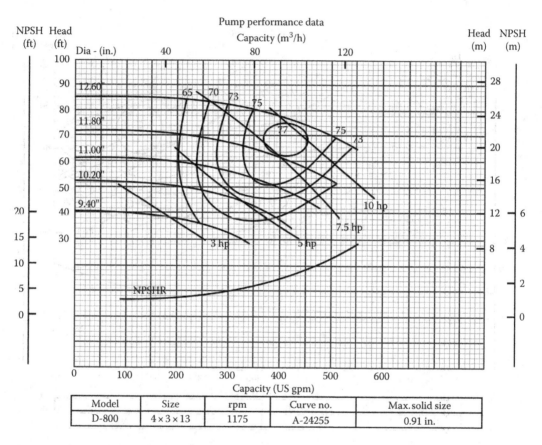

Model	Size	rpm	Curve no.	Max. solid size
D-800	4 × 3 × 13	1175	A-24255	0.91 in.

FIGURE H.14 Typical pump characteristic curves.

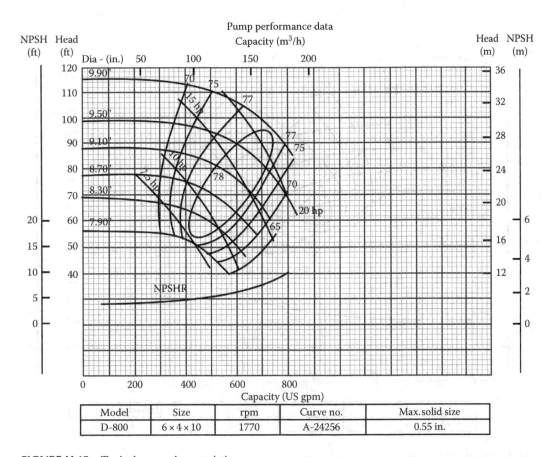

FIGURE H.15 Typical pump characteristic curves.

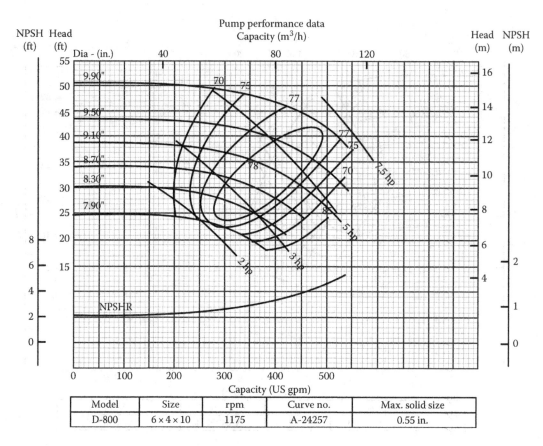

FIGURE H.16 Typical pump characteristic curves.

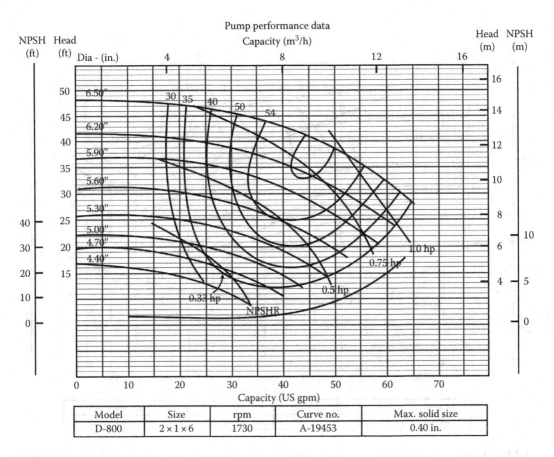

FIGURE H.17 Typical pump characteristic curves.

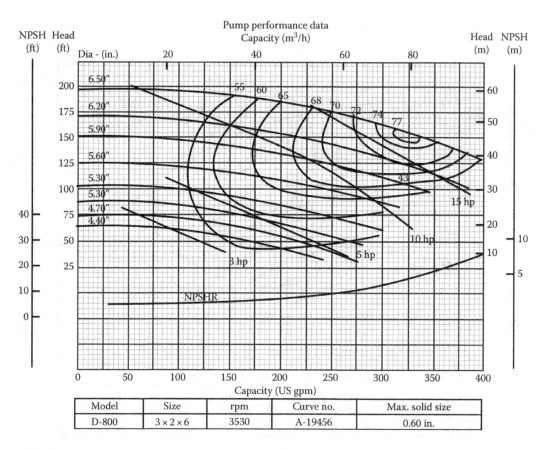

Model	Size	rpm	Curve no.	Max. solid size
D-800	3 × 2 × 6	3530	A-19456	0.60 in.

FIGURE H.18   Typical pump characteristic curves.

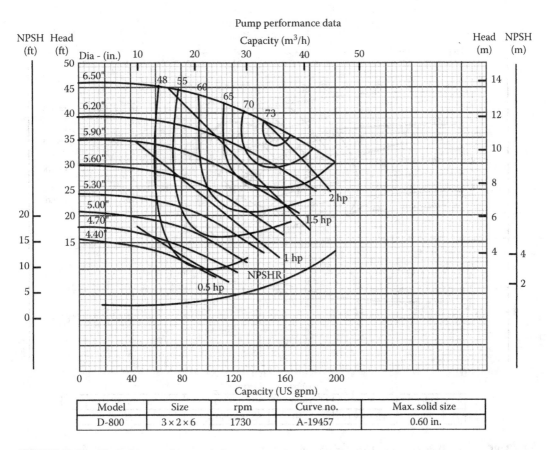

FIGURE H.19 Typical pump characteristic curves.

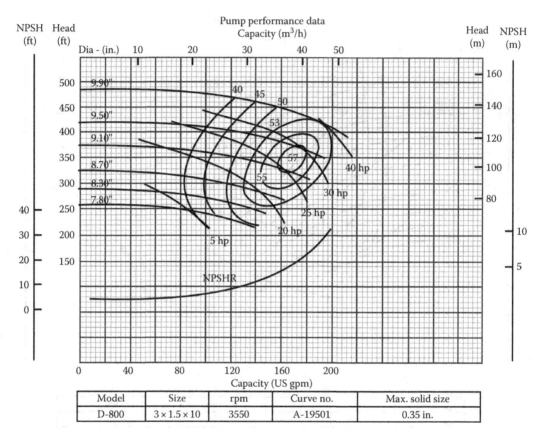

Model	Size	rpm	Curve no.	Max. solid size
D-800	3 × 1.5 × 10	3550	A-19501	0.35 in.

FIGURE H.20 Typical pump characteristic curves.

Appendix I: Fanno Line Tables for Adiabatic Flow of Air in a Constant Area Duct

---

TABLE I.1

Fanno Line: Adiabatic, Constant Area Flow ($k = 1.4$)

N_{Ma}	$\dfrac{T}{T^*}$	$\dfrac{P}{P^*}$	$\dfrac{P^0}{P^{0*}}$	$\dfrac{\tilde{V}}{\tilde{V}^*}$	$\dfrac{F}{F^*}$	$\dfrac{4fL}{D}$
0	1.2000	∞	∞	0	∞	∞
0.01	1.2000	10,9.544	5,7.874	0.01095	4,5.650	7,134.40
0.02	1.1999	5,4.770	2,8.942	0.02191	22,834	1,778.45
0.03	1.1998	3,6.511	1,9.300	0.03286	15,232	7,87.08
0.04	1.1996	27,382	14,482	0.04381	11,435	4,40.35
0.05	1.1994	21,903	11,5914	0.05476	9,1584	2,80.02
0.06	1.1991	18,251	9,6659	0.06570	7.6,428	19,3.03
0.07	1.1988	15.6,42	8.2,915	0.07664	6.5,620	14,0.66
0.08	1.1985	13.6,84	7.2,616	0.08758	5.7,529	10,6.72
0.09	1.1981	12.1,62	6.4,614	0.09851	5.1,249	8,3.496
0.10	1.1976	10.9,435	5.8,218	0.10943	4.6,236	66,922
0.11	1.1971	9.9,465	5.2,992	0.12035	4.2,146	54,688
0.12	1.1966	9.1,156	4.8,643	0.13126	3.8,747	45,408
0.13	1.1960	8.4,123	4.4,968	0.14216	3.58,80	38,207
0.14	1.1953	7.8,093	4.1,824	0.15306	3.34,32	32,511
0.15	1.1946	7.2,866	3.91,03	0.16395	3.13,17	27,932
0.16	1.1939	6.82,91	3.67,27	0.17482	2.94,74	24,198
0.17	1.1931	6.42,52	3.46,35	0.18568	2.78,55	21,115
0.18	1.1923	6.06,62	3.27,79	0.19654	2.64,22	18.5,43
0.19	1.1914	5.74,48	3.11,23	0.20739	2.51,46	16.3,75
0.20	1.1905	5.45,55	2.96,35	0.21822	2.40,04	14.5,33
0.21	1.1895	5.19,36	2.82,93	0.22904	2.29,76	12.9,56
0.22	1.1885	4.95,54	2.70,76	0.23984	2.20,46	11.5,96
0.23	1.1874	4.73,78	2.59,68	0.25063	2.12,03	10.4,16
0.24	1.1863	4.53,83	2.49,56	0.26141	2.04,34	9.3,865
0.25	1.1852	4.35,46	2.40,27	0.27217	1.97,32	8.4,834
0.26	1.1840	4.18,50	2.31,73	0.28291	1.90,88	7.6,876
0.27	1.1828	4.02,80	2.23,85	0.29364	1.84,96	6.9,832
0.28	1.1815	3.88,20	2.16,56	0.30435	1.795,0	6.3,572
0.29	1.1802	3.74,60	2.09,79	0.31504	1.744,6	5.7,989
0.30	1.1788	3.61,90	2.035,1	0.32572	1.697,9	5.2,992
0.31	1.1774	3.50,02	1.976,5	0.33637	1.654,6	4.8,507
0.32	1.1759	3.38,88	1.921,9	0.34700	1.614,4	4.44,68
0.33	1.1744	3.28,40	1.870,8	0.35762	1.576,9	4.08,21

(Continued)

TABLE I.1 (*Continued*)
Fanno Line: Adiabatic, Constant Area Flow (*k* = 1.4)

N_{Ma}	$\dfrac{T}{T^*}$	$\dfrac{P}{P^*}$	$\dfrac{P^0}{P^{0*}}$	$\dfrac{\tilde{V}}{\tilde{V}^*}$	$\dfrac{F}{F^*}$	$\dfrac{4fL}{D}$
0.34	1.1729	3.18,53	1.822,9	0.36822	1.542,0	3.75,20
0.35	1.1713	3.09,22	1.778,0	0.37880	1.509,4	3.45,25
0.36	1.1697	3.004,2	1.735,8	0.38935	1.478,9	3.18,01
0.37	1.1680	2.920,9	1.696,1	0.39988	1.450,3	2.93,20
0.38	1.1663	2.842,0	1.658,7	0.41039	1.423,6	2.70,55
0.39	1.1646	2.767,1	1.623,4	0.42087	1.398,5	2.49,83
0.40	1.1628	2.695,8	1.590,1	0.43133	1.374,9	2.30,85
0.41	1.1610	2.628,0	1.558,7	0.44177	1.352,7	2.13,44
0.42	1.1591	2.563,4	1.528,9	0.45218	1.331,8	1.97,44
0.43	1.1572	2.501,7	1.500,7	0.46257	1.312,2	1.82,72
0.44	1.1553	2.442,8	1.473,9	0.47293	1.293,7	1.69,15
0.45	1.1533	2.386,5	1.448,6	0.48326	1.276,3	1.56,64
0.46	1.1513	2.332,6	1.424,6	0.49357	1.259,8	1.45,09
0.47	1.1492	2.280,9	1.401,8	0.50385	1.244,3	1.34,42
0.48	1.1471	2.231,4	1.380,1	0.51410	1.229,6	1.24,53
0.49	1.1450	2.183,8	1.359,5	0.52433	1.215,8	1.15,39
0.50	1.1429	2.138,1	1.339,9	0.53453	1.202,7	1.06,908
0.51	1.1407	2.094,2	1.321,2	0.54469	1.190,3	0.99,042
0.52	1.1384	2.051,9	1.303,4	0.55482	1.178,6	0.91,741
0.53	1.1362	2.011,2	1.286,4	0.56493	1.167,5	0.84,963
0.54	1.1339	1.971,9	1.270,2	0.57501	1,157,1	0.786,62
0.55	1.1315	1.934,1	1.254,9	0.58506	1.147,2	0.728,05
0.56	1.1292	1.897,6	1.240,3	0.59507	1.137,8	0.673,57
0.57	1.1266	1.862,3	1.226,3	0.60505	1.128,9	0.622,86
0.58	1.1244	1.828,2	1.213,0	0.61500	1.120,5	0.575,68
0.59	1.1219	1.795,2	1.200,3	0.62492	1.112,6	0.531,74
0.60	1.1194	1.763,4	1.188,2	0.63481	1.1050,4	0.490,81
0.61	1.1169	1.732,5	1.176,6	0.64467	1.0979,3	0.452,70
0.62	1.1144	1.702,6	1.165,6	0.65449	1.0912,0	0.417,20
0.63	1.1118	1.673,7	1.155,1	0.66427	1.0848,5	0.384,11
0.64	1.1091	1.645,6	1.145,1	0.67402	1.0788,3	0.353,30
0.65	1.10650	1.618,3	1.135,6	0.68374	1.0731,4	0.324,60
0.66	1.10383	1.591,9	1.126,5	0.69342	1.0677,7	0.297,85
0.67	1.10114	1.566,2	1.117,9	0.70306	1.0627,1	0.272,95
0.68	1.09842	1.541,3	1.109,7	0.71267	1.0579,2	0.249,78
0.69	1.09567	1.517,0	1.101,8	0.72225	1.0534,0	0.228,21
0.70	1.09290	1.493,4	1.0943,6	0.73179	1.0491,5	0.208,14
0.71	1.09010	1.470,5	1.0872,9	0.74129	1.0451,4	0.189,49
0.72	1.08727	1.448,2	1.0805,7	0.75076	1.0413,7	0.172,15
0.73	1.08442	1.426,5	1.0741,9	0.76019	1.0378,3	0.156,06
0.74	1.08155	1.405,4	1.0681,5	0.76958	1.0345,0	0.141,13
0.75	1.07856	1.384,8	1.0624,2	0.77893	1.0313,7	0.127,28
0.76	1.07573	1.364,7	1.0570,0	0.78825	1.0284,4	0.114,46
0.77	1.07279	1.345,1	1.0518,8	0.79753	1.0257,0	0.102,62
0.78	1.06982	1.326,0	1.0470,5	0.80677	1.0231,4	0.091,67
0.79	1.06684	1.307,4	1.0425,0	0.81598	1.0207,5	0.081,59

(*Continued*)

TABLE I.1 (*Continued*)

Fanno Line: Adiabatic, Constant Area Flow ($k = 1.4$)

N_{Ma}	$\dfrac{T}{T^*}$	$\dfrac{P}{P^*}$	$\dfrac{P^0}{P^{0*}}$	$\dfrac{\tilde{V}}{\tilde{V}^*}$	$\dfrac{F}{F^*}$	$\dfrac{4fL}{D}$
0.80	1.06383	1.2892	1.0382,3	0.82514	1.0185,3	0.072,29
0.81	1.06080	1.2715	1.0342,2	0.83426	1.0164,6	0.063,75
0.82	1.05775	1.2542	1.0304,7	0.84334	1.0145,5	0.055,93
0.83	1.05468	1.2373	1.0269,6	0.85239	1.0127,8	0.048,78
0.84	1.05160	1.2208	1.0237,0	0.86140	1.0111,5	0.042,26
0.85	1.04849	1.2047	1.0206,7	0.87037	1.0096,6	0.036,32
0.86	1.04537	1.1889	1.0178,7	0.87929	1.0082,9	0.030,97
0.87	1.04223	1.1735	1.0152,9	0.88818	1.0070,4	0.026,13
0.88	1.03907	1.1584	1.0129,4	0.89703	1.0059,1	0.021,80
0.89	1.03589	1.1436	1.0108,0	0.90583	1.0049,0	0.017,93
0.90	1.03270	1.1291,3	1.0088,7	0.91459	1.0039,9	0.0145,13
0.91	1.02950	1.1150,0	1.0071,4	0.92332	1.0031,8	0.0115,19
0.92	1.02627	1.1011,4	1.0056,0	0.93201	1.0024,8	0.0089,16
0.93	1.02304	1.0875,8	1.0042,6	0.94065	1.0018,8	0.0066,94
0.94	1.01978	1.0743,0	1.0031,1	0.94925	1.0013,6	0.0048,15
0.95	1.01652	1.0612,9	1.0021,5	0.95782	1.0009,3	0.0032,80
0.96	1.01324	1.0485,4	1.0013,7	0.96634	1.0005,9	0.0020,56
0.97	1.00995	1.0360,5	1.0007,6	0.97481	1.0003,3	0.0011,35
0.98	1.00664	1.0237,9	1.0003,3	0.98324	1.0001,4	0.0004,93
0.99	1.00333	1.0117,8	1.0000,8	0.99164	1.0000,3	0.0001,20
1.00	1.00000	1.0000,0	1.0000,0	1.00000	1.0000,0	0
1.01	0.99666	0.9884,4	1.0000,8	1.00831	1.0000,3	0.0001,14
1.02	0.99331	0.9771,1	1.0003,3	1.01658	1.0001,3	0.0004,58
1.03	0.98995	0.9659,8	1.0007,3	1.02481	1.0003,0	0.0010,13
1.04	0.98658	0.9550,6	1.0013,0	1.03300	1.0005,3	0.0017,71
1.05	0.98320	0.9443,5	1.0020,3	1.04115	1.0008,2	0.0027,12
1.06	0.97982	0.9338,3	1.0029,1	1.04925	1.0011,6	0.0038,37
1.07	0.97642	0.9235,0	1.0039,4	1.05731	1.0015,5	0.0051,29
1.08	0.97302	0.9133,5	1.0051,2	1.06533	1.0020,0	0.0065,82
1.09	0.96960	0.9033,8	1.0064,5	1.07331	1.0025,0	0.0081,85
1.10	0.96618	0.8935,9	1.0079,3	1.08124	1.00305	0.0099,33
1.11	0.96276	0.8839,7	1.0095,5	1.08913	1.00365	0.0118,13
1.12	0.95933	0.8745,1	1.0113,1	1.09698	1.00429	0.0138,24
1.13	0.95589	0.8652,2	1.0132,2	1.10479	1.00497	0.0159,49
1.14	0.95244	0.8560,8	1.0152,7	1.11256	1.00569	0.0181,87
1.15	0.94899	0.8471,0	1.0174,6	1.1203	1.00646	0.0205,3
1.16	0.94554	0.8382,7	1.0197,8	1.1280	1.00726	0.0229,8
1.17	0.94208	0.8295,8	1.0222,4	1.1356	1.00810	0.0255,2
1.18	0.93862	0.8210,4	1.0248,4	1.1432	1.00897	0.0281,4
1.19	0.93515	0.8126,3	1.0275,7	1.1508	1.00988	0.0308,5
1.20	0.93168	0.8043,6	1.0304,4	1.1583	1.01082	0.0336,4
1.21	0.92820	0.7962,3	1.0334,4	1.1658	1.01178	0.0365,0
1.22	0.92473	0.7882,2	1.0365,7	1.1732	1.01278	0.394,2
1.23	0.92125	0.7803,4	1.0398,3	1.1806	1.01381	0.0424,1
1.24	0.91777	0.7725,8	1.0432,3	1.1879	1.01486	0.0454,7

(*Continued*)

TABLE I.1 (*Continued*)

Fanno Line: Adiabatic, Constant Area Flow *(k = 1.4)*

N_{Ma}	$\dfrac{T}{T^*}$	$\dfrac{P}{P^*}$	$\dfrac{P^0}{P^{0*}}$	$\dfrac{\tilde{V}}{\tilde{V}^*}$	$\dfrac{F}{F^*}$	$\dfrac{4fL}{D}$
1.25	0.91429	0.7649,5	1.0467,6	1.1952	1.01594	0.04858
1.26	0.91080	0.7574,3	1.0504,1	1.2025	1.01705	0.05174
1.27	0.90732	0.7500,3	1.0541,9	1.2097	1.01818	0.05494
1.28	0.90383	0.7427,4	1.0580,9	1.2169	1.01933	0.05820
1.29	0.90035	0.7355,6	1.0621,3	1.2240	1.02050	0.06150
1.30	0.89686	0.7284,8	1.0663,0	1.2311	1.02169	0.06483
1.31	0.89338	0.7215,2	1.0706,0	1.2382	1.02291	0.06820
1.32	0.88989	0.7146,5	1.0750,2	1.2452	1.02415	0.07161
1.33	0.88641	0.7078,9	1.0795,7	1.2522	1.02540	0.07504
1.34	0.88292	0.7012,3	1.0842,4	1.2591	1.02666	0.07850
1.35	0.87944	0.6946,6	1.0890,4	1.2660	1.02794	0.08199
1.36	0.87596	0.6881,8	1.0939,7	1.2729	1.02924	0.08550
1.37	0.87249	0.6818,0	1.0990,2	1.2797	1.03056	0.08904
1.38	0.86901	0.6755,1	1.1041,9	1.2864	1.03189	0.09259
1.39	0.06554	0.6693,1	1.1094,8	1.2932	1.03323	0.09616
1.40	0.86207	0.6632,0	1.1149	1.2999	1.03458	0.09974
1.41	0.85860	0.6571,7	1.1205	1.3065	1.03595	0.10333
1.42	0.85514	0.6512,2	1.1262	1.3131	1.03733	0.10694
1.43	0.85168	0.6453,6	1.1320	1.3197	1.03872	0.11056
1.44	0.84822	0.6395,8	1.1379	1.3262	1.04012	0.11419
1.45	0.84477	0.6338,7	1.1440	1.3327	1.04153	0.11782
1.46	0.84133	0.6282,4	1.1502	1.3392	1.04295	0.12146
1.47	0.83788	0.6226,9	1.1565	1.3456	1.04438	0.12510
1.48	0.83445	0.6172,2	1.1629	1.3520	1.04581	0.12875
1.49	0.83101	0.6118,1	1.1695	1.3583	1.04725	0.13240
1.50	0.82759	0.6064,8	1.1762	1.3646	1.04870	0.13605
1.51	0.82416	0.6012,2	1.1830	1.3708	1.05016	0.13970
1.52	0.82075	0.5960,2	1.1899	1.3770	1.05162	0.14335
1.53	0.81734	0.5908,9	1.1970	1.3832	1.05309	0.14699
1.54	0.81394	0.5858,3	1.2043	1.3894	1.05456	0.15063
1.55	0.81054	0.5808,4	1.2116	1.3955	1.05604	0.15427
1.56	0.80715	0.5759,1	1.2190	1.4015	1.05752	0.15790
1.57	0.80376	0.5710,4	1.2266	1.4075	1.05900	0.16152
1.58	0.83038	0.5662,3	1.2343	1.4135	1.06049	0.16514
1.59	0.79701	0.5614,8	1.2422	1.4195	1.06198	0.16876
1.60	0.79365	0.5567,9	1.2502	1.4254	1.06348	0.17236
1.61	0.79030	0.5521,6	1.2583	1.4313	1.06498	0.17595
1.62	0.78695	0.5475,9	1.2666	1.4371	1.06648	0.17953
1.63	0.78361	0.5430,8	1.2750	1.4429	1.06798	0.18311
1.64	0.78028	0.5386,2	1.2835	1.4487	1.06948	0.18667
1.65	0.77695	0.5342,1	1.2922	1.4544	1.07098	0.19022
1.66	0.77363	0.5298,6	1.3010	1.4601	1.07249	0.19376
1.67	0.77033	0.5255,6	1.3099	1.4657	1.07399	0.19729
1.68	0.76703	0.5213,1	1.3190	1.4713	1.07550	0.20081
1.69	0.76374	0.5171,1	1.3282	1.4769	1.07701	0.20431
1.70	0.76046	0.5129,7	1.3376	1.4825	1.07851	0.20780

(Continued)

TABLE I.1 (*Continued*)

Fanno Line: Adiabatic, Constant Area Flow (*k* = 1.4)

N_{Ma}	$\dfrac{T}{T^*}$	$\dfrac{P}{P^*}$	$\dfrac{P^0}{P^{0*}}$	$\dfrac{\tilde{V}}{\tilde{V}^*}$	$\dfrac{F}{F^*}$	$\dfrac{4fL}{D}$
1.71	0.75718	0.5088,7	1.3471	1.4880	1.08002	0.21128
1.72	0.75392	0.5048,2	1.3567	1.4935	1.08152	0.21474
1.73	0.75067	0.5008,2	1.3665	1.4989	1.08302	0.21819
1.74	0.74742	0.4968,6	1.3764	1.5043	1.08453	0.22162
1.75	0.74419	0.4929,5	1.3865	1.5097	1.08603	0.22504
1.76	0.74096	0.4890,9	1.3967	1.5150	1.08753	0.22844
1.77	0.73774	0.4852,7	1.4070	1.5203	1.08903	0.23183
1.78	0.73453	0.4814,9	1.4175	1.5256	1.09053	0.23520
1.79	0.73134	0.4777,6	1.4282	1.5308	1.09202	0.23855
1.80	0.72816	0.47407	1.4390	1.5360	1.09352	0.24189
1.81	0.72498	0.47042	1.4499	1.5412	1.09500	0.24521
1.82	0.72181	0.46681	1.4610	1.5463	1.09649	0.24851
1.83	0.71865	0.46324	1.4723	1.5514	1.09798	0.25180
1.84	0.71551	0.45972	1.4837	1.5564	1.00946	0.25507
1.85	0.71238	0.45623	1.4952	1.5614	1.1009	0.25832
1.86	0.70925	0.45278	1.5069	1.5664	1.1024	0.26156
1.87	0.70614	0.49937	1.5188	1.5714	1.1039	0.26478
1.88	0.70304	0.44600	1.5308	1.5763	1.1054	0.26798
1.89	0.69995	0.44266	1.5429	1.5812	1.1068	0.27116
1.90	0.69686	0.43936	1.5552	1.5861	1.1083	0.27433
1.91	0.69379	0.43610	1.5677	1.5909	1.1097	0.27748
1.92	0.69074	0.43287	1.5804	1.5957	1.1112	0.28061
1.93	0.68769	0.42967	1.5932	1.6005	1.1126	0.28372
1.94	0.68465	0.42651	1.6062	1.6052	1.1141	0.28681
1.95	0.68162	0.42339	1.6193	1.6099	1.1155	0.28989
1.96	0.67861	0.42030	1.6326	1.6146	1.1170	0.29295
1.97	0.67561	0.41724	1.6461	1.6193	1.1184	0.29599
1.98	0.67262	0.41421	1.6597	1.6239	1.1198	0.29901
1.99	0.66964	0.41121	1.6735	1.6824	1.1213	0.30201
2.00	0.66667	0.40825	1.6875	1.6330	1.1227	0.30499
2.01	0.66371	0.40532	1.7017	1.6375	1.1241	0.30796
2.02	0.66076	0.40241	1.7160	1.6420	1.1255	0.31091
2.03	0.65783	0.39954	1.7305	1.6465	1.1269	0.31384
2.04	0.65491	0.39670	1.7452	1.6509	1.1283	0.31675
2.05	0.65200	0.39389	1.7600	1.6553	1.1297	0.31965
2.06	0.64910	0.39110	1.7750	1.6597	1.1311	0.32253
2.07	0.64621	0.38834	1.7902	1.6640	1.1325	0.32538
2.08	0.64333	0.38562	1.8056	1.6683	1.1339	0.32822
2.09	0.64047	0.38292	1.8212	1.6726	1.1352	0.33104
2.10	0.63762	0.38024	1.8369	1.6769	1.1366	0.33385
2.11	0.63478	0.37760	1.8528	1.6811	1.1380	0.33664
2.12	0.63195	0.37498	1.8690	1.6853	1.1393	0.33940
2.13	0.62914	0.37239	1.8853	1.6895	1.1407	0.34215
2.14	0.62633	0.36982	1.9018	1.6936	1.1420	0.34488
2.15	0.62354	0.36728	1.9185	1.6977	1.1434	0.34760
2.16	0.62076	0.36476	1.9354	1.7018	1.1447	0.35030

(*Continued*)

TABLE I.1 (*Continued*)

Fanno Line: Adiabatic, Constant Area Flow (*k* = 1.4)

N_{Ma}	$\dfrac{T}{T^*}$	$\dfrac{P}{P^*}$	$\dfrac{P^0}{P^{0*}}$	$\dfrac{\tilde{V}}{\tilde{V}^*}$	$\dfrac{F}{F^*}$	$\dfrac{4fL}{D}$
2.17	0.61799	0.36227	1.9525	1.7059	1.1460	0.35298
2.18	0.61523	0.35980	1.9698	1.7099	1.1474	0.35564
2.19	0.61249	0.35736	1.9873	1.7139	1.1487	0.35828
2.20	0.60976	0.35494	2.0050	1.7179	1.1500	0.36091
2.21	0.60704	0.35254	2.0228	1.7219	1.1513	0.36352
2.22	0.60433	0.35017	2.0409	1.7258	1.1526	0.36611
2.23	0.60163	0.34782	2.0592	1.7297	1.1539	0.36868
2.24	0.59895	0.34550	2.0777	1.7336	1.1552	0.37124
2.25	0.59627	0.34319	2.0964	1.7374	1.1565	0.37378
2.26	0.59361	0.34091	2.1154	1.7412	1.1578	0.37630
2.27	0.59096	0.33865	2.1345	1.7450	1.1590	0.37881
2.28	0.58833	0.33641	2.1538	1.7488	1.1603	0.38130
2.29	0.58570	0.33420	2.1733	1.7526	1.1616	0.38377
2.30	0.58309	0.33200	2.1931	1.7563	1.1629	0.38623
2.31	0.58049	0.32983	2.2131	1.7600	1.1641	0.38867
2.32	0.57790	0.32767	2.2333	1.7637	1.1653	0.39109
2.33	0.57532	0.32554.	2.2537	1.7673	1.1666	0.39350
2.34	0.57276	0.32342	2.2744	1.7709	1.1678	0.39589
2.35	0.57021	0.32133	2.2953	1.7745	1.1690	0.39826
2.36	0.56767	0.31925	2.3164	1.7781	1.1703	0.40062
2.37	0.56514	0.31720	2.3377	1.7817	1.1715	0.40296
2.38	0.56262	0.31516	2.3593	1.7852	1.1727	0.40528
2.39	0.56011	0.31314	2.3811	1.7887	1.1739	0.40760
2.40	0.55762	0.31114	2.4031	1.7922	1.1751	0.40989
2.41	0.55514	0.30916	2.4254	1.7956	1.1763	0.41216
2.42	0.55267	0.30720	2.4479	1.7991	1.1775	0.41442
2.43	0.55021	0.30525	2.4706	1.8025	1.1786	0.41667
2.44	0.54776	0.30332	2.4936	1.8059	1.1798	0.41891
2.45	0.54533	0.30141	2.5168	1.8092	1.1810	0.42113
2.46	0.54291	0.29952	2.5403	1.8126	1.1821	0.42333
2.47	0.54050	0.29765	2.5640	1.8159	1.1833	0.42551
2.48	0.53810	0.29579	2.5880	1.8192	1.1844	0.42768
2.49	0.53571	0.29395	2.6122	1.8225	1.1856	0.42983
2.50	0.53333	0.29212	2.6367	1.8257	1.1867	0.43197
2.51	0.53097	0.29031	2.6615	1.8290	1.1879	0.43410
2.52	0.52862	0.28852	2.6865	1.8322	1.1890	0.43621
2.53	0.52627	0.28674	2.7117	1.8354	1.1901	0.43831
2.54	0.52394	0.28498	2.7372	1.8386	1.1912	0.44040
2.55	0.52163	0.28323	2.7630	1.8417	1.1923	0.44247
2.56	0.51932	0.28150	2.7891	1.8448	1.1934	0.44452
2.57	0.51702	0.27978	2.8154	1.8479	1.1945	0.44655
2.58	0.51474	0.27808	2.8420	1.8510	1.1956	0.44857
2.59	0.51247	0.27640	2.8689	1.8541	1.1967	0.45059
2.60	0.51020	0.27473	2.8960	1.8571	1.1978	0.45259
2.61	0.50795	0.27307	2.9234	1.8602	1.1989	0.45457
2.62	0.50571	0.27143	2.9511	1.8632	1.2000	0.45654

(*Continued*)

TABLE I.1 (*Continued*)

Fanno Line: Adiabatic, Constant Area Flow (*k* = 1.4)

N_{Ma}	$\dfrac{T}{T^*}$	$\dfrac{P}{P^*}$	$\dfrac{P^0}{P^{0*}}$	$\dfrac{\tilde{V}}{\tilde{V}^*}$	$\dfrac{F}{F^*}$	$\dfrac{4fL}{D}$
2.63	0.50349	0.26980	2.9791	1.8662	1.2011	0.45850
2.64	0.50127	0.26818	3.0074	1.8691	1.2021	0.46044
2.65	0.49906	0.26658	3.0359	1.8721	1.2031	0.46237
2.66	0.49687	0.26499	3.0647	1.8750	1.2042	0.46429
2.67	0.49469	0.26342	3.0938	1.8779	1.2052	0.46619
2.68	0.49251	0.26186	3.1234	1.8808	1.2062	0.46807
2.69	0.49035	0.26032	3.1530	1.8837	1.2073	0.46996
2.70	0.48820	0.25878	3.1830	1.8865	1.2083	0.47182
2.71	0.48606	0.25726	3.2133	1.8894	1.2093	0.47367
2.72	0.48393	0.25575	3.2440	1.8922	1.2103	0.47551
2.73	0.48182	0.25426	3.2749	1.8950	1.2113	0.47734
2.74	0.47971	0.25278	3.3061	1.8978	1.2123	0.47915
2.75	0.47761	0.25131	3.3376	1.9005	1.2133	0.48095
2.76	0.47553	0.24985	3.3695	1.9032	1.2143	0.48274
2.77	0.47346	0.24840	3.4017	1.9060	1.2153	0.48452
2.78	0.47139	0.24697	3.4342	1.9087	1.2163	0.48628
2.79	0.46933	0.24555	3.4670	1.9114	1.2173	0.48803
2.80	0.46729	0.24414	3.5001	1.9140	1.2182	0.48976
2.81	0.46526	0.24274	3.5336	1.9167	1.2192	0.49148
2.82	0.46324	0.24135	3.5674	1.9193	1.2202	0.49321
2.83	0.46122	0.23997	3.6015	1.9220	1.2211	0.49491
2.84	0.45922	0.23861	3.6359	1.9246	1.2221	0.49660
2.85	0.45723	0.23726	3.6707	1.9271	1.2230	0.49828
2.86	0.45525	0.23592	3.7058	1.9297	1.2240	0.49995
2.87	0.45328	0.23458	3.7413	1.9322	1.2249	0.50161
2.88	0.45132	0.23326	3.7771	1.9348	1.2258	0.50326
2.89	0.44937	0.23196	3.8133	1.9373	1.2268	0.50489
2.90	0.44743	0.23066	3.8498	1.9398	1.2277	0.50651
2.91	0.44550	0.22937	3.8866	1.9423	1.2286	0.50812
2.92	0.44358	0.22809	3.9238	1.9448	1.2295	0.50973
2.93	0.44167	0.22682	3.9614	1.9472	1.2304	0.51133
2.94	0.43977	0.22556	3.9993	1.9497	1.2313	0.51291
2.95	0.43788	0.22431	4.0376	1.9521	1.2322	0.51447
2.96	0.43600	0.22307	4.0763	1.9545	1.2331	0.51603
2.97	0.43413	0.22185	4.1153	1.9569	1.2340	0.51758
2.98	0.43226	0.22063	4.1547	1.9592	1.2348	0.51912
2.99	0.43041	0.21942	4.1944	1.9616	1.2357	0.52064
3.00	0.42857	0.21822	4.1346	1.9640	1.2366	0.52216
3.50	0.34783	0.16850	6.7896	2.0642	1.2743	0.58643
4.00	0.28571	0.13363	10.719	2.1381	1.3029	0.63306
4.50	0.23762	0.10833	16.562	2.1936	1.3247	0.66764
5.00	0.20000	0.08944	25.000	2.2361	1.3416	0.69381
6.00	0.14634	0.06376	53.180	2.2953	1.3655	0.72987
7.00	0.11111	0.04762	104.14	2.3333	1.3810	0.75281
8.00	0.08696	0.03686	190.11	2.3591	1.3915	0.76820
9.00	0.06977	0.02935	327.19	2.3772	1.3989	0.77898
10.00	0.05714	0.02390	535.94	2.3905	1.4044	0.78683
∞	0	0	∞	2.4495	1.4289	0.82153

Note: Numbers following the comma are less precise than others.

Index